Holographic Interferometry

WILEY SERIES IN PURE AND APPLIED OPTICS

Advisory Editor

Stanley S. Ballard, University of Florida

ALLEN AND EBERLY • *Optical Resonance and Two-Level Atoms*
BABCOCK • *Silicate Glass Technology Methods*
BOND • *Crystal Technology*
CATHEY • *Optical Information Processing and Holography*
EBERLY AND LAMBROPOULOS • *Multiphoton Processes*
GASKILL • *Linear Systems, Fourier Transforms, and Optics*
GERRARD AND BURCH • *Introduction of Matrix Methods in Optics*
HUDSON • *Infrared System Engineering*
JUDD AND WYSZECKI • *Color in Business, Science, and Industry*, Third Edition
KNITTL • *Optics of Thin Films*
LENGYEL • *Lasers*, Second Edition
LEVI • *Applied Optics, A Guide to Optical System Design*, Volume I
    Volume II
LOUISELL • *Quantum Statistical Properties of Radiation*
MALACARA • *Optical Shop Testing*
MCCARTNEY • *Optics of the Atmosphere; Scattering by Molecules and Particles*
MIDWINTER • *Optical Fibers for Transmission*
MOLLER AND ROTHSCHILD • *Far-Infrared Spectroscopy*
ROGERS • *Noncoherent Optical Processing*
SHULMAN • *Optical Data Processing*
VEST • *Holographic Interferometry*
WILLIAMS AND BECKLUND • *Optics*
ZERNIKE AND MIDWINTER • *Applied Nonlinear Optics*

Holographic Interferometry

CHARLES M. VEST
Department of Mechanical Engineering
The University of Michigan

John Wiley & Sons, New York / Chichester / Brisbane / Toronto

TA
1555
V47

Copyright © 1979 by John Wiley & Sons, Inc.

All rights reserved. Published simultaneously in Canada.

Reproduction or translation of any part of this work beyond that permitted by Sections 107 or 108 of the 1976 United States Copyright Act without the permission of the copyright owner is unlawful. Requests for permission or further information should be addressed to the Permissions Department, John Wiley & Sons, Inc.

Library of Congress Cataloging in Publication Data

Vest, Charles M
   Holographic interferometry.

   (Wiley series in pure and applied optics)
   Includes bibliographical references and index.
   1. Interferometer. 2. Holography. I. Title.
TA1555.V47   774   78-14883
ISBN 0-471-90683-2

Printed in the United States of America

10 9 8 7 6 5 4 3 2 1

*To M. L. Vest
and my family, Becky, Kemper, and John*

Preface

In writing this book I have attempted to provide a unified, self-contained treatment of the theory, practice, and application of holographic interferometry and related coherent optical measurement techniques. Emphasis is placed on quantitative evaluation of holographic interferograms of both opaque and transparent objects. The technical level and the scope of the book are such that it should be of interest to workers and students in coherent optics and to engineers and scientists from other disciplines who wish to evaluate or use holographic interferometry as a measurement or diagnostic technique. Clear, simple, physical reasoning is applied in a manner that circumvents the use of complicated mathematics wherever possible. Practical information and data regarding light sources, recording media, the design and construction of holographic systems, and physical properties such as refractive index are given throughout the book and are made readily accessible by careful indexing. The book serves as a guide to the literature of the subject and contains almost 700 cited references. It could be used as a textbook or supplementary reference in senior-graduate level courses in holographic interferometry, optical measurement techniques, engineering measurements, or holography.

The principles of holographic interferometry were discovered very soon after the introduction of off-axis holography by Leith and Upatnieks in 1961. Much of the basic theory, practice, and future potential of the method were presented clearly in the early papers of Powell and Stetson, Burch, and Heflinger, Wuerker, and Brooks. This stage was followed by a surge of activity, some of which suggested that holographic interferometry was viewed by some as a panacea for problems encountered in engineering and scientific metrology. Activity diminished as the practical limitations of the method were recognized, but work continued; the basic theory was refined, and more emphasis was placed on quantitative interpretation of interferograms. During the last 5 years a steady diffusion of the literature of holographic interferometry from optics journals into the journals of other disciplines reflects a new, more mature period of application and growth. It is my belief that this growth in the importance of holographic interferometry as a scientific and engineering tool will continue as specific applications for which it is uniquely suited are recognized. I hope that this book will contribute to the understanding required for future accomplishments. The current range of applications is remarkable. Readers of this book will learn that holographic interferometry is used in plasma experiments associated with the laser-fusion program, and has served to increase understanding of the mechanics of hearing, to measure relaxation rates of structural materials, to observe the vibration of turbine blades in operating jet engines, to design musical instruments, and to detect subsurface damage in fifteenth-century panel paintings.

Work in holographic interferometry has also stimulated the development, extension, or adaptation of several closely related measurement techniques based on the use of laser light. These include holographic contour generation, speckle photography and interferometry, projected fringe techniques, techniques incorporating television systems, and holographic photoelasticity. Because of their close, and often complementary, relationship to holographic interferometry, the theory, practice, and application of these techniques are also discussed in the book.

In organizing and presenting the material I have assumed that many readers will approach the topic without the benefit of a formal background in coherent optics or Fourier transform theory. For this reason I have written an introductory chapter in which the elements of coherent optics, holography, and holographic interferometry are presented. Although readers already conversant with coherent optics may wish to bypass all but the last section of this chapter, others will find that it contains virtually all the basic concepts required for the developments in the rest of the book. Also, simple physical and geometrical arguments are emphasized throughout the book. This is particularly evident in the analysis of fringe localization. Although all but the most elementary formalisms of Fourier optics and the mathematics associated with diffraction theory are bypassed, quantitative aspects of the subject are emphasized.

I have provided a substantial number of references to the published literature for those who wish to pursue particular topics in more detail. References are restricted almost exclusively to journal articles because I have found that most important results of research published in reports and symposia proceedings usually reappear in article form. There are, of course, exceptions and time lags, but that is a risk encountered in writing about a field in which applications are still developing.

I wish to acknowledge my debt to Emmett Leith, who, together with my colleagues in the Department of Mechanical Engineering at The University of Michigan, has encouraged my activities in engineering applications of coherent optics over the past 10 years. Special thanks go also to Joseph

Goodman, Albert Macovski, and Daniel Bershader of Stanford University, who provided the opportunity and professional atmosphere in which the initial research and writing were done. Karl Stetson contributed immeasurably to this book through his excellent published work and his generous personal cooperation and assistance. Research carried out with Don Sweeney, Predrag Radulovic, Ron Boyd, and Soyoung Cha have contributed enormously to my understanding of this topic. Vicki Rothhaar produced many of the interferograms used to illustrate this book, and Karl Stetson, Soyoung Cha, and Bruce Hansche carefully read major portions of the manuscript and provided valuable criticisms. Madelyn Hudkins greatly eased the effort of writing by producing hundreds of pages of error-free typing with remarkable efficiency and cheerfulness. Finally, I want to express my sincere thanks to my family for their patience, interest, and encouragement.

CHARLES M. VEST

Ann Arbor, Michigan
December 1978

Contents

Chapter 1. Coherent Optics, Holography, and Holographic Interferometry   1

   1.1 Introduction   1
   1.2 Light, Interference, and Coherence   1
   1.3 Diffraction and Spatial Filtering   15
   1.4 Laser Speckle   30
   1.5 Holography   36
   1.6 Holographic Interferometry   57
   References   64

Chapter 2. Opaque Objects: Measurement of Displacement and Deformation   67

   2.1 Introduction   67
   2.2 Equations for Fringe Interpretation   68
   2.3 Accuracy of Measurements   77
   2.4 The Holodiagram   90
   2.5 Alternative Methods of Fringe Readout   97
   References   104

Chapter 3. Opaque Objects: Formation and Localization of Fringes   107

   3.1 Introduction   107
   3.2 Analysis of Fringe Localization   108
   3.3 Fringe Localization with Collimated Illumination   111
   3.4 Fringe Localization with Spherical Wave Illumination   122
   3.5 Examples of Fringe Formation and Localization   123
   3.6 Other Approaches to Fringe Formation and Localization   142
   References   143

Chapter 4. Opaque Objects: Measurement of Strain, Stress, Bending Moments, and Vibration; Applications — 146

4.1	Introduction	146
4.2	Measurement of Strain, Stress, and Bending Moments	147
4.3	Measurement of Mechanical Vibrations	177
	4.3.1 Sinusoidal Vibrations	178
	4.3.2 Separable Motions	183
	4.3.3 Multiple Modes and Nonseparable Motions	187
	4.3.4 Real-Time and Stroboscopic Interferometry	197
	4.3.5 Temporally Modulated Holography	210
4.4	Applications	217
	4.4.1 Nondestructive Testing	217
	4.4.2 Medical and Dental Research	231
	4.4.3 Solid Mechanics	233
	References	245

Chapter 5. Transparent Objects: Fringe Formation and Localization — 254

5.1	Introduction	254
5.2	Phase Objects and Refraction in Transparent Media	255
5.3	Holography and Interferometry with Phase Objects	264
5.4	Fringe Localization in Diffuse-Illumination Interferometry	284
5.5	Nonlinear Holographic Interferometry	295
	References	308

Chapter 6. Transparent Objects: Fringe Interpretation, Special Techniques, and Applications — 311

6.1	Introduction	311
6.2	Fringe Interpretation	311
	6.2.1 Inversion Techniques	315
	6.2.2 Errors Due to Refraction	329
6.3	Special Techniques: Multiple-Beam, Multiple-Wavelength, and Multiple-Pass Interferometry	334
6.4	Applications	344
	6.4.1 Aerodynamics and Flow Visualization	344
	6.4.2 Plasma Diagnostics	353
	6.4.3 Heat and Mass Transfer	363
	6.4.4 Stress Analysis	373
	References	377

Chapter 7. Related Measurement Techniques — **387**

- 7.1 Introduction — 387
- 7.2 Holographic Contour Generation — 387
- 7.3 Speckle Photography and Speckle Interferometry — 396
  - 7.3.1 Speckle Photography — 396
  - 7.3.2 Speckle Interferometry — 413
- 7.4 Projected Fringe Techniques — 428
- 7.5 Techniques Incorporating Television Systems — 435
- 7.6 Holographic Photoelasticity — 438
- References — 452

Author Index — **457**

Subject Index — **461**

1

Coherent Optics, Holography, and Holographic Interferometry

1.1 INTRODUCTION

In this introductory chapter a variety of concepts and procedures which are required for development of the theory, practice, and application of holographic interferometry are presented. The subject matter has been chosen so that the treatment of holographic interferometry in this book will be selfcontained and accessible to persons having little or no prior knowledge of coherent optics. To further this goal, descriptive physical approaches are used in preference to detailed analysis wherever possible. In particular, the topics of coherence, diffraction, and holography are discussed without the use of the Fourier transform.

After a general disscussion of light, coherence, and interferometry, the topic of diffraction is introduced in order to develop the concept and nomenclature of spatial frequencies and the basic idea of spatial filtering. The phenomenon of laser speckle is discussed, and the manner in which it is analyzed quantitatively is described. This material is useful for understanding the holographic interferometry of diffusely scattering objects and is required for the discussion of speckle photography and interferometry in Chapter 7.

The theory and practice of holography are discussed briefly. Some emphasis is given to experimental considerations such as properties of photographic emulsions and selection of reference-to-object-beam ratios. Finally, the basic concepts of holographic interferometry are presented in order to establish nomenclature and prepare the reader for the detailed presentations in Chapters 2 to 6.

1.2 LIGHT, INTERFERENCE, AND COHERENCE

Light is a form of electromagnetic radiation. It is characterized by its amplitude, wavelength (or frequency), phase, polarization, speed of propagation, and direction of propagation. When light is scattered or reflected by the surface of an opaque object, or when it is transmitted through a transparent medium, any or all of these characteristics may be altered. By measuring the changes in these characteristics, we obtain information about the state of the object, for example, its size, shape, temperature, velocity, density, or state of stress. Holographic interferometry is one important method for carrying out measurements of this type. Some basic concepts regarding light, interference, and coherence are required for development of the theory and practice of holographic interferometry. A brief introduction to these topics is given in this section.

An electromagnetic wave such as light can be described by specifying the temporal and spatial dependence of its electric intensity vector **E**. A more complete description requires specification of the magnetic intensity **H**, the electric displacement **D**, and the magnetic induction **B**, which are interrelated by the Maxwell equations. We restrict our attention to **E** because we are interested in the form of the wave rather than its basic physics, and because the photographic recording materials used in holographic interferometry respond primarily to the **E** field.

The simplest type of electromagnetic wave is the linearly polarized plane wave. If such a wave is polarized in the y direction and propagates in the z direction, the three components of **E** are

$$E_x = 0,$$
$$E_y = A\cos(\omega t - kz), \qquad (1.1)$$
$$E_z = 0.$$

Here A is the amplitude of the wave, and its *circular frequency* ω and *wave number* k are given by

$$\omega = 2\pi\nu \qquad (1.2)$$

and

$$k = \frac{2\pi}{\lambda}, \qquad (1.3)$$

where ν is the temporal frequency and λ is the wavelength. The frequency of light is on the order of 10^{15} Hz, and visible light has wavelengths in the range $0.38 < \lambda < 0.76$ μm. The light wave travels at its phase speed $v = \omega/k$. This speed depends on the medium in which the light propagates. Its maximum value, 3×10^8 m/s, occurs in a vacuum and is denoted by c.

The wave described by equations 1.1 was termed a plane wave because at any instant of time **E** has the same value at all points lying in the same plane, z = constant, normal to the direction of propagation. It was termed

1.2 LIGHT, INTERFERENCE, AND COHERENCE

linearly polarized because **E** at any point is always directed along the same line parallel to the y axis. More generally, we describe the direction in which a light wave travels by its *propagation vector* **k**, which has magnitude $k = 2\pi/\lambda$ and points in the direction of propagation. A *plane wave* is a wave whose phase at any instant of time is constant at all points on any plane normal to **k**. If $\mathbf{r} = \hat{\mathbf{i}}x + \hat{\mathbf{j}}y + \hat{\mathbf{k}}z$ is the position vector of any point in space, as shown in Figure 1.1, the equation of a linearly polarized plane wave is

$$E_{x'} = 0,$$
$$E_{y'} = A\cos(\omega t - \mathbf{k}\cdot\mathbf{r}), \qquad (1.4)$$
$$E_{z'} = 0.$$

A surface over which phase is constant, in this case the planes $\mathbf{k}\cdot\mathbf{r} = $ constant, is called a *wavefront*. Another wave of simple form which is important in optics is the *spherical wave* arising from a point source of light. In this case the wavefronts are concentric spheres, $r = $ constant, about the point source. The amplitude of **E** decreases in inverse proportion to the distance from the source.

To discuss the concept of polarization further, we consider the time dependence of **E** at a point in space, such as that located at **r** in Figure 1.1. If x' and y' are any mutually orthogonal axes lying in the plane tangent to

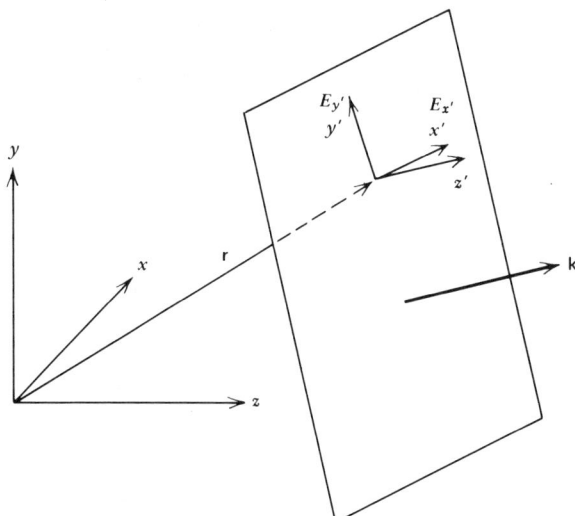

Figure 1.1 Plane wave of light propagating in the direction specified by the propagation vector **k**.

the wavefront at **r**, then

$$E_{x'} = A_{x'}\cos(\omega t - \mathbf{k}\cdot\mathbf{r}),$$
$$E_{y'} = A_{y'}\cos(\omega t - \mathbf{k}\cdot\mathbf{r} + \phi), \qquad (1.5)$$
$$E_{z'} = 0,$$

where ϕ denotes the phase difference between the x' and y' components of **E**. Since $E_{x'}$ and $E_{y'}$ both vary harmonically with time, the tip of the vector **E**, which is their resultant, traces out a closed curve in the x'-y' plane. Equations 1.5 are a parametric representation of this curve. A bit of algebra discloses that in general this equation describes an ellipse, as shown in Figure 1.2a. Such light is said to be *elliptically polarized*. Two

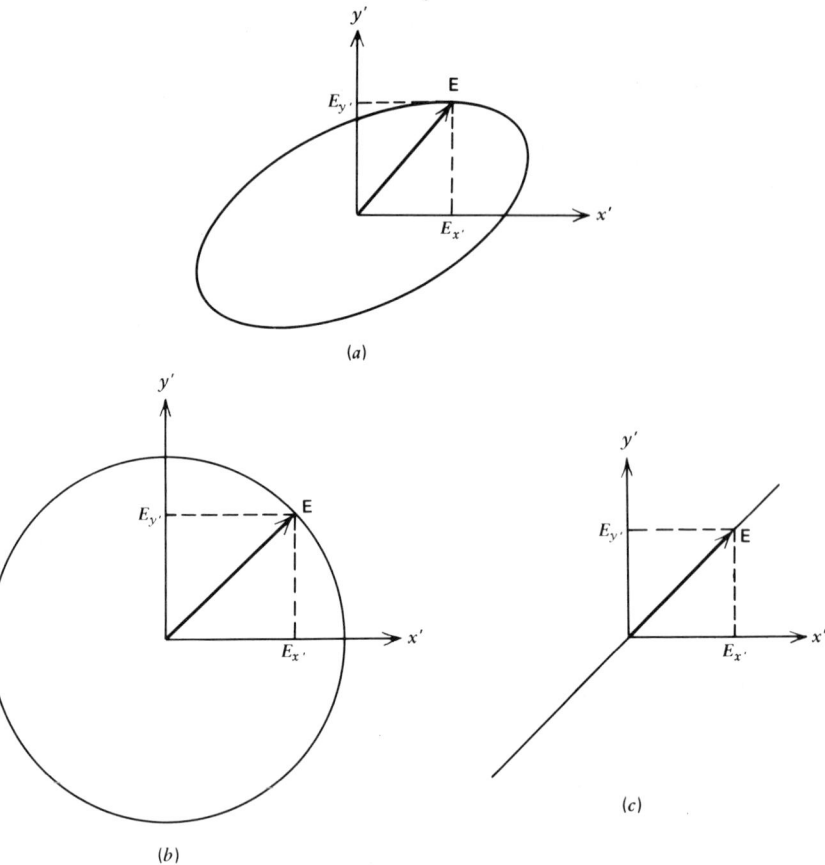

Figure 1.2 Polarization defined by the curve traced by the tip of the **E** vector: (*a*) elliptical polarization; (*b*) circular polarization; (*c*) linear polarization.

1.2 LIGHT, INTERFERENCE, AND COHERENCE

cases of great practical importance arise when the relative phase ϕ and the amplitudes $A_{x'}$ and $A_{y'}$ have special values. The first is *circularly polarized* light, for which $A_{x'} = A_{y'}$ and $\phi = \pm(2N+1)\pi/2$, where N is any positive integer or zero. The second is *linearly polarized* light, for which $\phi = \pm N\pi$. The term "plane polarization" is used synonymously with "linear polarization." The corresponding curves traced out by **E** in these cases are shown in Figures 1.2*b* and 1.2*c*, respectively.

In the vast majority of applications of holographic interferometry a laser is used as the light source. Lasers emit light waves of unusually simple form, having parameters which are quite constant with time and can be measured with high precision. Lasers emit narrow beams of nearly monochromatic light with almost perfectly plane wavefronts. Most lasers emit linearly polarized light. Light from a typical He-Ne continuous wave (cw) laser used for holographic interferometry has a wavelength $\lambda = 632.8$ nm, which is constant to within about 5×10^{-4} nm. It is emitted in a beam of about 2 mm diameter, which diverges at an angle less than 0.7 mrad and is linearly polarized to better than 1 part in 10^3. The most important characteristic of laser light for the applications discussed in this book is its high coherence. Coherence is discussed later in this section.

Spherical and plane waves used in holographic interferometry can be produced from a narrow beam of laser light, as shown in Figure 1.3. The beam is passed through a small positive lens such as a microscope objective of short focal length f_1. After passing through the focal point, the rays diverge to form a spherical wave. If desired, this wave can be *collimated* by using a second lens of larger focal length f_2. If this lens is placed a distance f_2 from the origin of the spherical wave, a plane wave is formed, as shown in the figure. Typically, laser light is linearly polarized in the vertical direction. If desired, this can be converted to circularly polarized light by passing it through a *quarter-wave plate*. (See Section 7.6.)

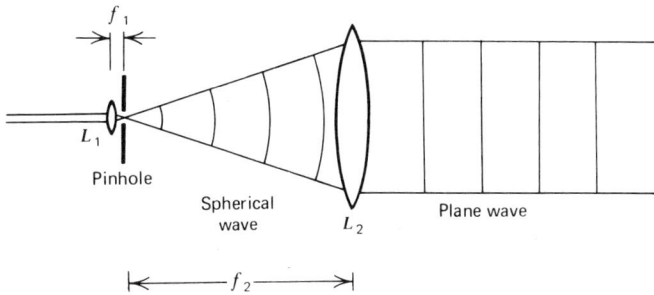

Figure 1.3 Thin collimated wave is expanded by lens L_1 of focal length f_1 to form a spherical wave. A plane wave is then formed by a second lens L_2 of focal length f_2.

The phenomenon of *interference* is central to the subject matter of this book. In the rest of this section we discuss interference and the related property of light referred to as coherence. It was noted above that the frequency of light is approximately 10^{15} Hz. Practical detectors such as photographic film, photodiodes, or the retina of an eye are not capable of responding to such extremely rapid variations. Rather, they respond to *irradiance*, which is the time-average energy flux of the light wave. We denote irradiance by I. Using electromagnetic theory, it can be shown that

$$I = \epsilon v \langle \mathbf{E}^2 \rangle, \tag{1.6}$$

where ϵ is the electrical permittivity of the medium in which the light travels, and v is the speed of propagation. The key point is that I is proportional to the time average of $\mathbf{E}^2$, so the proportionality constant ϵv in equation 1.6 will be dropped in the rest of our discussion. To initiate consideration of interference, we suppose that two different light waves $\mathbf{E}_1$ and $\mathbf{E}_2$ of the same frequency are superimposed. Since $\mathbf{E} = \mathbf{E}_1 + \mathbf{E}_2$, the irradiance will be

$$I = \langle \mathbf{E}^2 \rangle = \langle \mathbf{E}_1^2 \rangle + \langle \mathbf{E}_2^2 \rangle + 2 \langle \mathbf{E}_1 \cdot \mathbf{E}_2 \rangle. \tag{1.7}$$

For simplicity we assume that both waves are linearly polarized in the same direction. We then have a simple scalar computation involving

$$E_1 = A_1 \cos(\omega t - \mathbf{k}_1 \cdot \mathbf{r}) \tag{1.8}$$

and

$$E_2 = A_2 \cos(\omega t - \mathbf{k}_2 \cdot \mathbf{r} + \phi), \tag{1.9}$$

where ϕ is a constant relative phase between the two waves. Combining equations 1.7 to 1.9 and carrying out the averaging, we find that

$$I = I_1 + I_2 + 2\sqrt{I_1 I_2} \cos \delta \tag{1.10}$$

where $I_1 = A_1^2, I_2 = A_2^2$, and

$$\delta = \mathbf{k}_2 \cdot \mathbf{r} - \mathbf{k}_1 \cdot \mathbf{r} - \phi \tag{1.11}$$

is the phase difference between the two waves at any location. The irradiance varies from a minimum value $I_{\min} = I_1 + I_2 - 2(I_1 I_2)^{1/2}$ at points where $\delta = 2N + 1$ to a maximum value $I_{\max} = I_1 + I_2 + 2(I_1 I_2)^{1/2}$ at points where $\delta = 2N\pi$, N being an integer. The irradiance pattern in any plane can be recorded simply by exposing a sheet of photographic film to the

1.2 LIGHT, INTERFERENCE, AND COHERENCE

light. It can also be viewed on a diffusing screen such as a plate of ground glass. In either case a pattern consisting of alternate light and dark fringes will be observed. This fringe pattern enables one to measure the spatial distribution of phase difference between the two waves.

A specific, important example of interference is that of two spherical waves emanating from two point sources of light, S_1 and S_2, as shown in Figure 1.4. Assume that S_1 and S_2 radiate in phase, that is, $\phi=0$. The irradiance at any point P in space is given by equation 1.10 with the phase difference

$$\delta = k(r_1 - r_2).$$

The locus of points forming a surface of maximum irradiance is determined by setting $\delta = 2\pi N$:

$$r_1 - r_2 = \frac{2\pi N}{k} = N\lambda, \qquad N = 0, 1, 2, \ldots . \tag{1.12}$$

This is the equation of a family of hyperboloids of revolution about the axis $\overline{S_1 S_2}$ connecting the two point sources. The interference fringes which would be observed by placing a detector surface such as a sheet of film into the field is the intersection of these hyperboloids with that surface. For example, the intersections with a plane containing $\overline{S_1 S_2}$ are shown in Figure 1.4. The fringes formed on a plane will always be straight lines, or arcs of circles or hyperbolas, depending on the location and orientation of the plane.

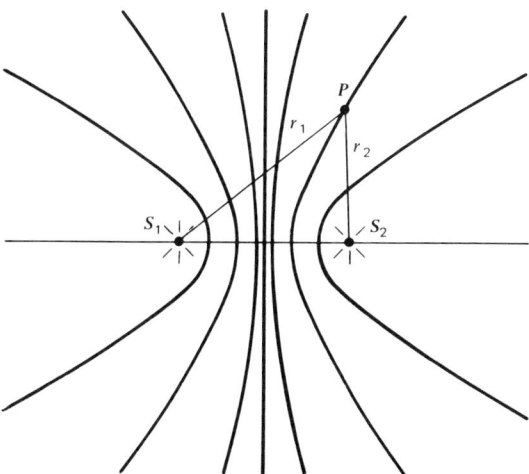

Figure 1.4 Interference of light emitted by two point sources, S_1 and S_2.

Two light waves which are capable of interfering with each other are said to be *coherent* (a term which will be defined more precisely below). Because of coherence requirements most interference experiments are conducted using two images of the same physical source. These images are produced by an instrument called an *interferometer*. There are two basic types of interferometers: *division of wavefront* interferometers and *division of amplitude* interferometers. We will consider briefly one example of each.

A simple division of wavefront interferometer is that used to form *Young's fringes*. This is shown in Figure 1.5. It is simply an opaque screen in which two small holes (or parallel slits) separated by a distance b have been cut. This screen is illuminated by a small point source located a distance l_s behind the screen and a small distance y_s above the axis of symmetry. The light diffracted by the two holes (or slits) forms an interference pattern which can be observed on a screen placed some distance l_o away. In practice, y_s, y, and b are much smaller than l_s and l_o. The irradiance of each of the light waves at y will therefore be nearly equal, $I_1 = I_2 = I_0$, so that equation 1.10 becomes

$$I = 2I_0(1 + \cos\delta)$$
$$= 4I_0 \cos^2\left(\frac{\delta}{2}\right). \tag{1.13}$$

Here the phase difference $\delta = (2\pi/\lambda)\Delta l$, where Δl is the difference in distance the light travels from the source S to the observation point at y:

$$\Delta l = \left\{\left[l_s^2 + \left(\frac{b}{2} - y_s\right)^2\right]^{1/2} + \left[l_0^2 + \left(\frac{b}{2} - y\right)^2\right]^{1/2}\right\}$$
$$- \left\{\left[l_s^2 + \left(\frac{b}{2} + y_s\right)^2\right]^{1/2} + \left[l_0^2 + \left(\frac{b}{2} + y\right)^2\right]^{1/2}\right\},$$

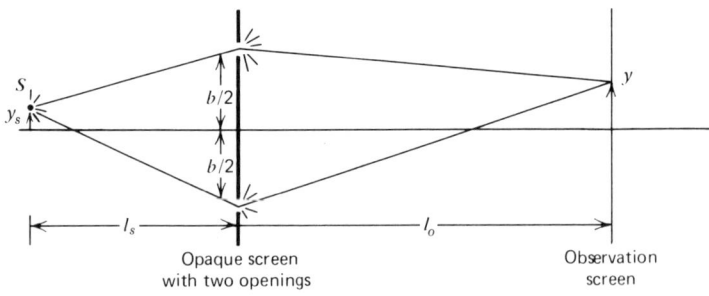

Figure 1.5 Interferometer to form Young's fringes.

1.2 LIGHT, INTERFERENCE, AND COHERENCE

which can be rearranged as

$$\Delta l = l_s \left\{ \left[1 + \frac{(b/2 - y_s)^2}{l_s^2} \right]^{1/2} - \left[1 + \frac{(b/2 + y_s)^2}{l_s^2} \right]^{1/2} \right\}$$

$$+ l_0 \left\{ \left[1 + \frac{(b/2 - y)^2}{l_0^2} \right]^{1/2} - \left[1 + \frac{(b/2 + y)^2}{l_0^2} \right]^{1/2} \right\}. \quad (1.14)$$

Because y_s, y, and b are much smaller than l_0 and l_s, the relation $(1 \pm \epsilon^2)^{1/2} \cong 1 \pm \frac{1}{2}\epsilon^2$ can be used to approximate equation 1.14 as

$$\Delta l = -\left(\frac{by_s}{l_s} + \frac{by}{l_0} \right). \quad (1.15)$$

The irradiance is then proportional to

$$I = I_0 \cos^2\left[\frac{\pi b}{\lambda} \left(\frac{y}{l_0} + \frac{y_s}{l_s} \right) \right]. \quad (1.16)$$

Equation 1.16 represents parallel fringes with a spacing of $\lambda l_0 / b$.

The *Michelson interferometer*, which is a division of amplitude interferometer, is shown in Figure 1.6. It consists of two plane mirrors oriented perpendicularly to each other and a beamsplitter bisecting the angle between the mirrors. A beamsplitter is a glass plate with a partially reflecting coating. In this case we assume that it reflects 50 percent of the incident light and transmits the other 50 percent. A plane wave incident from the left is divided into two plane waves by the beamsplitter. As shown in Figure 1.6a, these waves travel separate paths and are then recombined at the output of the interferometer. The interferometer can be carefully adjusted so that the optical pathlengths traversed by the two waves are identical. A uniform plane wave of light then leaves the interferometer. It is identical to that which entered the instrument.

Now suppose that mirror M_2 is tilted slightly, as shown, somewhat exaggerated, in Figure 1.6b. If the angle of tilt is ϵ, the wavefront reflected by mirror M_2 will be at an angle 2ϵ relative to the wavefront reflected by mirror M_1 when they leave the interferometer. For small angles ϵ the spacing of fringes viewed or photographed at the output plane of the interferometer will be $\lambda/2\epsilon$.

So far in our discussion of interference and interferometers it has been assumed that the illumination is derived from a monochromatic point source. Real sources are never perfectly monochromatic, and they have a

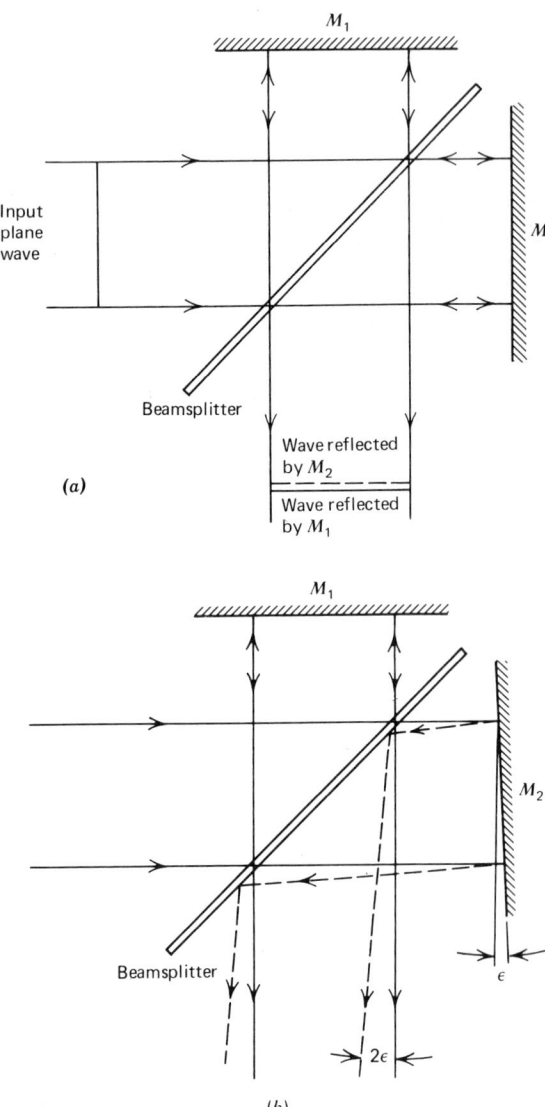

Figure 1.6 Michelson interferometer. (*a*) Mirrors are mutually perpendicular. (*b*) Mirror $M2$ is tilted by an angle ϵ.

finite spatial extent. Investigation of the effects of imperfect monochromaticity and finite source extent leads to a deeper understanding of the meaning of coherence. If, as indicated in Figure 1.7a, the Michelson interferometer is illuminated by a plane wave derived from a very small, but real, source, and if the distances traveled by the two waves are equal, the resulting fringes will have excellent contrast. The dark fringes formed

1.2 LIGHT, INTERFERENCE, AND COHERENCE

by destructive interference will be perfectly black. The contrast of fringes is quantified by defining *fringe visibility* v:

$$v \equiv \frac{I_{max} - I_{min}}{I_{max} + I_{min}}. \tag{1.17}$$

Since the fringes shown in Figure 1.7a have $I_{min} = 0$, their visibility is unity, $v = 1$.

Now suppose that mirror M_1 is translated to extend one arm of the interferometer by a distance $\Delta l_1 / 2$, as shown in Figure 1.7b. The resulting fringes will have only moderate visibility, for example, $v = 0.5$. If mirror M_1 is moved even further to an extension $\Delta l_2 / 2$, as in Figure 1.7c, the fringe visibility will degrade further until no fringes can be seen, $v \to 0$. The decrease in fringe visibility with increasing pathlength difference in the Michelson interferometer is a measure of the *temporal coherence* of the light source. Light is emitted from real sources in discrete wave packets (Figure 1.8). A wave packet of finite extent must be composed of light of a variety of wavelengths; a perfectly monochromatic wave must be of infinite extent. The concept can be understood in terms of a simple model. Assume that the source emits light of two different wavelengths, λ_1 and λ_2, which differ only slightly from the mean wavelength $\lambda_0 = (\lambda_1 + \lambda_2)/2$, and that the light of each wavelength contributes half the irradiance. The light of each wavelength will independently form a fringe pattern. At any point in the fringe pattern the difference $\Delta \delta$ in phase shift between the light of one wavelength and that of the other will be approximately

$$\Delta \delta = \frac{2\pi \Delta l}{\lambda_2} - \frac{2\pi \Delta l}{\lambda_1}$$
$$= \frac{2\pi \Delta l (\lambda_1 - \lambda_2)}{\lambda_1 \lambda_2} \cong 2\pi \Delta l \left(\frac{\Delta \lambda}{\lambda_0^2} \right), \tag{1.18}$$

where $\Delta l / 2$ is the difference in length of the two arms of the interferometer. If fringe visibility is to remain greater than 0.7, $\Delta \delta$ must be less than $\pi / 2$. This yields the criterion

$$\Delta l \leq \frac{1}{4} \left(\frac{\lambda_0^2}{\Delta \lambda} \right). \tag{1.19}$$

Thus, if the mismatch in pathlength exceeds a length of the order of $\lambda_0^2 / \Delta \lambda$, fringe visibility will decay to zero. This length is called the *coherence length*, l_c, of the light source. Temporal coherence is in reality more complicated than is implied in this example. Wave packets are emitted by real sources at random intervals; furthermore, they have randomly different lengths and contain many wavelength components, rather than just two. Nonetheless, if $\Delta\lambda$ is a measure of the spectral width of a source, its coherence length will be of the order of $\lambda_0^2/\Delta\lambda$.

Long coherence length is highly desirable for interferometry in general, and for holography and holographic interferometry in particular. Before the development of the laser, most interferometers had filtered mercury vapor lamps as sources; these have coherence lengths on the order of 0.03

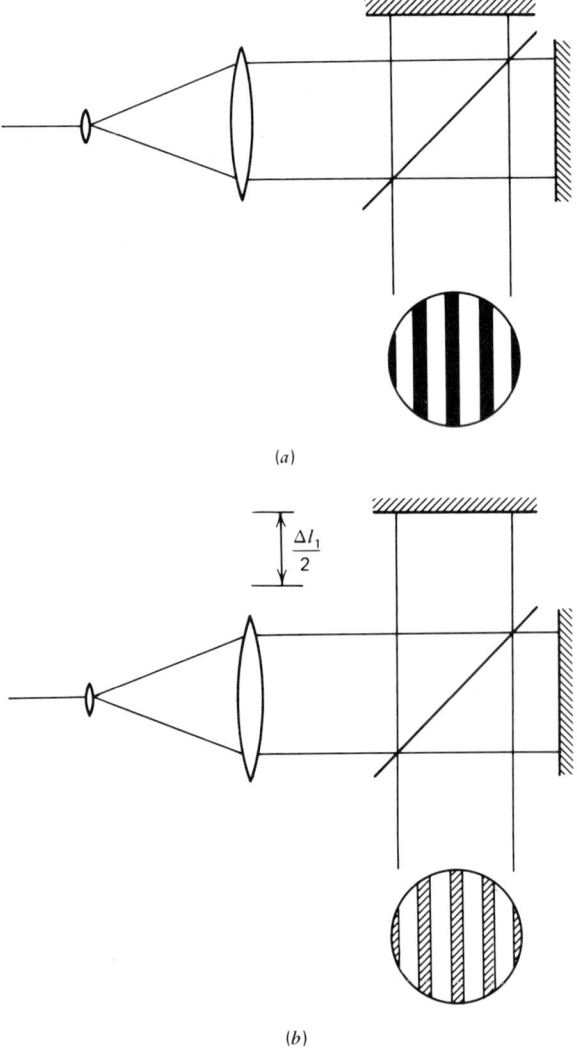

Figure 1.7 Temporal coherence effects displayed by using a Michelson interferometer. (a) Matched pathlengths, $\nu = 1$. (b) Pathlength mismatch Δl_1, $\nu \cong 0.5$. (c) Pathlength mismatch Δl_2, $\nu \cong 0$.

1.2 LIGHT, INTERFERENCE, AND COHERENCE

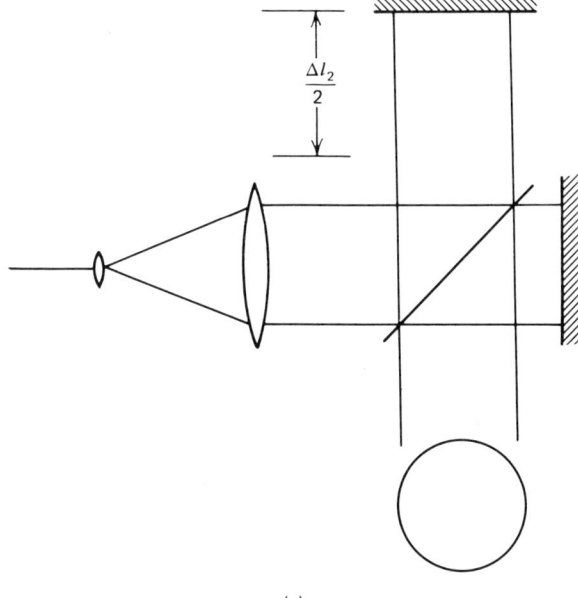

(c)

Figure 1.7 Cont.

to 0.2 mm. Lasers, on the other hand, have vastly greater coherence lengths. Ordinary continuous wave He-Ne lasers have coherence lengths on the order of 20 cm. Furthermore, at the expense of some loss of power, they can be modified to produce a single axial mode, thereby increasing the coherence length to several meters.

Spatial coherence is associated with finite source size. This concept also can be understood in terms of a simple model. Suppose that we perform the Young's fringe experiment using an idealized source that is monochromatic but finite in extent, as shown in Figure 1.9. Each point of the source independently emits light. Although all the light has wavelength λ, there is a random phase variation from point to point on the source. In this model we assume that the light from each point is incoherent with that from the others. The light from each point thus independently forms an interference pattern on the observation screen. The fringe pattern we observe is the superposition of all these patterns. The irradiance at any point y on the

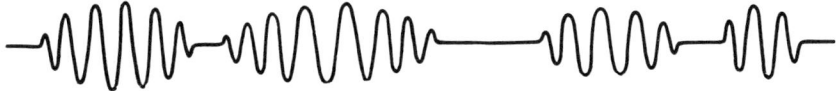

Figure 1.8 Schematic representation of wave packets emitted successively by a light source.

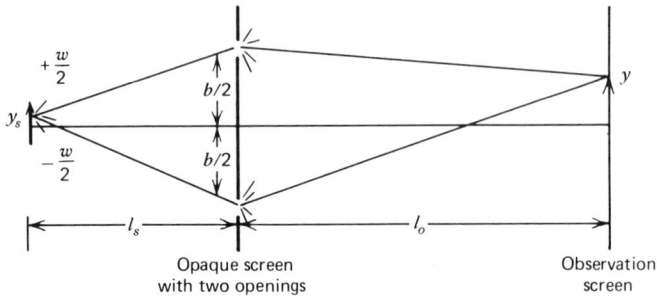

Figure 1.9 Young's fringes formed with a source of finite width w.

observation screen due to light emitted by a source point at y_s is given by equation 1.16. The total irradiance at any point y on the observation screen is found by integrating equation 1.16 over the width of the source:

$$I = \frac{1}{w} \int_{-w/2}^{w/2} I_0 \cos^2\left[\frac{\pi b}{\lambda}\left(\frac{y}{l_0} + \frac{y_s}{l_s}\right)\right] dy_s. \tag{1.20}$$

This integral can be computed after lengthy but straightforward algebra to yield

$$I = \tfrac{1}{2} I_0 \left[1 + \frac{\sin(\pi bw/\lambda l_s)}{\pi bw/\lambda l_s} \cos\left(\frac{2\pi by}{\lambda l_o}\right)\right], \tag{1.21}$$

where I_o is the maximum irradiance in the observation plane. The visibility of these fringes is

$$\nu = \left|\frac{\sin(\pi bw/\lambda l_s)}{\pi bw/\lambda l_s}\right|. \tag{1.22}$$

A plot of this function is shown in Figure 3.5. The fringes approach full visibility, $\nu = 1$, only when the width of the source approaches zero. As the source width increases, the visibility decreases to zero (Figure 1.10). If the width is increased still further, the visibility periodically increases slightly but never exceeds 0.217. The spatial coherence of thermal and gas discharge sources is associated primarily with the spatial extent of the source, as in our idealized model. The spatial coherence of lasers, on the other hand, is associated with the transverse mode structure of the cavity resonance. Most cw lasers are capable of resonating in the TEM_{00} mode, in which all points on the wavefront have essentially the same phase, and therefore they have extremely good spatial coherence.

1.3 DIFFRACTION AND SPATIAL FILTERING

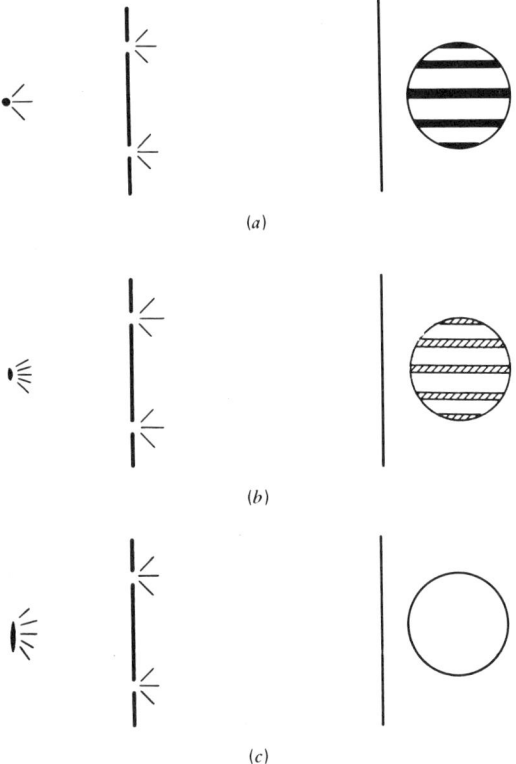

Figure 1.10 Spatial coherence effects displayed by forming Young's fringes with light from monochromatic sources of various widths. Source width increases from (a) to (c).

This cursory discussion of coherence is sufficient for the purposes of this book. A more detailed investigation of the phenomenon reveals that temporal and spatial coherence are really interrelated concepts, and that the inability to produce sources of perfect coherence is a result of the Heisenberg uncertainty principle. Readers interested in pursuing the study of coherence further will find a clear discussion in the textbook by Hecht and Zajac[1] and an in-depth development in the article by Mandel and Wolf.[2]

1.3 DIFFRACTION AND SPATIAL FILTERING

The formation of holograms is an interferometric process and will be described in terms of the concepts presented in Section 1.2. The reconstruction of holographically recorded optical waves involves diffraction, which is the deviation of the propagation direction of a wave due to interaction with an obstacle that changes its amplitude or phase. As we have seen, laser light, which is used for most work in holography and holographic interferometry, is highly monochromatic and coherent. Because of these nearly ideal properties, the diffraction of laser light can be described adequately in terms of elementary diffraction theory. In this section we introduce several concepts and results from diffraction theory which will be used in this book.

To initiate the discussion of diffraction, the representation of light waves presented in Section 1.2 will be modified. In most of our discussions it will be assumed that all light waves being considered are linearly polarized in the same direction, so that only one vector component of **E** need be considered. In other words, we essentially can neglect the vector nature of the electric intensity and treat light as a scalar. For the development of *scalar diffraction theory*, **E** is replaced by a scalar quantity $u(x,y,z,t)$, which we refer to simply as the *optical disturbance*. For example, a monochromatic plane wave is denoted by

$$u(x,y,z,t) = a(x,y,z)\cos(\omega t - \mathbf{k}\cdot\mathbf{r}). \tag{1.23}$$

Similarly, the expression for any monochromatic wave can be written in the form

$$u(x,y,z,t) = a(x,y,z)\cos[\omega t - \phi(x,y,z)], \tag{1.24}$$

where $a(x,y,z)$ is the real amplitude of the light wave, and $\phi(x,y,z)$ is its phase. Of course, the plane wave is a special case for which $\phi(x,y,z) = \mathbf{k}\cdot\mathbf{r}$ is constant over the planes $\mathbf{k}\cdot\mathbf{r} = \text{constant}$. Equation 1.24 can also be written as

$$u(x,y,z,t) = \text{Re}\{U(x,y,z)\exp(i\omega t)\}, \tag{1.25}$$

where

$$U(x,y,z) = a(x,y,z)\exp[-i\phi(x,y,z)] \tag{1.26}$$

is called the *complex amplitude* of the light. Because the temporal frequency of light is so high ($\sim 10^{15}$ Hz), we will have no need to consider the time dependence $\exp(-i\omega t)$ explicitly. The complex amplitude $U(x,y,z)$ contains all the information about the spatial structure of light waves that is essential for our purposes. This convenient notation therefore is used in the rest of this book.

The complex amplitude of a plane wave with propagation vector **k** is

$$U(x,y,z) = a(x,y,z)\exp(-i\mathbf{k}\cdot\mathbf{r}) \tag{1.27}$$

1.3 DIFFRACTION AND SPATIAL FILTERING

or

$$U(x,y,z) = a(x,y,z)\exp\left[-i2\pi\left(\frac{\cos\alpha_1}{\lambda}x + \frac{\cos\alpha_2}{\lambda}y + \frac{\cos\alpha_3}{\lambda}z\right)\right], \tag{1.28}$$

where, as shown in Figure 1.11, α_1, α_2, and α_3 are the angles between the direction of propagation of the plane wave and the x, y, and z axes, respectively. This complex amplitude can also be written in terms of the angles

$$\theta_1 = \frac{\pi}{2} - \alpha_1, \qquad \theta_2 = \frac{\pi}{2} - \alpha_2, \qquad \theta_3 = \frac{\pi}{2} - \alpha_3$$

as

$$U(x,y,z) = a(x,y,z)\exp\left[-i2\pi\left(\frac{\sin\theta_1}{\lambda}x + \frac{\sin\theta_2}{\lambda}y + \frac{\sin\theta_3}{\lambda}z\right)\right]. \tag{1.29}$$

Equations 1.28 and 1.29 are completely equivalent, but the latter notation is generally used in the literature because angles θ_1 and θ_2 are often small and this notation is particularly convenient when $\sin\theta_1 \cong \theta_1$ and $\sin\theta_2 \cong \theta_2$.

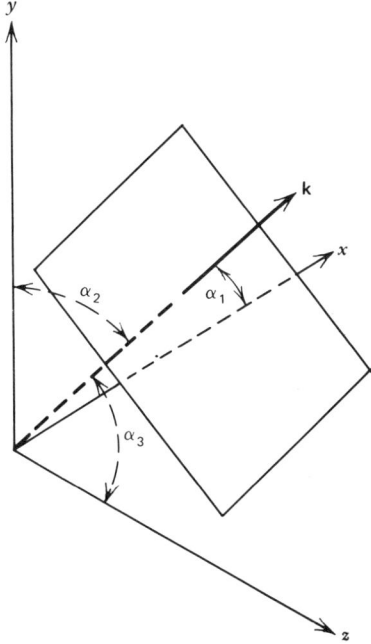

Figure 1.11 Plane wave with propagation vector **k**.

Consider a plane wave whose propagation vector **k** lies in the y-z plane. The intersections of several successive wavefronts with this plane are shown in Figure 1.12. The wavefronts depicted are separated by one wavelength. They intersect the y axis at periodic intervals $\lambda/\sin\theta_2$ and intersect the z axis at periodic intervals $\lambda/\sin\theta_3$. The reciprocals of these periods are called *spatial frequencies*. In general, the spatial frequencies of a plane wave are

$$f_x = \frac{\sin\theta_1}{\lambda}, \qquad f_y = \frac{\sin\theta_2}{\lambda}, \qquad f_z = \frac{\sin\theta_3}{\lambda} \tag{1.30}$$

and are the reciprocals of the period with which successive wavefronts intersect each coordinate axis. The units of spatial frequencies are reciprocal millimeters (mm^{-1}). Specifying the spatial frequencies of a plane wave

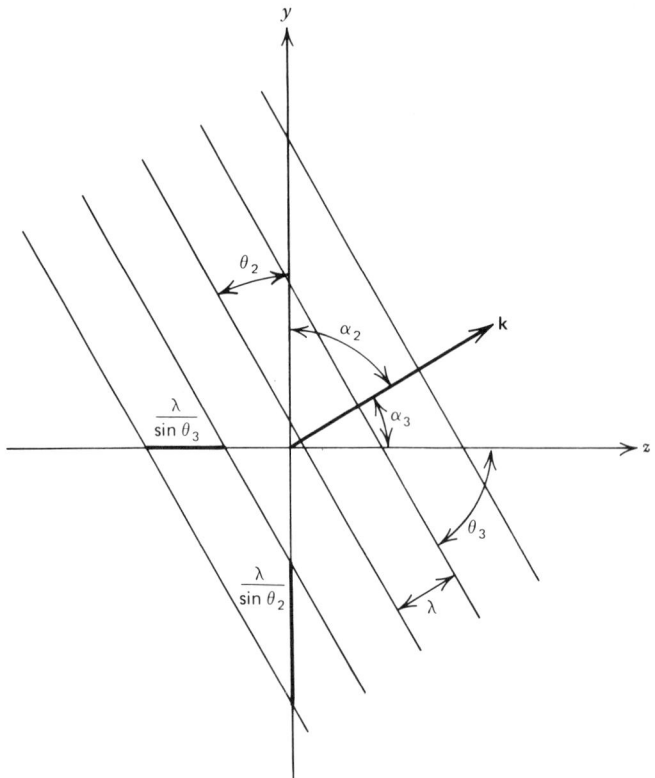

Figure 1.12 Intersections of successive wavefronts with the $y-z$ plane. The propagation vector **k** lies in this plane.

1.3 DIFFRACTION AND SPATIAL FILTERING

with known wavelength determines its direction of propagation. This notation is convenient for our purposes. Interference fringes whose period is the reciprocal of a spatial frequency can often be displayed. This gives a direct physical significance to spatial frequency. By combining equations 1.29 and 1.30, yet another useful notation for a plane wave is obtained:

$$U(x,y,z) = a(x,y,z)\exp\left[-i2\pi(f_x x + f_y y + f_z z)\right]. \qquad (1.31)$$

Frequencies f_x, f_y, and f_z are not independent. The direction cosines of the propagation vector must sum to unity. Correspondingly,

$$f_x^2 + f_y^2 + f_z^2 = \frac{1}{\lambda^2}. \qquad (1.32)$$

When a light wave travels through a transmission object such as a grating, a photographic transparency, or a hologram, its complex amplitude is altered. The *amplitude transmittance* **t** of an object is a function that describes this alteration; it is the ratio of the amplitude leaving to the amplitude entering. With reference to Figure 1.13,

$$\mathbf{U}_o(x,y) = \mathbf{t}(x,y)\mathbf{U}_i, \qquad (1.33)$$

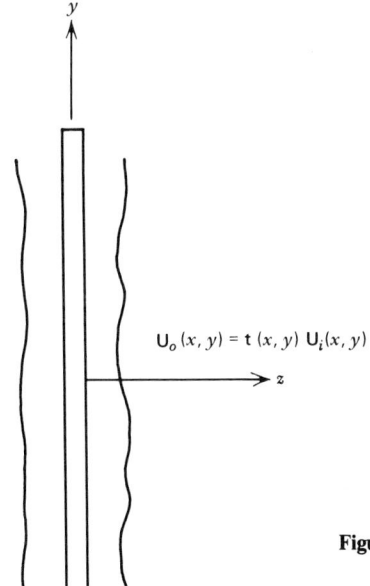

Figure 1.13 Amplitude transmittance $t(x,y)$ describes the alteration of complex amplitude $U_i(x,y)$ due to propagation through an object.

where the surface of the object is parallel to the x-y plane. If the object affects only the real amplitude of the transmitted light, it is termed an *amplitude object* and the transmittance is a real quantity, t. If the object affects only the phase of the transmitted light, it is termed a *phase object* and the transmittance is a complex quantity, **t**. In general, a transmission object such as a hologram affects both amplitude and phase and has a complex transmittance **t**.

An elementary but extremely important example of diffraction occurs when a plane wave of monochromatic light strikes a *sinusoidal amplitude grating*. The grating, which is depicted in Figure 1.14a, has an amplitude transmittance which can be expressed as

$$t(x,y) = t_0 + t_1 \cos(2\pi f_y y), \tag{1.34}$$

where f_y is the reciprocal of the period of the grating fringes. A grating of this type can be manufactured simply by exposing a sheet of photographic film to the light leaving the Michelson interferometer shown in Figure 1.6b. Suppose that a monochromatic plane wave traveling in the z direction impinges on the grating. The complex amplitude of this wave is

$$\mathbf{U}_i(x,y,z) = a_1 \exp\left(-\frac{i2\pi z}{\lambda}\right), \tag{1.35}$$

and at the plane of the grating

$$\mathbf{U}_i(x,y,0) = a_1. \tag{1.36}$$

Equations 1.33 to 1.36 can be used to calculate the complex amplitude of light leaving the grating. It is

$$\mathbf{U}_o(x,y,0) = a_1 t_0 + a_1 t_1 \cos(2\pi f_y y).$$

The physical significance of this amplitude becomes apparent if we express the cosine as the sum of two exponentials:

$$\mathbf{U}_o(x,y,0) = a_1 t_0 + \tfrac{1}{2} a_1 t_1 \exp(i2\pi f_y y) + \tfrac{1}{2} a_1 t_1 \exp(-i2\pi f_y y). \tag{1.37}$$

Equation 1.37 indicates that the sinusoidal amplitude grating divides the incident plane wave into three separate plane waves. One is directly transmitted, and the other two are diffracted into directions determined by the period of the grating fringes. This is depicted in Figure 1.14b.

Another example of diffraction is that which occurs when a plane wave of monochromatic light strikes an amplitude grating whose transmittance

1.3 DIFFRACTION AND SPATIAL FILTERING

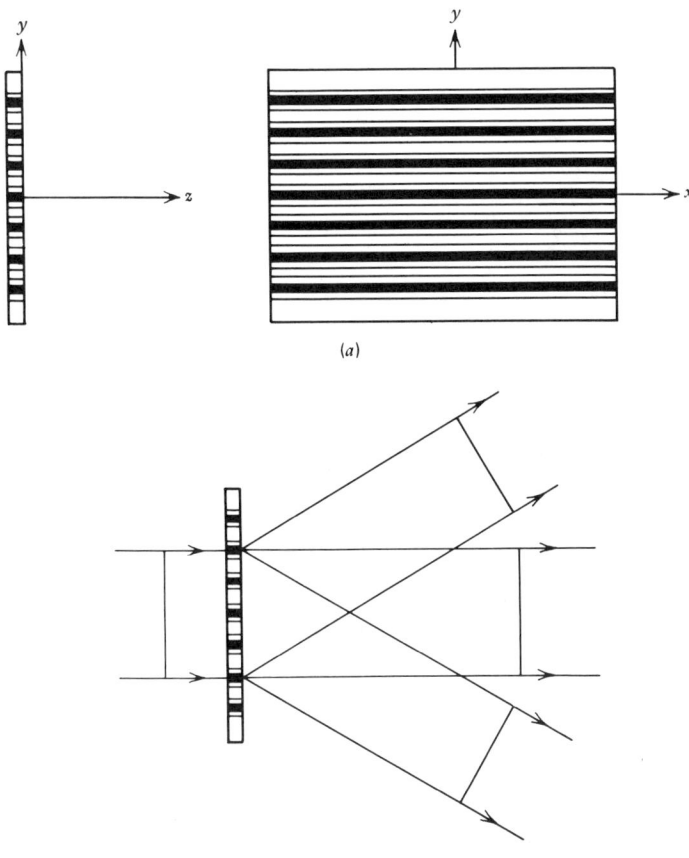

Figure 1.14 Diffraction by a sinusoidal amplitude grating. (*a*) The grating. (*b*) Diffraction of an incident plane wave into three components.

is

$$t(x,y) = t_0 + t_1 \cos\left[\frac{\pi(x^2+y^2)}{\lambda L}\right], \quad (1.38)$$

where L is a characteristic length whose significance will become apparent. This grating is in the form of concentric circular fringes whose spacing decreases with increasing radius, as shown in Figure 1.15*a*. A grating of this type can be manufactured by photographing the fringe pattern formed when a spherical light wave interferes with a plane wave. Its structure is similar to the well-known Fresnel zone plate grating. If a plane wave with

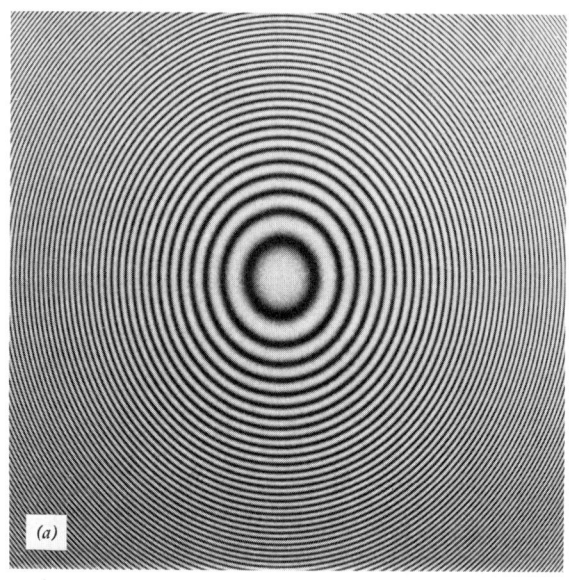

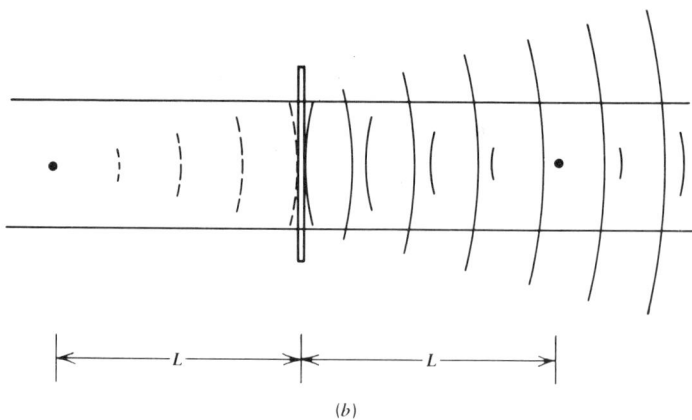

Figure 1.15 Diffraction by a zone-plate-like circular amplitude grating. (*a*) The grating. (*b*) Diffraction of an incident plane wave into three components.

1.3 DIFFRACTION AND SPATIAL FILTERING

real amplitude a_1 traveling in the z direction impinges on the grating, the complex amplitude of light leaving the grating is

$$\begin{aligned}U_o(x,y,0) &= a_1 t_0 + a_1 t_1 \cos\left[\frac{\pi(x^2+y^2)}{\lambda L}\right] \\ &= a_1 t_0 + \tfrac{1}{2} a_1 t_1 \exp\left[\frac{i\pi(x^2+y^2)}{\lambda L}\right] \\ &\quad + \tfrac{1}{2} a_1 t_1 \exp\left[-\frac{i\pi(x^2+y^2)}{\lambda L}\right]. \end{aligned} \qquad (1.39)$$

Equation 1.39 indicates that the plane wave has been divided into three waves. The first represents a directly transmitted plane wave; the second term, a spherical wave that appears to diverge from a point located a distance L to the left of the grating; and the third term, a spherical wave that converges to a point located a distance L to the right of the grating, as shown in Figure 1.15b. This interpretation follows from the expression for the complex amplitude of a spherical wave originating at $x=0, y=0, z=0$:

$$\begin{aligned}\mathbf{U}(\mathbf{r}) &= \frac{a}{r} \exp(-ikr) \\ &= \frac{a}{r} \exp\left[-i\left(\frac{2\pi}{\lambda}\right)(x^2+y^2+z^2)^{1/2}\right]. \end{aligned} \qquad (1.40)$$

If we evaluate this amplitude in a plane normal to the z axis a distance L from the origin,

$$U(x,y,L) = \frac{a}{r} \exp\left[-i\left(\frac{2\pi}{\lambda}\right)(x^2+y^2+L^2)^{1/2}\right].$$

If we consider points relatively close to the axis, $L \gg x,y$ and the following approximations can be made:

$$\begin{aligned}(x^2+y^2+L^2)^{1/2} &= L\left[1+\left(\frac{x}{L}\right)^2+\left(\frac{y}{L}\right)^2\right]^{1/2} \\ &\cong \left\{1+\frac{1}{2}\left[\left(\frac{x}{L}\right)^2+\left(\frac{y}{L}\right)^2\right]\right\} L. \end{aligned} \qquad (1.41)$$

Also, $r \cong L$, so the complex amplitude of the spherical wave is

$$\begin{aligned}U(x,y,L) &\cong \frac{a}{L} \exp\left(-\frac{i2\pi L}{\lambda}\right) \\ &\quad \exp\left[-\frac{i\pi(x^2+y^2)}{\lambda L}\right]. \end{aligned} \qquad (1.42)$$

The first exponential term in equation 1.42 represents a uniform phase over the plane $z = L$ and does not describe the shape of the wavefront. Comparison of the form of equations 1.39 and 1.42 verifies our interpretation of the diffracted waves. Waves such as those represented by the last two terms in equation 1.39, which differ only in the sign of their phase, are said to be *conjugate waves*. Another example of conjugate waves are those represented by the last two terms of equation 1.37.

As a final example, consider a *sinusoidal phase grating*. A phase grating is a transmission object whose refractive index or surface contour varies periodically, as indicated in Figures 1.16a and 1.16b. The amplitude transmittance of a sinusoidal phase grating is

$$\mathbf{t}(x,y) = \exp\left[i\left(\frac{M}{2}\right)\sin(2\pi f_y y)\right], \qquad (1.43)$$

where M is called the *modulation depth* of the grating and represents the maximum phase change introduced when light passes through the grating, and f_y is the reciprocal of the grating period. When a plane wave impinges on a grating of this type, it is divided into a number of plane waves traveling in directions for which the spatial frequencies are integer multiples of f_y of the grating (see Section 5.3). These plane waves are shown in Figure 1.16c. The wave diffracted at the smallest angle is called the *first-order* diffracted wave, the next is referred to as *second-order*, and so on. Order numbers are positive if the spatial frequency is positive, and negative if the spatial frequency is negative. Multiple diffraction orders also arise when amplitude gratings have nonsinusoidal fringes, for example, if they are formed from alternate black and transparent strips.

We have considered diffraction by periodic structures only, which often arise in the theory and practice of holographic interferometry. In addition, these examples are of fundamental importance because an arbitrary thin transmission, or reflection, object can be described by analyzing it into an equivalent set of gratings of various spatial frequencies, orientations, amplitudes, and modulation depths. The problem is a linear one, so the wave diffracted by an arbitrary object can be expressed as a sum of the various plane waves diffracted by the individual gratings. Furthermore, the propagation of complicated light waves can be understood in terms of the increasing spatial separation of the various component plane waves with distance. Image formation with coherent light is described in terms of the recombining of those component waves that pass through a lens or imaging system. These matters are best discussed in terms of the Fourier transform, and constitute what is known as *Fourier optics*. The interested reader can refer to the book by Goodman[3] for a detailed discussion of Fourier optics.

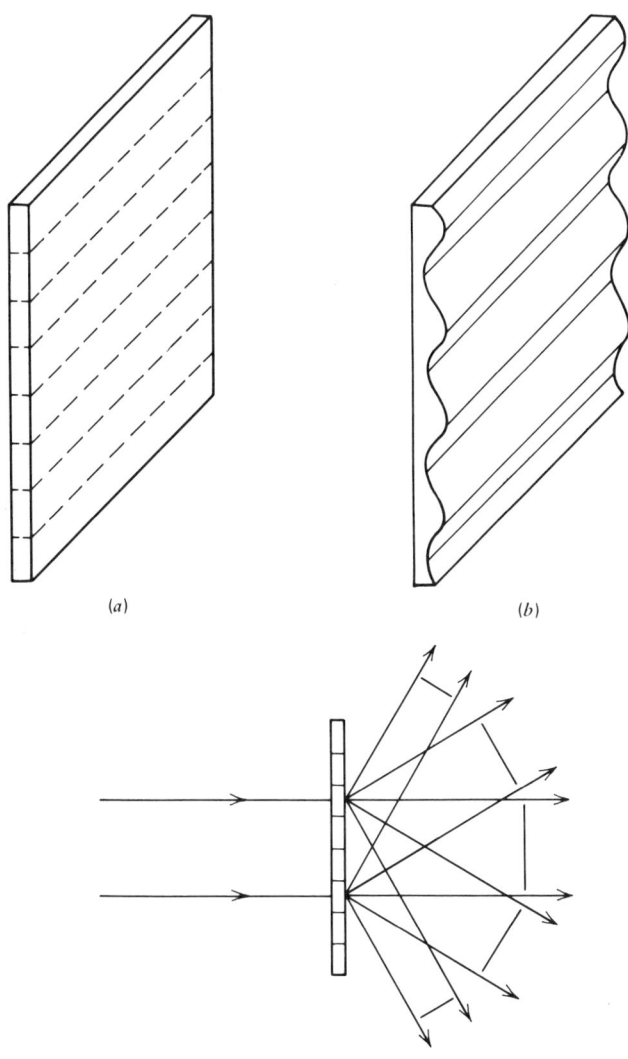

Figure 1.16 Diffraction by a sinusoidal phase grating. (*a*) Phase grating with sinusoidal variation of refractive index. (*b*) Phase grating with sinusoidal variation of surface contour. (*c*) Diffraction of an incident plane wave into three components.

We conclude the present section by briefly discussing *spatial filtering*, a technique which is important in many applications of holographic interferometry. Before doing so, we will review some elementary terminology regarding lenses and imaging systems. A *thin lens* is one whose thickness is small in comparison with other dimensions in the optical system. We restrict our brief review to thin lenses whose surfaces are spherical. The elementary aspects of imaging are most easily described in terms of geometric optics, that is, in terms of light rays rather than wavefronts.

All rays which enter a thin lens parallel to the optical axis are deflected so that they pass through a single point, as shown in Figure 1.17a. This point is the *focal point*, and its distance f from the center of the lens is the *focal length* of the lens. The plane normal to the optical axis that contains the focal point is called the *focal plane*. All rays which pass through the center of a thin lens are undeflected. These facts lead to the simple geometric construction shown in Figure 1.17b, which suggests the *thin-lens imaging equation*:

$$\frac{1}{l_o} + \frac{1}{l_i} = \frac{1}{f}, \qquad (1.44)$$

where l_o is the distance from the lens to the object plane, and l_i is the distance from the lens to the image plane. This equation is approximately valid for real lenses, and is quite accurate when an image is formed using only rays which are close to the optical axis. When an image is formed by rays physically converging to points on the observer's side of the lens, as in Figure 1.17b, the image is said to be *real*. When the rays appear to diverge from image points on the other side of the lens, as shown in Figure 1.17c, the image is said to be *virtual*. The distance l_i in equation 1.44 is negative for virtual images.

The nature of images in general, and the formation of interferograms in particular, are dependent on what portion of the light wave leaving the object passes through the imaging system. The element of an imaging system which determines the amount of light that reaches the image is called the *aperture stop*. In the case of a simple thin lens this is just the diameter of the lens, or of a diaphragm placed in front of the lens. The *entrance pupil* is a parameter of special importance in holographic interferometry. It is the image of the aperture stop as seen from an axial point on the object. The entrance pupil determines the solid angle, or extent of the cone of rays, that enters the imaging system. Similarly, the *exit pupil* is the image of the aperture stop as seen from an axial point on the image; it determines the extent of the cone of rays that leaves the system. For the case of a single-lens imaging system the aperture stop, entrance pupil, and exit pupil are all identical and can be referred to as the *aperture*.

1.3 DIFFRACTION AND SPATIAL FILTERING

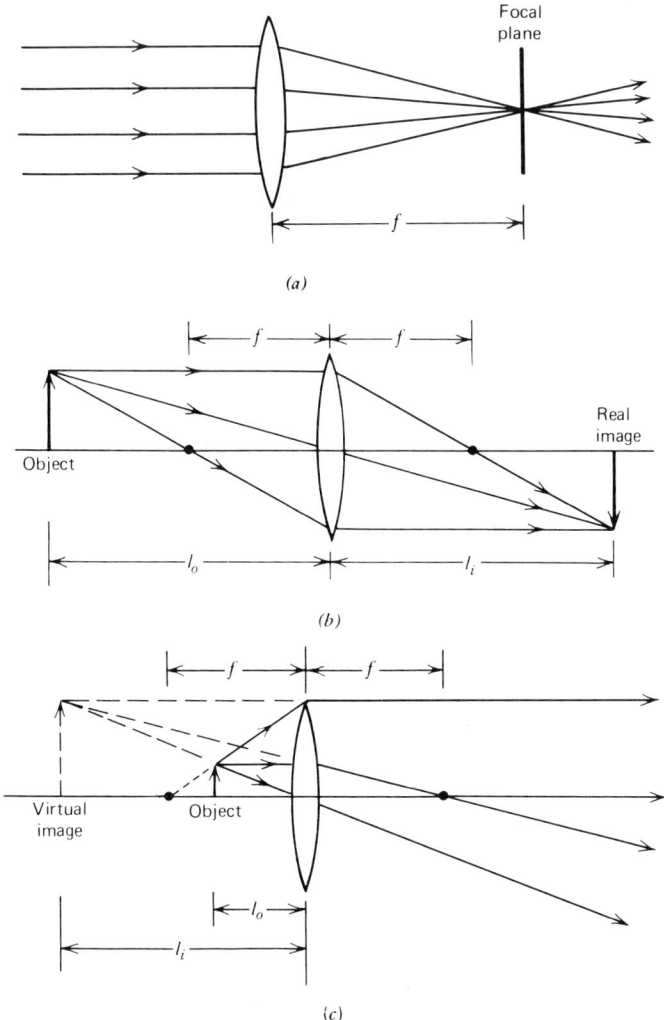

Figure 1.17 Imaging by a thin, spherical lens. (*a*) Lens of focal length f. (*b*) Formation of a real image. (*c*) Formation of a virtual image.

Imaging systems or lenses are usually specified in terms of their relative aperture, or *f number*. This is the ratio of the focal length to the diameter of the entrance pupil, f/D. For example, a lens with a 100 mm focal length and a 25 mm diameter is said to be an $f/4$ lens. A lens with a 200 mm focal length and a 50 mm diameter is also an $f/4$ lens. If the same distant object is imaged by each of these lenses, the average irradiance in their image planes will be identical.

To proceed with our discussion of spatial filtering, we consider the most fundamental spatial filtering operation, which is the isolation of individual plane wave components of light which have been diffracted by some object. As shown in Figure 1.17a, the light in a plane wave traveling in the axial direction will be concentrated by a thin lens to the spot where the optical axis intersects the back focal plane. A plane wave incident from an off-axis direction can be thought of as light from an off-axis object point located a large distance away, that is, $l_o \to \infty$. This light will be focused to an off-axis spot in the back focal plane. As indicated in Figure 1.18a, if the plane wave propagates at an angle with respect to the optical axis it will

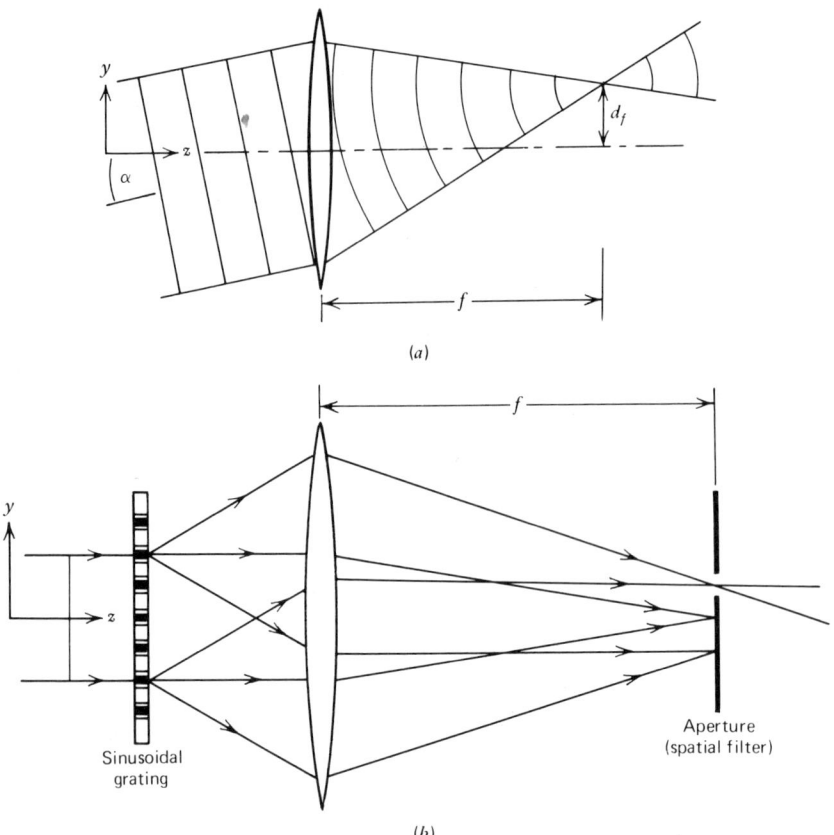

Figure 1.18 Spatial filtering. (a) Light of an off-axis plane wave is focused to an off-axis spot by a thin lens. (b) Example of spatial filtering to isolate a single component plane wave.

1.3 DIFFRACTION AND SPATIAL FILTERING

focus to a point a distance d_f from the axis, where

$$d_f = f \tan \alpha. \tag{1.45}$$

We are concerned primarily with the paraxial case; hence α is small, and

$$\tan \alpha \cong \sin \alpha \cong \alpha.$$

Equation 1.45 can also be expressed in terms of spatial frequency as $d_f = f(\lambda f_y)$. In general, if d_{f_x} and d_{f_y} represent x and y coordinates in the back focal plane,

$$d_{f_x} = f(\lambda f_x) \tag{1.46a}$$

and

$$d_{f_y} = f(\lambda f_y). \tag{1.46b}$$

Complicated wavefronts can be analyzed into a set of component plane waves of various spatial frequencies. Since there is a one-to-one correspondence between spatial frequency (direction of propagation) and location in the back focal plane of a lens, which is referred to as a *transforming lens*, these individual components can be physically isolated. This is done by placing a small aperture in the back focal plane which passes light of the desired spatial frequency, but blocks all other light. This is an elementary but extremely important type of spatial filtering. As an example, consider the system shown in Figure 1.18b. A sinusoidal amplitude grating diffracts an incident plane wave into three directions. This light enters a thin spherical lens, which focuses it to three separate spots in the back focal plane. As shown in the figure, an aperture can be used to transmit any one of the individual plane waves while blocking the others.

As noted in Section 1.2, typical cw laser beams are collimated and have diameters on the order of 1 to 2 mm. To expand these beams for use in holographic interferometry they are passed through a lens of short focal length, usually a microscope objective with a focal length in the range 4 to 32 mm. Because of the great expansion, small dust particles or lens imperfections generate large diffraction patterns in the form of concentric interference rings at various locations in the expanded beam. This optical noise is at best annoying and can make quantitative evaluation of interferograms difficult or impossible. A spatial filter, which is simply a small circular pinhole, can be located at the spot where the laser beam is focused. This passes the desired light of nearly zero spatial frequency but blocks the higher frequency diffracted light. In this manner a clean spherical wave free of large diffraction rings can be produced. Since the diameter of these filters is quite small (typically 5 to 25 μm), mechanically stable fixtures with precision adjustments must be used to locate them accurately in the focal plane of the expanding lens. Many such lens-pinhole assemblies are available commercially.

Spatial filtering is a much more general and powerful technique than is indicated by this brief discussion. In addition to apertures, spatial filters can be constructed that alter the complex amplitude distribution in the focal plane in order to carry out various transformations in optical image and information processing.[3]

1.4 LASER SPECKLE

When one views or photographs a diffusely reflecting (or transmitting) object in laser light, its image has a granular appearance. It seems to be covered with fine, randomly distributed light and dark speckles. If one focuses in front of or behind the object, this speckle pattern is still visible. If the observer moves, the speckles appear to twinkle and move relative to the object. This phenomenon of speckle is inherent in the use of highly coherent light. Since one of the most important features of holographic interferometry is that diffuse objects can be studied, it is useful to be familiar with a few basic characteristics of laser speckle. In this section we consider the irradiance, contrast, and characteristic size of speckle. The reader desiring a more detailed discussion of this phenomenon can consult Dainty.[4]

Figure 1.19 is a photograph of a speckle pattern produced by scattering laser light from a rough surface. In this context "rough" means that the surface has random, microscopic height variations whose scale is greater than a wavelength of light. This encompasses most surfaces other than those which are polished to high optical quality. Speckle can also be produced by transmission through scattering objects such as ground glass or opal glass.

The physical origin of speckle is quite simple. Each point on the object scatters some light to the observer. Because of its high coherence, the laser light scattered by one object point interferes with the light scattered by each of the other object points. When a detector such as film or the retina of an eye is placed in the optical field, it will observe a random pattern of interference fringes which is termed "speckle." The randomness is caused by the surface roughness because the phase of light scattered will vary from point to point in proportion to the local surface height.

To describe speckle quantitatively, we consider its formation when laser light is scattered by a surface whose microscopic contour is shown, greatly exaggerated, in Figure 1.20. We wish to determine the statistical characteristics of the irradiance in the observation plane. Rather than directly describing the variation from point to point in this plane, it is convenient

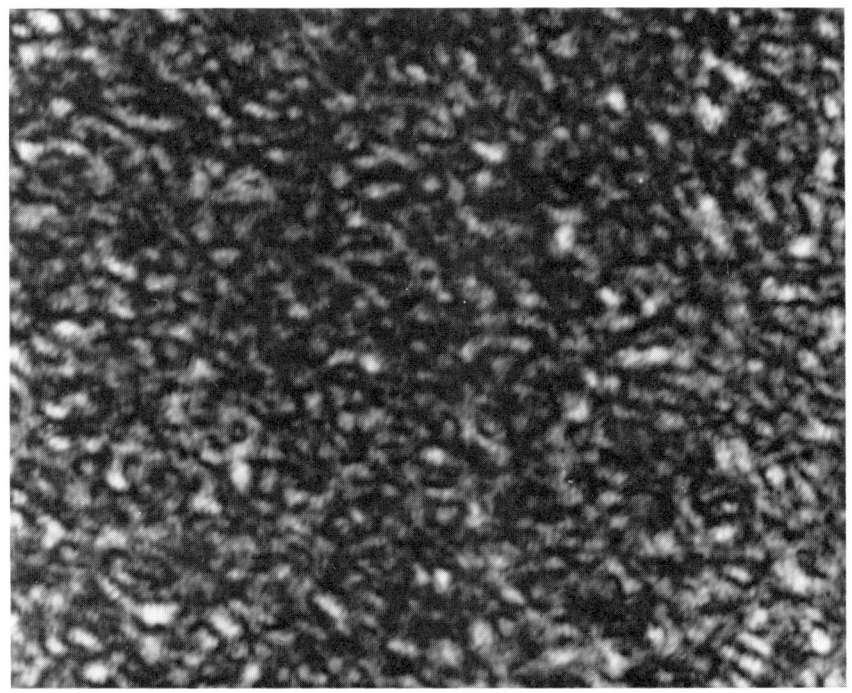

Figure 1.19 Typical laser speckle pattern.

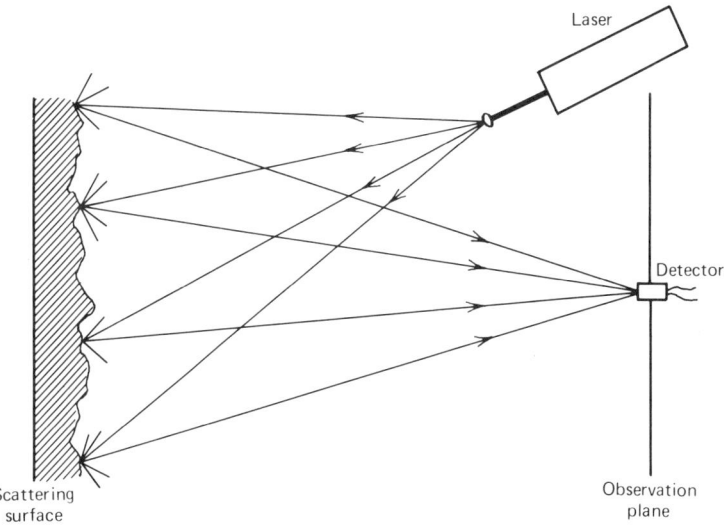

Figure 1.20 Surface with randomly varying microscopic contour scattering laser light to a detector.

to consider the *ensemble average* at a single detection point. In other words, we imagine that the detector remains fixed at location (x,y,z) and that we measure the irradiance produced there by each of a very large collection, or ensemble, of diffusers that are macroscopically identical, but microscopically different. The microscopic surface structures of all of the diffusers have the same statistical characteristics such as rms height. We assume that the incident light is monochromatic and linearly polarized, and that its polarization is not changed by scattering.

Using the notation of equation 1.26, we can write, the complex amplitude at the detector as

$$U(x,y,z) = a(x,y,z)\exp[-i\phi(x,y,z)]$$

and can represent it as a vector in the complex plane, as shown in Figure 1.21. This complex amplitude is actually the sum of a large number N of components which represent the light received from the vicinity of each point of the diffuse surface. If we denote the kth component by $N^{-1/2}U_k(x,y,z)$, then

$$U(x,y,z) = \frac{\sum_{k=1}^{N} U_k(x,y,z)}{\sqrt{N}}$$

$$= \left(\frac{1}{\sqrt{N}}\right) \sum_{k=1}^{N} a_k \exp(-i\phi_k). \qquad (1.47)$$

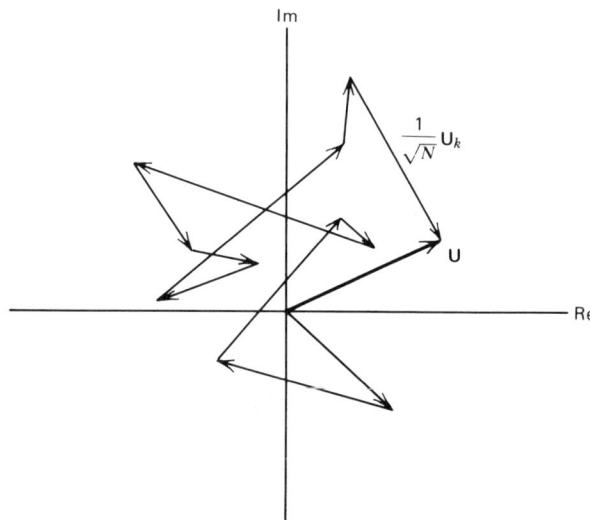

Figure 1.21 Addition of random complex amplitudes.

1.4 LASER SPECKLE

This summation is represented in Figure 1.21.

If it is assumed that (1) the amplitude and phase of each component are statistically independent and are further statistically independent of the amplitudes and phases of all other components, and (2) the phases ϕ_k are uniformly distributed over all values between $-\pi$ and $+\pi$, Goodman[5] has shown that the basic statistical properties of the field can be determined. In fact, the analysis is basically equivalent to the classical problem of a two-dimensional random walk.[6] In this case the random walk is in the complex plane of Figure 1.21. The complex amplitude turns out to obey Gaussian statistics. In particular, the joint probability density function of the real and imaginary parts U is

$$P_{r,i}(U^r, U^i) = \frac{1}{2\pi\sigma^2} \exp\left\{ -\frac{[(U^r)^2 + (U^i)^2]}{2\sigma^2} \right\}, \qquad (1.48)$$

where

$$\sigma = \lim_{N \to \infty} \left(\frac{1}{N}\right) \sum_{k=1}^{N} \tfrac{1}{2} \langle |U_k|^2 \rangle$$

and $P_{r,i}$ is defined so that $P_{r,i} dU^r dU^i$ is the probability that the complex amplitude at the detector will be infinitesimally close to the value specified in its argument by U^r and U^i.

We are more concerned with the irradiance $I = UU^*$, since this is the quantity which is actually detected. Goodman[5] shows that equation 1.48 can be transformed to yield the probability density function for irradiance, $P_I(I)$:

$$P_I(I) = \frac{1}{\langle I \rangle} \exp\left(-\frac{I}{\langle I \rangle}\right) \qquad (1.49)$$

The irradiance of a speckle pattern thus obeys *negative exponential statistics*. Figure 1.22 shows a plot of $P_I(I)$. The most probable irradiance in a speckle pattern is seen to be zero, that is, black. Results of experiments involving tens of thousands of irradiance measurements agree very closely with equation 1.49.[7]

A measure of the *contrast* of speckle is the ratio $C = \sigma_I / \langle I \rangle$, where σ_I is the standard deviation of the irradiance, and $\langle I \rangle$ is its mean value. For the distribution given by equation 1.49 the speckle contrast is unity.

For our purposes the most important statistical characteristic of laser speckle is its size. To estimate the size of a typical speckle, suppose that the speckle pattern is formed by uniformly illuminating a diffuser of width L, as shown in Figure 1.23a. For simplicity, we consider only the y dependence of the irradiance. The speckle pattern formed at a plane located a

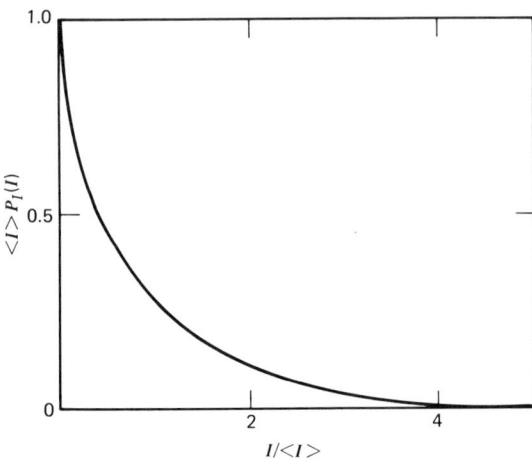

Figure 1.22 Probability density distribution of irradiance in a speckle pattern.

distance z from the diffuser is a superposition of the fringe patterns formed by light scattered by each pair of points on the diffuser. Any two points separated by distance l give rise to fringes of frequency $f = l/(\lambda z)$, as can be seen by inspection of equation 1.16. The finest possible fringes, that is, those of the highest spatial frequency, will be formed by the two end points and will have frequency $f_{max} = L/(\lambda z)$. For smaller separations l, there will be a large number of pairs of points giving rise to fringes of the corresponding frequency. Specifically, the number of pairs of points separated by l is proportional to $(L - l)$. The various fringes will have random phase with respect to each other; therefore, when the ensemble average of irradiance is formed, the contribution of fringes of each frequency will be proportional to the corresponding number of pairs of scattering points. Since this number is proportional to $(L - l)$, which in turn is proportional to $(f_{max} - f)$, the distribution of irradiance over fringe frequency is linear, as shown in Figure 1.24. The average fringe frequency will be

$$\langle f \rangle = \tfrac{1}{3} f_{max} = \tfrac{1}{3}\left(\frac{L}{\lambda z}\right); \tag{1.50}$$

hence the irradiance distribution across a "typical speckle" will be like

$$I(y) = 1 + \cos\left[2\pi\left(\frac{Ly}{3\lambda z}\right)\right].$$

The width of this speckle can be described as the distance between points where I drops to one half of its maximum value. This width is $1.5(\lambda z/L)$,

1.4 LASER SPECKLE

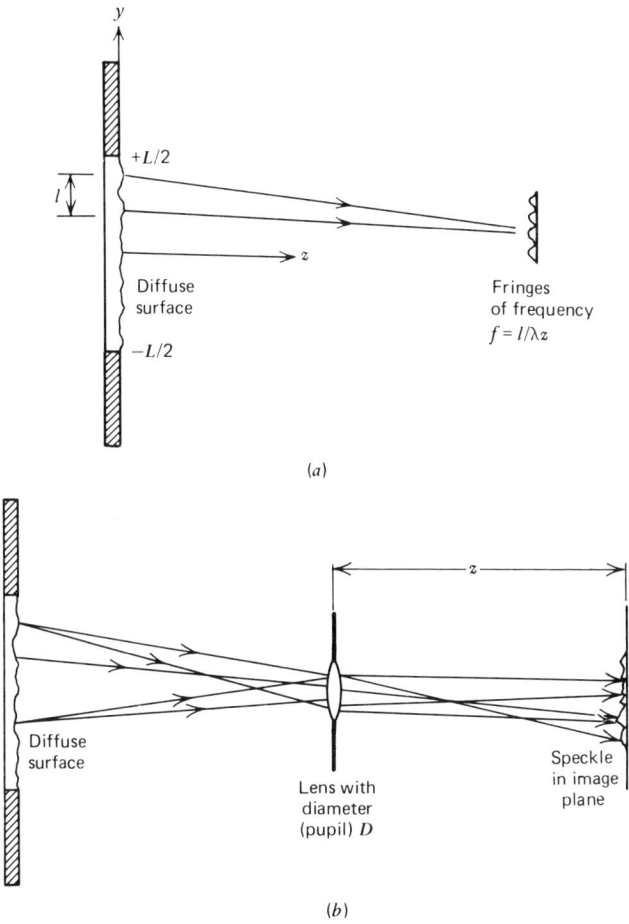

Figure 1.23 Formation of a speckle pattern. (a) Light from two points separated by distance l forms interference fringes of frequency $f = l/\lambda z$. (b) Formation of speckle, using a lens with entrance pupil D.

so we consider a typical speckle width b_s to be

$$b_s \cong 1.5\left(\frac{\lambda z}{L}\right). \tag{1.51}$$

In many cases of interest in this book a diffusely reflecting or transmitting object will be viewed through a lens or imaging system, as indicated in Figure 1.23b. To estimate the speckle size in this case, we need only treat the disk enclosed by the pupil of the lens as a uniformly illuminated

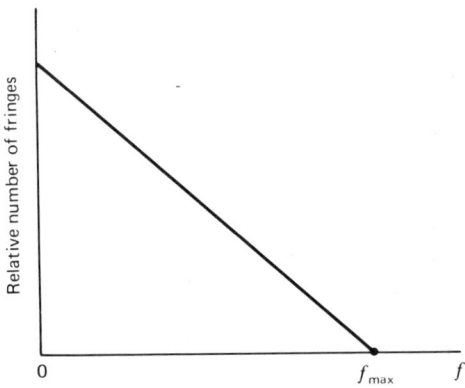

Figure 1.24 Spatial frequency distribution of interference fringes which contribute to speckle pattern.

diffuse surface. If the diameter of the lens pupil is D and the image is formed a distance z from the lens, $b_s \cong 1.5\lambda z/D$. A more rigorous analysis leads to a very similar expression:

$$b_s \cong 1.22\left(\frac{\lambda z}{D}\right). \qquad (1.52)$$

If the imaging system is focused on a relatively distant plane, $z \cong f =$ focal length of the lens; therefore

$$b_s \cong 1.22\lambda\left(\frac{f}{D}\right), \qquad (1.53)$$

where f/D is the f number of the lens. Typical imaging systems vary from about $f/1.4$ to $f/32$. If the speckle pattern is formed by imaging scattered He-Ne laser light ($\lambda = 632.8$ nm), the corresponding speckle size varies from 1 to 24 μm.

1.5 HOLOGRAPHY

Holography is a technique for recording and reconstructing light waves. The wave which is to be recorded is called the *object wave*. In order to reconstruct, that is, produce a facsimile of the object wave, it is sufficient to reproduce its *complex amplitude*, U_o, at one plane in space. Once this has been reproduced, the light propagating away from this plane will be identical to the original object wave. The distributions of both real amplitude and phase in the plane must be recorded; however, photographic film or any other detector responds only to irradiance. The object wave

1.5 HOLOGRAPHY

irradiance is $I_o = U_o U_o^*$, which is a real quantity, so film exposed to U_o can record the distribution of real amplitude, but the distribution of phase will be lost.

It is clear from the discussion in Section 1.2 that interferometry can be used to convert a phase distribution into an irradiance pattern, which can be recorded on photographic film. This is the basis of Gabor's invention of holography, which he described in detail in 1949.[8,9] He proposed to form an interference pattern by adding a coherent *reference wave* to the object wave. This interference pattern can be recorded on film. When the film is developed and illuminated appropriately, it diffracts light in a manner such that the complex amplitude U_o is reproduced at the plane of the film. All this can be accomplished using the simple system shown in Figure 1.25a. A plane wave of monochromatic light passes through a photographic transparency, which is the object. Part of the light will be diffracted by whatever image is recorded on the transparency, and part of the light will pass through the transparency without being scattered. The light which is diffracted is the object wave, and the undiffracted portion of the light serves as the reference wave. The complex amplitudes of these waves at the

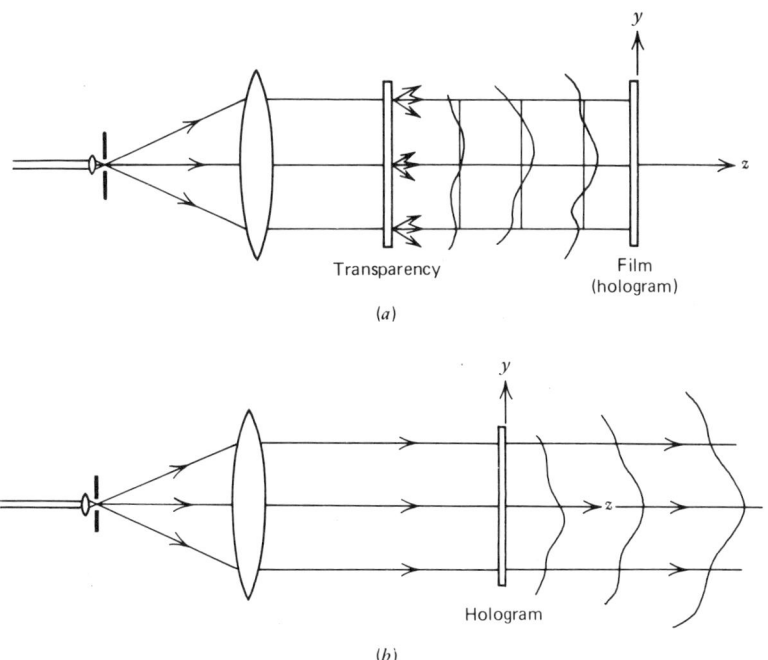

Figure 1.25 In-line (Gabor) holography. (*a*) Recording the hologram. (*b*) Reconstructing the object wave.

film plane, $z=0$, are as follows:

object wave: $\quad\quad\quad \mathbf{U}_0(x,y) = a_0(x,y)\exp[-i\phi_0(x,y)],\quad\quad$ (1.54)

reference wave: $\quad\quad \mathbf{U}_R(x,y) = a_R.\quad\quad$ (1.55)

The irradiance at the film plane is

$$I(x,y) = |a_R + \mathbf{U}_o|^2$$
$$= a_R^2 + |\mathbf{U}_o|^2 + a_R\mathbf{U}_o + a_R\mathbf{U}_o^*. \quad\quad (1.56)$$

We assume that the film is exposed to the irradiance pattern described by equation 1.56, and then is developed so that its amplitude transmittance $t(x,y)$ is proportional to $I(x,y)$:

$$t(x,y) = t_b + \beta(a_R\mathbf{U}_o + a_R\mathbf{U}_o^*). \quad\quad (1.57)$$

The term t_b in equation 1.57 is proportional to the first two terms in equation 1.56. If the object transparency is only weakly scattering, t_b will be nearly uniform over the film and therefore represents a bias level in the exposure. The constant of proportionality β is a property of the film and will be discussed in more detail below.

The developed film with transmittance described by equation 1.57 is referred to as a *Gabor hologram* or, alternatively, as an *in-line hologram*. The object wave is reconstructed by illuminating the hologram with a uniform plane wave of laser light, as shown in Figure 1.25b. This is referred to as the *reconstruction wave*; at the plane $z=0$ its complex amplitude is

$$\mathbf{U}_c(x,y) = a_c. \quad\quad (1.58)$$

When the hologram is illuminated by this wave, the complex amplitude just to the right of the hologram, $z=0+$, will be

$$\mathbf{U}_I(x,y) = ta_c$$
$$= a_c t_b + \beta a_c a_R \mathbf{U}_o + \beta a_c a_R \mathbf{U}_o^*. \quad\quad (1.59)$$

Since the term $\beta a_c a_R \mathbf{U}_o$ is the desired facsimile of $\mathbf{U}_o$, the goal of recording and reconstructing the object wave has been attained. The term $a_c t_b$ represents the portion of the reconstruction wave that is just attenuated and transmitted by the hologram, and $\beta a_c a_R \mathbf{U}_o^*$ is a wave that is proportional to the conjugate of the object wave.

Holographic reconstruction is a process which can be understood physically in terms of the discussion of diffraction presented in Section 1.3.

1.5 HOLOGRAPHY

Suppose that we record a hologram of the light diffracted by a single object point located a distance L from the film plane. The recording system is identical to that shown in Figure 1.25a, except that the transparency is replaced by the single point scatterer. The scattered light will form a spherical wave diverging from this point, so the hologram will be formed by the interference between this spherical object wave and the undiffracted portion of the plane wave. At the film plane the complex amplitudes of the object and reference waves are

$$\mathbf{U}_o(x,y) = \frac{a}{(x^2+y^2+z^2)^{1/2}} \exp\left[-i\left(\frac{2\pi}{\lambda}\right)(x^2+y^2+z^2)^{1/2}\right]$$
$$\cong a_o \exp\left[-\frac{i\pi(x^2+y^2)}{\lambda L}\right] \tag{1.60}$$

and

$$\mathbf{U}_R(x,y) = a_R. \tag{1.61}$$

The transmittance of the developed hologram is

$$t(x,y) = t_b + \beta a_o \exp\left[-\frac{i\pi(x^2+y^2)}{\lambda L}\right]$$
$$+ \beta a_o \exp\left[\frac{i\pi(x^2+y^2)}{\lambda L}\right]. \tag{1.62}$$

This hologram is a zone-plate-like diffraction grating of the type discussed in Section 1.3. When it is illuminated by a plane reconstruction wave, three component waves are produced. One is an attenuated transmitted plane wave. The second component is a spherical wave diverging from an apparent source at the location of the original object point; this corresponds to the reconstructed object wave $\beta a_c a_R \mathbf{U}_o$ in equation 1.59. The third component is a spherical wave converging to a point a distance L to the right of the hologram; this corresponds to the conjugate object wave $\beta a_c a_R \mathbf{U}_o^*$ in equation 1.59. These three components are shown in Figure 1.15b.

Although Gabor holograms are used extensively for forming three-dimensional images of aerosols, sprays, and other distributions of small particles,[10] they are rarely used in holographic interferometry. (A few exceptions to this statement will be noted in Chapter 6.) The primary reason is that the three diffracted waves overlap and are detected simultaneously by an observer's eye or recorded by a camera. This overlapping creates a low-contrast, noisy image, rather than the clean three-dimensional image which is necessary for interferometry.

Generally, *off-axis holography*, developed by Leith and Upatnieks,[11,12] is used in holographic interferometry. This scheme enables one to spatially separate the three waves produced by diffraction in the reconstruction process. It is based on a common procedure in communication theory, namely, the coding and decoding of signals by modulation of a high-frequency carrier wave. In holography, spatial frequencies are used rather than temporal frequencies, but the concept remains the same. The reference wave used to record an off-axis hologram propagates in a different angular direction from the object wave. Figure 1.26a illustrates the formation of an off-axis hologram using a plane reference wave. The complex amplitudes at the film plane, $z = 0$, are as follows:

object wave: $\quad \mathbf{U}_o(x,y) = a_o(x,y) \exp[-i\phi_o(x,y)] \quad$ (1.63)

reference wave: $\quad \mathbf{U}_R(x,y) = a_R \exp(i2\pi f_y y), \quad$ (1.64)

where $f_y = \sin\theta_R/\lambda$ is the spatial frequency of the reference wave. The irradiance at the film plane is

$$\begin{aligned} I(x,y) &= |\mathbf{U}_o + a_R \exp(i2\pi f_y y)|^2 \\ &= |\mathbf{U}_o|^2 + a_R^2 + a_R \mathbf{U}_o \exp(-i2\pi f_y y) \\ &\quad + a_R \mathbf{U}_o^* \exp(i2\pi f_y y). \end{aligned} \quad (1.65)$$

The film is exposed to the irradiance pattern, equation 1.65, and then developed so that its amplitude transmittance $\mathbf{t}(x,y)$ is proportional to $I(x,y)$:

$$t(x,y) = t_b + \beta\big[|\mathbf{U}_o|^2 + a_R \mathbf{U}_o \exp(-i2\pi f_y y) + a_R \mathbf{U}_o^* \exp(i2\pi f_y y)\big]. \quad (1.66)$$

Although equation 1.66 expresses $\mathbf{t}(x,y)$ in a convenient form, it is useful to substitute equation 1.63 and then combine the exponential terms to obtain

$$t(x,y) = t_b + \beta a_o^2(x,y) + 2\beta a_R a_o \cos\big[2\pi f_y y - \phi(x,y)\big]. \quad (1.67)$$

Equation 1.67 clearly shows that the hologram consists of a set of "carrier" interference fringes of spatial frequency f_y which are modulated in amplitude by $a_o(x,y)$ and in phase by $\phi(x,y)$.

To reconstruct the object wave, the hologram is illuminated by a plane wave traveling in the same direction as the original reference wave. Its

1.5 HOLOGRAPHY

complex amplitude at the hologram plane is

$$\mathbf{U}_c(x,y) = a_c \exp(i2\pi f_y y). \quad (1.68)$$

The resulting complex amplitude of light just to the right of the hologram, $z = 0+$, is

$$\mathbf{U}_I(x,y) = (t_b + \beta|\mathbf{U}_o|^2)a_c \exp(i2\pi f_y y) + \beta a_c a_R \mathbf{U}_o + \beta a_c a_R \mathbf{U}_o^* \exp(i4\pi f_y y). \quad (1.69)$$

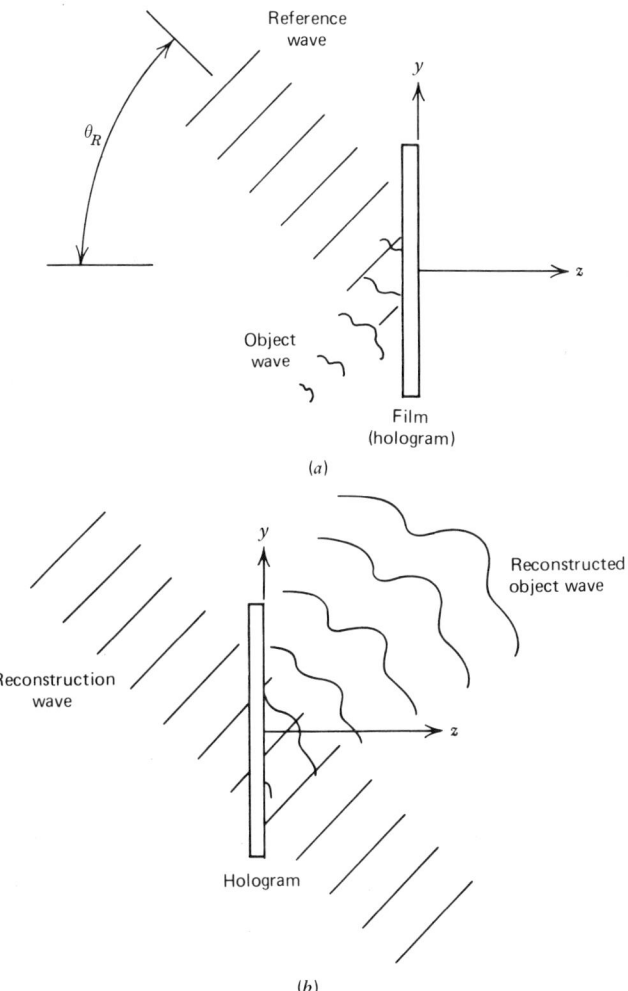

Figure 1.26 Off-axis (Leith-Upatnieks) holography. (*a*) Recording the hologram. (*b*) Reconstructing the object wave.

The first term of equation 1.69 represents a portion of the reconstruction wave which is transmitted by the hologram with attenuation and some irradiance modulation. The second term represents a diffracted wave which is a replica of the original object wave. A virtual image of the object has been formed, and the reconstructed object wave is spatially separated from the transmitted portion of the reconstruction wave (see Figure 1.26b). The third term in equation 1.69 represents a wave which is the conjugate of the original object wave. It propagates at an angle $-\sin^{-1}(2\theta)$ with respect to the z axis. Holograms are usually recorded on emulsions which are from 6 to 35 wavelengths thick. Diffraction effects occurring in holograms of this thickness suppress the conjugate wave unless the offset angle θ is very small; therefore only the reconstructed object wave and the transmitted component are seen.

A typical optical system for recording off-axis holograms is shown in Figure 1.27a. The system is quite simple. A *beamsplitter* is used to divide the incoming laser beam into an object beam and a reference beam. It is possible to use a small piece of plate glass, or perhaps a photographic plate from which the emulsion has been removed, as a beamsplitter. Unless the plate is tilted so that the laser beam strikes it at near-grazing incidence, approximately 90 percent of the light will be transmitted and the rest reflected. If the object is diffusely reflecting, the stronger (transmitted) beam should be used as the object beam because, after scattering, only a small fraction of it will irradiate the film. Better quality beamsplitters are available commercially; they are made by vacuum deposition of a thin metallic film on a glass substrate which is polished flat to $\lambda/4$ or better. The front (metallic) surface may be overcoated with a protective dielectric layer, and the back surface has a dielectric antireflection coating. Beamsplitters can be manufactured with any desired ratio of transmission to reflection, but are commonly supplied with ratios of 90/10, 70/30, and 50/50. The ratio varies somewhat with angle of incidence. A very convenient device called a variable beamsplitter is also available commercially. *Variable beamsplitters* are usually glass disks with reflective coatings whose thickness varies circumferentially. By rotating the disk about its axis a variety of transmission/reflection ratios can be obtained. This device is particularly useful for holographic interferometry. The *front-surface mirrors* used in systems such as the one in Figure 1.27a are similarly manufactured by vacuum deposition of metallic or multilayer dielectric coatings on optically flat glass substrates.

Both the object and reference beams are expanded by passing through a short focal length lens, usually a microscope objective, and filtered by a pinhole spatial filter. To attain uniform irradiance of the object, or across the reference wave, it is desirable to choose the focal length f_1 of the expanding lens so that the diameters of the expanded beams are about

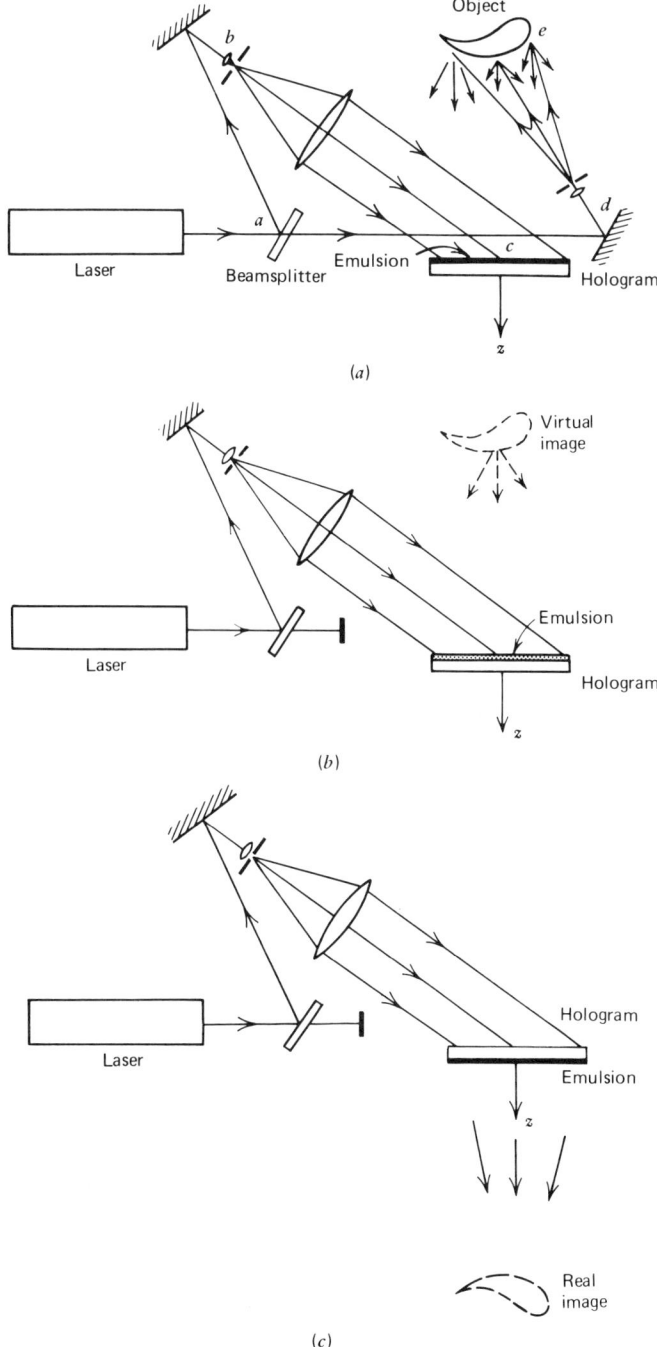

Figure 1.27 Off-axis holography. (*a*) System for recording the hologram. (*b*) Reconstruction of the true, virtual image. (*c*) Reconstruction of the conjugate, real image.

twice the diameter of the object or the collimating lens which they illuminate. The required focal length is

$$f_1 \cong f_2\left(\frac{d}{2D}\right), \qquad (1.70)$$

where f_2 is the focal length of the collimating lens or the distance from the pinhole to the object, d is the diameter of the unexpanded laser beam, and D is the diameter of the collimating lens or of the illuminated object area. If efficiency of light usage is important in a given experiment, it may not be possible to satisfy this condition.

Holograms can be recorded on any photographic emulsion whose resolution and transmittance versus exposure characteristics are appropriate. Commercial holographic emulsions are available on acetate film or on glass plates. The glass plates are usually preferred in holographic interferometry because of their dimensional stability. The plates are mounted in a plate holder which is mechanically stable and holds the plates firmly. For some purposes it is helpful to use a plate holder which can be tilted slightly by turning a micrometer screw. Special-purpose plate holders for holographic interferometry are discussed in Section 4.3.4.

The holographic apparatus should be mounted so that the *relative* motions of the components do not exceed $\lambda/4$. This is easiest to accomplish if all components, including the laser, are mounted on the same rigid surface. If the laser is mounted separately, the pinholes, object, and holographic plate should be rigidly attached to a common surface. The most frequent origin of relative motion of components is building vibration, which is transmitted through the optical table. This can be combatted by using tables of high mass on resilient mountings with low spring constants, thereby achieving a very low natural frequency and filtering out most vibrations. Traditionally, massive, granite optical tables mounted on inflated inner tubes have been used, but these have been superseded by pneumatically isolated tables constructed with honeycomb cores and stainless steel working surfaces. These achieve high rigidity and vibration isolation with lower mass and expense. Satisfactory, inexpensive isolation can be obtained by using sand tables or even bedsprings.

The precise geometric layout of the holographic system is not critical; however, the optical pathlengths of the reference and object beams must differ by less than the coherence length of the laser (see Section 1.2). For example, in the system shown in Figure 1.27a distance $\overline{abc}$ should equal distance $\overline{adec}$. The visibility of the interference fringes that form the hologram decreases as the pathlength imbalance increases, thereby reducing the brightness of the reconstructed holographic image. When the object is large, or the coherence length of the laser is small, the holodiagram,

1.5 HOLOGRAPHY

which is discussed in Section 2.4, is a useful aid to laying out the system within coherence constraints.

Considerable practical information regarding holographic systems can be found in the publication by Kallard.[13]

Once the hologram is formed by exposing and developing the plate, the object wave can be reconstructed by replacing the plate in the holder and blocking off the object wave. The reconstruction wave is then identical to the original reference wave. If a variable beamsplitter is used, it can be adjusted to pass the maximum amount of light into the reconstruction wave. An observer located as in Figure 1.27b will see a realistic, three-dimensional *virtual image* of the object. If the hologram is rotated 180° about a vertical axis so that the emulsion is placed on the observer's side of the plate, the reconstruction wave will be

$$U_c(x,y) = a_c \exp(-i2\pi f_y y), \tag{1.71}$$

which is conjugate to the reference wave. The resulting complex amplitude at $z = 0+$ will be

$$U_I(x,y) = \left(t_b + \beta |U_o|^2\right) a_c \exp(-i2\pi f_y y)$$
$$+ \beta a_c a_R U_0^* + \beta a_c a_R U_o^* \exp(i4\pi f_y y). \tag{1.72}$$

The second term in this equation represents a reconstructed wave that is conjugate to the original object wave. This forms a *real image* in space on the observer's side of the hologram. The image is inverted with respect to the z axis, as shown in Figure 1.27c. Such an image is termed *pseudoscopic*. Since this is a real image, it can be photographed conveniently without the use of lenses by placing a sheet of photographic film in the real image space. If a spherical reference wave is used, the real image cannot be reconstructed simply by rotating the plate. In this case the plate must be rotated and then illuminated by a converging spherical wave whose wavefront curvature at the plate is the negative of that of the reference wave. This is inconvenient experimentally, so one should use a plane reference wave when the use of real images is anticipated.

Most holograms are recorded on high-resolution photographic emulsions which consist of minute (~ 0.4 μm) crystals of silver halide suspended in a gelatin layer. This emulsion is usually coated on a glass plate, although acetate films are also used in some applications. When exposed to light, the crystals absorb optical energy; this results in the formation on the crystal of tiny patches of metallic silver, which are called *development centers*. The number of crystals with development centers will be high in regions where the irradiance to which the emulsion was exposed was high, and low where the irradiance was low. The emulsion is therefore said to contain a *latent image* of the irradiance distribution. When the film is placed in an appropriate chemical developer, a complex reaction occurs which results in complete conversion of silver to metallic form in crystals with development centers, while crystals without development centers are unaffected. Processing is completed by placing the developed emulsion in a chemical fixer which removes the remaining, unexposed silver halide. The grains of metallic silver are opaque; hence, macroscopically, the regions of emulsion which were exposed to high irradiance will be dark, and those exposed to low irradiance will be nearly transparent. The developed emulsion therefore contains a negative image of the irradiance distribution. Some quantitative knowledge of the characteristics of photographic emulsions is needed in order to record holograms of good quality. In particular, resolution requirements must be set, and exposure and development procedures must be chosen so that the assumed linear relationship between irradiance and amplitude transmittance, equation 1.66, is approximately satisfied.

Obviously the emulsion must be capable of resolving fringes at the carrier frequency of the hologram. A typical offset angle (θ_R in Figure 1.26a) might be 30°, so if an He-Ne laser with wavelength $\lambda = 632.8$ nm is used, the carrier frequency is

$$f_y = \frac{\sin \theta}{\lambda} = \frac{\sin 30°}{632.8 \times 10^{-6}}$$
$$= 790 \text{ lines/mm}.$$

This value represents a rather minimal resolution requirement, although even lower resolution can be used in a few applications involving interferometry of transparent objects. If we allow for larger offset angles and for large, diffuse objects which scatter light over a wide range of directions, a resolution of 1000 to 2000 lines/mm is desirable. This criterion is met by several emulsions designed for spectroscopy or holography, a number of which are listed in Table 1.1.

The transmittance of a photographic transparency such as a hologram is a function of exposure, that is, the energy density per unit area of light to which the film is subjected. Exposure is the product of irradiance and exposure time t_e:

$$E = It_e. \tag{1.73}$$

The transmission characteristics of films can be displayed on sensitometric curves, the most common of which is the Hurter-Driffield curve. This is a

1.5 HOLOGRAPHY

TABLE 1.1 NOMINAL CHARACTERISTICS OF HOLOGRAPHIC EMULSIONS

Type	Thickness (μm)	Resolution (lines/mm)	Nominal[a] Sensitivity ($\mu J/cm^2$)	Wavelength (nm)
Kodak emulsions				
649F	17	2000	70	632.8
120-02	6	2000	30	632.8
125	3	1250	2	441.6
			5	514.5
131	9	1250	0.5–0.8	632.8
Agfa Gavaert emulsions				
8E75	7	5000	10	632.8
10E75	7	2800	2	632.8
			1.8	514.5
8E56	7	5000	15	476
			25	521
10E56	7	2800	1.4	476
			1.9	521
14C70		1500	0.3	700
14C75		1500	0.3	700
Illingford emulsion				
He-Ne 1		—	5	632.8
USSR emulsions				
ФПГВ	—	2800	1-2	632.8
Mikrat-900	—	2800	5-10	632.8
SO-243	—	500	0.2	632.8
ВРП	—	2800	5-10	632.8

[a] These values may be used for *estimation* of exposure times. There is considerable variation in reported data, and the sensitivities may correspond to various densities in the range 0.4 to 1.0. Also, there are variations in reported values for resolution.

plot of density versus log (exposure). *Density* is defined by

$$D \equiv \log\left(\frac{1}{|t|^2}\right), \tag{1.74}$$

where **t** is the *amplitude transmittance*; $|t|^2$ is called the *intensity transmittance* and is a relevant parameter for incoherent as well as coherent light. A typical D-log (E) curve is shown in Figure 1.28a. A substantial portion of this curve is nearly linear. Photographic films are often characterized by

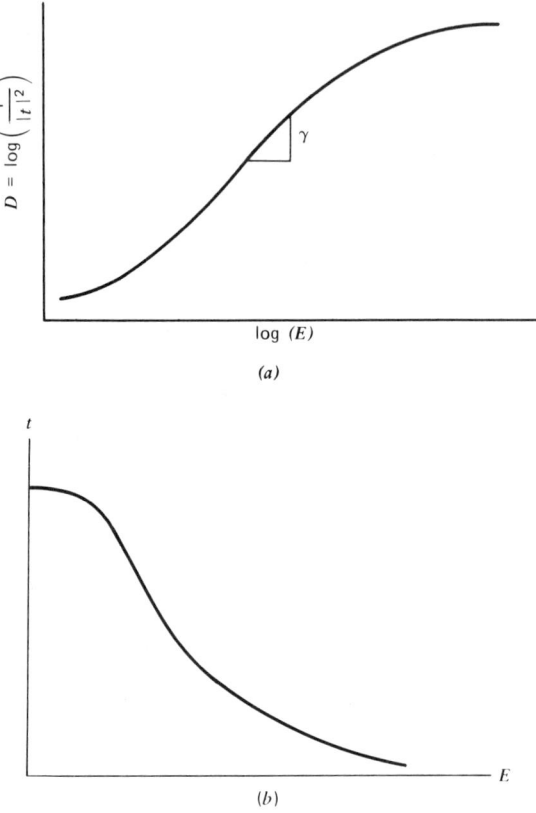

Figure 1.28 Sensitometric curves for photographic emulsion. (*a*) Hurter-Driffield curve $[D - \log(E)]$. (*b*) Transmittance-exposure curve.

the slope γ of this linear portion. The γ of a film can be controlled to some extent by the development procedure.

Density is a natural parameter for ordinary photography because human beings apparently judge brightness differences on an approximately logarithmic scale. Also, the number of silver grains per unit area in a developed emulsion has been found to be proportional to $\log(1/|t|^2)$. In holography, amplitude transmittance is of more direct usefulness than density, so sensitometric data are often displayed as real amplitude transmittance versus exposure.[14] A t-E curve is shown in Figure 1.28*b*. A third plot, t-$\log(E)$, is also used by many workers.

The transmittance-exposure characteristics of holographic film are used to set two important experimental parameters: exposure time t_e, and the ratio of reference-to-object-wave irradiance I_R/I_o. An ideal hologram

1.5 HOLOGRAPHY

should be linear, as in equation 1.66, have a high signal/noise ratio, and have a high diffraction efficiency η_{diff}, so that a bright reconstruction will result. *Diffraction efficiency* is defined by

$$\eta_{\text{diff}} \equiv \frac{I_I}{I_c}, \tag{1.75}$$

where I_I is the average irradiance of the reconstructed holographic object wave, and I_c is the irradiance of the reconstruction wave. An additional criterion may be good light economy or, equivalently, short exposure time. Unfortunately, these criteria are not mutually compatible, so some compromises must be made.

To obtain linearity, the average exposure of a hologram should be set at the center of the linear portion of the t-E curve [which does not coincide with the linear portion of the D-log (E) curve]. Fortunately, for most holographic films maximum diffraction efficiency also occurs at an amplitude transmittance that is close to the center of the linear portion of the t-E curve. There is some disagreement in the literature regarding the precise transmittance which yields maximum diffraction efficiency. The value $t = 0.5$, which corresponds to $D = 0.6$, is often quoted, but the correct value is apparently closer to $D = 1$, because films have a base density of approximately 0.3. In any event, holograms should be exposed and developed to attain a density slightly less than 1. By ordinary photographic standards this is an underexposure, that is, the plate will appear lighter than a typical photographic negative, which would be centered on the linear portion of the D-log(E) curve. The exposure E at which maximum diffraction efficiency occurs is called the *sensitivity* of the emulsion. Nominal values of this parameter are tabulated for several emulsions in Table 1.1.

In off-axis holography, exposure is basically spatially periodic:

$$E = E_0 + E_1 \cos(2\pi f_y y). \tag{1.76}$$

Here E_0 is the mean exposure, which we assume is set close to the center of the linear portion of the t-E curve. The modulation, $m = E_1/E_0$, of exposure is inversely proportional to the reference-to-object-beam ratio:

$$m = \frac{E_1}{E_0} = \frac{I_o}{I_R}. \tag{1.77}$$

Maximum diffraction efficiency of the hologram will result if the modulation is unity, so that the carrier fringes of the hologram have maximum contrast. On the other hand, if $m = 1$, amplitude transmittance will not vary linearly with exposure. The reason for this is indicated in Figure 1.29a, which depicts the distortion, due to the t-E characteristics of the film, of a fringe pattern which has modulation $m = 1$. A nonlinear response like this one gives rise to additional diffraction components which create a noisy holographic image. Better linearity can be achieved by using a larger reference-to-object-beam ratio so that $m < 1$, as indicated in Figure 1.29b. Reference-to-object-beam ratios from 3 : 1 to 10 : 1 are often recommended in order to achieve linearity.

An additional complication is that the response of a film to a modulated pattern is spatial frequency dependent. This effect is due to scattering of light in the emulsion during exposure. It is described quantitatively by the *modulation transfer function*,[15] or MTF, which we denote by $\mathbf{M}(f)$:

$$\mathbf{M}(f) = M(f)\exp\left[-i\Omega(f)\right]. \qquad (1.78)$$

A real film given the spatially periodic exposure described by equation 1.76 will, after development, transmit light as if it were a linear film which had been given a somewhat different exposure E_{eff}:

$$E_{\text{eff}} = E_0 + M(f)E_1\cos\left[2\pi f_y y - \Omega(f)\right]. \qquad (1.79)$$

Films appropriate for holography, where high spatial frequencies must be recorded, have MTFs which are high and nearly independent of f. A plot of $M(f)$ versus spatial frequency for such films exhibits a long, nearly horizontal, and rather straight curve, which sometimes is called the *Frieser plateau*. These curves are given by Johansson and Biedermann[16] for several holographic films.

Biedermann[17] has found that diffraction efficiency can be correlated with several of the film characteristics by the following equation:

$$\eta_{\text{diff}} = \left[\tfrac{1}{2}\log(e)\cdot m\cdot M(f)\cdot\alpha(\overline{E}\,)\right]^2, \qquad (1.80)$$

where $\alpha = \tfrac{1}{2}ln(10)\cdot t\cdot\gamma$ is measured at the mean exposure $\overline{E}$. The parameter α is the slope of the t-$\log(E)$ curve, which many workers prefer to the t-E curve to characterize holographic recording media.

In holographic interferometry the ultimate objective usually is to produce interference fringes of high visibility in a photograph of the reconstructed object wave. Experimental evidence indicates that this objective will be achieved at a beam ratio of approximately 1 : 1.[18,19]

It is recommended that a reference-to-object-beam ratio of 1 : 1 *be used for holographic interferometry, but this ratio can be increased if it is necessary to reduce exposure time. Nominal exposure times can be determined by dividing*

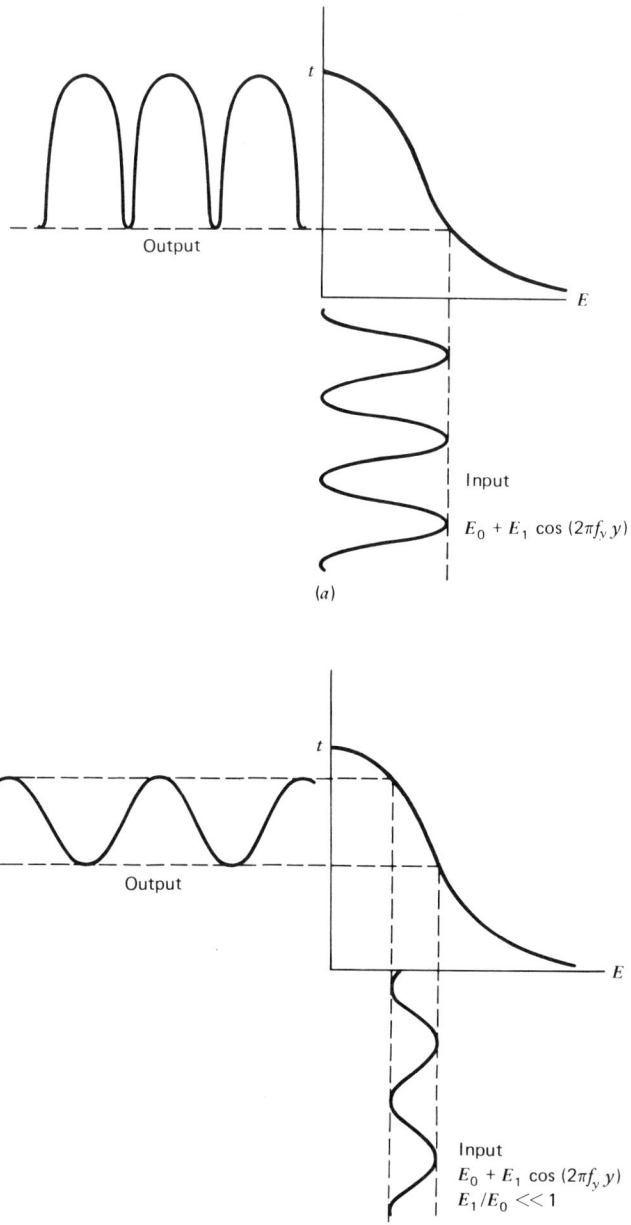

Figure 1.29 Effect of emulsion characteristics when recording periodic irradiance patterns. (*a*) Distortion of pattern due to nonlinear emulsion characteristics. (*b*) Minimization of distortion achieved by reducing modulation of the exposure.

the sensitivity listed in Table 1.1 by the irradiance measured in the hologram plane.

Once the photographic plate has been exposed to form a hologram, it must be developed. The following procedure is satisfactory for holograms recorded on either Kodak or Agfa–Gavaert plates:

1. Develop for 5 min in Kodak D-19 with mild agitation, and then rinse for 30 s in water at 20°C.
2. Fix for 3 to 5 min in Kodak Rapid Fixer with hardener.
3. Wash for 10 min in running water.
4. Dip plate in dilute Photo-Flo solution and then air-dry slowly at room temperature.

Some deviations from this simple procedure are recommended in later chapters for specific applications.

Sometimes it is desirable to form a *phase hologram* because the diffraction efficiency can thereby be increased, typically to 20 percent or even as high as 75 percent, whereas the maximum for amplitude holograms is only about 4 percent. Phase holograms are also used in nonlinear holographic interferometry, which is discussed in Chapters 5 and 6. A phase hologram can be made by bleaching an ordinary amplitude hologram. *Bleaching* is a process by which the metallic silver grains in the developed emulsion are converted to transparent silver salts whose refractive index differs from that of gelatin—for example, silver bromide ($n=2.25$) or silver chloride ($n=2.07$). In this manner the density distribution is converted to a refractive index distribution. In other processes the metallic silver grains are removed, thereby leaving a variation in the surface relief of the emulsion. Many bleaching processes have been developed and described in the literature. We describe two simple processes here.

The simplest bleaching procedure is the bromine vapor bleach developed by Graube.[20] The plate is overexposed to achieve a high density ($D \cong 3$) and then processed and dried as described above. A thin layer of liquid bromine is placed in the bottom of a flat glass tray. The developed hologram is suspended, emulsion side down, approximately 1 cm above the bromine. In approximately 15 min the plate will be observed to clear to essentially uniform transparency. It is then removed from the bromine vapor and allowed to stand overnight in flowing air under a fume hood to remove a yellowish bromine stain from the emulsion. It must be emphasized that bromine is highly toxic. *The bleach should be applied in a fume hood, and protective goggles and rubber gloves should be used.* Plates bleached in this manner have been observed to become clouded after several months' exposure to air.

A simple wet bleaching procedure is to expose the plate to a high density ($D \cong 2$) and then process it in the usual manner. After fixing and rinsing (but not drying), the plate is bleached until it clears in a solution of 15 g of potassium ferricyanide dissolved in 1000 ml of distilled water.[21] Finally, the plate is rinsed for 5 min in running water and air dried. This bleach solution is toxic and caustic, so protective goggles and rubber gloves should be used.

Other bleaching procedures are described in refs. 22 to 31.

Holograms can be recorded in many other materials such as dichromated gelatin, photoresist, photopolymers, photochromic materials, electro-optic crystals, magneto-optic materials, and, in the case of infrared lasers, various metallic films. Currently, most holographic interferometry is done with silver halide emulsions, but in the future other materials are likely to play an important role. In particular, erasable, reusable recording media will be especially important in applications to commercial testing. The reader may wish to consult the comprehensive monograph edited by Smith[32] for discussion of the properties of various recording materials and for an extensive bibliography on the subject.

In some applications of holographic interferometry a parameter such as wavelength or the position of the point source of the reference wave is changed between recording and reconstruction of the hologram. For example, a hologram may be recorded with a pulsed ruby laser ($\lambda = 694$ nm) but viewed using a continuous wave He-Ne laser ($\lambda = 632.8$ nm), or the object wave may be tilted between two exposures in order to introduce reference fringes into an interferogram. We now proceed to develop the *holographic imaging equations* which provide a convenient method for calculating the effects of such changes.

Let the hologram lie in the x-y plane, as shown in Figure 1.30, and consider a single *object point* a distance R_o from the center of the hologram in a direction defined by the angles α_o and β_o. Light is scattered as a spherical wave from this object point, giving rise to the complex amplitude

$$U_o = \frac{a_o'}{r_o} \exp\left[-ik_o(r_o + S_o)\right]$$

at point (x,y) on the hologram. Here $k_o = 2\pi/\lambda_o$ is the wave number of the laser light in the object beam, r_o is the distance from the object point to (x,y), and S_o is the distance from the illumination source to the object point.

If the $1/r_o$ variation of real amplitude is insignificant over the hologram, a_o'/r_o, to first approximation, can be replaced by a constant amplitude a_o:

$$U_o = a_o \exp\left[-ik_o(r_o + S_o)\right]. \tag{1.81}$$

Expressions like equation 1.81 can be written for the complex amplitudes

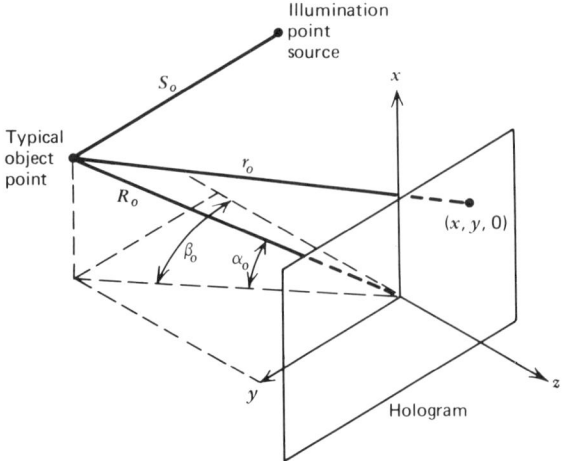

Figure 1.30 Coordinate system and notation for the holographic imaging equations.

of the *reference wave* (subscript R) and *reconstruction wave* (subscript c):

$$U_R = a_R \exp(-ik_R r_R), \tag{1.82}$$

$$U_c = a_c \exp(-ik_o r_c). \tag{1.83}$$

The reference and object waves originate at point sources located at (R_R, α_R, β_R) and (R_o, α_o, β_o), respectively. Because of coherence requirements $k_R = k_o$, but k_c may be different.

The hologram is formed by exposing a film plate to the complex amplitude:

$$U_o + U_R = a_o \exp[-ik_R(r_o + S_o)] + a_R \exp(-ik_R r_R).$$

After the plate is developed, the object wave is reconstructed by illuminating the plate with the reconstruction wave, which has complex amplitude U_c. From the discussion of off-axis holography we know that the hologram will diffract the reconstruction wave into three components. We are interested in the component whose complex amplitude is proportional to

$$\begin{aligned} U_1 &= U_c U_R^* U_o \\ &= a_1 \exp(-ik_c r_c) \exp[-ik_R(r_o + S_o - r_R)]. \end{aligned} \tag{1.84}$$

Equation 1.84 represents the complex amplitude in the hologram plane of the *primary image*; if the reconstruction wave is identical to the reference

1.5 HOLOGRAPHY

wave, $U_c = U_R$, this is a virtual image of the original object point. If the holographic emulsion is sufficiently thin, or if the angles $\alpha_o - \alpha_R$ and $\beta_o - \beta_R$ are sufficiently small, one also can observe the *conjugate image* corresponding to the complex amplitude

$$U_2 = U_c U_R U_o^*$$
$$= a_2 \exp(-ik_c r_c) \exp[ik_R(r_o + S_o - r_R)]. \tag{1.85}$$

The complex amplitudes, equations 1.84 and 1.85, each represent light which appears to radiate as a spherical wave from some *image point* (subscript *I*), so they can be written in the form

$$U_I = a_I \exp(-ik_c r_I). \tag{1.86}$$

The wave number k_c is used in this expression because the image is formed with light of wavelength λ_c. Now the apparent location of the primary image point can be determined by equating the exponent of equation 1.86 with the exponent of equation 1.84:

$$k_c r_I = k_c r_c + k_R(r_o + S_o - r_R). \tag{1.87}$$

The distance r_o from the object point to point (x, y) on the hologram can be expressed as

$$r_o = [z_o^2 + (x - x_o)^2 + (y - y_o)^2]^{1/2}$$
$$= (x^2 + y^2 - 2x_o x - 2y_o y + R_o^2)^{1/2}. \tag{1.88}$$

Assuming that R_o^2 is large compared with the quantity $(x^2 + y^2 - 2x_o x - 2y_o y)$, we can expand r_o in a binomial series about R_o:

$$r_o = R_o + \frac{x^2 + y^2}{2R_o} - \frac{x_o x + y_o y}{R_o} + \cdots. \tag{1.89}$$

Similar expressions can be written for r_R, r_c, and r_I. It should be noted that if this series is truncated after three terms the resulting approximation is valid under conditions which are less stringent than the usual paraxial approximation. Introducing approximations of the form of equation 1.89

into equation 1.87 yields

$$k_c\left(R_I + \frac{x^2+y^2}{2R_I} - \frac{x_I x + y_I y}{R_I} + \cdots\right)$$

$$= k_c\left(R_c + \frac{x^2+y^2}{2R_c} - \frac{x_c x + y_c y}{R_c} + \cdots\right)$$

$$+ k_R\left(R_o + S_o - R_R + \frac{x^2+y^2}{2R_o} - \frac{x^2+y^2}{2R_R}\right.$$

$$\left. - \frac{x_o x + y_o y}{R_o} + \frac{x_R x + y_R y}{R_R} + \cdots\right). \quad (1.90)$$

Equation 1.90 will be satisfied if we equate the coefficients of like powers of x and y appearing on the left- and right-hand sides. Matching the coefficients of (x^2+y^2) yields

$$k_c\left(\frac{1}{R_I}\right) = k_c\left(\frac{1}{R_c}\right) + k_R\left(\frac{1}{R_o}\right) - k_R\left(\frac{1}{R_R}\right). \quad (1.91)$$

Similarly, matching the coefficients of x and y gives

$$k_c\left(\frac{x_I}{R_I}\right) = k_c\left(\frac{x_c}{R_c}\right) + k_R\left(\frac{x_o}{R_o}\right) - k_R\left(\frac{x_R}{R_R}\right), \quad (1.92)$$

$$k_c\left(\frac{y_I}{R_I}\right) = k_c\left(\frac{y_c}{R_c}\right) + k_R\left(\frac{y_o}{R_o}\right) - k_R\left(\frac{y_R}{R_R}\right). \quad (1.93)$$

Equations 1.91 to 1.93 can be written in a more convenient form by noting that $k_R/k_c = \lambda_c/\lambda_R$, $x_I/R_I = \sin\alpha_I$, $y_I/R_I = \cos\alpha_I \sin\beta_I$, and so on. Also, it should be noted that a similar analysis of the conjugate image would result in matching conditions that differ from equations 1.91 to 1.93 only in the sign of the second and third terms on the right-hand side of each equation; therefore they can be written as

$$\frac{1}{R_I} = \frac{1}{R_c} \pm \frac{\lambda_c}{\lambda_R}\left(\frac{1}{R_o} - \frac{1}{R_R}\right), \quad (1.94)$$

$$\sin\alpha_I = \sin\alpha_c \pm \frac{\lambda_c}{\lambda_R}(\sin\alpha_o - \sin\alpha_R), \quad (1.95)$$

$$\cos\alpha_I \sin\beta_I = \cos\alpha_c \sin\beta_c \pm \frac{\lambda_c}{\lambda_R}$$

$$(\cos\alpha_o \sin\beta_o - \cos\alpha_R \sin\beta_R), \quad (1.96)$$

where the plus sign refers to the primary image, and the minus sign to the conjugate image. Equations 1.94 to 1.96 are the holographic imaging equations which can be used to determine the location (R_I, α_I, β_I) of the holographic image of the object point (R_o, α_o, β_o). Matching the higher order terms in equation 1.90 yields expressions for aberrations of the image,[33] which will not be needed in this book.

1.6 HOLOGRAPHIC INTERFEROMETRY

Through the use of off-axis holography, one can produce three-dimensional images of diffusely reflecting objects which appear to be overlaid by interference fringes that are indicative of deformation, displacement, or rotation of the object. Similarly, in the case of transparent objects, fringe patterns can be formed which are indicative of changes in refractive index or object thickness. This type of interferometry is possible because a light wave scattered by an object can be holographically recorded and reconstructed with such precision that it can be compared interferometrically with light scattered by the same object at another time. Alternatively, it can be compared interferometrically with a second holographic reconstruction of light scattered by the object. Accordingly, we define *holographic interferometry* as the interferometric comparison of two or more waves, at least one of which is holographically reconstructed. The composite of these two or more waves will be referred to as a *holographic interferogram*. The term *interferogram* with no modifying adjective will denote a pattern of interference fringes recorded on photographic film or formed on a two-dimensional viewing screen or the retina of the eye.

This book deals with holographic interferometry for which the illumination source is a laser emitting optical, infrared, or ultraviolet radiation. It may be noted, however, that holographic interferometry has also been demonstrated using microwave[34] and ultrasonic[35] radiation.

The application of holography to interferometry was first suggested by Horman,[36] who described the inclusion of a hologram in lieu of a test section in a Mach-Zehnder interferometer. Holographic interferometry of diffusely scattering objects was introduced in the context of vibration analysis by Powell and Stetson in a paper submitted for publication in March 1965[37] and was subsequently expanded upon in refs. 38 to 41. The development of two-exposure and real-time holographic interferometry occurred independently in several laboratories and was reported in papers read, or submitted for publication, during the summer and fall of 1965. This work includes that of Burch,[40] Collier, Doherty, and Pennington,[41] Stetson and Powell,[42] and Haines and Hildebrand,[43] who studied the deformation and displacement of diffusely reflecting objects, and that of Heflinger, Wuerker and Brooks[44,45] who used the method for aerodynamic measurements. The technique of holographic addition and subtraction was also introduced in 1965 by Gabor et al.[46] and subsequently was noted to be equivalent to holographic interferometry. Finally, it is interesting that the much earlier work of Burch and Palmer[47] dealing with the formation of moiré patterns by two photographically recorded gratings is very closely related to the concept of holographic interferometry.

We consider here briefly the technique of *two-exposure holographic interferometry*. Two other techniques, real-time and time-average holographic interferometry, are discussed in Chapter 4. The two-exposure technique requires the use of an off-axis system like that shown in Figure 1.27*a*. For illustrative purposes, suppose that the object in this figure is a turbine blade whose response to a certain mechanical force is to be determined. A holographic exposure of the object in its initial, unstressed state is recorded by exposing the photographic plate to the object and reference waves simultaneously. The turbine blade is then stressed by applying the desired force, and a second holographic exposure is made on the same plate. When this plate is developed and then illuminated by a reconstruction wave identical to the reference wave used to record it, an observer looking through the hologram will see a three-dimensional virtual image of the turbine blade overlaid with a pattern of interference fringes, as depicted in Figure 1.31. The observer will also note that the fringes appear to be localized in space, not necessarily on the surface of the blade, and that they appear to shift about and change form if the observer's viewing direction changes.

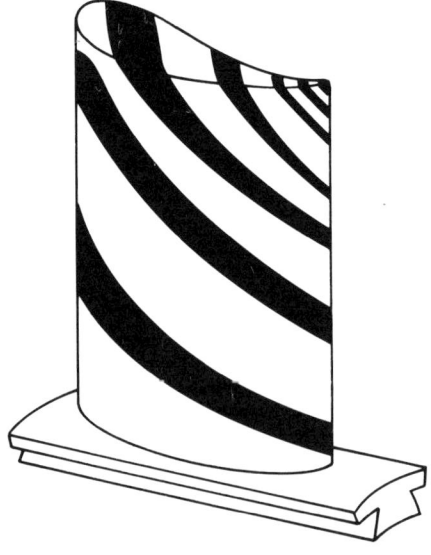

Figure 1.31 Sketch of fringe pattern formed by a two-exposure holographic interferometry of a turbine blade.

1.6 HOLOGRAPHIC INTERFEROMETRY

Having this qualitative description of the technique of holographic interferometry as background, we now proceed to discuss the underlying principle. Holography is a linear process in the sense that two or more optical waves can be recorded sequentially in time and later can be reconstructed simultaneously. Therefore the sum, difference, or even time average of a sequence of waves can be formed. For example, at some time t_1 the off-axis holographic system shown in Figure 1.27a can be used to record an optical wave whose complex amplitude in the hologram plane is $U_1(x,y)$. At time t_2 a second wave, $U_2(x,y)$, can be recorded on the same photographic plate. The process is simply holographic double exposure; the plate is exposed first to $U_1(x,y)$ together with a reference wave $U_R(x,y)$, and then to $U_2(x,y)$ together with $U_R(x,y)$. When the hologram formed by developing this plate is illuminated by $U_R(x,y)$, the complex amplitude of the reconstructed wave will be proportional to $[U_1(x,y) + U_2(x,y)]$, and the irradiance will be proportional to

$$I(x,y) = |U_1(x,y) + U_2(x,y)|^2. \tag{1.97}$$

In applications to interferometry, $U_1 = U_o(x,y)$ represents the light scattered or transmitted to the hologram plane by some object, and $U_2 = U'_o(x,y)$ represents light from the same object after it has been slightly deformed or changed in some manner. Slight deformation or changes of the object primarily affect the phase of U_o, so we write

$$U_o(x,y) = a(x,y)\exp[-i\phi(x,y)],$$
$$U'_o(x,y) = a(x,y)\exp\{-i[\phi(x,y)+\Delta\phi(x,y)]\}. \tag{1.98}$$

The irradiance of the reconstructed wave, equation 1.97, then becomes

$$\begin{aligned}I(x,y) &= |a(x,y)\exp[-i\phi(x,y)] \\ &\quad + a(x,y)\exp\{-i[\phi(x,y)+\Delta\phi(x,y)]\}|^2, \\ &= 2a^2(x,y)\{1+\cos[\Delta\phi(x,y)]\}.\end{aligned} \tag{1.99}$$

Equation 1.99 represents the irradiance of the object, $a^2(x,y)$, modulated by a fringe pattern $2\{1+\cos[\Delta\phi(x,y)]\}$. Dark fringes are contours of constant values of $\Delta\phi$ which are odd-integer multiples of π. Bright fringes are contours of constant values of $\Delta\phi$ which are even-integer multiples of π. In various applications $\Delta\phi$ may be related to physical quantities such as displacement, rotation, strain, bending moment, vibrational amplitude, temperature, pressure, mass concentration, electron density, or stress. The manner in which $\Delta\phi$ is evaluated and its quantitative relation to these quantities are discussed in Chapters 2 to 6.

The effectiveness, or uniqueness, of holographic interferometry for many applications is based on four key properties:

1. High information content.
2. Division of amplitude in time.
3. Permanent recording of events.
4. Temporal filtering.

The information content of holograms can be sufficiently high to permit the recording and reconstruction of fine details of a complicated wave with great fidelity. It is this property that makes it possible to study three-dimensional, diffusely reflecting objects interferometrically.

In Section 1.2 it was noted that classical interferometers are categorized as division of amplitude or division of wavefront devices, with "division" referring to a *spatial* division of light into two different paths. In holographic interferometry, amplitude is divided *temporally*; that is, the waves which interfere traverse essentially the same path in space at two different times. This is of particular importance in the interferometry of transparent media because it permits the use of test section windows and other optical elements of relatively poor quality.

A permanent record of optical waves is extremely useful for the study of transient events. Holographic interferograms can be recorded nearly instantaneously by using a pulsed laser, and then studied later by reconstruction with a continuous light source. Alignment and focusing of imaging optics and photographic recording of interferograms can be carried out with the cw reconstruction, thereby reducing experimental complexity.

The temporal filtering property refers to the formation of interferograms representing one temporal frequency component of a time-varying wave. The most common application is to time-average holographic interferometry for the study of mechanical vibrations, which is addressed in Chapter 4.

In preparation for the following chapters, a few comments regarding the nature of interference of two diffuse waves are in order. Figure 1.32a is a photograph of a two-exposure holographic interferogram. The object is an aluminum beam on which a rectangular reference grid has been painted. Between exposures a force was applied which caused the beam to deform slightly, giving rise to the distinct fringe pattern evident in the photograph. Figure 1.32b is a photographic enlargement of a small portion of the interferogram. It is evident that the interferogram has a fine structure which is a speckle pattern. The speckles in the regions which we identify as dark fringes are, *on the average*, darker than those in regions recognized as bright fringes.

The interferogram in Figure 1.32 is equivalent to that which would be formed if light were scattered simultaneously from the object in both its deformed and its undeformed configuration, as depicted in Figure 1.33.

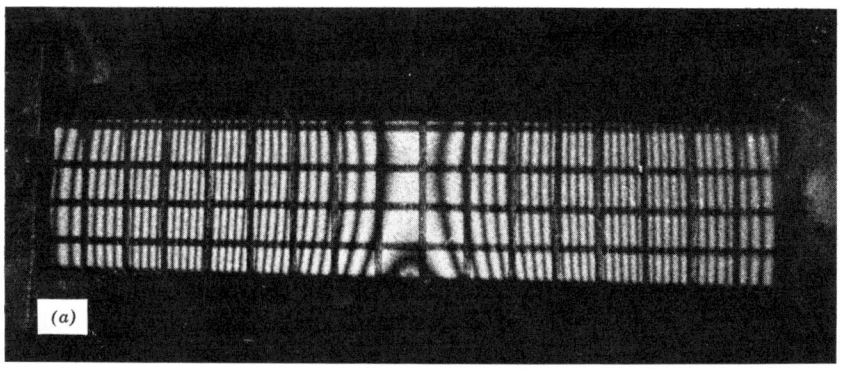

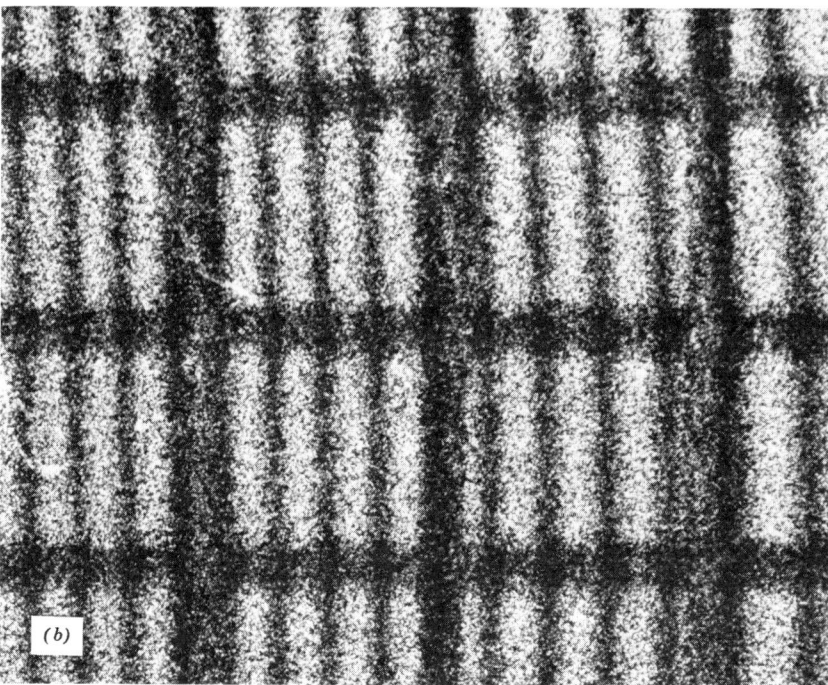

Figure 1.32 Two-exposure interferogram of a diffuse object. (*a*) The fringes of holographic interferometry appear distinct. (*b*) In an enlargment of a portion of (*a*) the underlying speckle structure is evident.

The light from each point of each surface interferes with light from every other point. From the discussion in Section 1.4 it is clear that this will give rise to a speckle pattern consisting of random irradiance variations of relatively high spatial frequency. This speckle pattern conveys information primarily about the microscopic contour of the object surface and the size of the viewing aperture. The systematic, low-frequency variation which constitutes the fringes of holographic interferometry conveys information about the displacement and deformation of the object between exposures. If a number of interferograms produced by identical deformations of surfaces which are macroscopically identical but differ in microstructure were examined, the systematic variations (fringes of holographic interferometry) would be the same in each case, but the fine structure (speckle pattern) would vary randomly. Walles[48,49] expanded this argument to develop a theory of fringe formation in which an ensemble average of a large number of holographic interferograms is formed in order to integrate out the fine structure, leaving a description sufficient for analysis of the formation of the fringes of holographic interferometry. Alternatively, Tanner[50] based an analysis on a single interferogram, but described the fringes in terms of irradiance averaged along fringe contours.

Let us consider further the manner in which fringes are formed when a diffusely reflecting surface is translated or deformed. Figure 1.33 shows such an object surface before and after displacement; P and P' are the same point on the surface before and after displacement. Light scattered by the surface in the neighborhood of point P gives rise to a complex

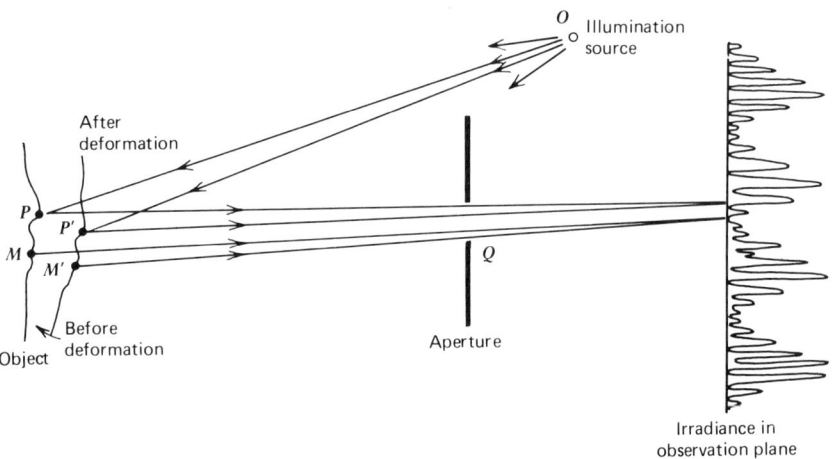

Figure 1.33 Schematic diagram of fringe formation. The irradiance in a plane where fringes are visible consists of a rapidly varying speckle pattern modulated by a slow, systematic variation which constitute the fringes of holographic interferometry.

1.6 HOLOGRAPHIC INTERFEROMETRY

amplitude $U_o(x,y;z)$ in a plane a distance z from the object. If the corresponding irradiance $I_o(x,y;z)$ were detected, after imaging through a finite aperture, it would be recognized as a speckle pattern created by the random variations in the microscopic surface contour of the object. In two-exposure holographic interferometry a second optical wave, scattered by the object surface in the neighborhood of P', is formed. This wave is nearly identical to the first, but is displaced and travels in a slightly different direction. At a given distance z from the object this wave will have a complex amplitude $U'_o(x,y;z)$, which differs in detail from $U_o(x,y;z)$. If we could compare the corresponding speckle patterns $I_o(x,y;z)$ and $I'_o(x,y;z)$ at an arbitrary plane in front of the object, their detailed structures would be completely different, that is, they would be uncorrelated. However, there exists one certain distance z from the object at which U_o and U'_o will be nearly identical except for a very small relative displacement and a variation in phase whose spatial scale is large in comparison with the variation giving rise to the speckle. The irradiance pattern formed when the field at this distance from the object is imaged has a fine speckle structure modulated by the broader, cosinusoidal variation which gives rise to the fringes of holographic interferometry. This is illustrated in Figure 1.32b. The fringes are said to be *localized*, for the given viewing direction, at this distance z from the object. To an observer the fringes indeed will appear to be located in this region of space. If the imaging system is focused on other planes in space, U_o and U'_o will differ in detail and will not include a systematic, large-scale, relative phase variation. Only a speckle pattern will be observed in such locations.

The phase difference δ, whose variation gives rise to the fringes of holographic interferometry, can be calculated by finding the change, due to object displacement, in optical pathlength from the illumination source to the object point P to the center of the observing aperture Q. In terms of Figure 1.33,

$$\delta = \frac{2\pi}{\lambda}\left[(\overline{OP} + \overline{PQ}) - (\overline{OP'} + \overline{P'Q})\right]. \tag{1.100}$$

This equation is the basis for a simple, geometric relation between object displacement and fringe patterns. *For purposes of computation of fringe patterns, we need consider only the change in optical pathlength of light scattered by corresponding points such as P and P'* in Figure 1.33. Interference of light scattered by noncorresponding points such as P and M, P and M', P' and M', or P' and M does not enter into these calculations. The relation between phase differences δ, determined by the use of holographic interferometry, and the motion of the object is discussed in detail in Chapter 2.

It will also be necessary for us to develop a scheme for determining the position in space at which fringes appear to localize for a given viewing direction. This position depends upon the nature and direction of object displacement or deformation, as well as on the directions of illumination and observation. Localization occurs where $U_o(x,y;z)$ and $U'_o(x,y;z)$ are most closely matched. Near such locations the corresponding speckle patterns I_o and I'_o nearly match and overlap, and U_o and U'_o differ only by a systematic, large-scale variation in phase. This position can be calculated using geometric optics by specifying that the phase difference for light scattered by P and P' must be equal to that for all neighboring corresponding points, such as M and M' in Figure 1.33. This approach is used in Chapter 3 to develop equations relating fringe localization to object displacement and deformation. The topic is discussed further in Chapter 4 in the context of strain and rotation measurement.

It is not possible to eliminate speckle from the interferograms. Even in the dark fringes it is inherent for three reasons. First, if the object is deformed between exposures, strain causes a change in the microscopic contour of the surface. This deforms the scattered waves so that U_o and U'_o will not match precisely even at locations where the fringes are localized. The second cause of speckle is the use of a finite viewing aperture. The aperture will sample slightly different portions of the waves leaving the neighborhood of P and P'. Hence U_o and U'_o cannot match perfectly in the image. This effect will cause an apparent decrease in fringe visibility at large strains. The third unavoidable cause of speckle is that light is always scattered by the grains in the photographic emulsion in which the hologram is recorded. (It is possible to reduce speckle in inteferograms by forming them with an imaging system having a moving aperture or random mask or, equivalently, by reimaging the interferogram in incoherent light.[51] Another approach is to record rainbow interferograms, which are reconstructed in white light.[52])

In summary, the formation of the fringes of holographic interferometry is a complicated phenomenon. Its detailed physical description requires a wave analysis describing diffraction by a random surface and a rigorous description of the effect of the viewing aperture. Such an analysis is given by Walles.[48,49] However, simple geometric optics can be used to develop computational schemes for predicting fringe localization due to a given motion, or to determine displacement, rotation, and strain from observation of holographic interferograms. The latter approach is emphasized in this book.

REFERENCES

1. E. Hecht and A. Zajac, *Optics*, Addison-Wesley, Reading, MA, 1974.
2. L. Mandel and E. Wolf, Coherence properties of optical fields, *Rev. Mod. Phys.*, **37**, 231–287 (1965).

REFERENCES

3. J. W. Goodman, *Introduction to Fourier Optics*, McGraw-Hill, New York, 1968.
4. J. C. Dainty (Ed.), *Laser Speckle and Related Phenomena*, Springer-Verlag, Berlin, 1975.
5. J. W. Goodman, Statistical properties of laser speckle patterns, pp. 9–75 of ref. 4.
6. J. W. Strutt (Lord Rayleigh), On the problem of random vibrations, and of random flights in one, two, or three dimensions, *Phil. Mag.*, **37**, 321–347 (1919).
7. T. S. McKechnie, Statistics of coherent light speckle produced by stationary and moving apertures, doctoral thesis, Univ. of London, 1974.
8. D. Gabor, Microscopy by reconstructed wavefronts, *Proc. Roy. Soc.*, **A197**, 454–487 (1949).
9. D. Gabor, Microscopy by reconstructed wavefronts, II, *Proc. Phys. Soc.*, **64**, 449–469 (1951).
10. B. J. Thompson, Holographic particle sizing techniques, *J. Phys. E: Sci. Instrum.*, **7**, 781–788 (1974).
11. E. N. Leith and J. Upatnieks, Reconstructed wavefronts and communication theory, *J. Opt. Soc. Am.*, **52**, 1123–1130 (1962).
12. E. N. Leith and J. Upatnieks, Wavefront reconstruction with diffused illumination and three-dimensional objects, *J. Opt. Soc. Am.*, **54**, 1295–1301 (1964).
13. T. Kallard, *Exploring Laser Light*, Optosonic Press, New York, 1977.
14. A. Kozma, Photographic recording of spatially modulated coherent light, *J. Opt. Soc. Am.*, **56**, 428–432 (1966).
15. Reference 3, pp. 157–159.
16. S. Johansson and K. Biedermann, Multiple-sine-slit microdensitometer and MTF evaluation for high resolution emulsions. 2: MTF data and other recording parameters of high resolution emulsions for holography, *Appl. Opt.*, **13**, 2288–2291 (1974).
17. K. Biedermann, A function characterizing photographic film that directly relates to brightness of holographic images, *Optik*, **28**, 160–176 (1968).
18. J. E. Sollid and J. B. Swint, A determination of the optimum beam ratio to produce maximum contrast photographic reconstruction from double-exposure holographic interferometry, *Appl. Opt.*, **9**, 2717–2719 (1970).
19. K. A. Stetson and K. Singh, Measurement of signal-to-noise ratio in hologram reconstructions, by vibration interferograms, *Opt. Laser Technol.*, **3**, 104–108 (1971).
20. A. Graube, Advances in bleaching methods for photographically recorded holograms, *Appl. Opt.*, **13**, 2942–2946 (1974).
21. J. Upatnieks and C. Leonard, Diffraction efficiency of bleached, photographically recorded interference patterns, *Appl. Opt.*, **8**, 85–89 (1969).
22. K. S. Pennington and J. S. Harper, Techniques for producing low-noise improved efficiency holograms, *Appl. Opt.*, **9**, 1643–1650 (1970).
23. J. Upatnieks and C. D. Leonard, Characteristics of dielectric holograms, *IBM J. Res. Devp.*, **14**, 527–532 (1970).
24. J. H. Altman, Pure relief images on Type 649-F plates, *Appl. Opt.*, **5**, 1689–1690 (1966).
25. V. Russo and S. Sottini, Bleached holograms, *Appl. Opt.*, **7**, 202 (1968).
26. H. M. Smith, Photographic relief images, *J. Opt. Soc. Am.*, **58**, 533–538 (1968).
27. W. J. Cathey, Jr., Three-dimensional wavefront reconstruction using a phase hologram, *J. Opt. Soc. Am.*, **55**, 457 (1965).
28. J. N. Latta, The bleaching of hologram diffraction gratings for maximum efficiency, *Appl. Opt.*, **7**, 2409–2416 (1968).
29. C. B. Burckhardt and E. T. Doherty, A bleach process for high-efficiency, low-noise holograms, *Appl. Opt.*, **8**, 2479–2482 (1969).

30. N. J. Phillips and D. Porter, An advance in the processing of holograms, *J. Phys. E: Sci. Instrum.*, **9**, 631–634 (1976).
31. R. L. van Renesse and F. A. Bouts, Efficiency of bleaching agents for holography, *Optik*, **38**, 156–168 (1973).
32. H. M. Smith (Ed.), *Holographic Recording Materials*, Springer-Verlag, Berlin, 1977.
33. E. B. Champagne, Nonparaxial imaging, magnification, and aberration properties in holography, *J. Opt. Soc. Am.*, **57**, 51–55 (1967).
34. R. W. Larson, J. S. Zelenka, and E. L. Johansen, Microwave radar imagery, *Proceedings of Symposium on Engineering Applications of Holography* (February 16–17, 1972, Los Angeles, CA), Society of Photo-optical Instrumentation Engineers, Redondo Beach, CA, 1972, p. 14.
35. K. Suzuki and B. P. Hildebrand, Holographic interferometry with acoustic waves, in *Acoustical Holography*, Vol. 6, N. Booth (Ed.), Plenum Press, New York, 1975, pp. 577–595.
36. M. H. Horman, An application of wavefront reconstruction to interferometry, *Appl. Opt.*, **4**, 333–336 (1965).
37. R. L. Powell and K. A. Stetson, Interferometric analysis by wavefront reconstruction, *J. Opt. Soc. Am.*, **55**, 1593–1598 (1965).
38. R. L. Powell and K. A. Stetson, Interferometric vibration analysis of three-dimensional objects by wavefront reconstruction, *J. Opt. Soc. Am.*, **55**, 612A (1965).
39. K. A. Stetson and R. L. Powell, Interferometric hologram evaluation and real-time vibration analysis of diffuse objects, *J. Opt. Soc. Am.*, **55**, 1694–1695 (1965).
40. J. M. Burch, The application of lasers in production engineering (The 1965 Viscount Nuffield Memorial Paper), *Prod. Eng.*, **44**, 431–442 (1965).
41. R. J. Collier, E. T. Doherty, and K. S. Pennington, Application of moiré techniques to holography, *Appl. Phys. Lett.*, **7**, 223–225 (1965).
42. K. A. Stetson and R. L. Powell, Hologram interferometry, *J. Opt. Soc. Am.*, **55**, 1570A (1965).
43. K. A. Haines and B. P. Hildebrand, Surface-deformation measurement using the wavefront reconstruction technique, *Appl. Opt.*, **5**, 595–602 (1966).
44. R. E. Brooks, L. O. Heflinger, and R. F. Wuerker, Interferometry with a holographically reconstructed comparison beam, *Appl. Phys. Lett.*, **7**, 248–249 (1965).
45. L. O. Heflinger, R. F. Wuerker, and R. E. Brooks, Holographic interferometry, *J. Appl. Phys.*, **37**, 642–649 (1966).
46. D. Gabor, G. W. Stroke, R. Restrick, A. Funkhouser, and D. Brumm, Optical image synthesis (complex amplitude addition and subtraction) by holographic Fourier transformation, *Phys. Lett.*, **18**, 116–118 (1965).
47. J. M. Burch and D. A. Palmer, Interferometric methods for the production of large gratings, *Opt. Acta*, **8**, 73–80 (1961).
48. S. Walles, Visibility and localization of fringes in holographic interferometry of diffusely reflecting surfaces, *Ark. Fys.*, **40**, 229–403 (1969).
49. S. Walles, On the concept of homologous rays in holographic interferometry of diffusely reflecting surfaces, *Opt. Acta*, **17**, 899–913 (1970).
50. L. H. Tanner, A study of fringe clarity in laser interferometry and holography, *J. Phys. E:Sci. Instrum.*, **1**, 517–522 (1968).
51. L. A. Östlund and K. Biedermann, Laser speckle reduction: equivalence of the moving aperture method and incoherent spatial filtering, *Appl. Opt.*, **16**, 685–690 (1977).
52. F. T. S. Yu and H. Chen, Rainbow holographic interferometry, *Appl. Opt.* (in press, 1978).

2
Opaque Objects: Measurement of Displacement and Deformation

2.1 INTRODUCTION

Perhaps the most remarkable application of optical holography is the interferometric comparison of diffuse wavefronts. It has made possible a vast generalization of interferometry as applied to the measurement of displacements and surface deformations of opaque objects. Classical interferometric techniques, such as the use of Michelson interferometers, are restricted to the comparison of wavefronts of simple geometric form, usually plane or spherical. Their application to metrology is confined to measuring the normal displacement of optically polished surfaces. Holographic interferometry, however, can be used to measure the *vector* displacement of points on diffusely reflecting surfaces of complicated shape. This is a consequence of the high information content of holograms, which makes possible the recording and reconstruction, with extreme fidelity, of the optical wavefront scattered by such a surface.

The production of a holographic interferogram is simple in concept and in practice. As an illustration, suppose that we wish to determine the response of some mechanical component to the application of a force. The component is mounted in a loading fixture and placed in the object position of an off-axis holographic system such as that shown in Figure 1.27a. With the component in an original, unstressed condition a holographic exposure of the film plate is made. The desired force is then applied, and a second holographic exposure of the same film plate is made. When such a doubly exposed hologram is illuminated with a duplicate of the reference wave, the holographic interferogram can be viewed. The result is a fascinating and appealing display. An observer looking through the hologram sees the three-dimensional image of the component overlaid with a pattern of interference fringes. The fringes appear to be localized in space—sometimes on the object, sometimes in front of it or behind it. Furthermore, the fringes have a dynamic character in that they appear to change position and be altered in shape as the observer moves his or her head about to vary the viewing direction.

Holographic interferometry has found many applications which require only qualitative interpretation of fringe patterns, particularly in the field of nondestructive testing.[1] Applications to metrology and stress analysis require quantitative evaluation of interferograms; this is the subject of this chapter. In it, procedures for determining the vector displacements of points on the surface of an object will be developed.

2.2 EQUATIONS FOR FRINGE INTERPRETATION

In Section 1.6 we learned that the irradiance distribution of the reconstructed wavefront from a two-exposure hologram is

$$I(x,y) = 2|a(x,y)|^2[1+\cos\Delta\phi(x,y)], \tag{2.1}$$

where $2|a(x,y)|^2$ represents the light from the image of the object. The term in square brackets describes the fringes; $\Delta\phi(x,y)$ is the change in optical phase due to the change in object position between exposures. We must develop procedures to determine this phase change by examining the fringe pattern, and to relate the phase change to the vector displacement of points on the surface of the object.

First, consider a simple example in which displacement occurs in only one dimension. The object is a cantilever beam with one end fixed to a rigid block. The beam is illuminated with a plane wave traveling in the $-z$ direction, normal to its surface (see Figure 2.1a). The light scattered by the cantilever beam is recorded during an initial exposure of an off-axis hologram. The free end of the beam is then displaced slightly in the z direction, and a second exposure is made. Figure 2.1b shows what a photograph of the virtual image from this two-exposure hologram would look like. The same pattern could be observed visually. Since displacements are small, each point on the object moves only in the z direction. Initially each point was in the plane $z=0$. After displacement each point moved to a new position described by $z=Z(x)$. If light travels a distance l_o from the source to a point on the object and back to the observer (or photograph) before the object is deformed, it will travel a distance $l_o - 2Z(x)$ after the object is deformed. The corresponding optical phase shift is $\Delta\phi(x) = (2\pi/\lambda)[2Z(x)]$, where λ is the wavelength of the laser light. Note that we have not mentioned the hologram explicitly—its function is simply to record the two wavefronts and reconstruct them simultaneously to form the interferogram shown in Figure 2.1b. We next assign fringe order numbers to the bright fringes in this figure. The base has not moved, so we assign the number $N=0$ to the bright fringe which covers it. The other

2.2 EQUATIONS FOR FRINGE INTERPRETATION

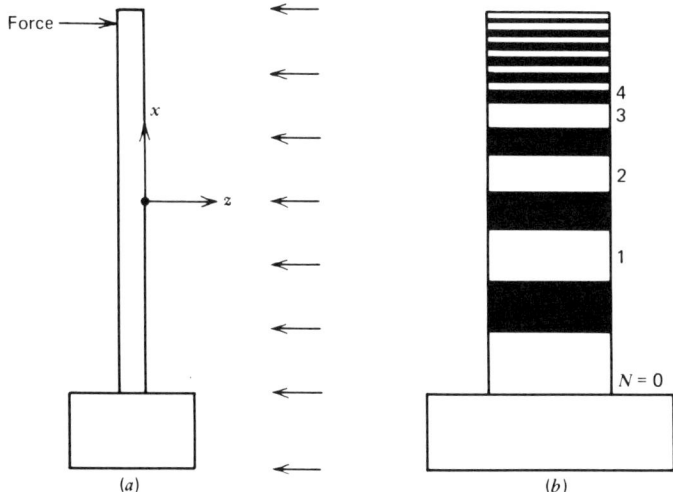

Figure 2.1 Normal deformation of a cantilever beam: (*a*) Side view of the beam. Displacement is in the $+z$ direction, and illumination in the $-z$ direction. (*b*) Fringe pattern observed on the front of the beam.

bright fringes are assigned the numbers $N = 1, 2, 3, \ldots$ consecutively. The Nth bright fringe corresponds to a phase change $\Delta\phi = 2\pi N = (2\pi/\lambda)[2Z(x)]$; hence

$$Z(x) = \frac{N\lambda}{2}. \tag{2.2}$$

By simply counting the fringes to a given location, the displacement can be calculated using equation 2.2.

In many practical applications a flat surface undergoes only normal deformation, and fringe interpretation is as simple as is indicated above. One obtains a direct full-field display of the normal deformation, which can be interpreted at a glance, as in classical interferometry. The general case of surfaces of complicated shape undergoing three-dimensional deformation, however, is not as simple. The price we pay for this generalization is that we can no longer interpret the interferogram at a glance, or from a single photograph. Vector displacements can be extracted from the hologram only by careful analysis on a point-by-point basis. We now develop the equations for doing so.

The approach to fringe interpretation given here follows the formulation of Sollid,[2] who generalized the earlier work of Aleksandrov and Bonch-Bruevich[3] and of Ennos[4] and presented the results in a unified manner and convenient notational system. We adopt the following model. The object surface consists of a collection of point scatterers. When an arbitrary surface point P undergoes a displacement $\mathbf{L}$ to a new position P', the interference between light scattered from P and light scattered from P' contributes to the fringes of holographic interferometry; interference between light scattered from P and light scattered from other surface points does not. This model is convenient and sufficient for our present purposes.

The role of the hologram is to reproduce the light scattered by the object both in its initial and in its displaced configuration. Once this light is accessible to us, no further consideration of the hologram is required in this analysis, except that its size and position limit the range of directions from which the object can be viewed. The relevant elements of the physical system are shown in Figure 2.2a. The object is illuminated by a point source located at O. Light is scattered by an object point P through the hologram to an observer at point Q. A two-exposure hologram has been recorded, and between exposures point P was displaced by $\mathbf{L}$. While looking at point P, the observer at Q discerns that the displacement resulted in an optical phase shift δ, which must now be related to $\mathbf{L}$. The notation δ is used here in preference to $\Delta\phi(x,y)$; δ represents the phase shift of light scattered by a particular object point in a particular direction.

In Figure 2.2b several vectors are defined for use in determining the relation between δ and $\mathbf{L}$. Vectors $\mathbf{R}$ and $\mathbf{r}_1$ lie in the plane defined by points O, P, and Q, and $\mathbf{k}_1$ and $\mathbf{k}_2$ are the propagation vectors of the light illuminating P and the light scattered toward the observer, respectively. Since the magnitude of a propagation vector is $2\pi/\lambda$, the phases of the two light rays which reach the observer are as follows:

$$\phi_1 = \mathbf{k}_1 \cdot \mathbf{r}_1 + \mathbf{k}_2 \cdot (\mathbf{R} - \mathbf{r}_1) + \phi_r, \tag{2.3}$$

$$\phi_2 = \mathbf{k}_3 \cdot \mathbf{r}_3 + \mathbf{k}_4 \cdot (\mathbf{R} - \mathbf{r}_3) + \phi_r. \tag{2.4}$$

Here ϕ_1 is the phase of the light scattered by P before displacement, ϕ_2 is the phase of the light scattered by P after displacement, and ϕ_r is the arbitrary phase assigned to these rays at the point source O. The phase difference measured by the observer is

$$\delta \equiv \phi_2 - \phi_1. \tag{2.5}$$

After displacement of P, the propagation vectors in the illumination and observation directions are $\mathbf{k}_3$ and $\mathbf{k}_4$. We define the small changes, $\Delta\mathbf{k}_1$ and $\Delta\mathbf{k}_2$, in these propagation vectors by

$$\mathbf{k}_3 = \mathbf{k}_1 + \Delta\mathbf{k}_1, \quad \mathbf{k}_4 = \mathbf{k}_2 + \Delta\mathbf{k}_2. \tag{2.6}$$

Combining the preceding equations, we have

$$\delta = (\mathbf{k}_2 - \mathbf{k}_1) \cdot (\mathbf{r}_1 - \mathbf{r}_3) + \Delta\mathbf{k}_1 \cdot \mathbf{r}_3 + \Delta\mathbf{k}_2 \cdot (\mathbf{R} - \mathbf{r}_3). \tag{2.7}$$

2.2 EQUATIONS FOR FRINGE INTERPRETATION

In real systems the magnitudes of $\mathbf{r}_1$ and $\mathbf{r}_3$ are much larger than $L = |\mathbf{r}_3 - \mathbf{r}_1|$, so for practical purposes $\Delta\mathbf{k}_1 \perp \mathbf{r}_3$ and $\Delta\mathbf{k}_2 \perp (\mathbf{R} - \mathbf{r}_3)$. Because of these relations the last two scalar products in equation 2.7 vanish, and we arrive at the relation

$$\delta = (\mathbf{k}_2 - \mathbf{k}_1) \cdot \mathbf{L}. \tag{2.8}$$

This simple relation forms the basis of quantitative interpretation of the fringes of holographic interferometry.

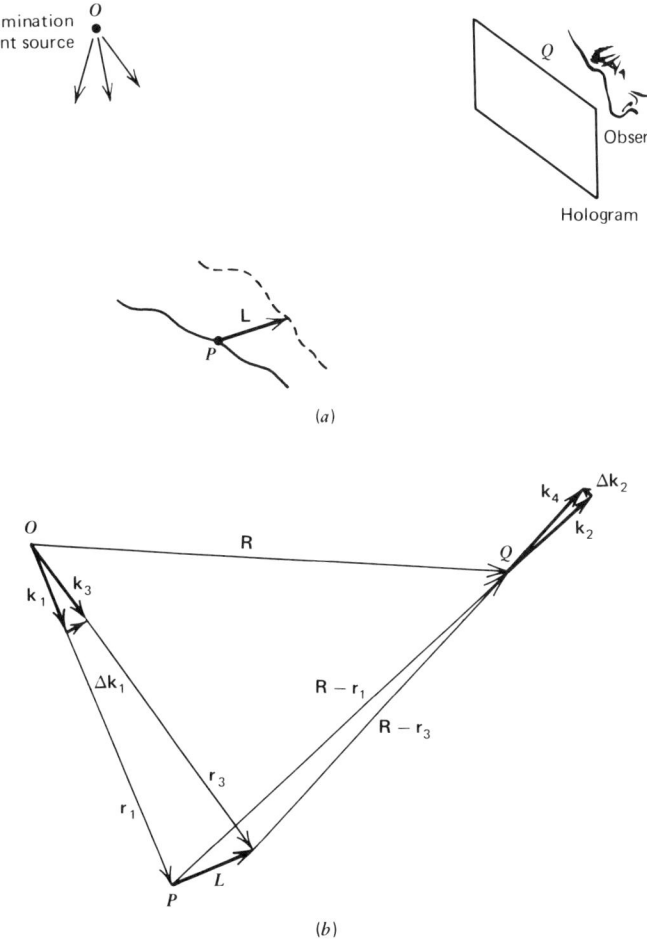

Figure 2.2 Nomenclature for fringe analysis (ref. 2). (*a*) Schematic diagram of the system. (*b*) Position and propagation vectors.

It is convenient to define the *sensitivity vector* **K** by

$$\mathbf{K} \equiv \mathbf{k}_2 - \mathbf{k}_1 \tag{2.9}$$

so that

$$\delta = \mathbf{K} \cdot \mathbf{L}. \tag{2.10}$$

Let 2θ be the angle between the illumination and viewing directions, as shown in Figure 2.3. Since both $\mathbf{k}_1$ and $\mathbf{k}_2$ have the magnitude $k = 2\pi/\lambda$, it is apparent that the sensitivity vector has the magnitude $2k\cos\theta$, and lies along the bisector of the angle between the illumination and viewing directions. A single observation of the interferogram thus determines a value of δ which yields a measurement of the component of **L** in this direction. The important concept of the sensitivity vector was implicit in early papers by Aleksandrov and Bonch-Bruevich,[3] Orr et al.,[5] and Ennos.[4]

Coplanar displacements are those for which **L** is known to lie in the *OPQ* plane. Such motions are of practical importance, and consideration of them yields insight into displacement measurement by holographic interferometry. Figure 2.4 depicts a case of coplanar displacement. Following the foregoing discussion, we see that

$$\delta = 2kL\cos\theta\cos\psi. \tag{2.11}$$

If we know a priori the direction of **L**, equation 2.11 determines the magnitude of the displacement. If this direction is not known, only the component parallel to the sensitivity vector has been determined. For most setups, **K** will have a direction close to the normal to the surface; hence δ, and therefore fringe order, is most sensitive to normal displacements, and

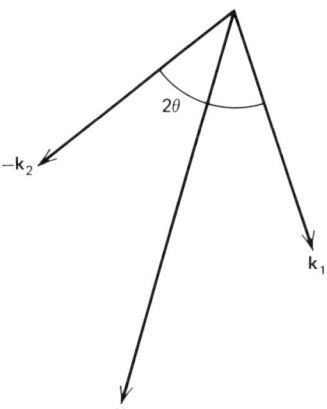

Figure 2.3 The sensitivity vector.

2.2 EQUATIONS FOR FRINGE INTERPRETATION

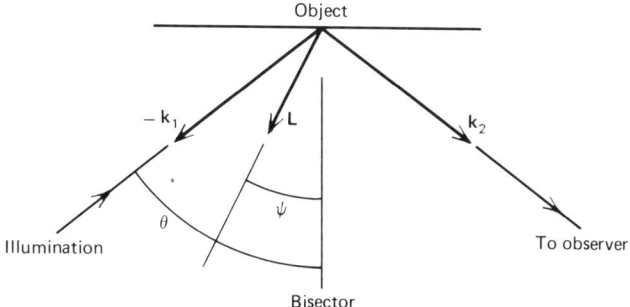

Figure 2.4 Coplanar displacement.

very insensitive to in-plane displacements. Although equation 2.11 yields the magnitude of the component of **L** along the direction of **K**, its sense is uncertain. The reason is that the sign of δ is not known. The value is assigned to δ according to $\delta = 2\pi N$ by counting white fringes; however, the fringes are described by $(1 + \cos\delta)$. Since the cosine is an even function, the same fringe pattern arises from positive or negative deformations. Fortunately, in most practical applications the sense of coplanar motions is understood from the physical configuration. In more complex measurement problems, however, this sign ambiguity is troublesome and must be confronted. This point will be discussed in later sections.

The approach to measuring a three-dimensional vector displacement is apparent. Since a single observation yields one vector component of **L** (equation 2.10), three observations must be made so that three independent vector components of **L** will be determined. One way to accomplish this is indicated in Figure 2.5a. Three different holographic interferograms are recorded, and phase measurements δ_1, δ_2, and δ_3 are made at the three points Q_1, Q_2, and Q_3, respectively. For each observation point Q_n, a propagation vector is defined which, along with $\mathbf{k}_1$, the propagation vector for the illumination, determines a sensitivity vector $\mathbf{K}_n$. These vectors are indicated in Figure 2.5b. An equation relating phase and displacement can be written for each observation:

$$\begin{aligned}
\delta_1 &= (\mathbf{k}_2 - \mathbf{k}_1) \cdot \mathbf{L} = \mathbf{K}_1 \cdot \mathbf{L}, \\
\delta_2 &= (\mathbf{k}_3 - \mathbf{k}_1) \cdot \mathbf{L} = \mathbf{K}_2 \cdot \mathbf{L}, \\
\delta_3 &= (\mathbf{k}_4 - \mathbf{k}_1) \cdot \mathbf{L} = \mathbf{K}_3 \cdot \mathbf{L}.
\end{aligned} \qquad (2.12)$$

If the three sensitivity vectors $\mathbf{K}_1$, $\mathbf{K}_2$, and $\mathbf{K}_3$ are noncoplanar, the system of equations 2.12 will determine the vector displacement **L**. This approach to displacement measurement is referred to as *multiple-hologram analysis*.

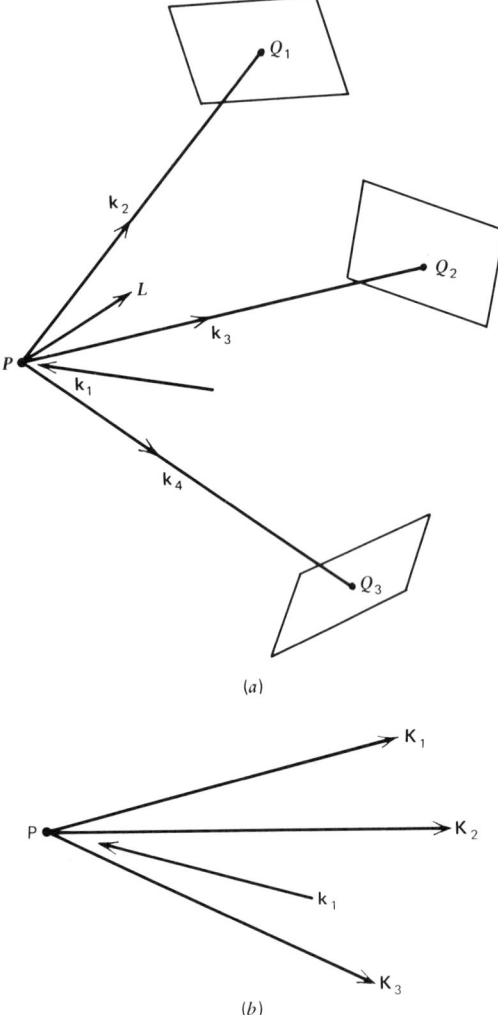

Figure 2.5 Multiple-hologram analysis. (*a*) Schematic diagram of the system. (*b*) Sensitivity vectors for the system.

It was first proposed by Ennos,[4] who used two holograms in conjunction with nearly grazing incident illumination to measure in-plane deformations of metal foils. Of course equations 2.12 are also valid if Q_1, Q_2, and Q_3 are three observation points on a single large hologram.

The phases δ_1, δ_2, and δ_3 are measured by counting fringes. Consider, for example, the determination of δ_1. As in the one-dimensional case discussed

2.2 EQUATIONS FOR FRINGE INTERPRETATION

above, the observer locates a point on the object which has not moved and assigns the bright fringe passing through this point the order number $N=0$. He or she then counts the number of bright fringes which lie between the point of zero motion and the point P under consideration in order to assign a fringe order number N_1 to the fringe which crosses P. Then $\delta_1 = 2\pi N_1$, and the process is repeated for the other observation points to determine δ_2 and δ_3. It is implied here that a point of zero motion exists and that the observer can locate it. This is not always the case. If the object undergoes certain rigid-body motions, or is not firmly connected to a stationary fixture, all points on the object move. Even if a point or region of zero motion exists, it may be difficult to identify. If the observer moves his or her eye from side to side while looking through the hologram, a zero-motion fringe remains in a fixed position on the object, while other fringes usually appear to move about. In practice this is not a very reliable way to locate the zero-motion fringe because the parallax effects involved may be small. A simple way to circumvent this problem is suggested in refs. 6 and 7. One end of a small flexible strip is attached to a fixed mass, which is in the field of view of the hologram. The strip is stretched slightly and its other end is attached to the object surface. A reliable zero-motion fringe is thus introduced into the interferogram. Köpf[8] has suggested a three-exposure holographic procedure in which the zero-motion fringe is easily identified because it is much brighter than the other fringes.

The most convenient way to determine **L** from equations 2.12 is to decompose all vectors into their orthogonal components in some x,y,z coordinate system. The orientation of this Cartesian system is arbitrary. If the object is planar, it may be convenient to let the x-y plane be parallel to the object. In other instances the coordinate system may be chosen for experimental convenience in determining the directions of the various vectors involved. In terms of the coordinate system chosen, equations 2.12 can be written as a system of linear algebraic equations. In matrix form this is as follows:

$$\begin{bmatrix} K_{1x} & K_{1y} & K_{1z} \\ K_{2x} & K_{2y} & K_{2z} \\ K_{3x} & K_{3y} & K_{3z} \end{bmatrix} \begin{bmatrix} L_x \\ L_y \\ L_z \end{bmatrix} = 2\pi \begin{bmatrix} N_1 \\ N_2 \\ N_3 \end{bmatrix}. \qquad (2.13)$$

The coefficient matrix is determined entirely by the geometry of the holographic system and the wavelength of light. The vector on the right-hand side comes from observation of the interference fringes. Equation 2.13 can be solved for the three orthogonal components of displacement, L_x, L_y, and L_z.

An alternative approach to measuring vector displacement is the *single-hologram analysis* introduced by Aleksandrov and Bonch-Bruevich.[3] Consider the system depicted in Figure 2.6. Three independent phase measurements can be made by viewing the object point P through three different points Q_1, Q_2, and Q_3 on the hologram. If the measurements of δ_1, δ_2, and δ_3 are made by counting fringes from a zero-motion fringe, all discussions and equations presented for multiple-hologram analysis apply. With a single large hologram, however, another technique can be used which does not require locating a zero-motion fringe. If the observer moves continuously from Q_1 to Q_2 while fixing his or her view through a small aperture or telescope on point P, fringes will appear to cross the field of view. The number of fringes, N_{12}, which appear to cross is a measure of $\delta_1 - \delta_2$; specifically, $\delta_1 - \delta_2 = 2\pi N_{12}$. With this in mind, equations of form 2.8 can be written for each of the four viewing directions shown in Figure 2.6:

$$\begin{aligned}
\delta_1 &= (\mathbf{k}_2 - \mathbf{k}_1) \cdot \mathbf{L}, \\
\delta_2 &= (\mathbf{k}_3 - \mathbf{k}_1) \cdot \mathbf{L}, \\
\delta_3 &= (\mathbf{k}_4 - \mathbf{k}_1) \cdot \mathbf{L}, \\
\delta_4 &= (\mathbf{k}_5 - \mathbf{k}_1) \cdot \mathbf{L}.
\end{aligned} \qquad (2.14)$$

Subtraction of pairs of these equations gives

$$\begin{aligned}
\delta_1 - \delta_2 &= (\mathbf{k}_2 - \mathbf{k}_3) \cdot \mathbf{L} = \overline{\mathbf{K}}_1 \cdot \mathbf{L}, \\
\delta_2 - \delta_3 &= (\mathbf{k}_3 - \mathbf{k}_4) \cdot \mathbf{L} = \overline{\mathbf{K}}_2 \cdot \mathbf{L}, \\
\delta_3 - \delta_4 &= (\mathbf{k}_4 - \mathbf{k}_5) \cdot \mathbf{L} = \overline{\mathbf{K}}_3 \cdot \mathbf{L}.
\end{aligned} \qquad (2.15)$$

The phase differences can be determined by scanning the hologram from

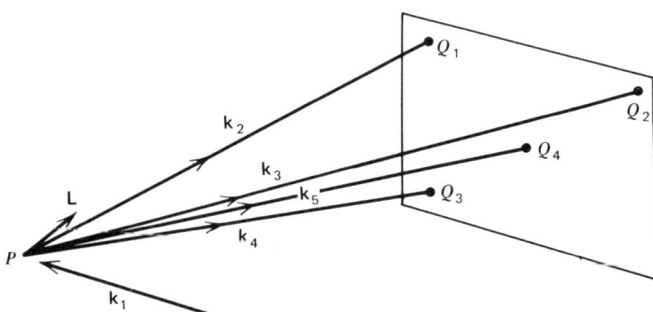

Figure 2.6 Single-hologram analysis. Observations are made from four locations on a single hologram.

point to point as just described, and a system of equations which determines the components of **L** can be written:

$$\begin{bmatrix} \overline{K}_{1x} & \overline{K}_{1y} & \overline{K}_{1z} \\ \overline{K}_{2x} & \overline{K}_{2y} & \overline{K}_{2z} \\ \overline{K}_{3x} & \overline{K}_{3y} & \overline{K}_{3z} \end{bmatrix} \begin{bmatrix} L_x \\ L_y \\ L_z \end{bmatrix} = 2\pi \begin{bmatrix} N_{12} \\ N_{23} \\ N_{34} \end{bmatrix}. \quad (2.16)$$

Equations of this form can be set up for a scan between any two viewing positions; normally, the pairs which give the largest phase changes would be chosen. We are again faced with the problem of sign ambiguity. In assigning values to changes of fringe order number such as N_{12}, an arbitrary sign convention may be chosen as long as it is followed consistently over the entire measurement. For example, N_{12} may be considered positive if the fringes appear to move toward the left relative to the observer as he or she scans from Q_1 to Q_2, and negative if they appear to move toward the right. If such a sign convention is followed, the correct magnitude and direction of **L** will be determined, but its sense will not be known unless it is physically apparent. This point will be considered further in Section 3.3.

2.3 ACCURACY OF MEASUREMENTS

Although the theory of vector displacement measurement, presented in Section 2.2, is straightforward, problems may arise in its experimental implementation which affect the accuracy of the measurement. The most obvious source of error is inherent in the sensitivity vector. If the general orientation of displacements is known a priori, the holographic system can be arranged so that the sensitivity vector will be nearly parallel to **L**, and an accurate determination of the magnitude of **L** can be made. If the direction of **L** is not known, the more general schemes leading to equations 2.13 or 2.16 must be applied. Consider the single-hologram analysis. Generally the solid angle subtended by the hologram is not large, so $\overline{\mathbf{K}}_1$, $\overline{\mathbf{K}}_2$, and $\overline{\mathbf{K}}_3$ are nearly parallel. As a result, the system of equations 2.16 will be poorly conditioned and small errors in the measurement of N_{12}, N_{23}, and N_{34} will be amplified and will lead to inaccurate values of L_x, L_y, and L_z. The solutions will also be sensitive to the accuracy with which the geometric measurements are made in order to specify the coefficient matrix. This can be discussed quantitatively using the techniques of linear algebra. To do so, we rewrite the system of equations 2.16 in matrix notation:

$$\overline{\mathbf{K}} \cdot \mathbf{L} = 2\pi \mathbf{N}, \qquad (2.17)$$

where

$$\overline{\mathbf{K}} = \begin{bmatrix} \overline{K}_{1x} & \overline{K}_{1y} & \overline{K}_{1z} \\ \overline{K}_{2x} & \overline{K}_{2y} & \overline{K}_{2z} \\ \overline{K}_{3x} & \overline{K}_{3y} & \overline{K}_{3z} \end{bmatrix}, \qquad \mathbf{L} = \begin{bmatrix} L_x \\ L_y \\ L_z \end{bmatrix},$$

and

$$\mathbf{N} = \begin{bmatrix} N_{12} \\ N_{23} \\ N_{24} \end{bmatrix}.$$

The analogous definitions can be made in terms of equation 2.13.

For the discussion of errors, we introduce the concept of the *norm*, or Euclidean length, of a vector such as **L**:

$$\|\mathbf{L}\| \equiv \left(|L_x|^2 + |L_y|^2 + |L_z|^2\right)^{1/2}. \qquad (2.18)$$

The norm of a matrix such as $\overline{\mathbf{K}}$ can be defined as [9]

$$\|\overline{\mathbf{K}}\| \equiv \max_{\mathbf{x} \neq \boldsymbol{\theta}} \frac{\|\overline{\mathbf{K}} \cdot \mathbf{X}\|}{\|\mathbf{X}\|}, \qquad (2.19)$$

where $\boldsymbol{\theta}$ is the null vector or, equivalently,

$$\|\overline{\mathbf{K}}\| = (\text{max. eigenvalue of } \overline{\mathbf{K}}^T \overline{\mathbf{K}})^{1/2}. \qquad (2.20)$$

First consider the effect of errors $\Delta \mathbf{N}$ in the measurement of fringe order numbers. The resulting error $\Delta \mathbf{L}$ in the displacement vector satisfies the equation

$$\overline{\mathbf{K}} \cdot (\mathbf{L} + \Delta \mathbf{L}) = 2\pi (\mathbf{N} + \Delta \mathbf{N}). \qquad (2.21)$$

Elementary inequalities satisfied by the norms can be used to show that[9]

$$\|\Delta \mathbf{L}\| \leq \|\overline{\mathbf{K}}^{-1}\| \cdot \|\Delta \mathbf{N}\| \cdot 2\pi. \qquad (2.22)$$

2.3 ACCURACY OF MEASUREMENTS

Similarly, if an error $\Delta\bar{\mathbf{K}}$ due to inaccurate measurements of system geometry is made in the coefficient matrix $\bar{\mathbf{K}}$, then [9]

$$(\bar{\mathbf{K}} + \Delta\bar{\mathbf{K}}) \cdot (\mathbf{L} + \Delta\mathbf{L}) = 2\pi\mathbf{N} \tag{2.23}$$

and

$$\frac{\|\Delta\mathbf{L}\|}{\|\mathbf{L}\|} \leq 2\pi \|\bar{\mathbf{K}}^{-1}\| \cdot \|\Delta\bar{\mathbf{K}}\|. \tag{2.24}$$

In deriving equation 2.24 it is assumed that $\|\Delta\mathbf{L}\|/\|\mathbf{L}\| \ll 1$. Equations 2.22 and 2.24 indicate the magnitude of errors in $\mathbf{L}$ which result from the experimental errors $\Delta\bar{\mathbf{K}}$ and $\Delta\mathbf{N}$. Matsumoto et al.[10] have applied inequalities 2.22 and 2.24 to the problem of displacement measurement by holographic interferometry. For a typical holographic system they calculated the errors in fringe readout and in measurement of propagation directions that could be allowed and still maintain a prespecified allowable error in $\mathbf{L}$. The curves these authors present confirm that as the angle between the sensitivity vectors decreases the accuracy demanded in the measurement of fringe order numbers increases.

A useful numerical measure of the effect of measurement errors is the *condition* of the matrix $\bar{\mathbf{K}}$, defined by

$$\mathrm{cond}(\bar{\mathbf{K}}) \equiv \|\bar{\mathbf{K}}\| \cdot \|\bar{\mathbf{K}}^{-1}\|. \tag{2.25}$$

In terms of the condition, we can rewrite inequalities 2.22 and 2.24 as[9]

$$\frac{\|\Delta\mathbf{L}\|}{\|\mathbf{L}\|} \leq \mathrm{cond}(\bar{\mathbf{K}}) \frac{\|\Delta\mathbf{N}\|}{\|\mathbf{N}\|} \tag{2.26a}$$

and

$$\frac{\|\Delta\mathbf{L}\|}{\|\mathbf{L}\|} \leq \mathrm{cond}(\bar{\mathbf{K}}) \frac{\|\Delta\bar{\mathbf{K}}\|}{\|\bar{\mathbf{K}}\|}. \tag{2.26b}$$

By means of computational procedures such as those given by Golub and Reinsch[12] the condition of $\bar{\mathbf{K}}$, or $\mathbf{K}$, can be computed for a given holographic system, and its value will aid in assessing the sensitivity of measurements of $\mathbf{L}$ to experimental errors.

Equations 2.22 and 2.24, or equations 2.26a and 2.26b, provide measures of the upper bounds of errors in measurements of $\mathbf{L}$. Such bounds are useful for studying general trends of measurement errors, but they usually give estimates that are much larger than the errors encountered in practice. A simple statistical analysis is more useful for estimating the expected error in a given experiment.[11] Let us assume that the experimental error, $\Delta \mathbf{N}$, in determining values of fringe order numbers can be represented by a random variable having a Gaussian distribution with zero mean. The standard deviation σ_N of this distribution is considered to be representative of the typical error range of fringe order number measurements. For a linear system like equation 2.21 the variance σ_L^2 of the corresponding values of $\mathbf{L}$ is related to σ_N^2 by

$$\frac{\sigma_L^2}{\sigma_N^2} = [\mathbf{K}^T\mathbf{K}]_{\text{diag}}^{-1}. \tag{2.27}$$

In equation 2.27 we use $\mathbf{K}$ if fringe-order measurements are *absolute*, that is measured with respect to a zero-motion fringe. $\bar{\mathbf{K}}$ is used when changes of fringe order are read by scanning (single-hologram analysis). To use equation 2.27 for error estimation, the matrix $\mathbf{K}$ is determined for the holographic system that is to be used. This matrix is then multiplied by its transpose and inverted. The element in the upper left corner of the resulting matrix represents $\sigma_{L_x}^2/\sigma_N^2$; the element in the center is $\sigma_{L_y}^2/\sigma_N^2$; and the element in the lower right corner is $\sigma_{L_z}^2/\sigma_N^2$. The results of such analyses for typical systems are displayed in Figures 2.7 to 2.9. The

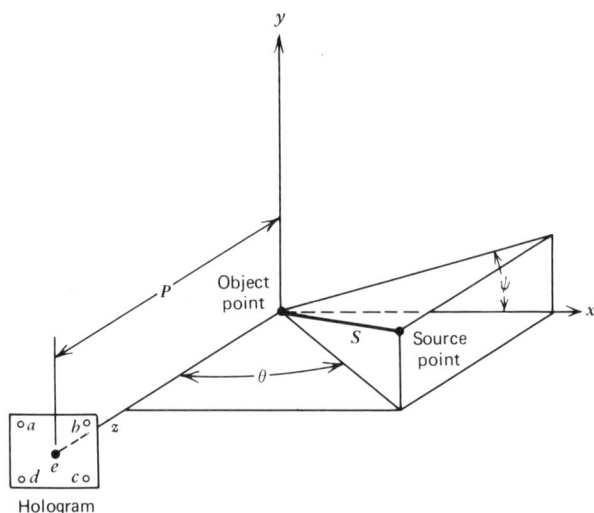

Figure 2.7 Holographic system geometry used in error analysis (ref. 11). Points a, b, c, d, and e are observation points. The plate, which has dimensions of 10×13 cm, is perpendicular to the z axis.

2.3 ACCURACY OF MEASUREMENTS

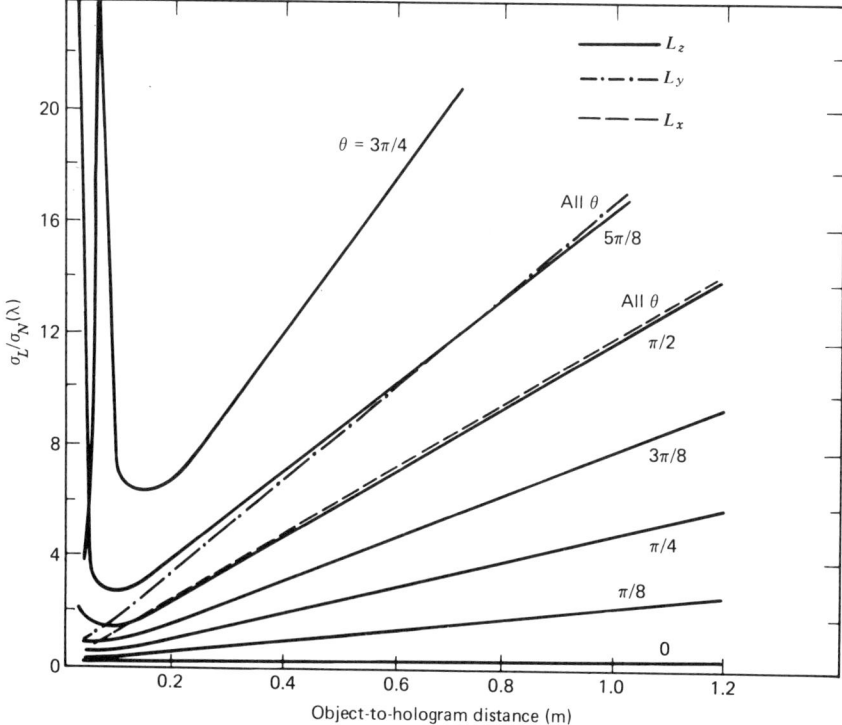

Figure 2.8 Measurement errors when observations are made from three points (ref. 11). Here σ_L is the standard deviation of errors in measuring component L_x, L_y, or L_z when the standard deviation of fringe order number errors is σ_N. System geometry is defined in Figure 2.7; observations are made at points a, b, and c.

geometry is defined in Figure 2.7. The hologram is assumed to be recorded on a 10×13 cm plate, and fringe order number measurements are made from three observation points, denoted as a, b, and c in the figure. The direction of incident illumination is specified by angles ψ and θ. Figure 2.8 is a plot of σ_L^2/σ_N^2 as a function of direction angle θ and object-to-hologram distance P when $\psi = 0$. When ψ is not zero, the results for σ_{L_x}/σ_N and σ_{L_y}/σ_N are nearly identical with those in Figure 2.8, and σ_{L_z}/σ_N differs at most by 50 percent of its value at $\psi = 0$.[11]

The observation points used in generating Figure 2.8 were chosen to represent a typical configuration. Of course the important factor is the *direction* of rays from the object point to each observation point. The errors estimated for a single hologram close to the object are equivalent to those for three separate holograms at a greater distance from the object if the observation directions are the same.

To illustrate the use of Figure 2.8, let us suppose that the distance from the object point under study to the hologram is 40 cm, and the illumination angle is 45°. From the figure, $\sigma_{L_x}/\sigma_N = 4.5\lambda$, $\sigma_{L_y}/\sigma_N = 6.8\lambda$, and $\sigma_{L_z}/\sigma_N = 2.0\lambda$, where λ is the wavelength of the laser light. As an example, suppose that the experimenter estimates that fringes can be read with an accuracy of ± 0.5 fringe, that is, $\sigma_N = 0.5$. The errors in displacement components would then be expected to be of the order of $\Delta L_x = \pm \frac{1}{2}(4.5\lambda)$, $\Delta L_y = \pm \frac{1}{2}(6.8\lambda)$, and $\Delta L_z = \pm \frac{1}{2}(2.0\lambda)$. If an He-Ne laser is used, $\lambda = 632.8$ nm, so the expected accuracy of computed values of L would be $\Delta L_x \cong \pm 1.4$ μm, $\Delta L_y \cong \pm 2.2$ μm, and $\Delta L_z = \pm 0.6$ μm. Calculations such as this provide useful *estimates* of accuracy for analyzing or designing measurement experiments.

The general trend predicted by Figure 2.8 is an increase in error with increasing object-to-hologram distance, that is, with decreasing solid angle subtended by the hologram at the object point. This is a manifestation of the poor condition of the matrix **K** when the viewing directions become nearly parallel. An exception is the drastic increase in error that can occur at small object-to-hologram distances. The physical significance of this singular behavior is that the sensitivity vector for one of the viewing directions is nearly normal to L_z; hence a small error in fringe order number will be highly amplified to give a large error ΔL_z.

Problems associated with poorly conditioned systems of equations are common in experimentation, and can be alleviated by using redundant data. Dhir and Sikora[13] showed that redundant measurements can significantly improve the accuracy with which L can be determined. Four or more phase measurements are made for each object point considered. This yields a system of equations similar to 2.13 or 2.16, but the number of equations is greater than the number of unknown vector components. This overdetermined system of equations can be solved in the least-mean-square sense by well-known numerical procedures.[14] The advantage of using redundant data is evident in Figure 2.9, which displays the results of an error analysis when five observations are made from the points labeled a, b, c, d, e in Figure 2.7. The use of redundant measurements leads to some improvement of accuracy, particularly in L_y, but the most important effect is the suppression of the singular behavior at small object-to-hologram distances. Clearly, redundant measurements should be used, especially when the direction of **L** is not known a priori.

In the experiments by Dhir and Sikora observations were made from several points on a single 10 × 13 cm hologram. Even the addition of one extra measurement resulted in significant improvement in the accuracy of the measurement of **L**. The use of four phase measurements per point was also shown to improve the accuracy of measurements reported in refs. 15

2.3 ACCURACY OF MEASUREMENTS

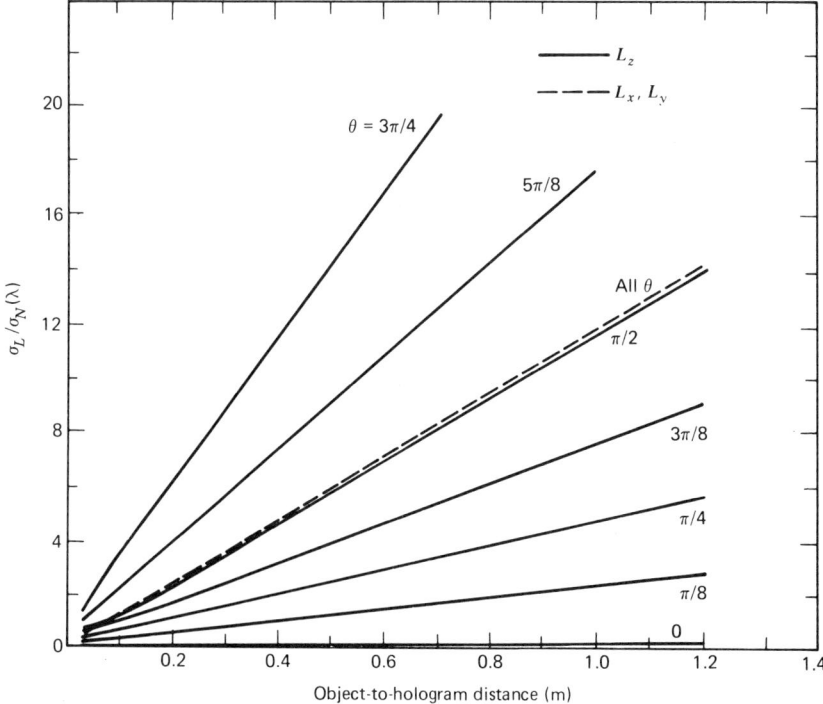

Figure 2.9 Measurement errors when observations are made from five points (ref. 11). Here σ_L, is the standard deviation of errors in measuring component L_x, L_y, or L_z when the standard deviation of fringe order number errors is σ_N. System geometry is defined in Figure 2.7; observations are made at points a, b, c, d, and e.

and 16. The effect of even greater redundancy was explored by Sciammerella and Gilbert[7] (14 observations per object point) and by King[17] (up to 20 observations per object point).

The effects of measurement errors when vector displacement is determined by scanning (single-hologram analysis) are shown in Figure 2.10. In this method changes of fringe order are measured as the hologram is scanned from point to point while the observer's view is fixed on object point P. Measurements of L_z by scanning are seen to be very error prone, and little improvement should be expected when redundant measurements are made. For object-to-hologram distances greater than about 20 cm in the system shown in Figure 2.7, σ_{L_z}/σ_N grows as the square of this distance, whereas σ_{L_x}/σ_N and σ_{L_y}/σ_N increase linearly.

It is apparent from the preceding discussion that a simple answer to the question of how accurately displacement vectors can be measured is not

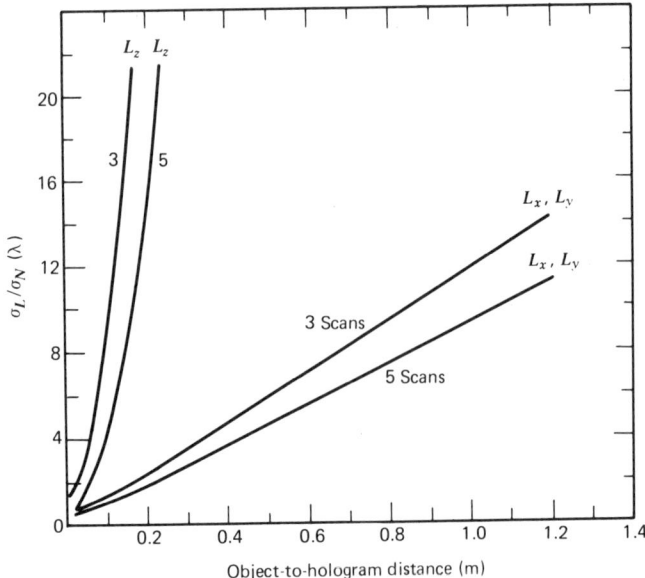

Figure 2.10 Measurement errors when observations are made by scanning a single hologram (ref. 11). Here σ_L is the standard deviation of errors in measuring L_x, L_y, or L_z when the standard deviation of fringe order number errors is σ_N. System geometry is defined in Figure 2.7; three-scan readout is along lines ab, bc, and ce, and five-scan readout is along lines ab, bc, ce, be, and ed.

possible. However, we have shown that the sources of experimental error can be divided into two categories: those associated with measuring *system geometry*, and those associated with measuring *optical phase*. These error sources and some of the methods of minimizing them are discussed below.

Errors associated with system geometry are those which affect the coefficient matrix of equations 2.13 or 2.16. The manner in which such errors affect the ultimate accuracy of displacement measurements depends upon two factors: the angular separation of the sensitivity vectors corresponding to each observation, and the orientation of the displacement vector relative to these sensitivity vectors. When the direction of the displacement vector is known a priori, the system can be designed so that one of the sensitivity vectors is nearly colinear with it; this will provide for an accurate determination of the magnitude of the displacement, but small deviations in orientation of **L** will not be measured accurately. In general, the best strategy is to maximize the angular separation of the sensitivity vectors. This usually necessitates the use of multiple-hologram systems. The separation can be increased even further if multiple illumination beams are used; this approach, of course, requires that each hologram be exposed separately.

2.3 ACCURACY OF MEASUREMENTS

Once the system geometry has been chosen, the coefficient matrix of equation 2.13 or equation 2.16 is strongly affected by the accuracy with which the orientations of the propagation vectors are determined. Careful measurement of the system geometry and its incorporation into equations or computer codes constitute one of the most crucial steps in measurements by holographic interferometry. If the object surface under study is not too large, it is advisable to use collimated object illumination, so that the vector $\mathbf{k}_1$ is a constant.

From the discussion above it is apparent that the use of several holograms, each with multiple observations, and the use of multiple collimated illumination beams[18] will tend to increase the experimental accuracy attainable. This increase is achieved, however, at the expense of additional system complexity and cost. It is advisable that the experimenter carry out at least a crude error analysis for the application of holographic interferometry to each particular measurement problem when designing the experimental system.

Once the system geometry has been determined, the remaining error sources are associated with the right-hand side of equations 2.13 or 2.16. They depend upon the magnitude of the displacement, the wavelength of the laser light, and the accuracy with which phase measurements can be made. The component of displacement parallel to the sensitivity vector for each observation, and its gradient along the object surface, together with the wavelength of the light, determine the number and spacing of interferometric fringes. Since the fringe shift caused by a given displacement is inversely proportional to λ, the measurement accuracy depends on the type of laser used. A wide range of wavelengths is available for holography, and this topic will be discussed in Chapter 6; currently, however, most experimenters use argon ion or He-Ne continuous wave lasers or pulsed ruby lasers with wavelengths in the range 450 to 700 nm. For measurement of very small displacements it is obviously desirable to use blue-green light from a source such as an argon ion or a He-Cd laser, although this yields less than a twofold increase in the number of fringes over that occurring in red light.

The accuracy with which fringe order numbers can be assigned depends on the accuracy with which phase can be measured, and on the accuracy with which such measurements can be assigned to a specific object point.

Fringe order numbers are usually determined by visual observation of the holographic interferogram, especially when single-hologram analysis is used. Such observations are limited to an accuracy of 0.5 fringe, with perhaps a crude interpolation to 0.25 fringe if the fringes are widely spaced. If the fringe pattern is photographed, fringe order numbers can be assigned either by counting, or by scanning the photograph with a microdensitometer. A microdensitometer trace is useful in estimating fractional fringes if the processing of the photograph is carefully controlled and if the microdensitometer is accurately calibrated. In this manner one generally estimates shifts to 0.1 fringe, although in careful experiments accuracy to 0.02 fringe has been reported.[7] It is difficult to assign a universal value to the accuracy with which a microdensitometer trace of holographic fringes can be read, because of the effect of laser speckle. As we will see in the next chapter, fringe localization effects may necessitate the use of a high-f-number camera system for photographing the fringes. This increases the characteristic speckle size, thus introducing a strong noise into the fringe pattern, so that fringe contours will not be smooth.

Determination of fractional fringe shifts is important if the total displacement yields a shift of less than one fringe, or if several fringes occur, but high accuracy is required. In the former case *holographic subtraction*[19-21] is useful and is accomplished by introducing a known phase shift β into either the reference or the object beam between the first and second holographic exposures. If $\beta = \pi$ radians, a dark-field interferogram with a fringe modulation of $(1 - \cos\delta)$ results. The zero-order fringe is dark, and small displacements yield a bright region which is easily detected against the dark background. For very small shifts, the irradiance is proportional to δ^2 and the slope of irradiance versus δ is zero. The dark-field hologram therefore offers no particular advantage for quantitative evaluation, although it is advantageous for qualitative observations. If $\beta = \pi/2$, however, the fringe modulation is $(1 - \sin\delta)$. For very small δ, the irradiance is proportional to δ and the derivative of irradiance with respect to δ has the maximum possible value, so that accuracy in detecting very small shifts is improved.

A controlled phase shift can be introduced into the laser beam by a variable-pressure gas cell.[20] It can also be accomplished by inserting a glass plate, such as a microscope coverslip, into the beam. The plate should be oriented at the Brewster angle to minimize losses. A small rotation of this plate by an angle $d\theta$ causes a phase shift

$$d\beta = d\theta\left[(1 - \alpha^{-1}\cos\theta)\sin\theta\right]t(2\pi/\lambda), \qquad (2.28)$$

where t is the thickness of the plate, $\alpha \equiv (n^2 - \sin^2\theta)^{1/2}$, n is the refractive index of the plate, and θ is the Brewster angle.[22] This method requires only a simple mechanical adjustment between holographic exposures.

Measurements with accuracy to a fraction of a fringe may be important even when the total displacement corresponds to a shift of several fringes. This is the case when the direction of **L** must be determined, and when displacements are to be differentiated to determine strains. When high accuracy is required, techniques which interface electronic phase detection with holographic interferometry can be used. Researchers at Brown-Boveri

2.3 ACCURACY OF MEASUREMENTS

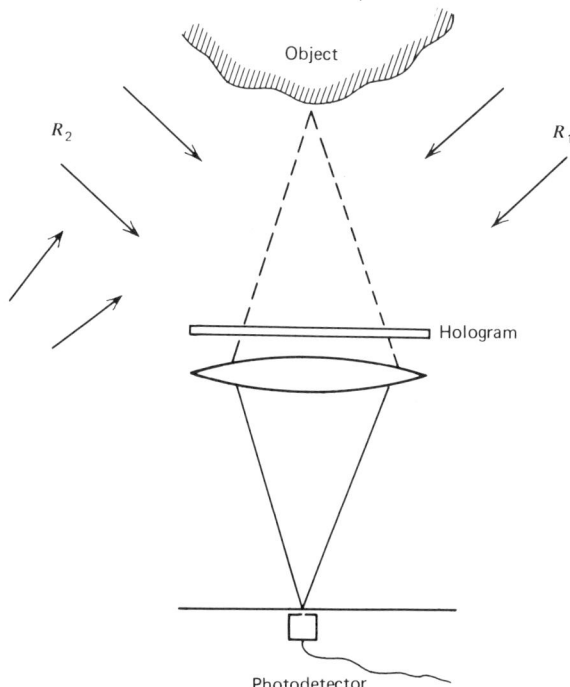

Figure 2.11 Holographic interferometry with electronic phase detection (ref. 23).

introduced such a technique,[23, 24] which uses the system shown schematically in Figure 2.11. The complex amplitude of the light, of temporal frequency ω_0, scattered by a point P and arriving at the photodetector is given by

$$\mathbf{U}_1(P) = a_1(P)\exp\left[i\omega_0 t + \phi_1(P)\right] \qquad (2.29)$$

before deformation, and by

$$\mathbf{U}_2(P) = a_2(P)\exp\left[i\omega_0 t + \phi_2(P)\right] \qquad (2.30)$$

after deformation. To understand this method we must retain the temporal as well as spatial variations of these fields. In Chapter 1 it was noted that optical frequencies are too high for detectors to follow, so phase differences must be converted to time-independent irradiance variations by interferometry. In the scheme considered here, the objective is to produce a signal of low frequency, whose phase can be measured electronically.

The low-frequency signal is produced by the "beating" of two optical waves having slightly different frequencies.

Two different reference beams, R_1 and R_2, are used in recording the hologram. Both have the same optical frequency ω_0. The first exposure, of the undeformed object, is recorded with reference beam R_1. The second exposure, of the deformed object, is recorded with reference beam R_2. During the reconstruction process the frequencies of R_1 and R_2 are slightly shifted; for example, let R_1' have frequency $\omega_1 = \omega_0 + \Delta\omega$, and let R_2' have frequency $\omega_2 = \omega_0 - \Delta\omega$, where $|\Delta\omega| \ll \omega_0$. When the hologram is illuminated by R_1' and R_2', two object waves are reconstructed simultaneously. They have complex amplitudes given by

$$\mathbf{U}_1'(P) = a_1(P)\exp[i\omega_1 t + \phi_1(P)] \qquad (2.31)$$

and

$$\mathbf{U}_2'(P) = a_2(P)\exp[i\omega_2 t + \phi_2(P)]. \qquad (2.32)$$

The irradiance measured at the detector is

$$I(t) = |\mathbf{U}_1' + \mathbf{U}_2'|^2$$

and can be expressed as

$$I(t) = a_1^2(P) + a_2^2(P) + 2a_1(P)a_2(P)\cos[\Omega t + \delta(P)]. \qquad (2.33)$$

Equation 2.33 indicates that the light irradiance leaving the hologram fluctuates sinusoidally at the beat frequency $\Omega = \omega_1 - \omega_2$, which can be much lower than the optical frequency ω_0; it has a phase $\delta(P)$ at the detector which is equal to the phase difference due to object displacement that we wish to measure. The frequency Ω can be low enough so that the photodetector can follow and convert $I(t)$ into an ac electric signal, whose phase $\delta(P)$ can be measured by standard electronic instruments. The necessary optical frequency shifts can be introduced into the reference beams by relatively simple devices such as rotating diffraction gratings[25] or half-wave plates,[26] or by electro-optical modulation.[27]

The Brown-Boveri workers have reported phase measurements accurate to better than 0.002 fringe. Their system used an argon ion laser. Phase shifts of ~100 kHz were introduced by a rotating grating. The system has been used to determine deflection, strain, change of curvature, and bending moments.[24] In theory it could be adapted to use in vibration analysis and real-time holographic interferometry.

2.3 ACCURACY OF MEASUREMENTS

Having discussed some of the individual factors which affect the accuracy of holographic interferometry, we conclude this section by citing experimental assessments of the precision with which displacement vectors can be measured. Sciammarella and Gilbert[7] have carried out a detailed experiment to measure the deformation of a Plexiglas disk, of 10 cm diameter, subjected to diametral compression. Their system had two collimated object beams; one was normal to the disk surface, and one was oriented at about 45°. Two holograms were recorded: one was parallel to the disk surface, and one was oriented at about 45°. For readout, each hologram was covered by a mask, one of which had 8 observation holes and the other, 6. Each hole had a diameter of 6.35 mm. The fringe pattern was photographed through each of the 14 observation holes. Each photograph was enlarged by a factor of 10 and scanned with a microdensitometer to assign fringe order numbers to a matrix of points on the surface of the disk. Fourteen equations like 2.12 were then solved by least-square analysis to determine three orthogonal components of displacement at each point. From these measurements surface strains ϵ_x and ϵ_y were computed and found to agree very well with theoretical computations of strain. The normal deformation w was then calculated using the relation

$$w = -\nu\left(\frac{t}{2}\right)(\epsilon_x + \epsilon_y), \qquad (2.34)$$

where ν is Poisson's ratio, and t is the thickness of the disk. Equation 2.34 was used in preference to the directly computed values of L_z because some rigid-body rotation of the object was detected and contributed to L_z. It was estimated that the measurements of w were accurate to within $\sim\frac{1}{8}\lambda \sim 0.1$ μm. This result agrees with the order-of-magnitude limit to accuracy suggested by the experiments of Sciammarella and Chang[16] and of King.[17] Other experimental assessments of the precision of displacement and deformation measurements for various configurations appear in refs. 28 to 30.

This section has dealt with the precision with which very small displacements and deformations can be measured. The related question of how large deformations and displacements measurable by holographic interferometry can be depends on quite different factors and is largely unexplored. Two factors which are obviously involved are loss of wavefront correlation and high fringe frequencies. When rigid-body motions are large or when a surface is strained excessively, the wavefronts of light scattered by the object at the times of the first and second holographic exposures lose correlation with each other. Observable fringes of holographic interferometry cannot be formed. Also, when the spacing of the fringes becomes as small as the characteristic size of the laser speckle, they cannot be observed and counted, except in special cases. Further discussion of these limiting factors requires the concepts introduced in the next chapter. Some techniques for measuring relatively large displacements are introduced in Chapter 7.

2.4 THE HOLODIAGRAM

The *holodiagram* is a concept and device developed by N. Abramson for describing and evaluating holography and holographic interferometry.[6, 31–37] It provides an alternative, geometrical formulation of many of the relations presented in this chapter. In this section we introduce the holodiagram and discuss its application to fringe interpretation, to a useful moiré analog of the fringes of holographic interferometry, and to the design of holographic systems.

Any holographic system is equivalent to that shown in Figure 2.12a. The solid ray represents laser light leaving the point source O and being scattered by the object point P to a point Q on the film (hologram). The dotted ray represents light from the same laser which serves as the portion of the reference beam impinging at Q. The fringes due to interference between the object and reference beam are referred to as *primary fringes*; they form the hologram. During the reconstruction process the primary fringes recorded on the hologram cause light to be diffracted toward the observer so that he or she sees the object point P. To form a holographic interferogram, P is displaced slightly to a new position P' and a second holographic exposure is made. The resulting hologram reconstructs the light scattered from both P and P'. An observer viewing point P through the hologram at Q simultaneously receives light which traveled path $\overline{OPQ}$ and that which traveled path $\overline{OP'Q}$. If these paths differ by an odd multiple of $\lambda/2$, the primary fringes at Q are washed out and no light will be diffracted toward the observer, who will see a dark fringe in the vicinity of P. If $\overline{OPQ}$ and $\overline{OP'Q}$ differ by an integral multiple of λ, the primary fringes will have full contrast at Q so that the hologram will scatter light toward the observer, who will see a bright fringe in the vicinity of P.

The basic holodiagram corresponding to the system of Figure 2.12a is shown in Figure 2.12b. It consists of a set of concentric ellipses having foci at points O and Q. These ellipses are the loci of points P for which the distance $\overline{OPQ}$ is a constant. We imagine this distance to differ by $\lambda/2$ from each ellipse to the next. In reality, these loci are three-dimensional surfaces, ellipsoids of revolution about the x axis. The diagram represents the intersection of these ellipsoids with the OPQ plane.

2.4 THE HOLODIAGRAM

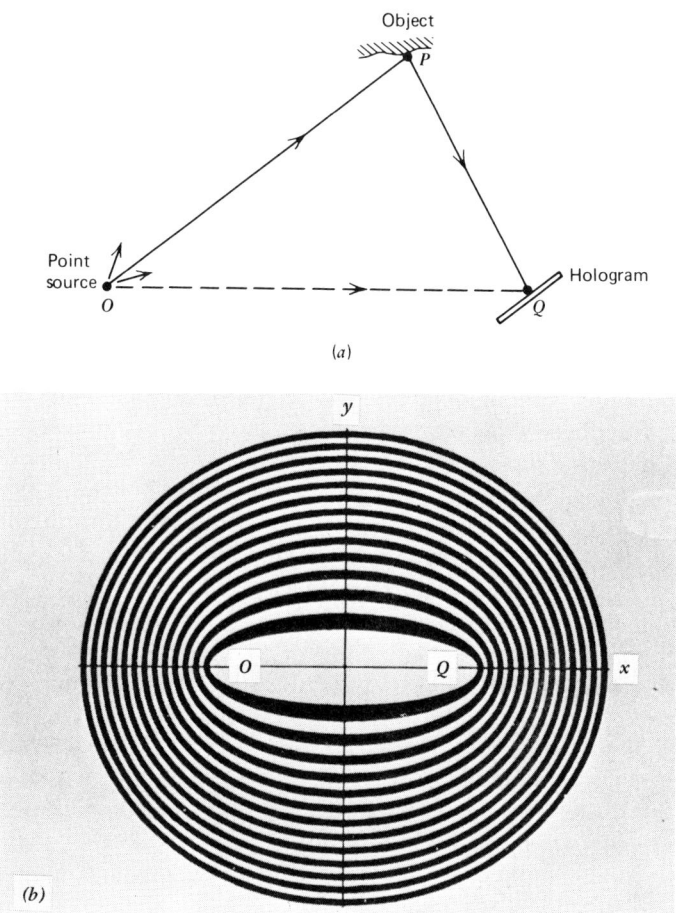

Figure 2.12 Nomenclature for the holodiagram. (*a*) Schematic diagram of a holographic system. (*b*) Ellipses of constant path length $\overline{OPQ}$.

Motion of an object point P along one of the ellipses maintains a constant optical path $\overline{OPQ}$; hence holographic interferometry is insensitive to such motions. Since a shift of one fringe corresponds to the displacement of P from one ellipse to another, holographic interferometry is most sensitive to motions normal to the ellipse passing through P. This is equivalent to the concept of the sensitivity vector discussed in Section 2.2. If an object point lies on the x axis to the right of point Q in Figure 2.12*a*, a shift of one fringe corresponds to displacement by a distance $\lambda/2$. At other locations in the OPQ plane a shift of one fringe corresponds to a larger displacement, say $k(\lambda/2)$. The holodiagram also includes curves of constant k. These loci, shown as solid curves in the holodiagram (Figure 2.13) are circles which pass through points O and Q. The centers of these circles are located on the y axis at points given by[32]

$$y_c = \tfrac{1}{2} \overline{OQ} \left[\frac{2-k^2}{(k^2-1)^{1/2}} \right]. \qquad (2.35)$$

The fringes of holographic interferometry can be evaluated as follows. The fringe order number N is assigned to the fringe passing through P by the usual procedure of counting fringes (Section 2.2). This number is multiplied by the k value corresponding to the point P on the holodiagram and by half the wavelength of light. The component of displacement *normal to the ellipse passing through P* is $kN(\lambda/2)$. Of course, we cannot really plot all the ellipses and k circles, but we can plot, for example, every 1000th curve. This would correspond to ellipses for which $\overline{OQ}$ differs by about 1 mm, and should result in sufficient resolution of the k values for practical measurement problems.

Abramson has used the holodiagram to discern the form of fringes which would be caused by a variety of object translations and rotations for

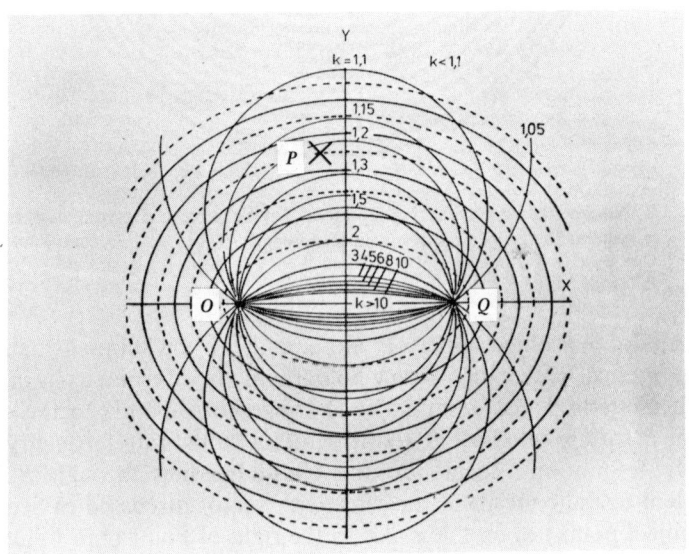

Figure 2.13 The holodiagram (ref. 6).

2.4 THE HOLODIAGRAM

various system geometries.[32] These results are especially useful for the qualitative understanding of fringes when the distances $\overline{OP}$ and $\overline{PQ}$ are comparable to $\overline{OQ}$. The sensitivity vector may then vary considerably over the surface of an object, resulting in fringe patterns whose significance is not obvious, until clarified by predictions like this. Abramson has also introduced a moiré analogy to holographic interferometry which is very useful for predicting fringe patterns due to simple motions.[33, 36]

The moiré analogy can be developed by imagining the ellipsoids of the holodiagram to exist as a set of concentric shells in space. Successive shells correspond to path lengths $\overline{OPQ}$, which differ by $\lambda/2$. The spaces between every other pair of shells are imagined to be filled with an ink which can be "activated" by the experimenter. The undeformed object is placed in its original position, and the experimenter activates the ink. This imprints a pattern of primary fringes on the object along the intersection of its surface with the ink-filled shells—a process analogous to the first exposure of a holographic interferogram. The ink is then deactivated, and the object is displaced or deformed. The experimenter activates the ink again, in analogy to a second holographic exposure. The object surface is now imprinted with two slightly different sets of primary fringes, which form a moiré pattern[38, 39] of coarse, *secondary fringes*. These secondary fringes are good analogs of the fringes of holographic interferometry.

The moiré analogy is especially convenient for rigid-body translations and rotations. If the object is a plane which is very close to the source O and observation point Q, the primary fringes imagined to be imprinted on the object will be nearly identical to the ellipses of Figure 2.12b. For this case the anticipated fringes of holographic interferometry for in-plane motions can be determined by placing one of two identical transparencies of the ellipses on top of the other, and translating or rotating one of them in the direction of interest. Figure 2.14 shows the moiré fringes, as well as the corresponding actual fringes of holographic interferometry, for a rotation and a translation of this type. Note that a casual observer might assume these fringe patterns to be due to deformations rather than rigid-body motions.

To extend the moiré analogy to objects located farther from the illumination and observation points, it must be recalled that the holodiagram is a planar section of a set of three-dimensional ellipsoids of revolution. Figure 2.15 shows the intersection of an object surface AB with the OPQ plane. Points where the ellipses intersect AB lie on dark primary fringes. The prediction of fringes of holographic interferometry is again accomplished by forming the moiré pattern between two identical transparencies of the set of ellipses. In Figure 2.15a the object is illuminated at nearly grazing incidence and viewed normally. The object has been translated parallel to its surface. It is seen that moiré fringes intersect the object

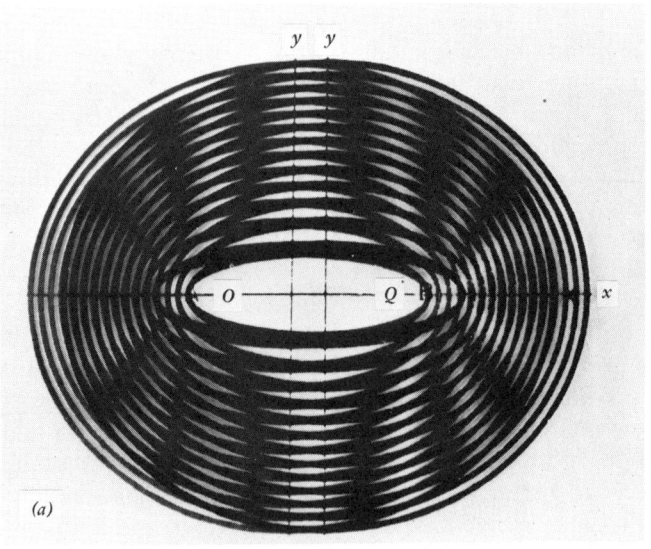

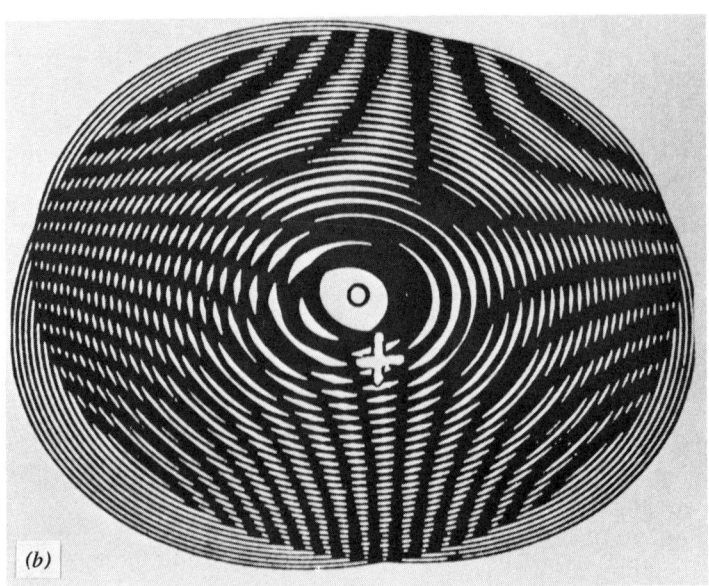

Figure 2.14 The moiré analogy (ref. 36). (*a*) In-plane translation moiré pattern. (*b*) Moiré pattern for rotation about O. (*c*) Holographic fringe pattern corresponding to (*b*). Fringe curvature similar to that indicated in (*a*) is shown in Figure 3.13.

2.4 THE HOLODIAGRAM

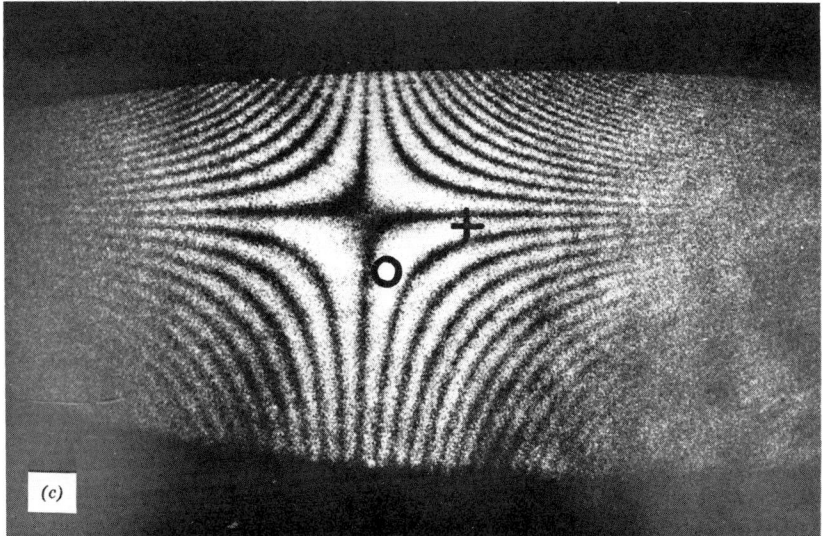

Figure 2.14 Cont.

along AB; hence, with appropriate scaling, we would predict that four fringes would be seen in a holographic interferogram. In Figure 2.15b illumination and viewing conditions are the same as in Figure 2.15a, but the object has been translated *normal* to its surface. In this case we would predict that only one fringe would be seen.

Sollid[40] conducted experiments which showed that for this configuration many more fringes are seen when the translation is transverse than when it is normal, and Abramson[33] showed that this was predicted by the holodiagram, as outlined above. This is a very interesting example, because, as the reader can readily confirm, equation 2.8 indicates that the phase shift δ is identical in the two cases. One could easily be misled into assuming that identical numbers of fringes would be observed. This example illustrates the usefulness of the holodiagram for predicting and understanding the fringes of holographic interferometry. It also indicates that further knowledge of the nature of fringe formation is required to supplement equation 2.8 when the inverse problem of *predicting* the fringes of holographic interferometry is to be solved. This is the subject matter of Chapter 3.

The holodiagram is also useful in the design of holographic systems. The most obvious application is to maximize or minimize sensitivity to certain motions. The holodiagram (Figure 2.12b) can be drawn to scale, or even sketched on the optical bench. We know that the system will be most sensitive to motions normal to the ellipses, and least sensitive to motions tangential to the ellipses. This knowledge can be used to maximize the

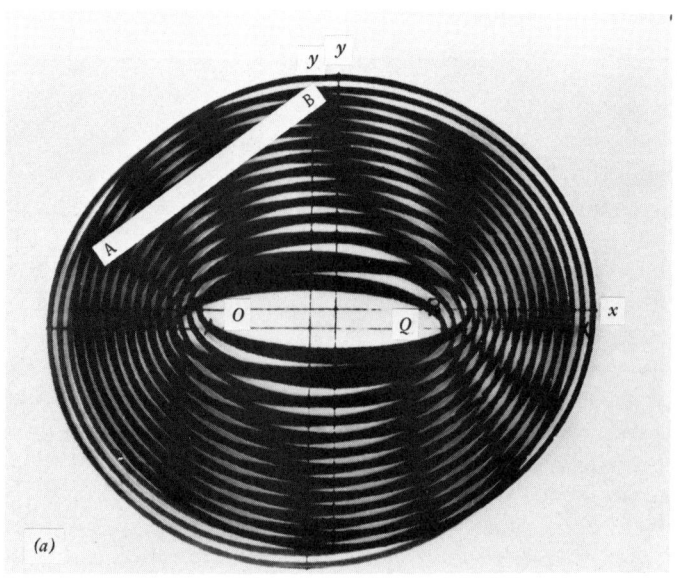

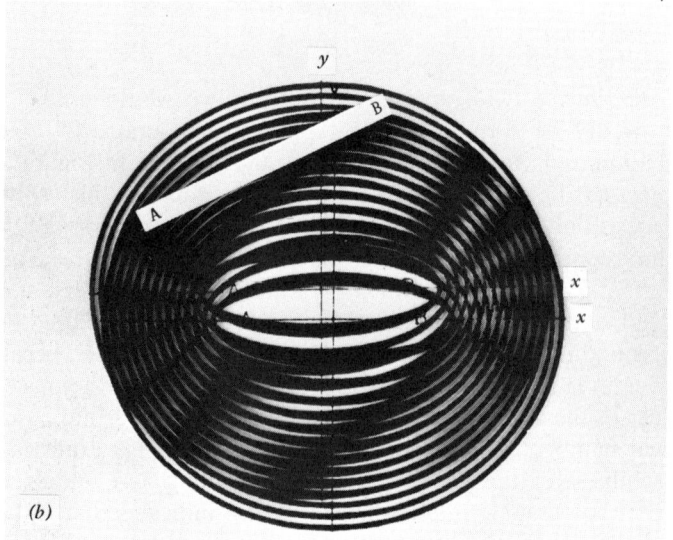

Figure 2.15 The moiré analogy. (*a*) In-plane translation of surface AB. (*b*) Normal translation of surface AB.

sensitivity of a holographic interferogram to a particular component of deformation. On the other hand, if a hologram of an object which is not completely stable is to be recorded, it can be oriented so that the undesirable motion is tangential to the ellipses. This maximizes the chances of recording a good hologram.

The holodiagram is also useful when a hologram of a very large object is to be recorded. In holography, object size is restricted by the coherence length l_c of the laser. If, for some object points, the pathlength difference between the object and reference beams exceeds this length, no hologram is formed. A holodiagram can be drawn in which the spacing of the ellipses corresponds to a pathlength difference l_c. If the object and the mirror which reflects light into the reference beam can be oriented to occupy the space enclosed by the same two neighboring ellipses, a hologram can be recorded. Using a system arranged according to this procedure, Abramson[31] has recorded holograms of an object 2 m long, using a cw gas laser with a coherence length of 30 cm. Similar reasoning is especially useful when a pulsed laser is used.

For many other applications of the holodiagram and for further details of its use, the reader is referred to the writings of Abramson.[6, 31-37] Reference 35 contains a summary of its various applications.

2.5 ALTERNATIVE METHODS OF FRINGE READOUT

In the preceding sections of this chapter we discussed the analysis of fringes, assuming that they were observed visually or by simple photography of the virtual image. Other methods of recording fringe data have been proposed and demonstrated. These methods are particularly important to the development of automated systems for the analysis of holographic interferograms. In this section three such techniques are described.

Let us begin by considering the multiple-hologram analysis (Section 2.2), in which fringes are counted in three different holograms. A photograph of the virtual image of the object is taken through each of the holograms. Since the perspective is different for each photograph, the object shape appears different in each view. This is illustrated by the three interferograms of a circular disk shown in Figure 2.16. Unless it is possible to scribe a grid on the object surface, it is difficult to identify corresponding object points in each photograph. This problem is especially complicated if fringes are to be counted by an automated device. Matsumoto et al.[41] have demonstrated a method for circumventing this difficulty by projecting the *real image* (Section 1.5) from the holographic interferogram back onto the object. A setup for this purpose is shown schematically in Figure 2.17. All reference waves are plane. Their conjugates, which are simply plane waves

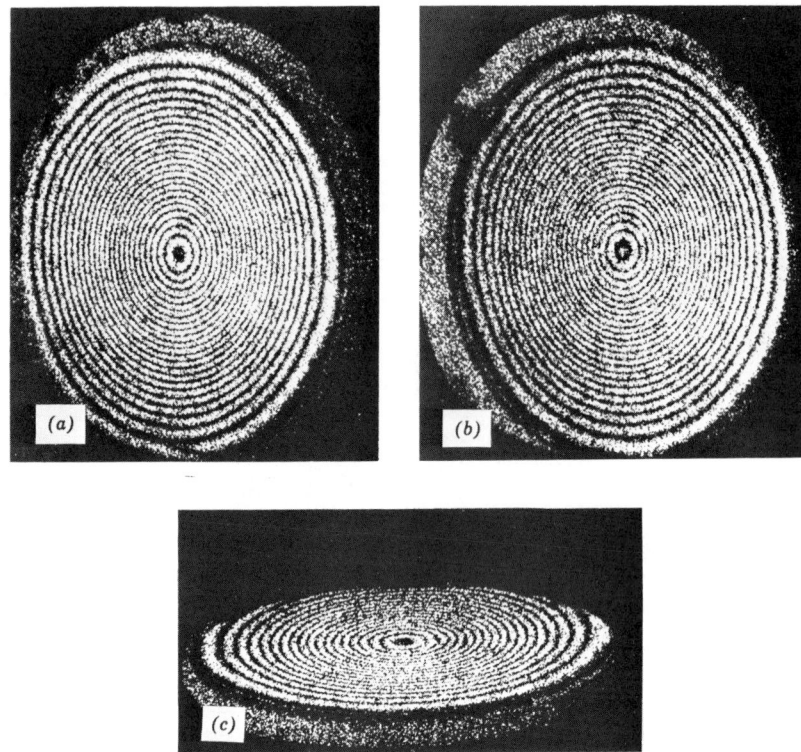

Figure 2.16 Three views of the virtual image of a two-exposure hologram of a deformed circular disk (ref. 41).

traveling in the opposite direction, are used for reconstructing the holograms. This causes the real image to form on the object surface. Since light is not scattered by the surface at points where destructive interference occurs, the fringe pattern can be observed on the object surface. A camera is used to photograph this pattern along the direction in which the illumination beam originally traveled. While the camera position remains fixed, each hologram is individually reconstructed and the corresponding fringe pattern on the object surface is photographed.

Figure 2.18 includes three views of the same disk which was shown in Figure 2.16. The apparent shape of the disk is identical in Figures 2.18a, 2.18b, and 2.18c, but the fringe pattern differs, since the sensitivity vectors are different for each view. It is easy to assign fringe numbers to corresponding points in each photograph. The resulting data can be analyzed using equation 2.13. Matsumoto et al.[41] also demonstrated that the fringe

2.5 ALTERNATIVE METHODS OF FRINGE READOUT

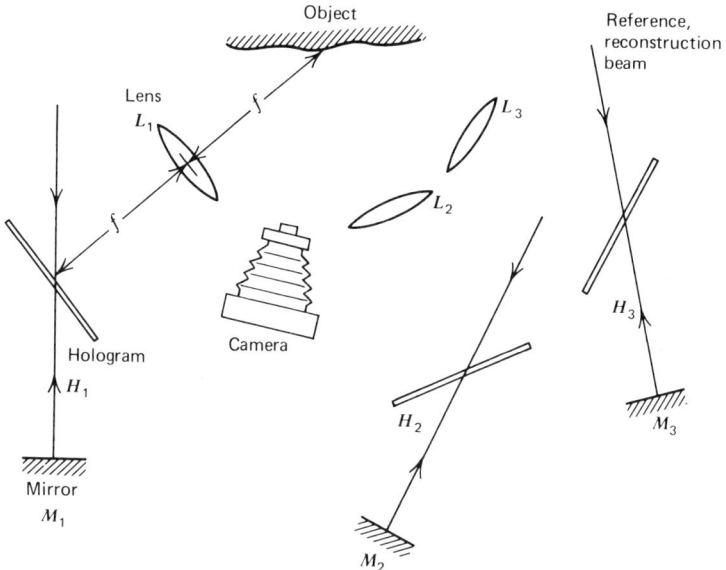

Figure 2.17 System for recording and projecting holographic interferograms onto the object surface (ref. 41).

patterns of Figure 2.18 can be recorded by placing a sheet of film directly on the object surface, if it is of a simple, flat shape.

It should be noted that in this method the fringes are detected at the object surface. In Chapter 3 it will be shown that object motions which have rigid-body translation components generate fringes that localize well off the object surface. Therefore this technique could not be used to study such motions. Nevertheless, it is convenient for the many applications in which object motion is primarily normal deformation and rotation.

A second method based on use of the real image was demonstrated by Bellani and Sona.[30] It is particularly well suited to an automated measurement system based on the fringe shift due to changing viewing angle, as in the single-hologram analysis of Section 2.2. Recall that in this method the observer notes the number of fringes which appear to cross an object point P as the viewing direction is continuously varied. With this method some difficulty may be encountered in maintaining proper registration between the fringe shift and point P. In the method reported in ref. 30, the fringes are observed at a fixed point P in the real image while the effective viewing angle is varied by scanning the hologram with a thin reconstruction beam.

Suppose that a two-exposure hologram is recorded using an off-axis system like the one shown in Figure 2.19a. The wavefront is reconstructed

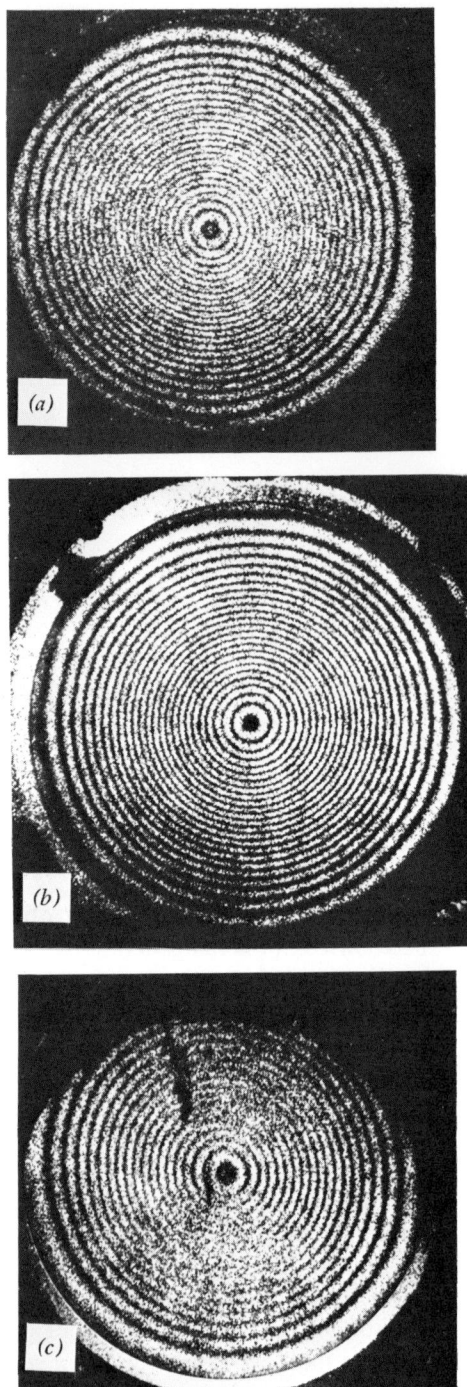

Figure 2.18 Real images of a two-exposure hologram projected onto the object (ref. 41). The hologram is identical to that used to produce the interferograms in Figure 2.16.

2.5 ALTERNATIVE METHODS OF FRINGE READOUT

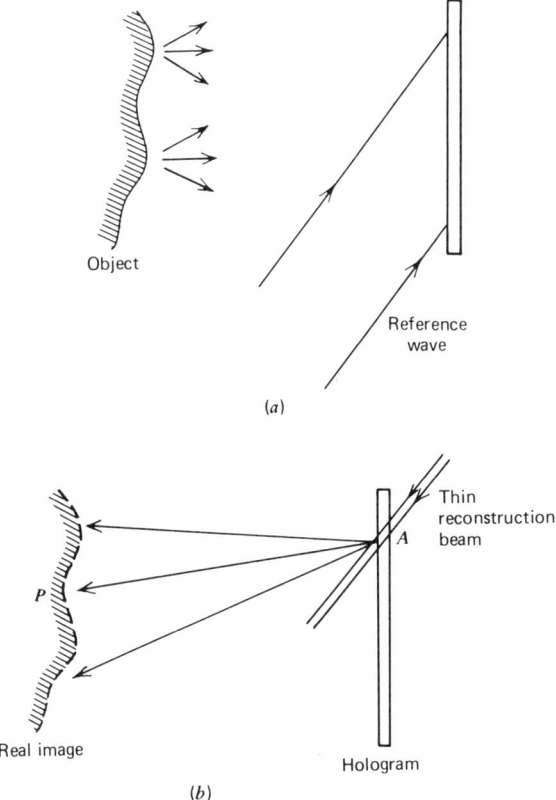

Figure 2.19 Formation of real holographic images. (*a*) Schematic diagram of the recording system. (*b*) Real image being formed with a thin reconstruction beam.

(Figure 2.19*b*) with a thin beam which is conjugate to the original reference wave but illuminates only a small region of the hologram near point *A*. In the object space, a real image is produced which is formed only from light passing through point *A* on the hologram. A fringe detected at point *P* in this real image is equivalent to the fringe one would observe at point *P* while looking at the virtual image through point *A*. The advantage of this reconstruction method is that it is easy to clearly define object point *P* and the equivalent viewing direction along $\overline{PA}$. A system for measuring fringe shifts in this manner[30] is shown schematically in Figure 2.20*a*. A thin beam is focused on a small plane mirror at the location of the point source of the original reference wave. A spherical mirror on the other side of the hologram is used to form the conjugate of the original spherical reference wave. The hologram is scanned by rotating the small mirror about two orthogonal axes. The corresponding fringe shift is detected by a

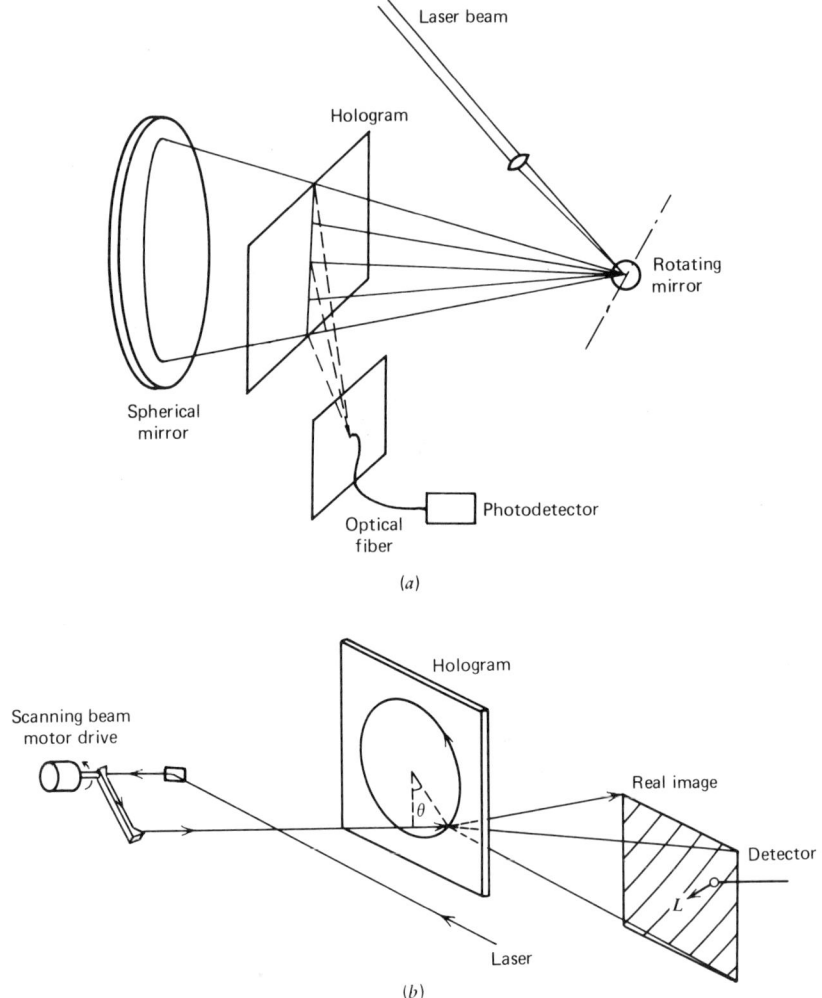

Figure 2.20 A system for scanning a two-exposure hologram with a thin beam. (a) Linear scanning (ref. 30). (b) Circular scanning (ref. 45).

photomultiplier tube which receives light through a fiber optic probe located at the object point P, whose displacement is to be measured. In this manner fringe shift data can be recorded for analysis using equations 2.16.

Because of the small illumination aperture at the hologram this system is less susceptible to problems associated with fringe localization than if the entire hologram were illuminated (unless the fringes localize near the hologram). However, speckle noise is a potential problem if large fringe shifts occur. A more sophisticated detector is required for the measurement of general deformations since the *direction*, as well as the magnitude, of fringe shifts must be detected.

Ek and Biedermann[42] have presented a detailed analysis of displacement measurement by scanning a two-exposure hologram with a thin reconstruction beam. Good accuracy can be attained if fringe orders can be measured with respect to a zero-motion fringe, or if two complete sets of scanning data are recorded from different directions. The upper limit to measurable displacements is set by laser speckle. The effect of speckle is minimized if the apertures of the system are matched to give a cosine fringe modulation of about 0.4. Errors due to optical elements in thin-beam readout systems have been discussed by Ovechkin et al.[43]

Fringe shift data can also be recorded by scanning photographs of the virtual image. Landry and Wise[44] have reported a system in which the virtual image of a two-exposure hologram was photographed at 0.75° intervals as the viewing direction was varied through a range of 29°. The optical density of each photograph was digitized by a commercial image dissector photomultiplier and processor. These data were processed by computer to follow the motion of each fringe across the surface, and ultimately to generate a plot of contours connecting points for which the fringe shift was identical. The displacement of each point was then computed. The system was used to measure the in-plane rigid-body translation of a flat surface. Errors were on the order of 0.3 μm for a 6 μm translation.

To date, automated systems have been demonstrated to measure only simple motions of planar objects, but the concepts form the basis for future development of more sophisticated systems capable of measuring deformations of objects of complicated shape. In fact, at the time of this writing, Ek and Biedermann[45] have reported the construction and operation of a system capable of measuring three-dimensional displacement components with an accuracy of approximately 0.04 μm. In their system a fixed detector in the real image space records irradiance variations as the hologram is scanned by a thin reconstruction beam along a circular path, as shown in Figure 2.20*b*.

In this chapter we have discussed the basic principles and techniques for determining vector displacements from measurements of fringe order. Other aspects of fringe interpretation which require explicit consideration of fringe localization effects are discussed in Chapters 3 and 4. For further discussion of this topic and alternative formulations the reader may wish to refer to the papers by Kohler,[46–48] the semiquantitative methodology described by Liu and Kurtz,[49] the technique of Boone,[50] and the review article by Briers.[51]

REFERENCES

1. R. K. Erf (Ed.), *Holographic Nondestructive Testing*, Academic Press, New York, 1974.
2. J. E. Sollid, Holographic interferometry applied to measurements of small static displacements of diffusely reflecting surfaces, *Appl. Opt.*, **8**, 1587–1595 (1969).
3. E. B. Aleksandrov and A. M. Bonch-Bruevich, Investigation of surface strains by the hologram technique, *Sov. Phys. Tech. Phys.*, **12**, 258–265 (1967).
4. A. E. Ennos, Measurements of in-plane surface strain by hologram interferometry, *J. Sci. Instrum.*, Ser. II, **1**, 731–746 (1968).
5. L. W. Orr, S. W. Tehon, and N. E. Barnett, Isophase surfaces in interference holography, *Appl. Opt.*, **7**, 202–203 (1968).
6. N. Abramson, The holo-diagram. II: A practical device for information retrieval in hologram interferometry, *Appl. Opt.*, **9**, 97–101 (1970).
7. C. A. Sciammarella and J. A. Gilbert, Strain analysis of a disk subjected to diametral compression by means of holographic interferometry, *Appl. Opt.*, **12**, 1951–1956 (1973).
8. U. Köpf, Fringe order determination and zero motion fringe identification in holographic displacement measurements, *Opt. Laser Technol.* **5**, 111–113 (1973).
9. G. Forsythe and C. B. Moler, *Computer Solution of Linear Algebraic Systems*, Prentice-Hall, Englewood Cliffs, NJ, 1967.
10. T. Matsumoto, K. Iwata, and R. Nagata, Measuring accuracy of three-dimensional displacements in holographic interferometry, *Appl. Opt.*, **12**, 961–967 (1973).
11. D. Nobis and C. M. Vest, Statistical analysis of errors in holographic interferometry, *Appl. Opt.*, **17**, 2198-2204 (1978).
12. G. H. Golub and C. Reinsch, Singular value decomposition and least squares solutions, *Numer. Math.*, **14**, 403–420 (1970).
13. S. K. Dhir and J. P. Sikora, An improved method for obtaining the general-displacement field from a holographic interferogram, *Exp. Mech.*, **12**, 323–327 (1972).
14. G. Golub, Numerical methods for solving linear least squares problems, *Numer. Math.*, **7**, 206–216 (1965).
15. J. P. Sikora and F. T. Mendenhall, Jr., Holographic vibration study of a rotating propellor blade, *Exp. Mech.*, **14**, 230–232 (1974).
16. C. A. Sciammarella and T. Y. Chang, Holographic interferometry applied to the solution of a shell problem, *Exp. Mech.*, **14**, 217–224 (1974).
17. P. W. King, Holographic interferometry technique utilizing two plates and relative fringe orders for measuring micro-displacements, *Appl. Opt.*, **13**, 231–233 (1974).
18. Y. Y. Hung, C. P. Hu, D. R. Henley, and C. E. Taylor, Two improved methods of surface-displacement measurements by holographic interferometry, *Opt. Commun.*, **8**, 48–51 (1973).
19. D. Gabor, G. W. Stroke, R. R. Restrick, A. Funkhouser, and D. Brumm, Optical image synthesis (complex addition and subtraction) by holographic fourier transformation, *Phys. Lett.*, **18**, 116–118 (1965).
20. L. F. Collins, Difference holography, *Appl. Opt.*, **7**, 203–204 (1968).
21. A. F. Metherell, S. Spinak, and E. J. Pisa, Subfringe interferometric holography for linearly recording small displacements, *J. Opt. Soc. Am.*, **59**, 1534 (1969).
22. E. E. Bergmann, Simple phase adjustment for "difference" holography, *Rev. Sci. Instrum.*, **44**, 1134 (1973).

REFERENCES

23. R. Dändliker, B. Ineichen, and F. M. Mottier, High resolution hologram interferometry by electronic phase measurement, *Opt. Commun.*, **9**, 412–416 (1973).
24. R. Dändliker, B. Elliasson, B. Ineichen, and F. M. Mottier, Accurate determination of the change of curvature (bending) of a surface through holographic interferometry, *J. Opt. Soc. Am.*, **64**, 1381 (1974).
25. W. H. Stevenson, Optical frequency shifting by means of a rotating diffraction grating, *Appl. Opt.*, **9**, 649–652 (1970).
26. R. Crane, Interference phase measurement, *Appl. Opt.*, **8**, 538–542 (1969).
27. I. P. Kaminov, *An Introduction to Electrooptic Devices*, Academic Press, New York, 1974.
28. K. Shibayama and H. Uchiyama, Measurement of three-dimensional displacements by hologram interferometry, *Appl. Opt.*, **10**, 2150–2154 (1971).
29. T. Matsumoto, K. Iwata, and R. Nagata, Measurement of deformation in a cylindrical shell by holographic interferometry, *Appl. Opt.*, **13**, 1080–1084 (1974).
30. V. F. Bellani and A. Sona, Measurement of three-dimensional displacements by scanning a double-exposure hologram, *Appl. Opt.*, **13**, 1337–1341 (1974).
31. N. Abramson, The holodiagram: A practical device for making and evaluating holograms, *Appl. Opt.*, **8**, 1235–1240 (1969).
32. N. Abramson, The holodiagram. III: A practical device for predicting fringe patterns in hologram interferometry, *Appl. Opt.*, **9**, 2311–2320 (1970).
33. N. Abramson, The holodiagram. IV: A practical device for simulating fringe patterns in hologram interferometry, *Appl. Opt.*, **10**, 2155–2161 (1971).
34. N. Abramson, The holodiagram. V: A device for practical interpreting of hologram interference fringes, *Appl. Opt.*, **11**, 1143–1147 (1972).
35. N. Abramson, The holodiagram. VI: A practical device in coherent optics, *Appl. Opt.*, **11**, 2562–2571 (1972).
36. N. Abramson, Practical interpretation of holographic interferograms, *Optik*, **37**, 337–346 (1973).
37. N. Abramson, Fundamental resolution of optical systems, *Optik*, **39**, 141–149 (1973).
38. G. Oster, *The Science of Moiré Patterns*, Edmund Scientific Co., Chicago, 1969.
39. O. Bryngdahl, Moiré: formation and interpretation, *J. Opt. Soc. Am.*, **64**, 1287–1294 (1974).
40. J. E. Sollid, Translational displacements versus deformation displacements in double-exposure holographic interferometry, *Opt. Commun.*, **2**, 282–288 (1970).
41. T. Matsumoto, K. Iwata, and R. Nagata, Distortionless recording in double-exposure holographic interferograms, *Appl. Opt.*, **12**, 1660–1662 (1973).
42. L. Ek and K. Biedermann, Analysis of a system for hologram interferometry with a continuously scanning reconstruction beam, *Appl. Opt.*, **16**, 2535–2542 (1977).
43. A. P. Ovechkin, S. Ya. Lovkov, and I. I. Dukhopel, Calculation of the instrument error in holographic interferometers having a narrow reference beam, *Sov. J. Opt. Technol.*, **43**, 258–261 (1976).
44. M. J. Landry and C. M. Wise, Automatic data reduction of certain holographic interferograms, *Appl. Opt.*, **12**, 2320–2327 (1973).
45. L. Ek and K. Biedermann, Implementation of hologram interferometry with a continuously scanning reconstruction beam, *Appl. Opt.*, **17**, 1727–1732 (1978).

46. H. Kohler, Interference fringe dynamics in the quantitative evaluation of holographic interferograms, *Optik*, **47**, 9–24 (1977).
47. H. Kohler, The measurement of holographic interferometric deformation fields, *Optik*, **47**, 271–282 (1977).
48. H. Kohler, General formulation of the holographic-interferometric evaluation methods, *Optik*, **47**, 469–475 (1977).
49. H. K. Liu and R. L. Kurtz, A practical method for holographic interference fringe assessment, *Opt. Eng.*, **16**, 176–186 (1977).
50. P. M. Boone, Determination of three orthogonal displacement components from one double exposure hologram, *Opt. Laser Technol.*, **4**, 162–166 (1972).
51. J. D. Briers, The interpretation of holographic interferograms, *Opt. Quant. Electron.*, **8**, 469–501 (1976).

3

Opaque Objects: Formation and Localization of Fringes

3.1 INTRODUCTION

In this chapter we investigate the physical nature of the fringes observed in holographic interferometry. Particular emphasis is placed on the phenomenon of *fringe localization*. Analysis of localization provides deeper understanding of the measurement techniques discussed in Chapter 2, and forms a basis for additional means of interpreting fringes due to displacement and deformation.

The fringes one observes in a two-exposure hologram of a diffuse object appear to be localized in space. They may seem to be on the object surface, or in some region of space behind, or in front of, it. The fringes exhibit a type of parallax as the observer shifts his or her viewing position. This phenomenon is illustrated by Figure 3.1, which shows three photographs of a single holographic interferogram. The object is a flat diffuse surface which was slightly rotated between holographic exposures. The resulting hologram was illuminated by the reference wave in order to reconstruct the optical wavefronts, and the photographs were made with a standard 35 mm camera. The camera lens was set at $f/2.8$ so that it would have a small depth of field. In Figure 3.1a the camera is focused on the object surface, and fringes are barely visible. In Figure 3.1b the camera is focused on a plane about 15 cm behind the object. The fringes are quite distinct in this second photograph, indicating that they were localized in this plane. If this hologram was viewed from a different direction, the observer would note that the fringes appear to localize in a different location. In this chapter we will analyze this phenomenon and develop mathematical relations which predict where fringes appear to localize in terms of the geometry of the holographic system and the vector displacement field of the object surface.

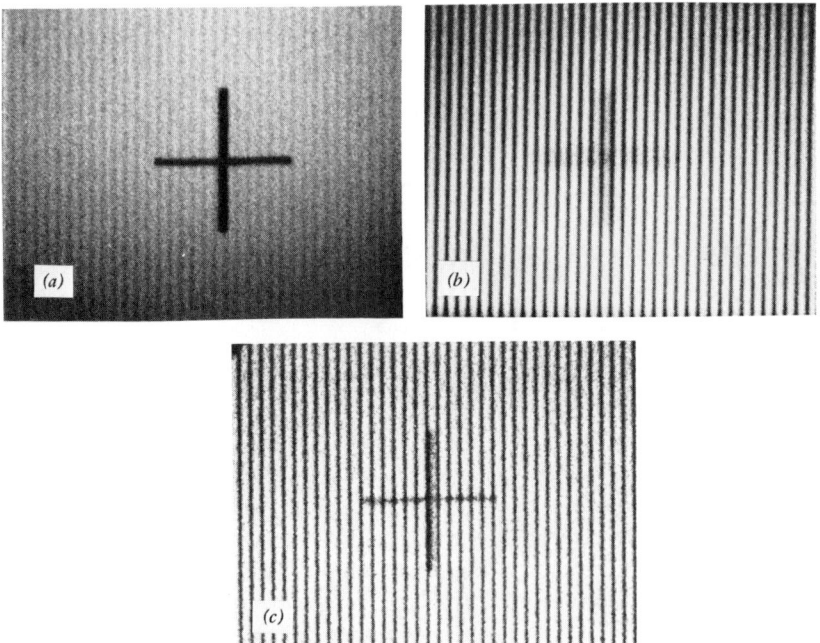

Figure 3.1 Photographs of a two-exposure holographic image showing the effect of fringe localization. (*a*) The object is in focus. (*b*) The fringe localization plane is in focus. (*c*) The object is in focus, but the aperture is small.

3.2 ANALYSIS OF FRINGE LOCALIZATION

In holographic interferometry two optical wavefronts, U_o and U'_o, are reconstructed simultaneously. Portions of these waves are imaged onto a detector surface, such as a sheet of film, a ground glass screen, or the retina of the eye. At this surface an irradiance pattern $I = |U_o + U'_o|^2$ is recorded or observed. It is important to recognize that interference fringes are detected only on this surface. When a person views a two-exposure hologram, the optical interference occurs on his or her retina, not at some other location in space. It follows that the imaging system plays a crucial role in forming the fringes of holographic interferometry. The most important parameter of the imaging system in this context is its *aperture*, or entrance pupil, which determines the angular extent of the ray cone forming the image of each point. The importance of the aperture is illustrated in Figure 3.1. We noted that when the object surface is photographed with a large aperture, as in Figure 3.1*a*, fringes are barely visible. However, if the lens is stopped down to a smaller aperture, as in Figure

3.2 ANALYSIS OF FRINGE LOCALIZATION

3.1c, distinct fringes are seen in the image of the object surface. The role of the aperture of the imaging system is central to the formation and localization of the fringes of holographic interferometry. In fact Stetson[1] has noted that without a finite viewing aperture no fringes of holographic interferometry could be observed.

A simple system for viewing a two-exposure hologram is shown in Figure 3.2. The composite optical field $(U_o + U'_o)$ is imaged by a lens focused on a plane in front of the object surface. The optical field at the detector surface is identical to that which would exist if both the undeformed object and the deformed object were present simultaneously. We can ignore the presence of the hologram and assume that U_o and U'_o propagate from the object surface. The image of the composite field at point Q is formed at Q' on the detector. Some of the rays which form this image are shown in the figure. If the ambient refractive index is homogeneous, the optical pathlengths of all rays connecting Q to Q' are equal, so the relative phase of each ray at Q' is identical to its relative phase at Q. The interference pattern at Q' is therefore identical to that which would be formed at Q by the same ray cone. This is an important conceptual point, because it allows us to analyze fringe formation by considering a ficticious observer inserted into the field $U_o + U'_o$ at point Q. This observer detects the light arriving in a ray cone centered on the axis $\overline{PQ}$ and having an angular extent determined by the aperture of the real viewing system. (If the imaged plane is behind the object, the ray cone emanates from a virtual point Q behind the object surface and the following analysis is still valid.)

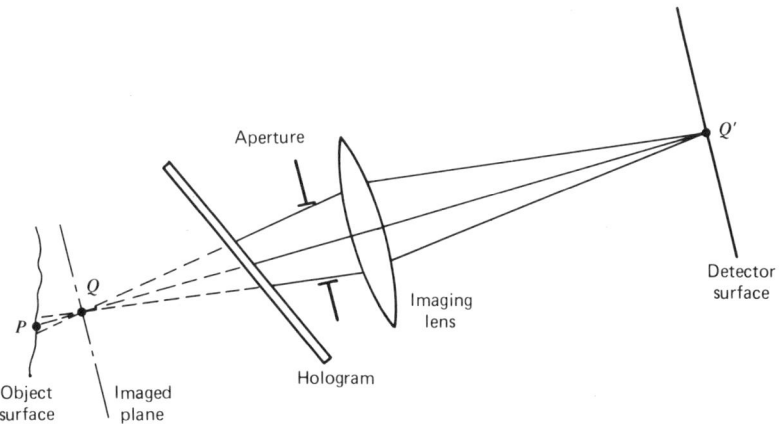

Figure 3.2 Formation of an interference pattern by imaging the composite optical field leaving a two-exposure hologram.

To analyze the interference at Q we apply the methodology introduced in Section 1.6, so that we consider only the interference of light from each surface point with light from the *same* point after it has been displaced by **L**.

A pair of rays travels along each ray path—one emanating from the undeformed object, and one from the deformed object. When a pair of rays arrive at Q, the two rays differ in phase by an amount δ, given by equation 2.8. The irradiance at Q is the sum of the irradiances over the cone of ray pairs, so the interference pattern in the neighborhood of Q is the sum of the interference patterns due to the motion of all surface points subtended by this cone. At an arbitrary location Q the value of δ varies significantly, and randomly, over the ray directions within the cone. No fringes of holographic interferometry can be observed with a large-aperture viewing system near such points. However, there may exist points Q such that the value of δ is nearly constant over the cone. There is a well-defined average irradiance near such a point, and a well-defined fringe pattern in its neighborhood. A real observer will find that the fringes of holographic interferometry appear to lie near a curve, or sometimes surface, in space which is the locus of such points.

Analytically, we determine localization by seeking the points along a given viewing direction for which the variation of δ over a small cone of ray pairs is minimized. We assume the object surface to be locally plane and attach a Cartesian coordinate system to it. As shown in Figure 3.3, the

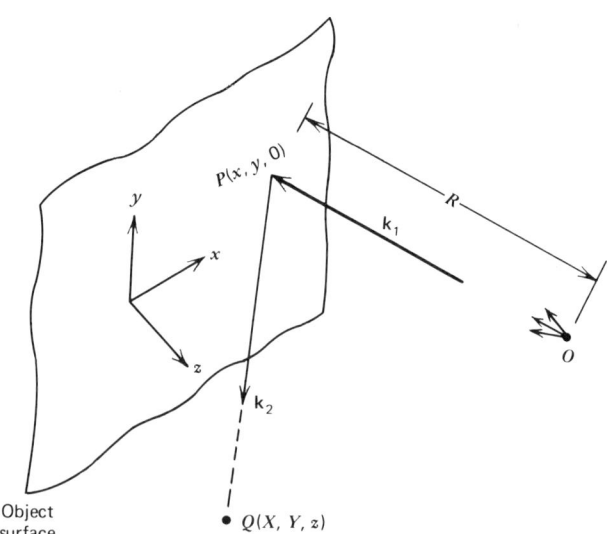

Figure 3.3 Nomenclature for analysis of fringe localization. Points P and Q correspond to points P and Q in Figure 3.2; O is the illumination source.

z direction is normal to the object surface. An arbitrary object point $(x,y,0)$ is illuminated by light with propagation vector $\mathbf{k}_1$, and scatters light with propagation vector $\mathbf{k}_2$ toward the observer at point $Q(X,Y,z)$, which is considered to be at a *fixed location* (X,Y) in a plane located a distance z from the object surface. Then $\mathbf{k}_1$ and $\mathbf{k}_2$ are functions of object point coordinates x,y. The condition for fringe localization is

$$d\delta = \frac{\partial \delta}{\partial x} dx + \frac{\partial \delta}{\partial y} dy = 0, \tag{3.1}$$

where dx and dy are differential changes in the object point which is viewed from Q. The distance z from the surface to Q is a parameter in the expression for δ. The values of z for which equation 3.1 is satisfied define the curve or surface fringe localization.

3.3 FRINGE LOCALIZATION WITH COLLIMATED ILLUMINATION

In most applications of holographic interferometry the curvature of the illumination wavefront at the object surface is quite small. Since this affords considerable analytical simplification, we first discuss the case of collimated illumination in detail. In Section 3.4 the analysis will be extended to the more general case of illumination with a spherical wave.

The basic expression for phase difference δ was derived in Chapter 2:

$$\delta = (\mathbf{k}_2 - \mathbf{k}_1) \cdot \mathbf{L} = \mathbf{K} \cdot \mathbf{L}. \tag{3.2}$$

It is convenient to factor out $2\pi/\lambda$ so that

$$\delta = \frac{2\pi}{\lambda} (\hat{\mathbf{k}}_2 - \hat{\mathbf{k}}_1) \cdot \mathbf{L} = \frac{2\pi}{\lambda} \tilde{\mathbf{K}} \cdot \mathbf{L}, \tag{3.3}$$

where $\hat{\mathbf{k}}_2$ and $\hat{\mathbf{k}}_1$ are unit vectors, and $\tilde{\mathbf{K}} \equiv \hat{\mathbf{k}}_2 - \hat{\mathbf{k}}_1$. The Cartesian components of these vectors are denoted by

$$\hat{\mathbf{k}}_1 = (k_{1x}, k_{1y}, k_{1z}),$$
$$\hat{\mathbf{k}}_2 = (k_{2x}, k_{2y}, k_{2z}),$$
$$\tilde{\mathbf{K}} = (k_x, k_y, k_z),$$
$$\mathbf{L} = (L_x, L_y, L_z).$$

For collimated illumination, $\hat{\mathbf{k}}_1$ is a constant, but $\hat{\mathbf{k}}_2$ and $\mathbf{L}$ in equation 3.3 are functions of surface coordinates x,y and observer position. We denote partial derivatives with respect to surface coordinates by superscripts, for example, $\partial k_{2x}/\partial y = k_{2x}^y$. With this notation, equation 3.1, for collimated illumination, becomes

$$(k_{2x}^x L_x + k_{2y}^x L_y + k_{2z}^x L_z + k_x L_x^x + k_y L_y^x + k_z L_z^x) dx$$
$$+ (k_{2x}^y L_x + k_{2y}^y L_y + k_{2z}^y L_z + k_x L_x^y + k_y L_y^y + k_z L_z^y) dy = 0. \quad (3.4)$$

Explicit expressions for the derivatives of the components of $\hat{\mathbf{k}}_2$ can be obtained by considering the geometry of Figure 3.3. The vector $\hat{\mathbf{k}}_2$ always points from the object point $P(x,y,0)$ to the fixed observation point $Q(X,Y,z)$. The distance $\overline{PQ}$ is denoted by ρ. The components of vector $\hat{\mathbf{k}}_2$ are as follows:

$$k_{2x} = \frac{X-x}{[(X-x)^2+(Y-y)^2+z^2]^{1/2}} = \frac{X-x}{\rho},$$

$$k_{2y} = \frac{Y-y}{[(X-x)^2+(Y-y)^2+z^2]^{1/2}} = \frac{Y-y}{\rho}, \quad (3.5)$$

$$k_{2z} = \frac{z}{[(X-x)^2+(Y-y)^2+z^2]^{1/2}} = \frac{z}{\rho}.$$

The derivative k_{2x}^x is easily evaluated after noting that X, Y, and z are parameters, not functions of x and y:

$$k_{2x}^x = \rho^{-2}[-\rho + \rho^{-1}(X-x)^2] = -\rho^{-1}(1-k_{2x}^2).$$

since $\hat{\mathbf{k}}_2$ is a unit vector,

$$k_{2x}^2 + k_{2y}^2 + k_{2z}^2 = 1. \quad (3.6)$$

Using equation 3.6 and the relation $\rho = z/k_{2z}$, we find that the expression for k_{2x}^x becomes

$$k_{2x}^x = -\frac{k_{2z}}{z}(1-k_{2x}^2) = -\frac{k_{2z}}{z}(k_{2y}^2 + k_{2z}^2).$$

All other derivatives can be evaluated in a similar manner. The results are as follows:

$$k_{2x}^x = -\frac{k_{2z}}{z}(k_{2y}^2 + k_{2z}^2), \quad k_{2x}^y = \frac{k_{2z}}{z} k_{2x} k_{2y},$$

$$k_{2y}^x = \frac{k_{2z}}{z} k_{2x} k_{2y}, \quad k_{2y}^y = -\frac{k_{2z}}{z}(k_{2x}^2 + k_{2z}^2), \quad (3.7)$$

$$k_{2z}^x = \frac{k_{2z}}{z} k_{2x} k_{2z}, \quad k_{2z}^y = \frac{k_{2z}}{z} k_{2y} k_{2z}.$$

3.3 FRINGE LOCALIZATION WITH COLLIMATED ILLUMINATION

Combining equations 3.4 and 3.7 yields the localization condition for collimated illumination:

$$\left\{ \frac{k_{2z}}{z} \left[(k_{2y}^2 + k_{2z}^2) L_x - k_{2x} k_{2y} L_y - k_{2x} k_{2z} L_z \right] \right.$$
$$\left. - (k_x L_x^x + k_y L_y^x + k_z L_z^x) \right\} dx$$
$$+ \left\{ \frac{k_{2z}}{z} \left[-k_{2x} k_{2y} L_x + (k_{2x}^2 + k_{2z}^2) L_y - k_{2y} k_{2z} L_z \right] \right.$$
$$\left. - (k_x L_x^y + k_y L_y^y + k_z L_z^y) \right\} dy = 0. \tag{3.8}$$

If the viewing aperture is of roughly the same dimension in all directions (e.g., a square or circular aperture), dx and dy can be varied independently in equation 3.8, so the coefficients of dx and dy must each be identically zero. This leads to the following two equations:

$$z = \frac{k_{2z} \left[(k_{2y}^2 + k_{2z}^2) L_x - k_{2x} k_{2y} L_y - k_{2x} k_{2z} L_z \right]}{k_x L_x^x + k_y L_y^x + k_z L_z^x}, \tag{3.9}$$

$$z = \frac{k_{2z} \left[-k_{2x} k_{2y} L_x + (k_{2x}^2 + k_{2z}^2) L_y - k_{2y} k_{2z} L_z \right]}{k_x L_x^y + k_y L_y^y + k_z L_z^y}. \tag{3.10}$$

When a particular geometry, $\hat{k}_1$ and $\hat{k}_2$, and a vector displacement field $L(x,y)$ are specified, each of these equations describes a surface $z = z(x,y,z)$ in space. In general, the surface described by equation 3.9 will be different from that described by equation 3.10. Since both equations must be satisfied, the fringes are localized along a *curve* which is the intersection of these two surfaces. In some special cases, equations 3.9 and 3.10 may predict that fringes will localize on a surface. (See Section 3.5.)

We have developed a theory which predicts that the fringes of holographic interferometry generally localize on some curve in space. Experiments indicate the fringes do indeed localize in the vicinity of the curves predicted by equations 3.9 and 3.10, but the *sharpness* of this localization depends on the aperture of the viewing system. This effect was illustrated in Figures 3.1a and 3.1c. It was shown that the region of localization was sharply defined when photographed at $f/2.8$, but could be extended to include the object surface when the aperture was reduced. To understand this phenomenon, consider Figure 3.4. A two-exposure hologram is being viewed through a rectangular aperture. On the viewing screen a fringe pattern is formed which is the image of the optical field in a plane containing point Q. The central ray passing through Q emanates from a

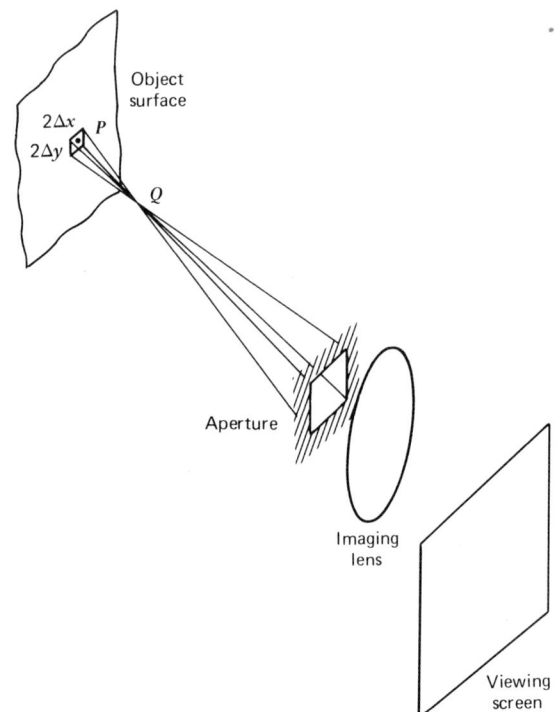

Figure 3.4 Formation of an interference pattern by an imaging system with a rectangular aperture.

point $P(x_o, y_o)$ on the object surface. The irradiance at Q is the integral of the irradiances of all the ray pairs within the ray cone defined by the aperture. This cone subtends a rectangular area of dimension $2\Delta x, 2\Delta y$ on the object surface; hence the irradiance at Q is

$$I(Q) = \int_{x_o - \Delta x}^{x_o + \Delta x} \int_{y_o - \Delta y}^{y_o + \Delta y} \{1 + \cos[\delta(x,y)]\} \, dx \, dy. \qquad (3.11)$$

Since $\Delta x, \Delta y$ are small, $\delta(x,y)$ can be approximated by the initial terms of a Taylor series expansion about x_o, y_o:

$$\delta(x,y) = \delta(x_o, y_o) + \frac{\partial \delta}{\partial x}\bigg|_{x_o, y_o} (x - x_o) + \frac{\partial \delta}{\partial y}\bigg|_{x_o, y_o} (y - y_o)$$

$$= \delta_o + \delta_o^x (x - x_o) + \delta_o^y (y - y_o). \qquad (3.12)$$

3.3 FRINGE LOCALIZATION WITH COLLIMATED ILLUMINATION

Substituting equation 3.12 into equation 3.11 and evaluating the integral yields

$$I(Q) = 4\Delta x \Delta y \left\{ 1 + \frac{\sin(\delta_o^x \delta x)}{\delta_o^x \Delta x} \cdot \frac{\sin(\delta_o^y \Delta y)}{\delta_o^y \Delta y} \cos[\delta(x_o, y_o)] \right\}. \quad (3.13)$$

This represents the fringe pattern, which, as pointed out in Section 1.6, modulates a speckle pattern. The *visibility* of interferometric fringes is defined by

$$v = \frac{I_{max} - I_{min}}{I_{max} + I_{min}}, \quad (3.14)$$

where I_{max} is the maximum irradiance occurring in the bright fringe nearest the point under consideration, and I_{min} is the minimum irradiance occurring in the nearest dark fringe.[2] (As noted in Section 1.6, the irradiance of a fringe must be interpreted as the irradiance averaged along a fringe contour, or as the average irradiance at a point of an ensemble of holographic inteferograms.) Fringes with the maximum possible contrast have $v = 1$. As visibility goes to zero, contrast becomes so low that fringes cannot be observed. The fringes described by equation 3.13 have a visibility given by

$$v = \left| \frac{\sin(\delta_o^x \Delta x)}{\delta_o^x \Delta x} \cdot \frac{\sin(\delta_o^y \Delta y)}{\delta_o^y \Delta y} \right|. \quad (3.15)$$

We can now give a precise definition of fringe localization. *Fringes of holographic interferometry are localized, for a given viewing direction, where their visibility is a maximum.* Maximization of fringe visibility is the same criterion which is applied in classical broad-source interferometry.[3] The function $\sin\xi/\xi$ is plotted in Figure 3.5. The visibility will have its maximum value $v = 1$ at points where $\delta_o^x = 0$ and $\delta_o^y = 0$. This is, of course, precisely the criterion which led to the localization conditions, equations 3.9 and 3.10. This pair of equations determines the location, specified by a value of z, along the viewing direction at which $v = 1$. If the value of $\delta_o^x \Delta x$

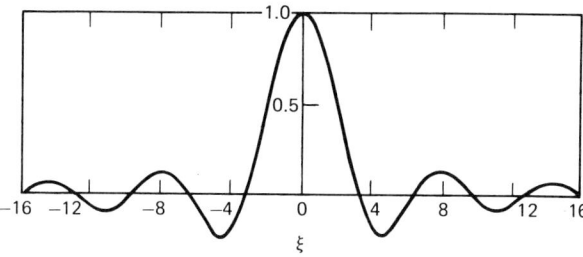

Figure 3.5 Plot of the function $(\sin\xi)/\xi$.

or $\delta_o^y \Delta y$ increases rapidly as z is changed from the localization value, the region of fringe localization is sharply defined. If $\delta_o^x \Delta x$ or $\delta_o^y \Delta y$ increases slowly as z is varied, the region of localization is broad. Since δ_o^x and δ_o^y are determined by the system geometry and object displacement field, the sharpness of localization is controlled by the viewing aperture. If the aperture is large, localization will be sharp, as illustrated in Figures 3.1a and 3.1b. If the aperture is small, relatively distinct fringes can be observed over an extended region; hence in Figure 3.1c fringes can be seen in the plane of the object, even though it is several centimeters from the center of the region of localization.

Assume that $\delta_o^x \approx \delta_o^y = \delta_o^l$ and $\Delta x \approx \Delta y = \Delta l$; then $v \approx |\sin^2(\delta_o^l \Delta l)/(\delta_o^l \Delta l)^2|$. This function is plotted in Figure 3.6. For fixed Δl, visibility decreases to zero with distance away from the localization region defined by equations 3.9 and 3.10, but it exhibits small periodic increases at even larger distances. This is illustrated by the photograph in Figure 3.7. The experimental configuration was such that fringes were localized in a plane which intersected the object surface in a line. The distance from the localization plane to a point on the object surface increases linearly with distance from this line. The photograph was made with a large-aperture lens focused on the object surface. There is a region of very high visibility near one line in the surface, and successive smaller local maxima are clearly visible. Klyberg[4] has carried out detailed experiments verifying this phenomonon.

According to equation 3.15, fringe visibility (and therefore localization) is the product of two terms, one a function of Δx and one a function of Δy. If one of these aperture dimensions, for example, Δy, is very small, then

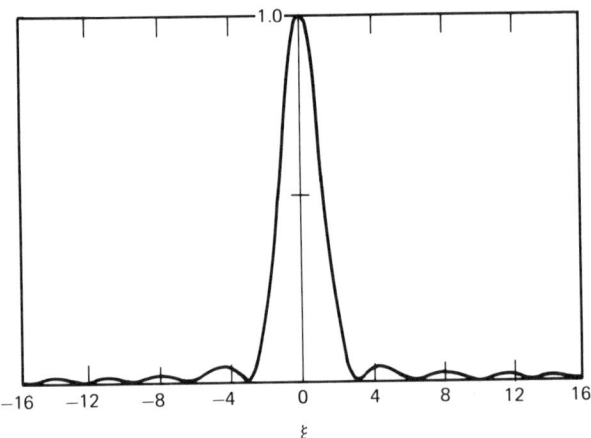

Figure 3.6 Plot of the function $(\sin \xi)^2/\xi^2$.

3.3 FRINGE LOCALIZATION WITH COLLIMATED ILLUMINATION

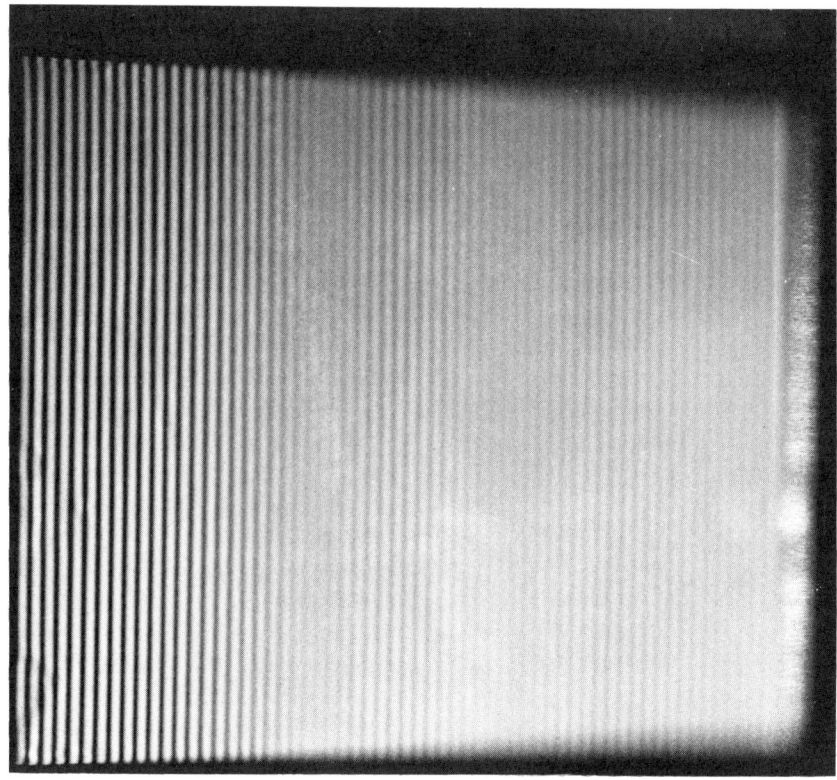

Figure 3.7 Fringe pattern with oscillating fringe visibility.

$\sin(\delta_o^y \Delta y)/(\delta_o^y \Delta y) \approx 1$, and the localization is determined primarily by the value of $\delta_o^x \Delta x$. This suggests the use of *slit apertures* in holographic interferometry.[5,6] If an opaque screen with a long, narrow slit parallel to the x axis is placed between an observer and a two-exposure hologram, fringe localization will be determined by equation 3.9 alone. More generally, a slit aperture may be tilted so that its projection, along the viewing direction, onto the object surface is a line of slope $(dy/dx)_{\text{ap}}$. In this case we must return to the basic localization condition, equation 3.4. Since dy and dx are no longer independent, fringes are localized where the following single relation is satisfied:

$$\left\{ k_{2z}\left[(k_{2y}^2 + k_{2z}^2)L_x - k_{2x}k_{2y}L_y - k_{2x}k_{2z}L_z \right] - z(k_x L_x^x + k_y L_y^x + k_z L_z^x) \right\}$$
$$+ \left(\frac{dy}{dx} \right)_{\text{ap}} \left\{ k_{2z}\left[-k_{2x}k_{2y}L_x + (k_{2x}^2 + k_{2z}^2)L_y - k_{2y}k_{2z}L_z \right] \right.$$
$$\left. - z(k_x L_x^y + k_y L_y^y + k_z L_z^y) \right\} = 0. \quad (3.16)$$

A useful application of equation 3.16 can be deduced by considering the viewing system depicted in Figure 3.8a. The observer looks through a slit aperture and focuses on the fringes in the center of his or her field of view. Fringes now localize in a *surface*. The fringe spacing which is observed can be related to the vector displacement **L** of the object surface by projecting the rate of change of optical phase onto the line imaged by the observer. If we define $\hat{\mathbf{k}}_{ap}$ to be the unit vector parallel to this line, and let x_{ap} be the coordinate along it, then

$$\frac{d\delta}{dx_{ap}} \cdot \left\{ \frac{\hat{\mathbf{k}}_{ap} \cdot [\hat{\mathbf{i}} + (dy/dx)_{ap}\hat{\mathbf{j}}]}{|\hat{\mathbf{i}} + (dy/dx)_{ap}\hat{\mathbf{j}}|} \right\} = \frac{d\delta}{dx_s}, \quad (3.17)$$

where $d\delta/dx_s$ is the rate of change of δ along the shadow of the slit on the surface,

$$\frac{d\delta}{dx_s} = \frac{\delta^x + (dy/dx)_{ap}\delta^y}{|\hat{\mathbf{i}} + (dy/dx)_{ap}\hat{\mathbf{j}}|}, \quad (3.18)$$

and

$$\delta^x = \frac{2\pi}{\lambda}(k_x L_x^x + k_y L_y^x + k_z L_z^x),$$

$$\delta^y = \frac{2\pi}{\lambda}(k_x L_x^y + k_y L_y^y + k_z L_z^y).$$

Combining equations 3.17 and 3.18 yields

$$\frac{d\delta}{dx_{ap}} = \frac{\delta^x + (dy/dx)_{ap}\delta^y}{\hat{\mathbf{k}}_{ap} \cdot [\hat{\mathbf{i}} + (dy/dx)_{ap}\hat{\mathbf{j}}]}. \quad (3.19)$$

Since $\hat{\mathbf{k}}_{ap}$ is a unit vector which is normal to $\hat{\mathbf{k}}_2$ and lies in the plane containing $\hat{\mathbf{k}}_2$ and the slit, it is defined by three conditions:

$$|\hat{\mathbf{k}}_{ap}| = 1, \quad (3.20)$$

$$\hat{\mathbf{k}}_{ap} \cdot \hat{\mathbf{k}}_2 = 0, \quad (3.21)$$

$$\hat{\mathbf{k}}_{ap} \times \hat{\mathbf{k}}_2 \cdot \left[\hat{\mathbf{i}} + \left(\frac{dy}{dx}\right)_{ap}\hat{\mathbf{j}}\right] = 0. \quad (3.22)$$

The solution of equations 3.20 to 3.22 is

$$\hat{\mathbf{k}}_{ap} = \frac{\mathbf{k}_a}{\left\{[\hat{\mathbf{i}} + (dy/dx)_{ap}\hat{\mathbf{j}}] \cdot \mathbf{k}_a\right\}^{1/2}}, \quad (3.23)$$

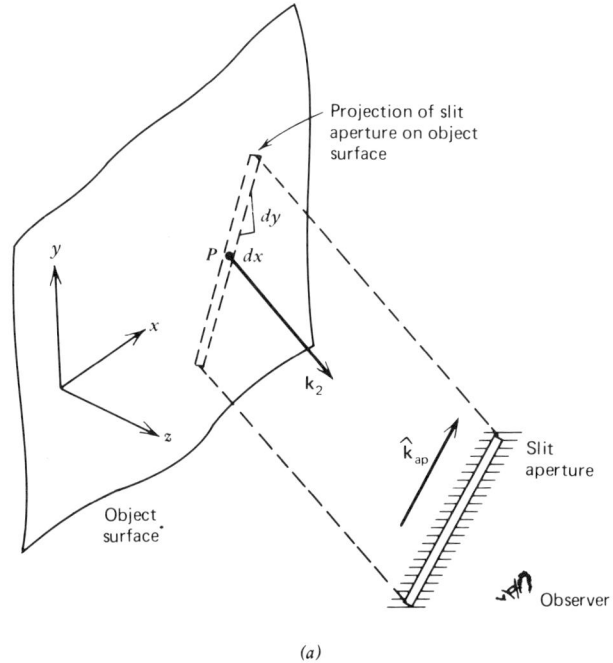

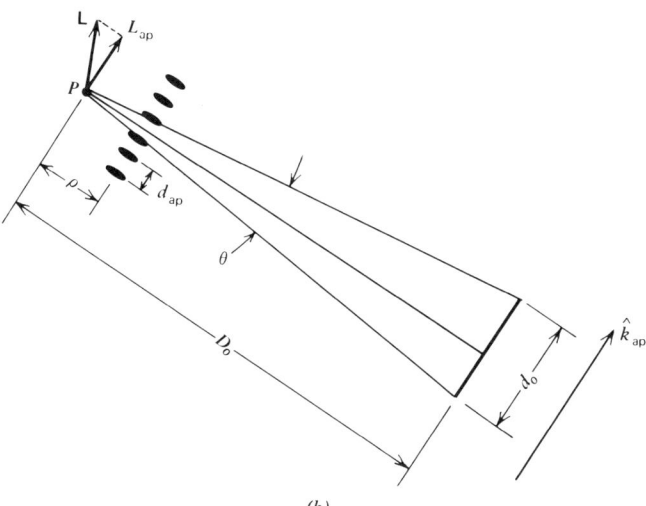

Figure 3.8 Observation of fringes through a slit aperture. (a) Observation geometry. (b) Localized fringes viewed through slit aperture.

where

$$k_a = \left\{ \hat{i}\left[(k_{2y}^2 + k_{2z}^2) - \left(\frac{dy}{dx}\right)_{ap} k_{2x}k_{2y}\right] \right.$$
$$+ \hat{j}\left[-k_{2x}k_{2y} + \left(\frac{dy}{dx}\right)_{ap}(k_{2x}^2 + k_{2z}^2)\right]$$
$$\left. + \hat{k}\left[-k_{2x}k_{2z} - \left(\frac{dy}{dx}\right)_{ap} k_{2y}k_{2z}\right] \right\}. \quad (3.24)$$

Using equations 3.23 and 3.24, we find that the localization condition for a slit aperture, equation 3.16, becomes

$$\frac{2\pi}{\lambda}\left\{\left[\hat{i} + \left(\frac{dy}{dx}\right)_{ap}\hat{j}\right]\cdot k_a\right\}^{1/2}(\hat{k}_{ap}\cdot L) = \frac{z}{k_{2z}}\left[\delta^x + \left(\frac{dy}{dx}\right)_{ap}\delta^y\right]; \quad (3.25)$$

however, equations 3.19 and 3.23 imply that

$$\delta^x + \left(\frac{dy}{dx}\right)_{ap}\delta^y = \left\{\left[\hat{i} + \left(\frac{dy}{dx}\right)_{ap}\hat{j}\right]\cdot k_a\right\}^{1/2}\frac{d\delta}{dx_{ap}},$$

so the localization equation for slit apertures becomes

$$\frac{2\pi}{\lambda}\hat{k}_{ap}\cdot L = \rho\frac{d\delta}{dx_{ap}}. \quad (3.26)$$

This relation indicates that the fringe spacing viewed through a slit aperture is a direct measure of L_{ap}, the component of displacement normal to the observation direction and parallel to the slit. Since a change of one fringe corresponds to a phase shift of magnitude 2π, the spacing between fringes d_{ap} is

$$\left(\frac{d\delta}{dx_{ap}}\right)d_{ap} = 2\pi,$$

and equation 3.26 leads to

$$L_{ap} = \lambda\frac{\rho}{d_{ap}}. \quad (3.27)$$

This interesting relation was derived by Stetson[5] for slit apertures, and is equivalent to a relation derived earlier by Haines and Hildebrand[7,8] in an analysis which did not explicitly involve a description of the aperture. For a given orientation of the slit, a measurement of the fringe spacing plus a measurement of the localization distance ρ determines one vector compo-

3.3 FRINGE LOCALIZATION WITH COLLIMATED ILLUMINATION

nent of **L**. If the procedure is repeated for three different slit orientations, **L** can be determined. This method has the disadvantage that in many types of motion ρ is very small and is difficult to measure with precision.

Equation 3.27 is closely related to *fringe parallax*, which is the basis of the single-hologram, or hologram scanning, methods discussed in Chapter 2. In these techniques the observer views an object point P through a telescope, and counts the fringes which pass his or her field of view as the viewing position moves from a point Q_1 to a point Q_2. By using equation 2.15, the number of fringes N which pass can be related to the displacement **L** by

$$(\mathbf{k}_3 - \mathbf{k}_2) \cdot \mathbf{L} = N(2\pi). \tag{3.28}$$

If the observer changes the viewing direction by an angle θ, as indicated in Figure 3.8b, so that one fringe crosses the object point, then

$$|\mathbf{k}_3 - \mathbf{k}_2| = \frac{2\pi}{\lambda} |\hat{\mathbf{k}}_3 - \hat{\mathbf{k}}_2| = \frac{2\pi}{\lambda} 2\sin\left(\frac{\theta}{2}\right). \tag{3.29}$$

The component L_p in the plane of this observer's motion and normal to the bisector of the angle between $\mathbf{k}_2$ and $\mathbf{k}_3$ is

$$L_p = \frac{(\mathbf{k}_3 - \mathbf{k}_2) \cdot \mathbf{L}}{|\mathbf{k}_3 - \mathbf{k}_2|} = \frac{N\lambda}{2\sin(\theta/2)}. \tag{3.30}$$

Stetson has pointed out the close relation between equations 3.27 and 3.30. For small scanning angle the geometry of Figure 3.8 can be used to relate these two equations as

$$\frac{L_{\mathrm{ap}}}{\lambda} = \frac{\rho}{d_{\mathrm{ap}}} \cong \frac{D_o}{d_o} = \frac{L_{\mathrm{p}}}{\lambda}. \tag{3.31}$$

This is an approximate equality because L_p and L_{ap} are generally in different directions.

Equation 3.31 provides the basis for the sign convention used when applying the single-hologram method. In Section 2.2 it was suggested that N be assigned a positive value if the fringes appear to move to the left of the observer's field of view as he or she scans to the right from Q_1 to Q_2. This occurs if the fringes are localized in front of the object surface, as can be deduced from the parallax effects depicted in Figure 3.8. Fringe motion in the direction opposite to observer motion corresponds to a positive distance ρ and, according to equation 3.31, a positive value of L_{ap}. The opposite sign is used when the fringes appear to move toward the right, that is, in the same direction as the observer's motion. This occurs when the fringes localize behind the object surface, in which case L_{ap} should be negative according to equations 3.31. Following this sign convention will yield a consistent set of vector components, from which the magnitude and direction of **L** can be determined. Its sense will not be known, but if it can be determined a priori at any point on the object, all ambiguity can be removed.

3.4 FRINGE LOCALIZATION WITH SPHERICAL WAVE ILLUMINATION

When holographic interferometry is used to study the motion of a large object, it is often necessary to use spherical, rather than plane, waves to illuminate the object. This is so because large-diameter collimating lenses or mirrors are generally expensive and increase the size of holographic systems. For most measurements of deformation the analysis of Section 3.3 is sufficient since the curvature of spherical illumination can usually be neglected over small regions of the object surface. However, to form a more general understanding of fringe systems, especially when rigid-body motions are present, an extension of the analysis of fringe localization to include wavefront curvature is desirable.

As before, the condition for fringe localization is derived by combining equations 3.1 and 3.2. Since the restriction that $\hat{\mathbf{k}}_1$ is constant does not apply, the result is

$$(k_x^x L_x + k_y^x L_y + k_z^x L_z + k_x L_x^x + k_y L_y^x + k_z L_z^x) dx$$
$$+ (k_x^y L_x + k_y^y L_y + k_z^y L_z + k_x L_x^y + k_y L_y^y + k_z L_z^y) dy = 0. \quad (3.32)$$

This differs from equation 3.4 in that all terms involve the vector $\tilde{\mathbf{K}}$ parallel to the sensitivity vector, rather than $\hat{\mathbf{k}}_2$. To evaluate the derivatives of $\tilde{\mathbf{K}}$, consider the configuration depicted in Figure 3.3. The object point $P(x,y,0)$ is illuminated by light propagating in the direction $\hat{\mathbf{k}}_1$ from a fixed point source at O, and scatters light in direction $\hat{\mathbf{k}}_2$ toward point Q; R is the distance from the point source to the object point, that is, R is the radius of the curvature of the illuminating wavefront. The various derivatives of $\hat{\mathbf{k}}_1$ can be evaluated by the same method used to derive equation 3.7. The result is as follows:

$$k_{1x}^x = -\frac{1}{R}(k_{1y}^2 + k_{1z}^2), \quad k_{1x}^y = \frac{1}{R}k_{1x}k_{1y},$$
$$k_{1y}^x = \frac{1}{R}k_{1x}k_{1y}, \quad k_{1y}^y = -\frac{1}{R}(k_{1x}^2 + k_{1z}^2), \quad (3.33)$$
$$k_{1z}^x = \frac{1}{R}k_{1x}k_{1z}, \quad k_{1z}^y = \frac{1}{R}k_{1y}k_{1z}.$$

3.5 EXAMPLES OF FRINGE FORMATION AND LOCALIZATION

Recalling that $\tilde{\mathbf{K}} \equiv \hat{\mathbf{k}}_2 - \hat{\mathbf{k}}_1$, we can substitute equations 3.7 and 3.33 into equation 3.32 to obtain

$$\left\{\left[\frac{k_{2z}}{z}\left(k_{2y}^2+k_{2z}^2\right)-\frac{1}{R}\left(k_{1y}^2+k_{1z}^2\right)\right]L_x - \left(\frac{k_{2z}}{z}k_{2x}k_{2y}-\frac{1}{R}k_{1x}k_{1y}\right)L_y\right.$$

$$\left.-\left(\frac{k_{2z}}{z}k_{2x}k_{2z}-\frac{1}{R}k_{1x}k_{1z}\right)L_z - \left(k_x L_x^x + k_y L_y^x + k_z L_z^x\right)\right\}dx$$

$$+\left\{-\left(\frac{k_{2z}}{z}k_{2x}k_{2y}-\frac{1}{R}k_{1x}k_{1y}\right)L_x + \left[\frac{k_{2z}}{z}\left(k_{2x}^2+k_{2z}^2\right)-\frac{1}{R}\left(k_{1x}^2+k_{1z}^2\right)\right]L_y\right.$$

$$\left.-\left(\frac{k_{2z}}{z}k_{2y}k_{2z}-\frac{1}{R}k_{1y}k_{1z}\right)L_z - \left(k_x L_x^y + k_y L_y^y + k_z L_z^y\right)\right\}dy = 0. \quad (3.34)$$

Equation 3.34 is the most general condition for fringe localization and can be applied in the same manner as equation 3.8, the localization condition for collimated illumination. When a square or circular aperture is used, the two terms which are coefficients of dx and dy, respectively, must vanish independently. This defines a curve of localization. When a slit aperture is used, these two terms are not independent, and equation 3.34 defines a surface of localization for each slit orientation.

3.5 EXAMPLES OF FRINGE FORMATION AND LOCALIZATION

In Chapter 2 we discussed in detail methods for making point-by-point measurements of the vector displacement field of a surface. This is the approach usually followed for measuring surface deformation. In this section we consider the significance of the fringe system as a whole, rather than on a point-by-point basis. This approach is useful for qualitative analysis of holographic interferograms. It also leads to quantitative results for certain rigid-body motions and special cases such as isotropic strain.[9] We must solve, therefore, the inverse problem of holographic interferometry—given the system geometry and object motion, characterize the fringe system. Some general conclusions regarding the form and localization of fringes will be drawn, and examples of specific motions will be considered in detail.

To interpret the results of these examples it must be noted that x,y are coordinates of an observed object point, and z is the distance from the object surface to the point of localization along the line between the observer and the object point $(x,y,0)$, that is, $z = \rho k_{2z}$. *If one wishes to write the equation of a surface or curve of localization in the (X,Y,z) coordinate* *system of Figure 3.3, the following transformation must be made:*

$$X = x - \left(\frac{k_{2x}}{k_{2z}}\right)z,$$

$$Y = y - \left(\frac{k_{2y}}{k_{2z}}\right)z. \tag{3.35}$$

In the initial discussions in this section, it is assumed that $\hat{\mathbf{k}}_1$ is constant (collimated illumination) and that the viewing direction $\hat{\mathbf{k}}_2$ is the same for all points. The latter condition can be realized if the object is observed with a telecentric viewing system.[10] For visual observation of holograms, this assumption is nearly satisfied if the object subtends a relatively small solid angle at the viewing position.

Our objective is to describe the geometric form of fringe patterns and to predict where fringes will localize. When the viewing aperture is made sufficiently small, fringes can be observed in the image of the object surface, and can be described as curves in an x,y coordinate system lying in the object surface. These curves are loci of points for which the phase difference δ, given by equation 2.8, is an even-integer multiple of π (a bright fringe), or an odd-integer multiple of π (a dark fringe). For constant $\hat{\mathbf{k}}_1$ and $\hat{\mathbf{k}}_2$, the explicit expression for δ is

$$\delta(x,y) = \frac{2\pi}{\lambda}\left[k_x L_x(x,y) + k_y L_y(x,y) + k_z L_z(x,y)\right], \tag{3.36}$$

so the equation for bright fringes is

$$k_x L_x + k_y L_y + k_z L_z = N\lambda, \qquad N = 0, 1, 2, \cdots. \tag{3.37}$$

As the aperture of the viewing system is increased, depth of focus decreases and the localization of fringes becomes apparent. The curve or surface in which fringes localize can be predicted by equations 3.9 and 3.10. We assume the object surface to be planar.

If an object undergoes *rigid-body translation*, L_x, L_y, and L_z are identical at all points on its surface, so their derivatives are identically zero, and equations 3.9 and 3.10 indicate that the fringes localize at $z = \infty$. They indeed can be observed with good visibility only at infinity, that is, in the back focal plane of a lens. Figure 3.9 is a photograph, taken through a two-exposure hologram, of an object surface which was translated parallel to itself between exposures. No fringes are visible. Figure 3.10a shows the fringe pattern, from the same hologram, in the back focal plane of a thin spherical lens. These fringes are perpendicular to the surface displacement

3.5 EXAMPLES OF FRINGE FORMATION AND LOCALIZATION

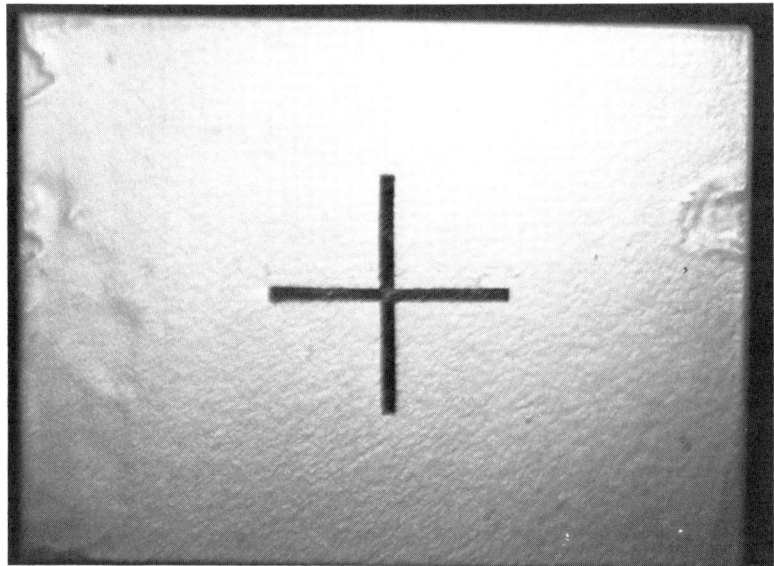

Figure 3.9 Photograph of a two-exposure hologram of an object which was translated in-plane. No fringes are visible at the object surface.

vector **L**. If the lens is placed at a distance of one focal length, f, from the object surface, equation 1.46 can be used to relate the fringe spacing to displacement. Phase difference $\delta = (2\pi/\lambda)(k_{2x} - k_{1x})L_x$ is constant for all ray pairs leaving the surface with a given value of k_{2x}, which is equal to λf_x. Using equation 1.46, we find that all such ray pairs impinge on a line $x_f = k_{2x}f$ in the back focal plane. A shift of one fringe corresponds to a change of k_{2x} by an amount λ/L_x, so the fringe spacing near the center of the back focal plane is

$$d_{\text{fp}} = \left(\frac{\lambda}{L_x}\right) f. \tag{3.38}$$

Figure 3.10b shows fringes in the back focal plane when the object is translated normal to its surface by an amount L_z. These circular fringes are fringes of equal inclination, known as *Haidinger fringes* in classical interferometry. An argument similar to that above can be used to determine the radial spacing of the fringes. The phase difference $\delta = (2\pi/\lambda)(k_{2z} - k_{1z})L_z$ is constant for all ray pairs leaving the surface in a given direction. For simplicity, we consider all ray pairs for which $k_{2y} = 0$. For a phase shift of

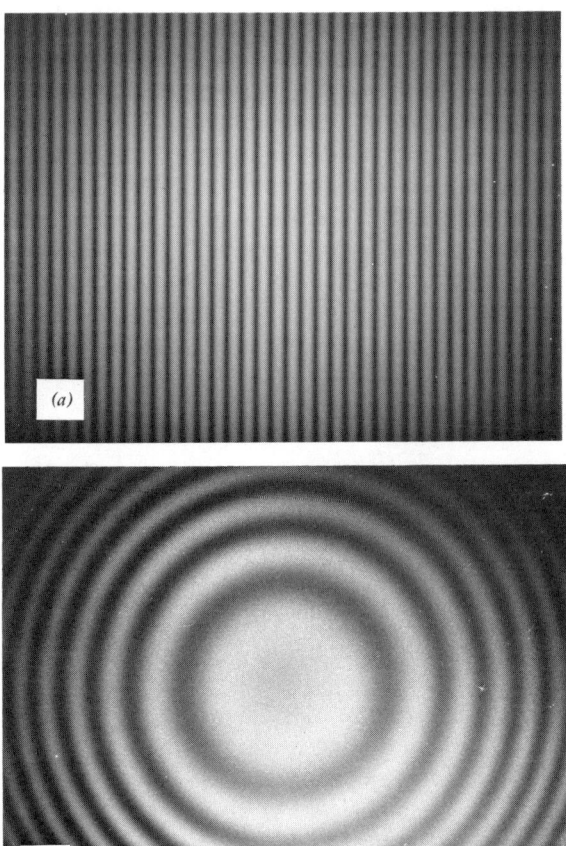

Figure 3.10 Fringes due to rigid-body translation observed in the back focal place of a Fourier-transforming lens. (*a*) In-plane translation. (*b*) Normal translation.

$m(2\pi)$,
$$k_{2z} = k_{1z} + m\left(\frac{\lambda}{L_z}\right),$$
but since $\hat{\mathbf{k}}_2$ is a unit vector,
$$k_{2x}^2 + k_{2z}^2 = 1,$$
so
$$k_{2x} = \left\{1 - \left[k_{1z} + m\left(\frac{\lambda}{L_z}\right)\right]^2\right\}^{1/2}.$$

3.5 EXAMPLES OF FRINGE FORMATION AND LOCALIZATION

The quantity $m(\lambda/L_z)$ will generally be small compared to k_{1z}, so

$$k_{2x} \cong \left[1 - k_{1z}^2 - 2mk_{1z}\left(\frac{\lambda}{L_z}\right)\right]^{1/2}.$$

The center fringe, for which $k_{2x} = 0$, therefore corresponds to

$$m_0 = \frac{1}{2k_{1z}}\left(\frac{L_z}{\lambda}\right)(1 - k_{1z}^2).$$

The Nth fringe from the center corresponds to

$$m = \frac{1}{2k_{1z}}\left(\frac{L_z}{\lambda}\right)(1 - k_{1z}^2) - N.$$

If this value of m is substituted into the expression for k_{2x}, we have

$$k_{2x} = \left[2Nk_{1z}\left(\frac{\lambda}{L_z}\right)\right]^{1/2}.$$

Hence the Nth fringe from the center is located at radius

$$r_{\text{fp}} = fk_{2x} \cong f\left[2Nk_{1z}\left(\frac{\lambda}{L_z}\right)\right]^{1/2}. \tag{3.39}$$

Equation 3.39 can be used to determine a normal displacement L_z.

When an object is rotated or deformed, the denominators of equations 3.9 and 3.10 usually do not vanish, so fringes localize at some finite distance from the surface. As an example, consider rotation about an axis lying in the surface of the object. If the object rotates by a small angle θ about the y axis,

$$L_x \cong 0, \qquad L_y \cong 0, \qquad L_z \cong \theta x,$$

and the localization conditions, equations 3.9 and 3.10, yield

$$z = -\frac{k_{2x}k_{2z}^2}{k_z}x \tag{3.40}$$

and

$$k_{2z}^2 k_{2y}^2 \theta x = 0. \tag{3.41}$$

The equations of the bright fringes are

$$k_z \theta x = N\lambda, \tag{3.42}$$

where $N = 0, 1, 2, \ldots$. If $k_{2y} = 0$, equation 3.41 is satisfied, so the fringes will appear to be localized in a plane defined by equation 3.40. Note that if the object is viewed in the normal direction $k_{2x} = 0$, so the fringes will localize on the object surface. In general, when the object rotates about an axis lying on the object surface, an observer will see straight fringes which are parallel to the rotation axis. They will appear to be on the object surface and will exhibit little parallax as the viewing position is varied slightly. If one looks at the object from above or below, so that $k_{2y} \neq 0$, equation 3.41 requires that $x = 0$. Equation 3.40 must also be satisfied, so the fringes will appear to localize along the y axis.

If the object is rotated by an angle θ about an axis parallel to the surface, but not lying in it, the fringes localize off the surface. For example, let the rotation axis be parallel to the y axis, but lie a distance r behind the surface in the x-z plane; then

$$L_x = r\theta, \qquad L_y = 0, \qquad L_z = -\theta x.$$

If $k_{2y} = 0$, equation 3.10 is satisfied and localization is determined by equation 3.9:

$$z = \frac{k_{2z}\left[(k_{2y}^2 + k_{2z}^2)r + k_{2x}k_{2z}x\right]}{-k_z}, \tag{3.43}$$

which is the equation of a plane. If viewed normally, the fringes are in a plane parallel to the object surface but behind it. The fringes are straight lines parallel to the rotation axis and are defined by

$$k_x r\theta - k_z \theta x = N\lambda. \tag{3.44}$$

Fringes of this type were depicted in Figure 3.1. The object was rotated slightly about an axis 22 cm behind the object. The photograph in Figure 3.1b was made with the camera focused on a plane 15 cm behind the object, which is the localization plane predicted by equation 3.43 for normal viewing.

Next we consider rotation by an angle θ about the z axis, which is normal to the object surface. Here

$$L_x = -\theta y, \qquad L_y = \theta x, \qquad L_z = 0.$$

3.5 EXAMPLES OF FRINGE FORMATION AND LOCALIZATION

The localization conditions are

$$z = \frac{k_{2z}\left[-\left(k_{2y}^2 + k_{2z}^2\right)y - k_{2x}k_{2y}x\right]}{k_y}, \quad (3.45)$$

$$z = \frac{k_{2z}\left[k_{2x}k_{2y}y + \left(k_{2x}^2 + k_{2z}^2\right)x\right]}{-k_x}, \quad (3.46)$$

and the bright fringes are described by

$$k_y \theta x - k_x \theta y = N\lambda. \quad (3.47)$$

The nature of this fringe system is highly dependent on the direction of illumination and viewing. If the object is illuminated and viewed in the normal direction, equation 3.2 indicates that no fringes are formed, since the displacement vector is normal to the sensitivity vector. The situation is quite different if the object is illuminated at an angle of 45° (see Figure 3.11a), while viewed normally. For this case,

$$k_{1x} = \frac{1}{\sqrt{2}}, \quad k_{1y} = 0, \quad k_{1z} = -\frac{1}{\sqrt{2}},$$
$$k_{2x} = 0, \quad k_{2y} = 0, \quad k_{2z} = 1.$$

This results in a system of horizontal fringes,

$$\frac{1}{\sqrt{2}} \theta y = N\lambda, \quad (3.48)$$

localized in the line

$$z = \sqrt{2}\, x, \quad y = 0. \quad (3.49)$$

The condition $y=0$ is required by equation 3.45 if fringes are to be observed at a finite distance z as k_y tends to zero in the denominator. Figure 3.11 illustrates this localization. It includes two photographs (Figures 3.11b and 3.11c) of the fringe system taken with a low f number in the two planes indicated in Figure 3.11a. If the virtual image of the object is viewed from above at an angle of 45°, the fringes are lines oriented at 45° to the x axis:

$$\frac{1}{\sqrt{2}} \theta(x+y) = N\lambda. \quad (3.50)$$

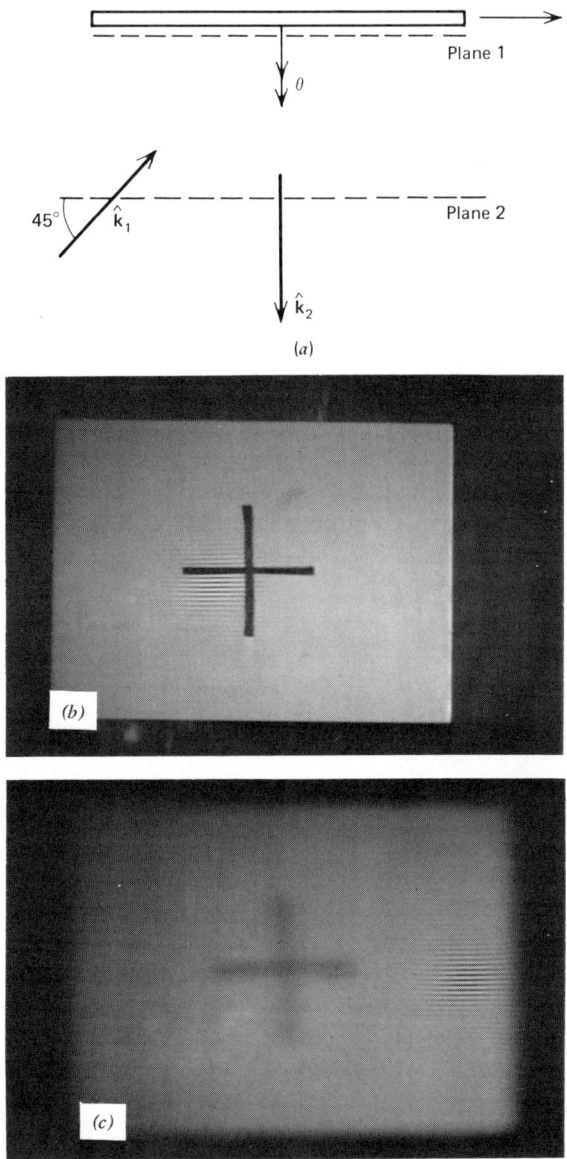

Figure 3.11 Fringes due to rotation about an axis normal to the object surface. (*a*) The geometrical configuration. (*b*) The fringes in plane 1. (*c*) The fringes in plane 2.

3.5 EXAMPLES OF FRINGE FORMATION AND LOCALIZATION

This fringe system is displayed in Figure 3.12. This interferogram was recorded with a high-f-number camera and confirms that, as k_y is varied, the fringes rotate about the z axis. This characteristic is useful for recognizing rotation about axes normal to the surface. The fringes in this system are localized in a line in space, which intersects the object at the origin. The line is the intersection of two planes:

$$z = -y, \quad z = \tfrac{1}{2}(x+y). \tag{3.51}$$

The foregoing examples indicate that a given rotational motion gives rise to a system of fringes whose orientation and localization depend strongly on the directions of illumination and viewing. A single photograph of a fringe pattern, or casual observation of a two-exposure hologram, is not sufficient to discern the nature of the motion which occurred. It is possible, however, to draw some useful conclusions regarding fringes due to rigid-body motions. The most general rigid-body motion consists of a translation $\mathbf{L}_t$ and a rotation $\mathbf{\Theta}$. The corresponding surface displacement is

$$\mathbf{L} = \mathbf{L}_t - \mathbf{r} \times \mathbf{\Theta}, \tag{3.52}$$

where $\mathbf{r} = \hat{\mathbf{i}}x + \hat{\mathbf{j}}y$ is the position vector of a point on the surface. The

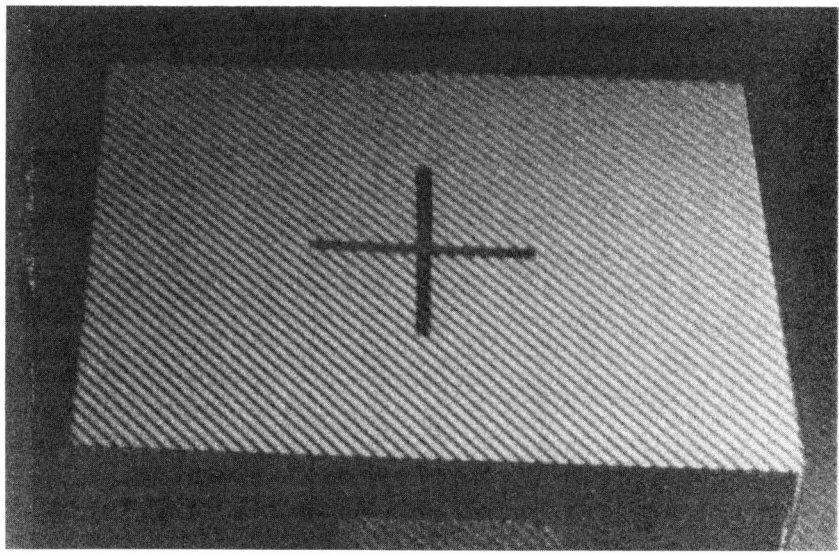

Figure 3.12 The fringes due to rotation about an axis normal to the object surface. The object is viewed from above at an angle of 45° to the horizontal with the object in focus.

equation of the resulting fringes is

$$-(k_x\Theta_z - k_z\Theta_x)y + (k_y\Theta_z - k_z\Theta_y)x + \tilde{\mathbf{K}}\cdot\mathbf{L}_t = N\lambda. \quad (3.53)$$

These fringes are equally spaced straight lines whose slope is

$$\left(\frac{dy}{dx}\right)_{\text{fringes}} = -\frac{(\tilde{\mathbf{K}}\times\boldsymbol{\Theta})_x}{(\mathbf{K}\times\boldsymbol{\Theta})_y}. \quad (3.54)$$

The spacing and orientation of these fringes are independent of the translation $\mathbf{L}_t$. Substitution of equation 3.52 into equations 3.9 and 3.10 indicates that fringes always localize in a straight line or in a plane. The orientation of the line of localization depends only on the rotation $\boldsymbol{\Theta}$, but its distance from the object surface is affected by $\mathbf{L}_t$. It can be shown in addition that for pure rotation the rotation axis, the line of localization, and the line of sight are all coplanar if the object is viewed *retroreflectively* ($\mathbf{k}_2 = -\mathbf{k}_1$). In this case the line of localization is midway between the object surface and the rotation axis.[9] Stetson[9,11] derived the interesting result that for rigid-body motions the fringes appear to the observer as the intersection of the object surface with equally spaced parallel planes in space. This result is independent of the shape of the object surface.

When deformations are involved, the displacement vector $\mathbf{L}$ may be a nonlinear function of coordinates on the object surface. In this case fringes will appear to localize along a curve in space, or in a curved surface. The formation of fringes due to surface deformation can be analyzed using equation 3.2, 3.9, and 3.10, just as for rigid-body motions. Table 3.1 is a catalog of fringe equations and localization for a few common system geometries and displacements or deformations. For simplicity this catalog has been constructed for the case in which $\mathbf{L}$, $\hat{\mathbf{k}}_1$, and $\hat{\mathbf{k}}_2$ are coplanar, which is nearly true in many applications. These results would also apply to motion with similar components in the plane of a slit aperture.

In the foregoing examples of fringe formation and localization it was assumed that $\hat{\mathbf{k}}_1$ and $\hat{\mathbf{k}}_2$ are constant. If spherical wave illumination is used, $\hat{\mathbf{k}}_1$ is not constant and curvature effects will not be negligible if the point source is sufficiently close to the object. These effects are most obvious for the case of rigid-body translations. Such motions lead to fringes at infinity when the illumination is collimated. If the illumination is a spherical wave, the fringes will be curved and may localize at a finite distance from the object. The form of fringes is easily deduced by writing the explicit expressions for the components of $\hat{\mathbf{k}}_1$. For a point source located at (x_o, y_o, z_o),

$$k_{1x} = \frac{x - x_o}{R}, \quad k_{1y} = \frac{y - y_o}{R}, \quad k_{1z} = -\frac{z_o}{R}, \quad (3.55)$$

3.5 EXAMPLES OF FRINGE FORMATION AND LOCALIZATION

where $R = [(x-x_o)^2 + (y-y_o)^2 + z_o^2]^{1/2}$. If the viewing direction $\hat{\mathbf{k}}_2$ is nearly constant over the object surface, the equation of bright fringes is

$$\left(k_{2x} - \left(\frac{x-x_o}{R}\right)\right)L_x + \left(k_{2y} - \frac{y-y_o}{R}\right)L_y$$
$$+ \left(k_{2z} + \frac{z_o}{R}\right)L_z = N\lambda. \tag{3.56}$$

The fringes are second-order curves in the object coordinates. An example is shown in Figure 3.13. The point source was located 25 cm in front of the object surface and 15 cm to the left of the center of the object. The viewing direction was normal to the object surface. Without knowledge that there is significant curvature of the illuminating wavefront, one would assume that this pattern represents a complex deformation of the object surface. In fact, the motion was a simple in-plane translation L_x. Fringe localization is also affected by the curvature of the illuminating wavefront. In this example, if we evaluate $\hat{\mathbf{k}}_1$ and $\hat{\mathbf{k}}_2$ at the center of the object, then

$$k_{1x} = \frac{1}{\sqrt{2}}, \quad k_{1y} = 0, \quad k_{1z} = -\frac{1}{\sqrt{2}},$$
$$k_{2x} = 0, \quad k_{2y} = 0, \quad k_{2z} = 1.$$

Substitution of these values into the general localization condition, equation 3.35, indicates that the fringes localize in a curved surface near $z = 2R$. This example is indicative of the primary effects of point source illumination. It introduces curvature into the fringes and into the surface or curve of localization, and can lead to localization at finite distances even if the object motion is a rigid-body translation.

If the illuminated object surface subtends a large solid angle at the aperture of the viewing system, as in Figure 3.14, $\hat{\mathbf{k}}_2$ cannot be considered constant over the surface. An important concept for considering the effect of variations of $\hat{\mathbf{k}}_2$ is summarized in what may be called the *observer-projection theorem*.[5,9] The theorem is implicit in the geometrical optics approach to fringe formation which is followed in this chapter, but can also be derived by wavefront analysis.[5] This theorem states that, if fringes of holographic interferometry are localized off the object surface, they can be projected onto the object surface radially from the center of the aperture of the viewing system. The theorem is easily visualized if we think of the loci of constant phase difference δ in the space depicted in Figure 3.14. These loci are fringe laminae or surfaces, extending radially outward from the viewing aperture. The fringes seen by an observer are the intersection of these surfaces with the region of localization. The intersection of these surfaces with the object surface are the fringe curves given by equation 3.37 or 3.56. The theorem is useful when fringes to be photographed are

TABLE 3.1 A CATALOG of FRINGE EQUATIONS AND LOCALIZATION

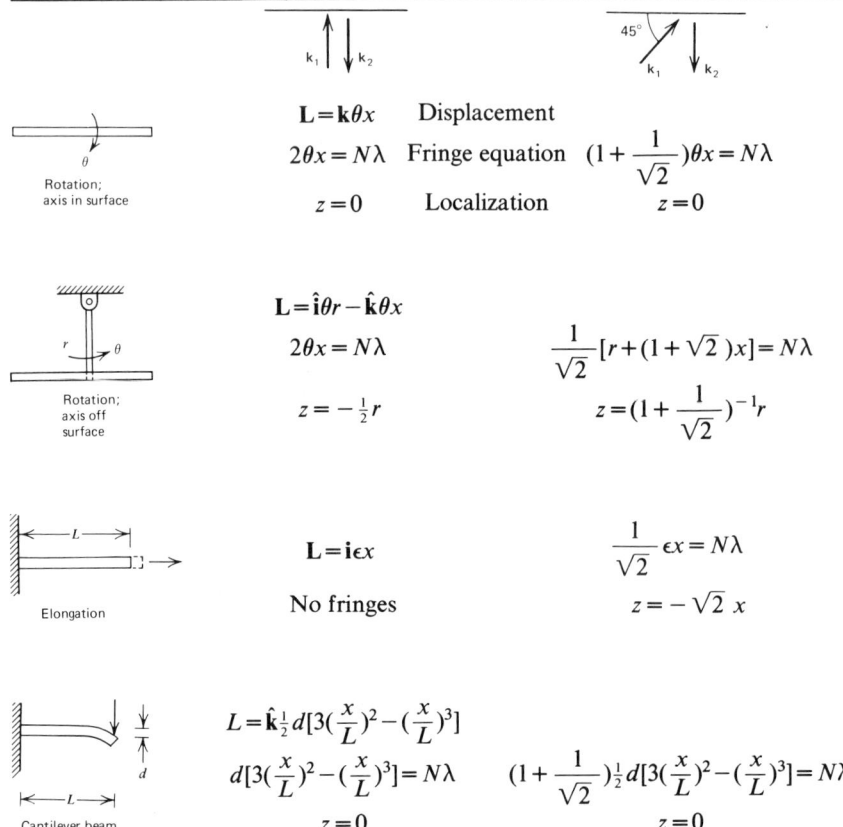

localized well off the object surface, especially if their spatial frequency is high. In order to photograph the fringes and object surface simultaneously, the aperture of the camera must be small. This small aperture may cause unacceptable noise, in the form of laser speckle, in the photograph of the fringes. Following the observer-projection theorem, however, one may photograph the fringes with a large aperture in their localization surface. A separate photograph of the object surface can be made from the same camera position. A conventional enlarger can then be used to superimpose these two photographs with appropriate relative magnification to yield the correct fringe curves on the object surface.

3.5 EXAMPLES OF FRINGE FORMATION AND LOCALIZATION

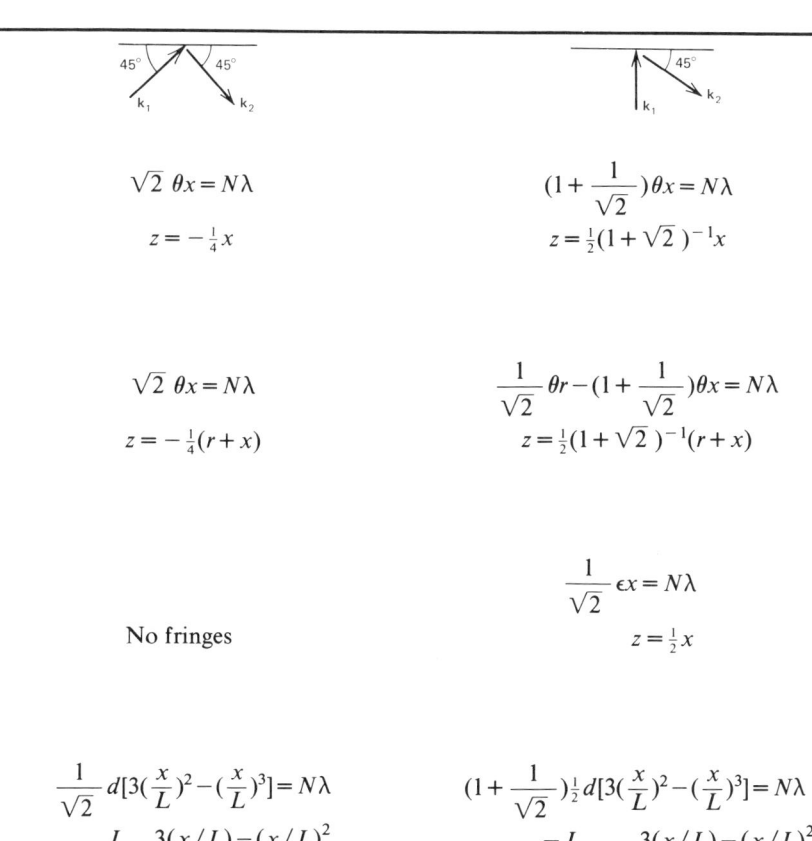

$\sqrt{2}\,\theta x = N\lambda$

$z = -\tfrac{1}{4}x$

$(1 + \dfrac{1}{\sqrt{2}})\theta x = N\lambda$

$z = \tfrac{1}{2}(1 + \sqrt{2})^{-1}x$

$\sqrt{2}\,\theta x = N\lambda$

$z = -\tfrac{1}{4}(r+x)$

$\dfrac{1}{\sqrt{2}}\theta r - (1 + \dfrac{1}{\sqrt{2}})\theta x = N\lambda$

$z = \tfrac{1}{2}(1 + \sqrt{2})^{-1}(r+x)$

$\dfrac{1}{\sqrt{2}}\epsilon x = N\lambda$

No fringes

$z = \tfrac{1}{2}x$

$\dfrac{1}{\sqrt{2}}d[3(\dfrac{x}{L})^2 - (\dfrac{x}{L})^3] = N\lambda$

$z = \dfrac{L}{6\sqrt{2}}\dfrac{3(x/L)-(x/L)^2}{2-(x/L)}$

$(1+\dfrac{1}{\sqrt{2}})\tfrac{1}{2}d[3(\dfrac{x}{L})^2 - (\dfrac{x}{L})^3] = N\lambda$

$z = \dfrac{-L}{6(1+1/\sqrt{2})}\dfrac{3(x/L)-(x/L)^2}{2-(x/L)}$

The concept of fringe surfaces, or laminae, extending outward from the object toward the observer is very useful for visualizing various aspects of fringe formation and localization, particularly those associated with perspective. For example, an observer viewing an object through a large-aperture imaging system will see fringes only in the neighborhood of the intersection of the localization curve with the image plane, which is normal to $\hat{k}_2$. The observed fringes will be the intersection of the fringe laminae with this imaged plane. We can now demonstrate that *in a region of localization the observed fringes are normal to the apparent displacement.* The apparent displacement, L_{ob}, is the component of displacement normal

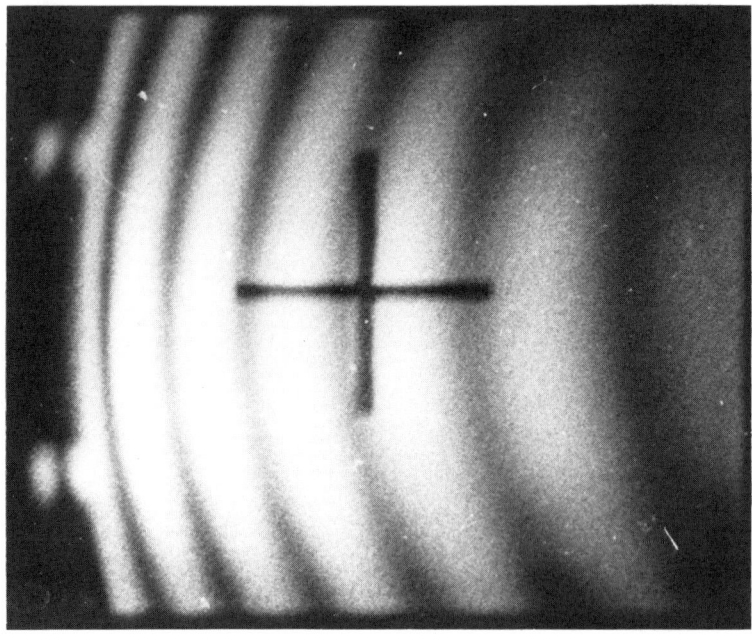

Figure 3.13 Fringes formed with spherical wave illumination. The object underwent an in-plane translation.

to the observation direction $\hat{\mathbf{k}}_2$:

$$L_{\text{ob}} = (\hat{\mathbf{k}}_2 \times L) \times \hat{\mathbf{k}}_2$$

$$= \hat{\mathbf{i}}\left[(k_{2y}^2 + k_{2z}^2)L_x - k_{2x}k_{2y}L_y - k_{2x}k_{2z}L_z\right]$$

$$+ \hat{\mathbf{j}}\left[-k_{2x}k_{2y}L_x + (k_{2x}^2 + k_{2z}^2)L_y - k_{2y}k_{2z}L_z\right]$$

$$+ \hat{\mathbf{k}}\left[-k_{2x}k_{2z}L_x - k_{2y}k_{2z}L_y + (k_{2x}^2 + k_{2z}^2)L_z\right]$$

$$= \hat{\mathbf{i}}L_{\text{ob}_x} + \hat{\mathbf{j}}L_{\text{ob}_y} + \hat{\mathbf{k}}L_{\text{ob}_z}. \tag{3.57}$$

Let us construct a vector Δ which lies in the observation plane and is normal to the observed fringes. To do this, we note that the expression

3.5 EXAMPLES OF FRINGE FORMATION AND LOCALIZATION

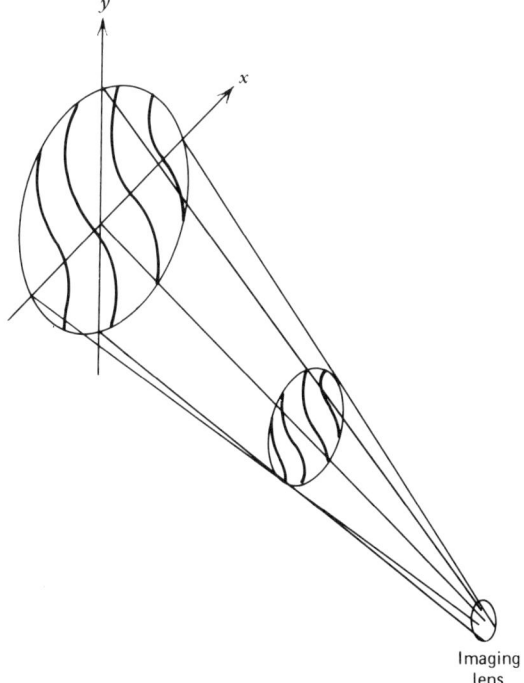

Figure 3.14 The observer-projection theorem.

$\delta = (2\pi/\lambda)\tilde{\mathbf{K}} \cdot \mathbf{L}$ associates a phase difference δ with each point on the object surface. The intersections of the fringe laminae with the object surface are contours of constant δ. These are the fringes observed when the object is imaged with a small aperture. Let $\nabla_n \delta$ be the two-dimensional gradient of δ in the object surface:

$$\nabla_n \delta = \hat{i}\left(\frac{\partial \delta}{\partial x}\right) + \hat{j}\left(\frac{\partial \delta}{\partial y}\right)$$

$$= \hat{i}\left(\frac{2\pi}{\lambda}\right)(k_x L_x^x + k_y L_y^x + k_z L_z^x)$$

$$+ \hat{j}\left(\frac{2\pi}{\lambda}\right)(k_x L_x^y + k_y L_y^y + k_z L_z^y). \tag{3.58}$$

Here $\nabla_n \delta$ is a vector which is normal to the fringes in the object surface, so the vector $\hat{\mathbf{k}} \times \nabla_n \delta$ lies along the fringes in the object surface. The vector $\mathbf{\Delta}$

that we wish to construct must be normal both to $\hat{k} \times \nabla_n \delta$ and to $\hat{k}_2$, that is,

$$\begin{aligned}\Delta &\equiv (\hat{k} \times \nabla_n \delta) \times \hat{k}_2 \\ &= \hat{i}\left[k_{2z}(k_x L_x^x + k_y L_y^x + k_z L_z^x)\right] \\ &\quad + \hat{j}\left[k_{2z}(k_x L_x^y + k_y L_y^y + k_z L_z^y)\right] \\ &\quad + \hat{k}\left[k_{2x}(k_x L_x^x + k_y L_y^x + k_z L_z^x)\right. \\ &\quad \left. - k_{2y}(k_x L_x^y + k_y L_y^y + k_z L_z^y)\right] \\ &= \hat{i}\Delta_x + \hat{j}\Delta_y + \hat{k}\Delta_z. \end{aligned} \qquad (3.59)$$

The observed fringes are normal to the apparent displacement if $\mathbf{L}_{ob}$ and Δ are parallel, that is, if

$$\frac{L_{ob_x}}{\Delta_x} = \frac{L_{ob_y}}{\Delta_y} = \frac{L_{ob_z}}{\Delta_z}. \qquad (3.60)$$

When the components of $\mathbf{L}_{ob}$ and Δ from equations 3.57 and 3.59 are substituted into equation 3.60, we find that the condition $L_{ob_x}/\Delta_x = L_{ob_y}/\Delta_y$ is identical to requiring that equations 3.9 and 3.10 be satisfied simultaneously, that is, that the fringes be localized. That L_{ob_z}/Δ_z is also equal to the ratios of the x and y components follows after some algebraic rearrangement. In summary, at the region of fringe localization

$$\frac{L_{ob_x}}{\Delta_x} = \frac{L_{ob_y}}{\Delta_y} = \frac{L_{ob_z}}{\Delta_z} = \frac{z}{k_{2z}^2} \qquad (3.61)$$

and the italicized statement on p. 135 is proved.

The last equality in equation 3.61 can be written explicitly as

$$\frac{z}{k_{2z}^2} = \frac{-k_{2x}k_{2z}L_x - k_{2y}k_{2z}L_y + (k_{2y}^2 + k_{2z}^2)L_z}{k_{2x}(k_x L_x^x + k_y L_y^x + k_z L_z^x) - k_{2y}(k_x L_x^y + k_y L_y^y + k_z L_z^y)}. \qquad (3.62)$$

Since the distance from the object surface to the localization region along the observation direction is $\rho = z/k_{2z}$, this equation can also be written as

$$\frac{\rho}{k_{2z}} = \frac{L_z - k_{2z}\hat{k}_2 \cdot \mathbf{L}}{(\lambda/2\pi)\hat{k}_2 \cdot \nabla_n \delta}. \qquad (3.63)$$

Similarly, the first two equalities, that is, the localization conditions 3.9 and

3.5 EXAMPLES OF FRINGE FORMATION AND LOCALIZATION

3.10, can be expressed in the following form

$$\rho = \frac{L_x - k_{2x}\hat{k}_2 \cdot L}{(\lambda/2\pi)\hat{i} \cdot \nabla_n \delta}, \tag{3.64}$$

$$\rho = \frac{L_y - k_{2y}\hat{k}_2 \cdot L}{(\lambda/2\pi)\hat{j} \cdot \nabla_n \delta}. \tag{3.65}$$

Equations 3.63 to 3.65 explicitly relate the localization distance to $\nabla_n \delta$, which can be expressed in terms of the surface strain and rotation. The use of these equations for the measurement of strain will be discussed in Section 4.2.

Viewing a holographic interferogram from a point close to the object surface has an effect on fringe curvature and localization similar to that due to curvature of the illuminating wavefront. Motions which cause systems of straight fringes when viewed with a telecentric system, or from a large distance, generate fringes which appear curved to an observer close to the surface. The analysis of fringe formation developed above is applicable to this case if the appropriate expressions for $\hat{k}_2$ are used. A simple example of such analysis can be developed for the case of rigid-body translation. In Section 2.4 we discussed an example in which an object was illuminated with light nearly parallel to its surface and was viewed in a direction normal to the surface. Let the observer be located a distance z from the center of the object; then

$$k_{1x} \cong 1, \quad k_{1y} = 0, \quad k_{1z} \cong 0,$$
$$k_{2x} = \frac{x}{\rho}, \quad k_{2y} = \frac{y}{\rho}, \quad k_{2z} = \frac{z}{\rho},$$

where $\rho = (x^2 + y^2 + z^2)^{1/2}$, and the coordinate origin is located at the center of the object surface. If the object is translated a distance L_x parallel to its surface, the equation of bright fringes is

$$\left[\left(1 + \frac{N\lambda}{L_x}\right)^{-2} - 1\right] x^2 - y^2 = z^2. \tag{3.66}$$

Near the center of the object the fringe spacing in the x direction is

$$d_{f_x} \cong \frac{\lambda z}{L_x}. \tag{3.67}$$

If the object is translated normal to its surface a distance L_z, the equation of bright fringes is

$$x^2 + y^2 = \left[\left(\frac{L_z}{N\lambda}\right)^2 - 1\right]z^2. \tag{3.68}$$

Near the center of the object the fringe spacing in the x direction is

$$d_{f_x} \cong \frac{\lambda z^2}{L_z x}. \tag{3.69}$$

Comparison of equation 3.67 with equation 3.69 confirms the earlier conclusion that this configuration will be much more sensitive to in-plane displacement than to normal displacement. A more general and important conclusion from this example is that rigid-body translations cause observable fringes in the virtual image if the observation point is sufficiently close to the object surface.

For purposes of analysis we have assumed so far that the object surface under consideration is flat. This is not a limitation on the fundamental equations 3.1 and 3.2 if they are expressed in an appropriate coordinate system; however, algebraic complexity is increased considerably if the object surface is not planar. Prikryl[12] has carried out such an analysis and developed the fringe localization equations for a surface whose shape can be expressed in analytical form. Here we consider the important example of a spherical object, for which the analysis is not difficult.

As a final example, suppose that a spherical object of radius r undergoes a small uniform expansion L_r between exposures of a two-exposure hologram. At all surface points the displacement vector points radially outward. If, as in Figure 3.15, the object is illuminated with collimated light and viewed along the same direction, the resulting phase shift is

$$\delta = 2L_r \cos\theta.$$

Since $\cos\theta = z/r$, the equation of bright fringes on the surface is

$$z = N\lambda \frac{r}{2L_r}. \tag{3.70}$$

Thus the fringes are circles which are the intersections of the object surface with a set of parallel planes normal to the direction of illumination and spaced according to equation 3.70. This example illustrates a result due to Stetson,[9,11] who showed that the fringes will appear to be the intersection of the object surface with a set of equally spaced parallel planes regardless of the shape of the object if it undergoes a homogeneous deformation. An

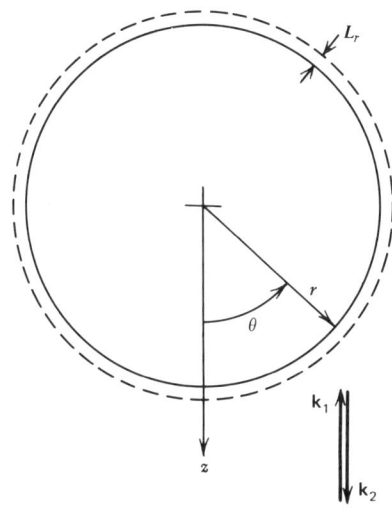

Figure 3.15 A spherical object which undergoes a uniform expansion. The sphere is illuminated with collimated light and viewed along the same direction.

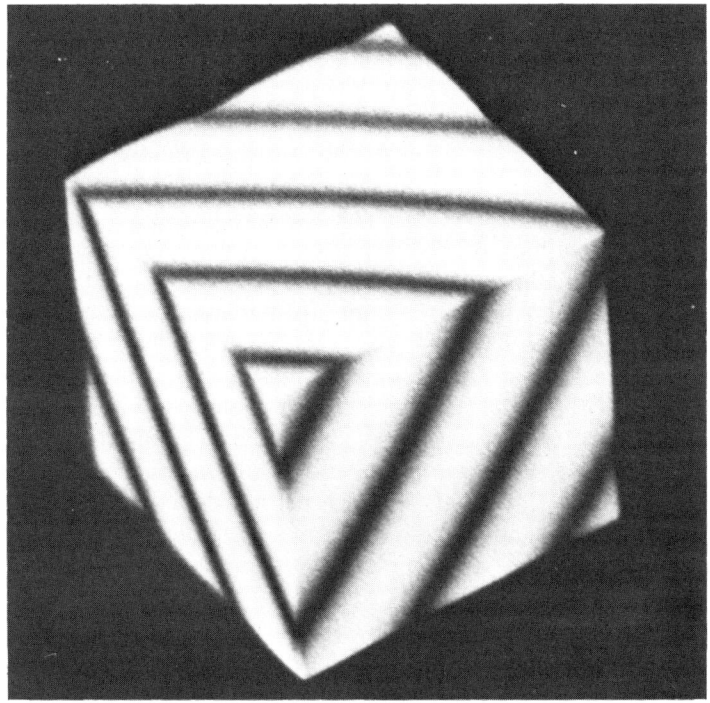

Figure 3.16 Fringes on an aluminum cube due to a uniform, thermally induced expansion.

example of such a fringe system is shown in Figure 3.16. These fringes were caused by a small, thermally induced expansion of an aluminum cube. Since the direction of the deformation vector is known a priori, the fringe system shown in Figure 3.16 can be evaluated from a single view. A simple method for measuring the thermal expansion coefficient of materials by evaluating fringes like these has been developed.[13] This concept has also been applied to the measurement of strain of a spherical object.[14,15]

3.6 OTHER APPROACHES TO FRINGE FORMATION AND LOCALIZATION

In this chapter we have presented a self-contained analysis of fringe localization based on geometric optics. Several alternative approaches to the problem of localization appear in the literature. The phenomonon of fringe localization was recognized in the earliest papers on holographic interferometry and was analyzed in the pioneering work of Haines and Hildebrand.[7,8] They developed a fringe analysis technique in which vector displacement was determined by measurements of fringe spacing and position of localization.

A long and thorough study of the general problem of fringe formation and localization has been reported in a series of papers by Stetson and his co-workers.[1,5,9,11,16–18] His theory is based on wavefront analysis and explicit consideration of the effect of the viewing aperture. The object surface is considered to be analyzed into small, perfectly blazed gratings which diffract incident light into the various components of the spatial spectrum of the object wavefront. The effect of translation, rotation, and strain of these gratings on the wavefront leaving the object surface is determined. The methods of Fourier optics are then applied to study the effect of aperturing and propagation of the interfering optical fields. Localization is defined to refer to the points in space where fringe visibility is maximized because of the vanishing of noise terms in the expression for the irradiance of the imaged, interfering wavefronts. The condition for fringe localization is the vanishing of the gradient in reciprocal space of a fringe function related to, but not identical with, the phase difference δ in (equations 29a and 29b of ref. 5) which defines the curve of localization. For collimated illumination these equations can be manipulated algebraically to show that they are identical to equations 3.9 and 3.10 in this chapter.

The theory was extended by Molin and Stetson[16] to apply to the case of spherical wave illumination. Several detailed experiments[16–18] were conducted by Molin and Stetson to verify the theory of fringe formation and localization. These experiments include study of the effects of curved object surfaces, and noncollimated illumination. Part of this work deals with the concept and application of fringe vectors, which are particularly useful for measurement of strain and rotation. Fringe vectors will be discussed in Section 4.2. Other Molin–Stetson experiments refer specifically to vibrating objects and will be discussed in Section 4.3.

Tsuruta et al.[19] also considered the effect of analyzing the object surface into gratings. This analysis provides a basis for a ray-tracing method to deduce fringe localization for certain rigid-body motions. Boone and Verbiest[20] and Froehly et al.[21] also presented theories of localization for specific cases of rigid-body motion.

Steel[22] discussed the relation of holographic interferometry to classical interferometry with broad sources. This analogy is quite complete if one considers a situation in which the illumination source is rigidly attached to the object in a holographic interferometer. Steel presented a graphical construction for determining fringe localization due to translation and rotation. It is simple and useful. Two rays are drawn, one each from identical points on the initial and displaced surfaces. The rays have identical inclinations to the surface. They will intersect at the point of localization for the viewing direction in question. High visibility requires equal distances of propagation along these rays. An approach which is conceptually similar, but more detailed and analytical, was taken by Walles.[23] He used the concept of homologous rays, and emphasized the close relation between localization theory and the theory of coherence, as did Klyberg.[4] References 4 and 6 both describe in detail the broadening of the region of localization by varying the viewing aperture. In the approach of Walles, localization occurs at the point of closest approach of homologous rays from the deformed and undeformed object surface.

Welford[24,25] utilized a different definition of localization. He carried out an analysis which sought positions in space at which fringes appear to be fixed, as judged by parallax. He concluded that such absolute localization does not occur. He also derived a formula for the aperture required to achieve good fringe visibility when the object surface is in focus. This was verified experimentally by Machado Gama.[26]

The reader may wish to refer to these various papers for more detailed treatments of specific aspects of fringe formation and localization in holographic interferometry.

REFERENCES

1. K. A. Stetson, A rigorous theory of the fringes of hologram interferometry, *Optik*, **29**, 386–400 (1969).
2. W. H. Steel, *Interferometry*, Cambridge University Press, Cambridge, 1967.

3. M. Born and E. Wolf, *Principles of Optics*, 2nd ed., Pergamon Press, New York, 1964, p. 292.
4. G. Klyberg, *Studies of Separation, Movements and Visibility of Holographic Interference Fringes Obtained by Turning and Translating a Rough Plane Object in Collimated Illumination and with Telecentric Observation*, Technical Report TR18.6, Institute for Optical Research, Stockholm, Sweden, 1968.
5. K. A. Stetson, The argument of the fringe function in hologram interferometry of general deformations, *Optik*, **31**, 576–591 (1970).
6. S. Walles, Visibility and localization of fringes in holographic interferometry of diffusely reflecting surfaces, *Ark. Fys.*, **40**, 299–403 (1969).
7. K. A. Haines and B. P. Hildebrand, Surface-deformation measurement using the wavefront reconstruction technique, *Appl. Opt.*, **5**, 595–602 (1966).
8. K. A. Haines and B. P. Hildebrand, Interferometric measurements on diffuse surfaces by holographic techniques, *IEEE Trans.*, **IM-15**, 149–161 (1966).
9. K. A. Stetson, Fringe interpretation for hologram interferometry of rigid-body motions and homogeneous deformations, *J. Opt. Soc. Am.*, **64**, 1–10 (1974).
10. Reference 3, pp. 186–187.
11. K. A. Stetson, Fringe vectors and observed-fringe vectors in hologram interferometry, *Appl. Opt.*, **14**, 272–273 (1975).
12. I. Prikryl, Localization of interference fringes in holographic interferometry, *Opt. Acta*, **21**, 675–681 (1974).
13. L. O. Heflinger, R. F. Wuerker, and H. Spetzler, Thermal expansion coefficient measurement of diffusely reflecting samples by holographic interferometry, *Rev. Sci. Instrum.*, **44**, 629–633 (1973).
14. B. D. Hansche and C. G. Murphy, Strain measurements by holometry, paper presented at International Conference and Exhibit, Instrument Society of America, Chicago, October 4–7, 1971.
15. O. J. Burchett, Analysis techniques for the inspection of structures by holographic interferometry, *Mater. Eval.*, **30**, 25–31 (1972).
16. N.-E. Molin and K. A. Stetson, Measurement of fringe loci and localization in hologram interferometry for pivot motion, in-plane rotation, and in-plane translation, Part I *Optik*, **31**, 157–177 (1970).
17. N.-E. Molin and K. A. Stetson, Measurement of fringe loci and localization in hologram interferometry for pivot motion, in-plane rotation, and in-plane translation, Part II, *Optik*, **31**, 281–291 (1970).
18. N.-E. Molin and K. A. Stetson, Fringe localization in hologram interferometry of mutually independent and dependent rotations around orthogonal, non-intersecting axes, *Optik*, **33**, 399–422 (1971).
19. T. Tsuruta, N. Shiotake, and Y. Itoh, Formation and localization of holographically produced interference fringes, *Opt. Acta*, **16**, 723–733 (1969).
20. P. M. Boone and R. Verbiest, Applications of hologram interferometry and translation measurement, *Opt. Acta*, **16**, 555–567 (1969).
21. C. Froehly, J. Monneret, J. Pasteur, and J. Ch. Vienot, Etude des faibles deplacements d'objets opaques et de la distorsion dans les lasers à solide par interférometrie holographie, *Opt. Acta*, **16**, 343–362 (1969).
22. W. H. Steel, Fringe localization and visibility in classical and hologram interferometers, *Opt. Acta*, **17**, 873–881 (1970).

23. S. Walles, On the concept of homologous rays in holographic interferometry of diffusely reflecting surfaces, *Opt. Acta*, **17**, 899–913 (1970).
24. W. T. Welford, Fringe visibility and localization in hologram interferometry, *Opt. Commun.*, **1**, 123–125 (1969).
25. W. T. Welford, Fringe visibility and localization in hologram interferometry with parallel displacement, *Opt. Commun.*, **1**, 311–314 (1970).
26. M. A. Machado Gama, Fringe localization in hologram and classical broad source interferometry, *Opt. Commun.*, **8**, 362–365 (1973).

4

Opaque Objects: Measurement of Strain, Stress, Bending Moments, and Vibration; Applications

4.1 INTRODUCTION

In the preceding chapters the basic optical physics associated with holographic interferometry was discussed, and methods for measuring displacement and surface deformation of opaque objects were described. In some applications of optical metrology this information is sufficient. More often, however, the experimenter needs to determine strain, stress, or bending moments, which are related to various derivatives of deformation. In this chapter the concepts of strain, stress, and bending moments are introduced and related to the components of deformation that can be measured by holographic interferometry. Numerical and optical techniques for differentiating deformation data are discussed. The measurement of the distribution of amplitude of mechanical vibration by holographic interferometry is also discussed. The methods considered include time-average, real-time, stroboscopic, and temporally modulated holographic interferometry. In addition to presenting basic theory and techniques, an effort has been made to provide practical information regarding experimental procedures and equipment.

Applications of holographic interferometry to nondestructive testing, medical and dental research, and solid mechanics are considered in the Section 4.4. The objectives of this portion of the chapter include discussion of both the scope and the limitations of applications of holographic interferometry. The presentation includes general discussions and descriptions of typical applications, as well as numerous references to relevant technical literature.

4.2 MEASUREMENT OF STRAIN, STRESS, AND BENDING MOMENTS

In Chapter 2 it was shown that holographic interferometry can be used to measure the vector displacement $\mathbf{L} = \hat{\mathbf{i}}L_x + \hat{\mathbf{j}}L_y + \hat{\mathbf{k}}L_z$ of each point on an opaque, diffusely reflecting object surface which undergoes small translations, rotations, or deformations. In experimental mechanics the deformation of test objects in response to mechanical or thermal loading is studied in order to determine strain, stress, or bending moments. These quantities are of interest because they affect the strength, safety, and lifetime of mechanical structures or components. Strain is a kinematic quantity related to the derivatives of displacement. Stress and bending moments can be inferred from strain measurements when the constituitive equations of the material, most commonly Hooke's law of elasticity, are known. In this section we discuss the measurement of strain, stress, and bending moments by holographic interferometry.

There are two basic approaches to determining strain. The first is an indirect method in which strains are calculated by differentiating measured displacements by computational or optical means. The second is a direct approach in which strains are determined by observation of fringe localization.

Let u, v, and w denote the x, y, and z components, respectively, of displacement of the points of a solid object. In this chapter this notation is used in preference to L_x, L_y, and L_z because it is standard in the literature of stress analysis, and because it serves to emphasize that we are concerned with the *deformation* and *rotation* of the object and not with rigid-body translation. Strain is a tensor quantity requiring nine components for its complete specification. Of these nine components, six are independent. At any point in a solid object there are three components of *normal strain*:

$$\epsilon_x = \frac{\partial u}{\partial x}, \tag{4.1a}$$

$$\epsilon_y = \frac{\partial v}{\partial y}, \tag{4.1b}$$

$$\epsilon_z = \frac{\partial w}{\partial z}, \tag{4.1c}$$

and three independent *shear strains*:

$$\gamma_{xy} = \frac{\partial u}{\partial y} + \frac{\partial v}{\partial x}, \tag{4.2a}$$

$$\gamma_{yz} = \frac{\partial v}{\partial z} + \frac{\partial w}{\partial y}, \tag{4.2b}$$

$$\gamma_{zx} = \frac{\partial w}{\partial x} + \frac{\partial u}{\partial z}. \tag{4.2c}$$

Physically, the normal strains are the change in length per unit length of a small element of the material in each coordinate direction. The shear strains are the decrease in angle between two line segments of material points which were initially parallel to the coordinate directions. The defining relations between strain and displacement can be developed for the simplified case of two-dimensional deformation by referring to Figure 4.1. In this figure lines $\overline{Pa}$, $\overline{ab}$, $\overline{bc}$, and $\overline{Pc}$ connect points of the undeformed solid. When the solid is deformed, point P moves to P' by translations u and v in the x and y directions, respectively. If deformation is very small, the translations of other points are given by the first two terms of Taylor series expansions, as indicated in Figure 4.1. Normal strain in the x direction is defined as

$$\epsilon_x = \frac{\overline{P'c'} - \overline{Pc}}{\overline{Pc}},$$

but

$$\overline{P'c'} = \Delta x \left[\left(\frac{\partial u}{\partial x} + 1 \right)^2 + \left(\frac{\partial v}{\partial x} \right)^2 \right]^{1/2}$$

$$\cong \left(1 + \frac{\partial u}{\partial x} \right) \Delta x.$$

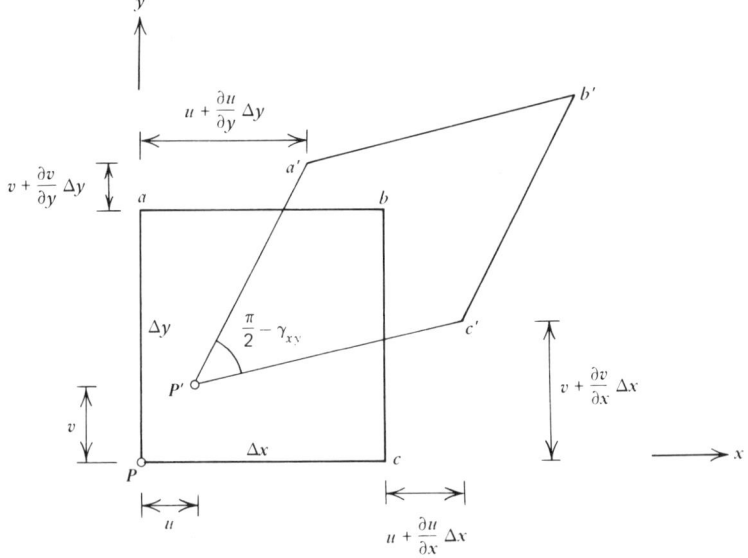

Figure 4.1 Small deformation of a solid. Point P translates to P' during deformation. Small displacements of other points are given by the first two terms of Taylor series expansions.

4.2 MEASUREMENT OF STRAIN, STRESS, AND BENDING MOMENTS

Since $\overline{Pc} = \Delta x$, $\epsilon_x = \partial u/\partial x$; similarly, $\epsilon_y = \partial v/\partial y$. The shear strain γ_{xy} is the sum of two angles:

$$\gamma_{xy} = \tan^{-1}\left[\frac{(\partial u/\partial y)\Delta y}{\Delta y}\right] + \tan^{-1}\left[\frac{(\partial v/\partial x)\Delta x}{\Delta x}\right].$$

Since the angles are assumed to be very small,

$$\gamma_{xy} = \frac{\partial u}{\partial y} + \frac{\partial v}{\partial x}.$$

Extension of this development to three-dimensional deformation yields the defining relations given in equations 4.1 and 4.2. More detailed discussions of strain are available in several texts such as that of Shames.[1]

It is appropriate at this point to note that for very small motions the *rotation* of an object also can be expressed in terms of derivatives of displacement. The components ω_x, ω_y, and ω_z of rotation about the x, y, and z axes, respectively, are as follows:

$$\omega_x = \tfrac{1}{2}\left(\frac{\partial w}{\partial y} - \frac{\partial v}{\partial z}\right), \tag{4.3a}$$

$$\omega_y = \tfrac{1}{2}\left(\frac{\partial u}{\partial z} - \frac{\partial w}{\partial x}\right), \tag{4.3b}$$

$$\omega_z = \tfrac{1}{2}\left(\frac{\partial v}{\partial x} - \frac{\partial u}{\partial y}\right). \tag{4.3c}$$

Holographic interferometry does not provide sufficient information about displacement to permit evaluation of all the derivatives appearing in equations 4.1 to 4.3. In particular, derivatives of displacement in the direction normal to the surface cannot be evaluated. If we assume that the object is at least locally plane, and that the x and y axes lie in this plane, then three or more fringe observations yield three components of displacement at each point on the surface: $u(x,y)$, $v(x,y)$, and $w(x,y)$. This is sufficient to calculate the in-plane strains ϵ_x, ϵ_y, and γ_{xy} in the object surface and the in-plane rotation ω_z about an axis normal to the surface. The out-of-plane rotation, or tilt, of the object surface can also be evaluated. To see this, note that equations 4.3a to 4.3c each represent an average of the rotation of two orthogonal faces of a cubic element; for example, ω_z is the average of the rotations of surfaces Pa and Pc in Figure 4.1. This averaging is not required to evaluate the out-of-plane rotation of the object surface, $z=0$, so its components are $\omega_x = \partial w/\partial y$ and $\omega_y = \partial w/\partial x$.

Knowledge of the displacement components $u(x,y)$, $v(x,y)$, and $w(x,y)$ is sufficient to describe the state of strain in the special, but important, case of plane stress. A thin, flat specimen is said to be in a state of *plane stress* when all stresses are parallel to its face. Examples are the stretching of thin sheets or membranes and the pulling of flat tensile specimens commonly used to measure the mechanical properties of materials.

Aleksandrov and Bonch-Bruevich[2] and Ennos[3] considered the measurement of in-plane strains. Examination of equation 2.11 or Figure 2.3 indicates that for in-plane displacements to be measured the object beam should be at a small angle to the object surface, that is, near grazing incidence, and should be viewed in the same direction. An example of a holographic system with low incidence angle and retroreflective viewing is shown in Figure 4.2. A similar system was used by Wilson[4] to measure in-plane displacements in a rectangular membrane with a hole in the center subjected to a uniform tensile stress in the longitudinal direction. The theoretical x-direction displacement for this configuration can be determined using the theory of elasticity and is shown in Figure 4.3a. Interferometric fringes which are contours of constant displacement u are shown in Figure 4.3b. This interferogram was recorded using a membrane specimen of white silastic rubber 0.06 mm thick. Good qualitative agreement with theory is evident. Wallach et al.[5] studied the same stress configuration in aluminum tensile specimens. By numerical differentiation of displacements they determined values of longitudinal strain ϵ_x at several

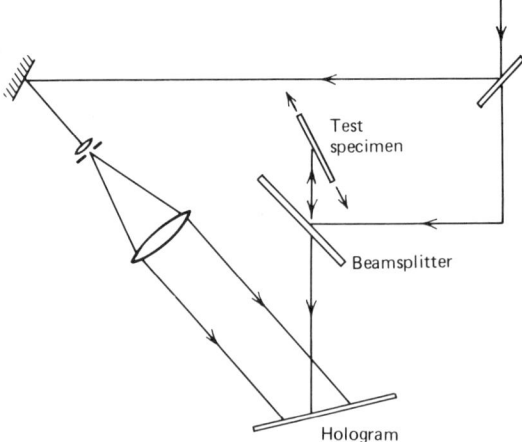

Figure 4.2 Holographic interferometer for measurement of in-plane displacements. The beamsplitter in the object beam permits retroreflective viewing and a low angle of incidence.

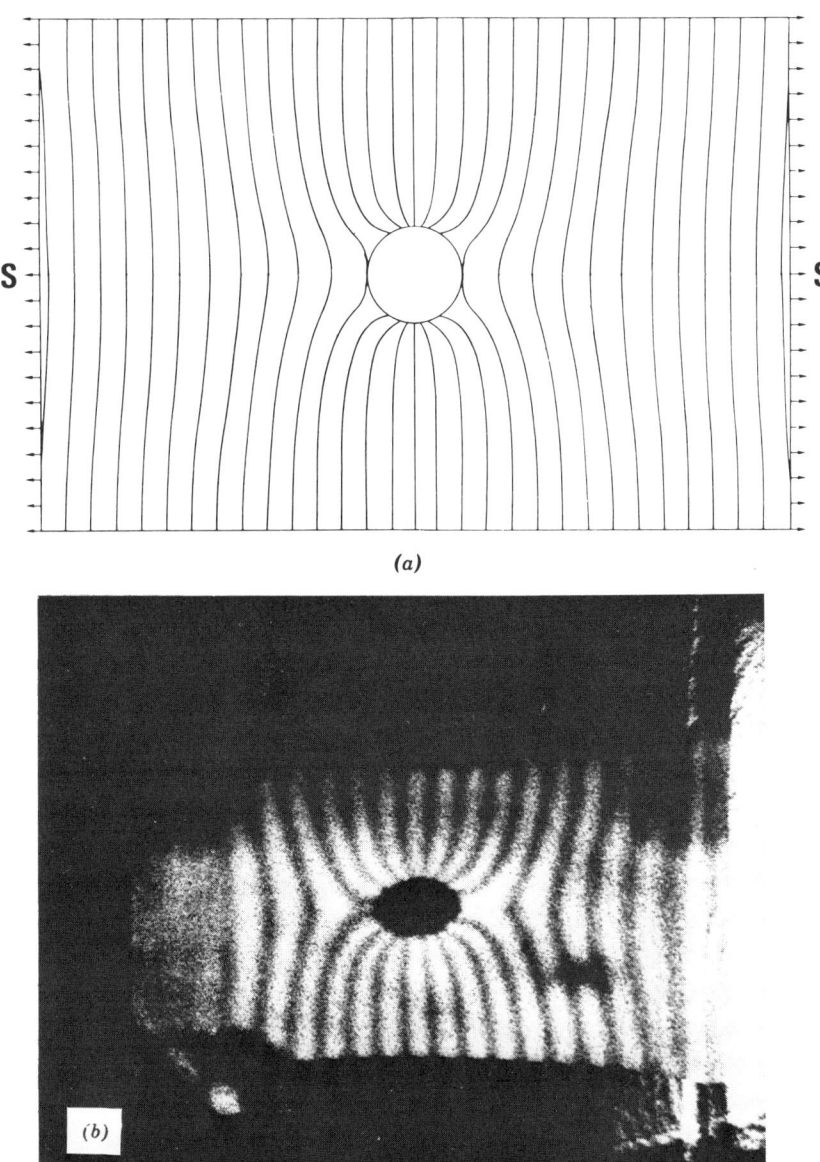

Figure 4.3 In-plane displacement distribution of a membrane with a circular hole subjected to tensile stress. (*a*) Theoretical distribution. (*b*) Holographic interferogram (Ref. 4). These displacement distributions can be differentiated to determine in-plane strains.

locations. Strains measured by holographic interferometry were compared with strain gage measurements made near the edges of the specimen; the maximum error was 3.5 percent. Also, the distribution of ϵ_x along the y axis was compared with the theoretical distribution; the maximum deviation was 12 percent.

We have seen that in-plane strains in flat specimens subjected to a plane-stress field can be determined by measuring displacements u and v parallel to the object surface. Strains in another class of objects—elastic beams, plates, and shells—can be determined by measuring displacements w normal to the object surface. Such measurements, however, depend on assumptions regarding the kinematics of deformation and on a knowledge of the constituitive equations (stress-strain relations) of the object.

A *beam* is a long, slender solid object with a constant cross section. This configuration is a common structural component and, to first approximation, models turbine blades, certain test specimens used to determine mechanical properties, and a variety of machine elements. An example of a beam is the cantilever shown in Figure 2.1. The discussion associated with that figure described the method of determining $w(x)$, the normal displacement, as a function of distance x from the base of the beam. If the material from which the beam is manufactured is elastic, stress and strain are proportional, for example,

$$\sigma_x = E\epsilon_x, \qquad (4.4)$$

where the *normal stress* σ_x is the x-direction force per unit area acting on a material surface which is normal to the x direction, and E is the *modulus of elasticity*, a property of the material. Shear stresses and strains are also proportional:

$$\tau_{zx} = G\gamma_{zx}, \qquad (4.5)$$

where τ_{zx} is the x-direction force per unit area acting on a material surface normal to the z axis, and G is the *shear modulus of elasticity*, a property of the material. It can be shown[6] that

$$G = \frac{E}{2(1+\nu)}, \qquad (4.6)$$

where ν is *Poisson's ratio*; ν is a material property which is the constant ratio of lateral strain to longitudinal strain, for example,

$$\epsilon_z = -\nu\epsilon_x$$

4.2 MEASUREMENT OF STRAIN, STRESS, AND BENDING MOMENTS

in a simple specimen subjected to a tensile or compressive load in the x direction.

If a beam, for example, that shown in Figure 4.4, is deflected by a small amount because of the action of transverse (z-direction) forces, longitudinal (x-direction) forces, and/or bending moments about axes parallel to the y axis, the longitudinal strain at the observable surface of the beam is

$$\epsilon_x = \frac{du_0}{dx} - \frac{1}{2}h\left(\frac{d^2w}{dx^2}\right), \tag{4.7}$$

where h is the thickness of the beam, and u_0 is the displacement of the center plane ($z=0$) of the beam. If the material is perfectly elastic and the

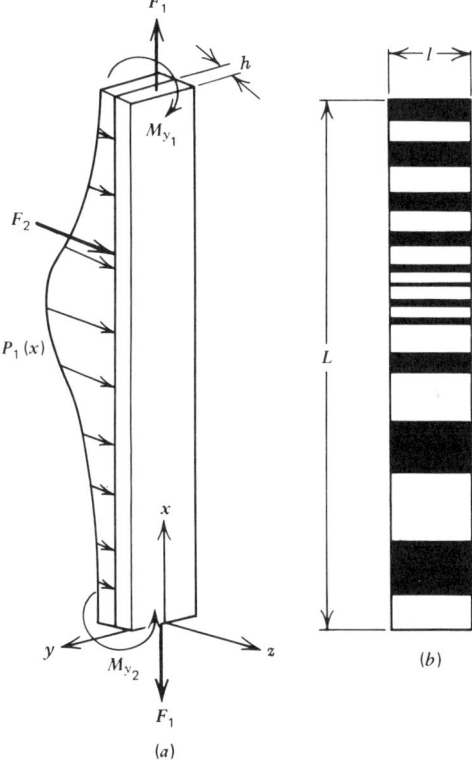

Figure 4.4 An elastic beam. (*a*) The beam is subjected to a combination of transverse loads ($F_2, p_1(x)$), axial loads (F_1) and bending moments (M_{y1}, M_{y2}) all of which are uniform in the y direction. (*b*) Interference pattern due to small deformations of the beam.

deflections are small, the bending moment at any location x is

$$M_y = -D\left(\frac{d^2w}{dx^2}\right), \tag{4.8}$$

where $D = Eh^3/12$. The longitudinal stress at the surface of the beam is

$$\sigma_x = E\left[\frac{du_0}{dx} - \tfrac{1}{2}h\left(\frac{d^2w}{dx^2}\right)\right]. \tag{4.9}$$

Equations 4.7 to 4.9 are approximate. Basically they assume that $l \ll L$ and $h \ll L$, that the material is elastic, that deformations are very small, and that plane material surfaces normal to the x axis remain plane after bending. Although restrictive, these assumptions are nearly satisfied in many situations for which sensitive interferometric measurements are feasible.

An object having width l comparable to its length L, but a uniform thickness h that is very small compared to L and l, is referred to as a *plate*.[7] If a thin plate is slightly deformed because of the action of distributed transverse (z-direction) loads over its surface, transverse or axial (x- or y-direction) forces applied at its edges, and moments (M_x, M_y) applied at its edges, the strains at its observable surface are as follows:

$$\begin{aligned}
\epsilon_x &= \frac{\partial u_0}{\partial x} - \tfrac{1}{2}h\left(\frac{\partial^2 w}{\partial x^2}\right), \\
\epsilon_y &= \frac{\partial v_0}{\partial y} - \tfrac{1}{2}h\left(\frac{\partial^2 w}{\partial y^2}\right), \\
\gamma_{xy} &= \left(\frac{\partial u_0}{\partial y} + \frac{\partial v_0}{\partial x}\right) - h\left(\frac{\partial^2 w}{\partial x \partial y}\right),
\end{aligned} \tag{4.10}$$

where u_0, v_0 are the components of in-plane translation of the central plane ($z=0$) of the plate. The corresponding stresses are

$$\begin{aligned}
\sigma_x &= E\left[\frac{\partial u_0}{\partial x} - \frac{h/2}{1-\nu^2}\left(\frac{\partial^2 w}{\partial x^2} + \frac{\nu \partial^2 w}{\partial y^2}\right)\right], \\
\sigma_y &= E\left[\frac{\partial v_0}{\partial y} - \frac{h/2}{1-\nu^2}\left(\frac{\partial^2 w}{\partial y^2} + \frac{\nu \partial^2 w}{\partial x^2}\right)\right], \\
\tau_{xy} &= G\left[\frac{\partial u_0}{\partial y} + \frac{\partial v_0}{\partial x} - \frac{h/2}{1+\nu}\left(\frac{\partial^2 w}{\partial x \partial y}\right)\right];
\end{aligned} \tag{4.11}$$

4.2 MEASUREMENT OF STRAIN, STRESS, AND BENDING MOMENTS

and the bending moments per unit length are

$$M_x = -D\left(\frac{\partial^2 w}{\partial x^2} + \frac{\nu \partial^2 w}{\partial y^2}\right),$$

$$M_y = -D\left(\frac{\partial^2 w}{\partial y^2} + \frac{\nu \partial^2 w}{\partial x^2}\right), \qquad (4.12)$$

$$M_{xy} = -M_{yx} = D(1-\nu)\left(\frac{\partial^2 w}{\partial x \partial y}\right),$$

where $D = Eh^3/[12(1-\nu^2)]$ is the *flexural rigidity* of the plate.

In many applications of beams and plates bending stresses are predominant, that is, $\partial u_0/\partial x$ and $\partial v_0/\partial y$ are negligible. Equations 4.10 to 4.12 indicate that in this case in-plane surface strains and stresses, and also bending moments, can be determined by measurement of the distribution of normal deformation $w(x,y)$. This is generally much more convenient than measuring $u(x,y)$ and $v(x,y)$ with sufficient accuracy to evaluate the derivatives in equations 4.1 and 4.2, and will usually yield more accurate results.

From the preceding discussion it is evident that the determination of strain, stress, and bending moments requires differentiation of the measured distribution of surface displacement. It is a very difficult task to obtain accurate derivatives from discrete displacement measurements because some degree of experimental error is unavoidable. The data obtained by holographic interferometry must be handled with extreme care if meaningful quantitative determinations of strain, stress, and bending moments are to be achieved. There are two basic approaches to differentiating displacement data to obtain strain: direct numerical differentiation by a finite difference approximation, and fitting an interpolating polynomial to the data, which can in turn be differentiated analytically. These two approaches can be discussed in terms of the bending of a simple beam. A third approach in which optical differentiation is used will be discussed later in this section. Figure 4.4b shows a holographic interferogram of a deformed beam. Let us assume that the beam is normally illuminated and viewed retroreflectively, so that at any point the normal displacement of the surface is

$$w = \frac{N\lambda}{2}. \qquad (4.13)$$

If the interferogram was obtained by two-exposure interferometry, N takes on integer values at the center of each bright fringe and half-integer values at the center of each dark fringe. A set of measured values of displacement

w_i at discrete locations x_i can be obtained from the interferogram. Usually the locations x_i will be the centers of bright and/or dark fringes. These values can be displayed graphically as in Figure 4.5.

In the *finite difference* approach, we assume that the normal displacement can be expanded in a Taylor series about the point x_i:

$$w(x) = w_i + \left(\frac{dw}{dx}\right)_{x_i}(x - x_i) + \frac{1}{2}\left(\frac{d^2w}{dx^2}\right)_{x_i}(x - x_i)^2. \tag{4.14}$$

The finite difference approximation to $(dw/dx)_{x_i}$ is obtained by evaluating the first two terms of this series at x_{i+1}:

$$\left(\frac{dw}{dx}\right)_{x_i} \cong \frac{w(x_{i+1}) - w(x_i)}{x_{i+1} - x_i} = \frac{w_{i+1} - w_i}{h_i}, \tag{4.15}$$

where $h_i = x_{i+1} - x_i$. This is called the *forward difference* approximation. Similarly, the first two terms of the series could be evaluated at x_{i-1} to give

$$\left(\frac{dw}{dx}\right)_{x_i} \cong \frac{w(x_i) - w(x_{i-1})}{x_i - x_{i-1}} = \frac{w_i - w_{i-1}}{h_{i-1}}, \tag{4.16}$$

which is called the *backward difference* approximation. At points other than x_0 and x_n, it is generally preferable to use the *central difference* approximation, which is simply the average of the forward and backward

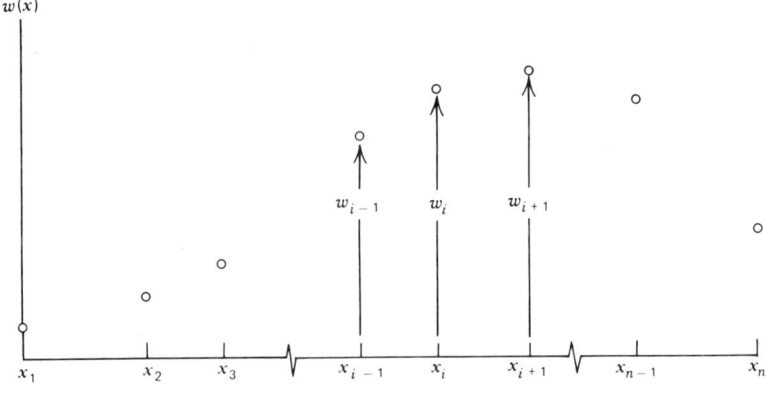

Figure 4.5 Discrete measurements of normal displacement of a beam obtained by holographic interferometry.

4.2 MEASUREMENT OF STRAIN, STRESS, AND BENDING MOMENTS

differences:

$$\left(\frac{dw}{dx}\right)_{x_i} \cong \frac{w_{i+1} - w_{i-1}}{h_i + h_{i-1}}. \tag{4.17}$$

To obtain a finite difference approximation to the second derivative, d^2w/dx^2, we evaluate equation 4.14 at $x = x_i$ and at $x = x_{i-1}$ and retain the first three terms:

$$w(x_{i+1}) = w(x_i) + \left(\frac{dw}{dx}\right)_{x_i} (x_{i+1} - x_i)$$

$$+ \frac{1}{2}\left(\frac{d^2w}{dx^2}\right)_{x_i} (x_{i+1} - x_i)^2,$$

$$w(x_{i-1}) = w(x_i) + \left(\frac{dw}{dx}\right)_{x_i} (x_i - x_{i-1})$$

$$+ \frac{1}{2}\left(\frac{d^2w}{dx^2}\right)_{x_i} (x_i - x_{i-1})^2.$$

When these two equations are added and equation 4.17 is substituted for $(dw/dx)_{x_i}$, the following approximation of the second derivative results:

$$\left(\frac{d^2w}{dx^2}\right)_{x_i} = \frac{\left(\frac{2h_{i-1}}{h_i + b_{i-1}}\right) w_{i+1} + \left(\frac{2h_i}{h_i + h_{i-1}}\right) w_{i-1} - 2w_i}{\frac{1}{2}(h_i^2 + h_{i-1}^2)}. \tag{4.18}$$

This equations has a very simple form if the intervals h_i and h_{i-1} are equal; however, this is not the usual case in interferometry because the intervals are determined by the unevenly spaced fringe locations.

If sufficient data of high accuracy are available, equations 4.15 to 4.18 can be used to evaluate first and second derivatives of displacement for use in computing strains, stresses, or bending moments. In practice, however, data points may be widely separated, slight distortions may be introduced by the optical system used to photograph the interferogram, and there is always some uncertainty in locating the center of each fringe. In addition the errors associated with measuring the system geometry discussed in Section 2.3 are always incurred. The effect of these errors is amplified when derivatives are calculated. For this reason it may be useful to smooth the data by fitting an appropriate curve to it. Computation of derivatives can then be based on this curve. *Cubic spline functions* are an appropriate choice for fitting displacement data with the objective of subsequent differentiation. Although other alternatives for representing and smoothing interferometric data are available, spline functions appear to be an attractive choice for three reasons: their definition is based on the theory of mechanical deformation; the behavior of their second derivatives is understood a priori; and their application to interferometric determination of strain has been studied.[8,9]

Consider the problem of fitting cubic splines to the discrete deformation data shown in Figure 4.5. A cubic spline is mathematically equivalent to fitting a draftsman's spline to these data points. A draftsman's spline is a thin, flexible metal strip which is interlaced among pins inserted into each data point such as those represented in Figure 4.5. The spline contacts each pin. In this manner the metal strip is formed into a smooth curve passing through each data point. The mathematical spline is a composite of n separate third-order polynomials, each of which represents the function $w(x)$ between adjacent data points, for example, in the interval between x_i and x_{i+1}. The second derivative of these polynomials varies linearly between each pair of adjacent data points—a characteristic it shares with the draftsman's spline. Let $P_i(x)$ denote the cubic polynomial which represents $w(x)$ in the interval between x_i and x_{i+1}. The spline function representation of $w(x)$ over the entire range from $x = x_1$ to $x = x_n$ consists of $(n-1)$ cubic polynomials $P_1(x), P_2(x), \ldots, P_{n-1}(x)$. Each cubic polynomial involves four arbitrary coefficients, so there is a total of $4(n-1)$ coefficients whose values must be determined in terms of the data.

Each polynomial must be equal to the known data values at each end of the interval. This requirement provides $2(n-1)$ relations:

$$\left.\begin{array}{l} P_i(x_i) = w(x_i), \\ P_i(x_{i+1}) = w(x_{i+1}) \end{array}\right\}, \quad i = 1, 2, \ldots, n-1. \tag{4.19}$$

It is further required that the first and second derivatives of successive polynomials must match each other at their common data point. This requirement providies an additional $2(n-2)$ relations:[10]

$$P_i'(x_i) = P_{i-1}'(x_i), \quad i = 2, 3, \ldots, n-1, \tag{4.20}$$

$$P_i''(x_i) = P_{i-1}''(x_i), \quad i = 2, 3, \ldots, n-1. \tag{4.21}$$

The remaining two relations must be obtained by *specifying* conditions on the first or second derivatives of $w(x)$ at the end points $x = x_1$ and $x = x_n$. These conditions should be based on the mechanics of the object being studied. For example, if normal displacement of a beam or plate is being measured, $P' = 0$ at a built-in end and $P'' = 0$ at a simply supported end:

$$P_1'(x_1) = 0 \quad \text{or} \quad P_1''(x_1) = 0, \tag{4.22}$$

$$P_{n-1}'(x_n) = 0 \quad \text{or} \quad P_{n-1}''(x_n) = 0. \tag{4.23}$$

4.2 MEASUREMENT OF STRAIN, STRESS, AND BENDING MOMENTS

Because the second derivative of each polynomial is a linear function of x, $P_i''(x)$ can be found at any point in the interval $x_i \leq x \leq x_{i+1}$ by linear interpolation:

$$P_i''(x) = \frac{x_{i+1} - x}{h_i} P_i''(x_i) + \frac{x - x_i}{h_i} P_i''(x_{i+1}). \tag{4.24}$$

To define the polynomials $P_i(x)$, equation 4.24 is integrated twice and conditions 4.19 are imposed. After straightforward but rather lengthy algebraic rearrangement, this results in the desired expression for the cubic polynomials:

$$\begin{aligned} P_i(x) = {} & \frac{P_i''(x_i)}{6h_i}(x_{i+1} - x)^3 + \frac{P_i''(x_{i+1})}{6h_i}(x - x_i)^3 \\ & + \left(\frac{w_{i+1}}{h_i} - \frac{h_i P_i''(x_{i+1})}{6} \right)(x - x_i) \\ & + \left(\frac{w_i}{h_i} - \frac{h_i P_i''(x_i)}{6} \right)(x_{i+1} - x), \\ & i = 1, 2, \ldots, n - 1. \end{aligned} \tag{4.25}$$

The second derivatives $P_i''(x_i) = P_{i-1}''(x_i)$ that appear as coefficients in equation 4.24 are determined by differentiating equation 4.25 and imposing the condition of continuity of second derivatives, equation 4.20:

$$\begin{aligned} & \frac{h_{i-1}}{h_i} P_{i-1}''(x_{i-1}) + 2\left(1 + \frac{h_{i-1}}{h_i}\right) P_i''(x_i) + P_i''(x_{i+1}) \\ & = \frac{6}{h_i}\left(\frac{w_{i+1} - w_i}{h_i} - \frac{w_i - w_{i-1}}{h_{i-1}}\right), \quad i = 2, 3, \ldots, n - 1. \end{aligned} \tag{4.26}$$

Equations 4.26, 4.22, and 4.23 constitute a set of n simultaneous linear algebraic equations in the n unknown values of $P_i''(x_i)$. The procedure to fit a cubic spline to interferometric data consists of solving these n equations and substituting the resulting values of $P_i''(x_i)$ into equation 4.25, which defines the cubic spline function:

$$\begin{aligned} w(x) = P_i(x) \quad & \text{for } x_i \leq x \leq x_{i+1}, \\ & i = 1, 2, 3, \ldots, n - 1. \end{aligned} \tag{4.27}$$

Equation 4.27 can be differentiated to provide first or second derivatives for the computation of strain, stress, or bending moments.

Our objective in introducing the cubic spline approximation was to smooth the data in order to reduce error amplification and avoid erratic second derivatives. If the end-point conditions are $P_0''(x_0) = P_{n-1}''(x_n) = 0$, it can be shown that the cubic spline is the smoothest possible interpolating function in the sense that the mean-square curvature is minimized:

$$\int_{x_1}^{x_n} |f''(x)|^2 dx \leq \int_{x_1}^{x_n} |g''(x)|^2 dx, \qquad (4.28)$$

where $f(x)$ is the cubic spline approximation to $w(x)$, and $g(x)$ is any other twice-differentiable interpolating function fitting the same data set w_i, $i = 1, 2, \ldots, n$.

Although the cubic spline approximation is simple to apply and succeeds in smoothing the data, some caution is advised for two reasons. First, if the spline is forced to fit closely spaced data points, rapid variation of the derivatives may result even though the displacement itself is represented accurately. Second, enforcing continuity of first derivatives and linearity of second derivatives is equivalent to assuming that the object being studied deforms like an elastic beam; thus an excessively smooth representation of the deformation of inelastic materials or test objects with complex cross sections could result. Of course, similar difficulties are inherent with any interpolating scheme and serve to emphasize that computing accurate derivatives from discrete experimental data is difficult.

Brandt and Taylor[8,9] have studied the use of cubic spline approximations to compute strains and bending moments in turbine blades, using holographic interferometry. An error analysis by computer simulation of experiments[8] indicated, as expected, that errors in strain and bending moment are one and two orders of magnitude higher than those in displacement, but fortunately errors appeared to be lowest in regions where strain and bending moment were highest. These workers further found that accuracy was enhanced by using only part of the available data points in order to avoid artificial oscillations of derivatives. On the basis of this study they developed an algorithm for obtaining good spline fits to interferometric data.[9] This algorithm is summarized as follows:

1. A cubic spline is fit to data points corresponding to each bright fringe. Preliminary smoothing is then accomplished by shifting the positions of the interior data points by a distance not exceeding 10 percent of the local fringe spacing until the mean-square curvature is minimized (see equation 4.28). The resulting spline serves as "data" for further operations.
2. A new spline is fit to "data" obtained from the initial spline at each point where its first derivative is a maximum, minimum, or zero; at each end point; and midway between any data points separated by more than 20 percent of the length of the object.

4.2 MEASUREMENT OF STRAIN, STRESS, AND BENDING MOMENTS

A computer code based on this algorithm was used to analyze experimental data, and the results were found to be in agreement with an error analysis which was derived separately. Detailed information regarding the theory and application of splines is given by Ahlberg et al.[11]

Our discussion of strain analysis by the spline approximation has dealt with cases in which differentiation in one direction of a single component of displacement is required. In more complicated problems derivatives in two directions of all three displacement components, u, v, and w, may be required. Sollid[12] has reported the existence of computer codes for this purpose that are based on the representation of u, v, and w over the entire surface by Bezier vector polynomials.[13] A discussion of these polynomials, which were originally developed for application to numerical control of machines, is beyond the scope of this book.

No simple statement of the expected accuracy of strain measurements by numerical differentiation of displacement data obtained by holographic interferometry is possible because of the many factors involved. Perusal of reported experimental values obtained by careful analysis of relatively simple strain fields suggests that average errors in the range of 5 to 35 percent are typical.

An alternative to numerical differentiation of displacement data is to determine strains, bending moments, and related quantities by *optical differentiation* of interferograms. In particular, first and second derivatives of displacement can be displayed as moiré patterns formed by overlaying identical photographic transparencies of an interferogram. Although generally less accurate than carefully computed numerical derivatives, moiré patterns can be obtained in a very simple manner, and they provide a useful visual display of slopes, strains, or bending moments. Although numerical differentiation of interference patterns by the moiré method has been known for some time, its use was restricted to the study of very simple configurations.[14-16] Holographic interferometry makes it possible to apply moiré differentiation to the study of diffusely reflecting surfaces.

Figure 4.6a is a two-exposure interferogram of a circular aluminum diaphragm which was deformed by a uniform pressure. As would be expected, the interference fringes, which are contours of constant normal displacement w, are concentric circles. If two identical photographic transparencies of this fringe pattern are made, one can be laid on top of the other and the two brought into exact registration. If one transparency is then translated in the x direction by a small amount with respect to the other one, the pattern shown in Figure 4.6b is produced. A set of relatively broad fringes of low visibility can be seen in this figure; they are caused by a "beating" between the two interferograms and represent contours of constant slope $\partial w/\partial x$ of the deformed diaphragm. This interpretation of the moiré pattern is explained in the analysis that follows.

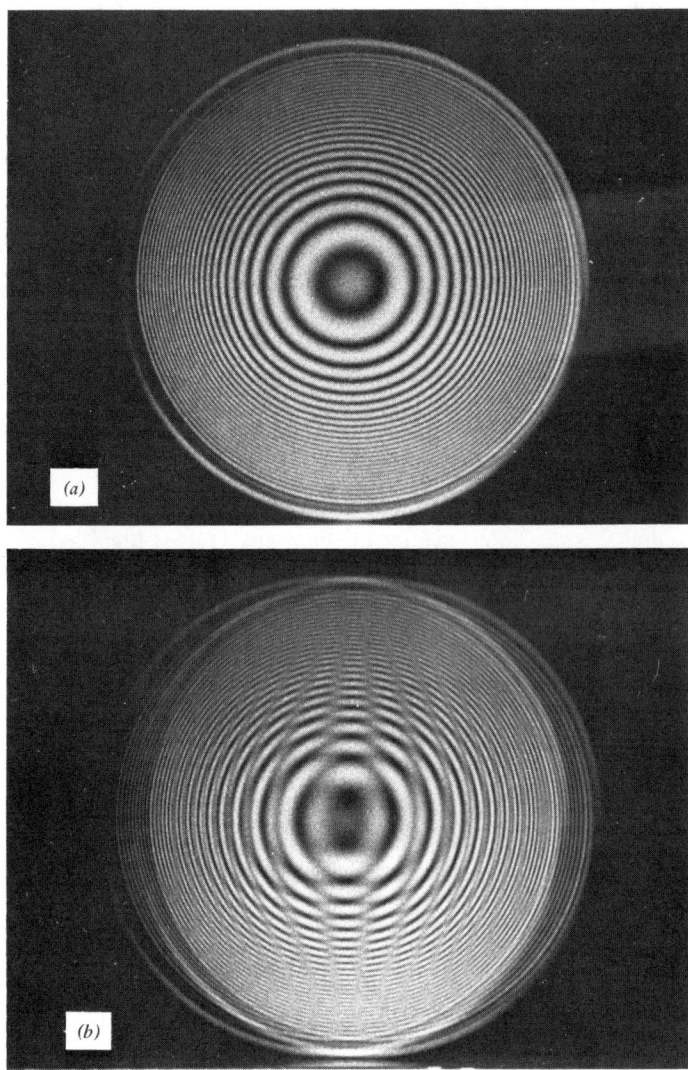

Figure 4.6 Determination of the slope of a circular diaphragm deformed by a uniform pressure. (*a*) Two-exposure holographic interferogram. (*b*) Moiré pattern formed by a small relative lateral displacement of identical photographic transparencies of interferogram (*a*). The moiré fringes are contours of constant slope differential $\partial w/\partial x$.

4.2 MEASUREMENT OF STRAIN, STRESS, AND BENDING MOMENTS

The transmittance of a photographic transparency of a double-exposure holographic interferogram is

$$t(x,y) = \tfrac{1}{2}\{1 + \cos[\Delta\phi(x,y)]\}$$
$$= \cos^2\left[\tfrac{1}{2}\Delta\phi(x,y)\right], \quad (4.29)$$

where the transmittance has been normalized and a zero bias level has been assumed for simplicity. The argument $\Delta\phi(x,y)$ was shown in Chapter 2 to be related to the object displacement $L(x,y)$ by

$$\Delta\phi(x,y) = \frac{4\pi}{\lambda}\cos\theta\cos\psi L(x,y), \quad (4.30)$$

where θ is half the angle between the illumination and observation directions, and ψ is the angle between the displacement vector and the bisector of the illumination and observation vectors. When one of two identical copies of this transparency is laid on top of the other and shifted laterally by an amount Δx with respect to it, the resulting transmittance is

$$t(x,y) = \cos^2\left[\tfrac{1}{2}\Delta\phi(x,y)\right]\cos^2\left[\tfrac{1}{2}\Delta\phi(x-\Delta x, y)\right]. \quad (4.31)$$

The trigonometric identity,

$$\cos a \cos b = \tfrac{1}{2}[\cos(a+b) + \cos(a-b)],$$

can be used to rewrite equation 4.31 as

$$t(x,y) = \tfrac{1}{4}\left\{\cos^2\left[\tfrac{1}{2}\Delta\phi(x,y) - \tfrac{1}{2}\Delta\phi(x-\Delta x, y)\right]\right.$$
$$+ \cos^2\left[\tfrac{1}{2}\Delta\phi(x,y) + \tfrac{1}{2}\Delta\phi(x-\Delta x, y)\right]\right\}$$
$$+ \tfrac{1}{2}\cos\left[\tfrac{1}{2}\Delta\phi(x,y) + \tfrac{1}{2}\Delta\phi(x-\Delta x, y)\right]$$
$$\cdot \cos\left[\tfrac{1}{2}\Delta\phi(x,y) - \tfrac{1}{2}\Delta\phi(x-\Delta x, y)\right]. \quad (4.32)$$

The last term in this equation describes the moiré fringes; the first two terms are related to the average transmittance over a large number of fringes, and their effect can be ignored in most instances. To interpret the term of interest we assume that the first two terms of a Taylor series expansion represent $\Delta\phi(x-\Delta x, y)$ adequately:

$$\Delta\phi(x-\Delta x, y) \cong \Delta\phi(x,y) - \frac{\partial\Delta\phi(x,y)}{\partial x}\Delta x;$$

then

$$\Delta\phi(x,y) + \Delta\phi(x-\Delta x,y) \simeq 2\Delta\phi(x,y) - \left(\frac{\partial \Delta\phi}{\partial x}\right)\Delta x, \quad (4.33)$$

$$\Delta\phi(x,y) - \Delta\phi(x-\Delta x,y) \simeq \left(\frac{\partial \Delta\phi}{\partial x}\right)\Delta x. \quad (4.34)$$

Retaining only the lowest order terms in equations 4.33 and 4.34, we find that the moiré pattern is described by

$$\tfrac{1}{2}\cos\left[\Delta\phi(x,y)\right]\cdot\cos\left\{\tfrac{1}{2}\left[\frac{\partial \Delta\phi(x,y)}{\partial x}\right]\Delta x\right\}.$$

This represents a fringe pattern whose period is comparable to that of the original interferogram, but it is multiplied by a slowly varying envelope function that defines the visibility of the fringe system. The visibility is a minimum at points where

$$\frac{1}{2}\left[\frac{\partial \Delta\phi(x,y)}{\partial x}\right]\Delta x = (2N_m+1)\left(\frac{\pi}{2}\right),$$

$$N_m = 0, 1, 2, \ldots. \quad (4.35)$$

Hence the regions of low visibility (moiré fringes) are contours of constant value of the derivative $\partial \Delta\phi/\partial x$.

If the object under study is flat, undergoes normal deformation $w(x,y)$, and is illuminated in the normal direction and viewed retroreflectively, the equation for the moiré fringes can be rewritten as

$$\frac{\partial w}{\partial x} = \frac{(2N_m+1)\lambda}{4\Delta x}, \quad (4.36)$$

where N_m is the fringe number of the moiré fringes and has an integer value at the center of each fringe of low visibility. Clearly, the transparencies also could be translated by an amount Δy in the orthogonal direction to form moiré contours of constant $\partial w/\partial y$. Boone and Verbiest[17] and Saito et al.[18] have used this method to measure the slope of deformation surfaces of plates. Agreement with theoretically predicted slopes was excellent.

If interferograms of pure in-plane deformation could be recorded, moiré differentiation could be used to determine directly the in-plane strains $\epsilon_x = \partial u/\partial x$ and $\epsilon_y = \partial v/\partial y$. Unfortunately, several practical difficulties arise. Sensitivity to in-plane deformation requires illumination and viewing at near-grazing incidence, so the resulting image is optically distorted. Furthermore, the fringes are significantly deformed even by very small

4.2 MEASUREMENT OF STRAIN, STRESS, AND BENDING MOMENTS

normal displacements, and the distinct fringes of relatively high frequency which are desirable for forming moiré patterns are difficult to attain with in-plane deformation.

Stetson[19] proposed and demonstrated a moiré technique which enables one to determine *second derivatives*; thus bending moments and in-plane strains of plates and shells can be measured. The procedure, which is quite fascinating, is illustrated in Figure 4.7. The object is a rectangular plate which was deformed slightly by application of a concentrated normal load at the center of the plate. A two-exposure interferogram is shown in Figure 4.7a. The viewing direction is normal to the plate, and the fringes are contours of constant normal displacement $w(x,y)$. Suppose that we wish to investigate the bending moments or strain at the location indicated by the small cross in this figure. Two identical photographic transparencies of the interferogram are prepared and overlaid, bringing the fringe patterns into registration. Then a pin is placed through both transparencies (literally or figuratively) at the point of interest, and one transparency is rotated 180°. The resulting fringe pattern is shown in Figure 4.7b. In the vicinity of the point about which the transparency was rotated an elliptical region of low visibility can be discerned. This is a moiré fringe. We will now show that the orientation of this ellipse indicates the principal strain directions and that the lengths of the major and minor axes of the ellipse are measures of the magnitudes of the principal strains.

The transmittance of each transparency is represented by equation 4.29. For simplicity, the coordinate origin $x=0$, $y=0$ can be placed at the point where strain is to be measured. After one transparency is rotated 180° about this point, the transmittance of the combined transparencies is

$$t(x,y) = \cos^2\left[\tfrac{1}{2}\Delta\phi(x,y)\right] \cdot \cos^2\left[\tfrac{1}{2}\Delta\phi(-x,-y)\right] \tag{4.37}$$

or, equivalently,

$$\begin{aligned}
t(x,y) = \tfrac{1}{4}\Big\{ &\cos^2\left[\tfrac{1}{2}\Delta\phi(x,y) - \tfrac{1}{2}\Delta\phi(-x,-y)\right] \\
+ &\cos^2\left[\tfrac{1}{2}\Delta\phi(x,y) + \tfrac{1}{2}\Delta\phi(-x,-y)\right]\Big\} \\
+ \tfrac{1}{2}&\cos\left[\tfrac{1}{2}\Delta\phi(x,y) + \tfrac{1}{2}\Delta\phi(-x,-y)\right] \\
\cdot &\cos\left[\tfrac{1}{2}\Delta\phi(x,y) - \tfrac{1}{2}\Delta\phi(-x,-y)\right].
\end{aligned} \tag{4.38}$$

As in the case of equation 4.32, the moiré pattern is described by the last term of equation 4.38. To interpret this term, we expand $\Delta\phi(x,y)$ in a

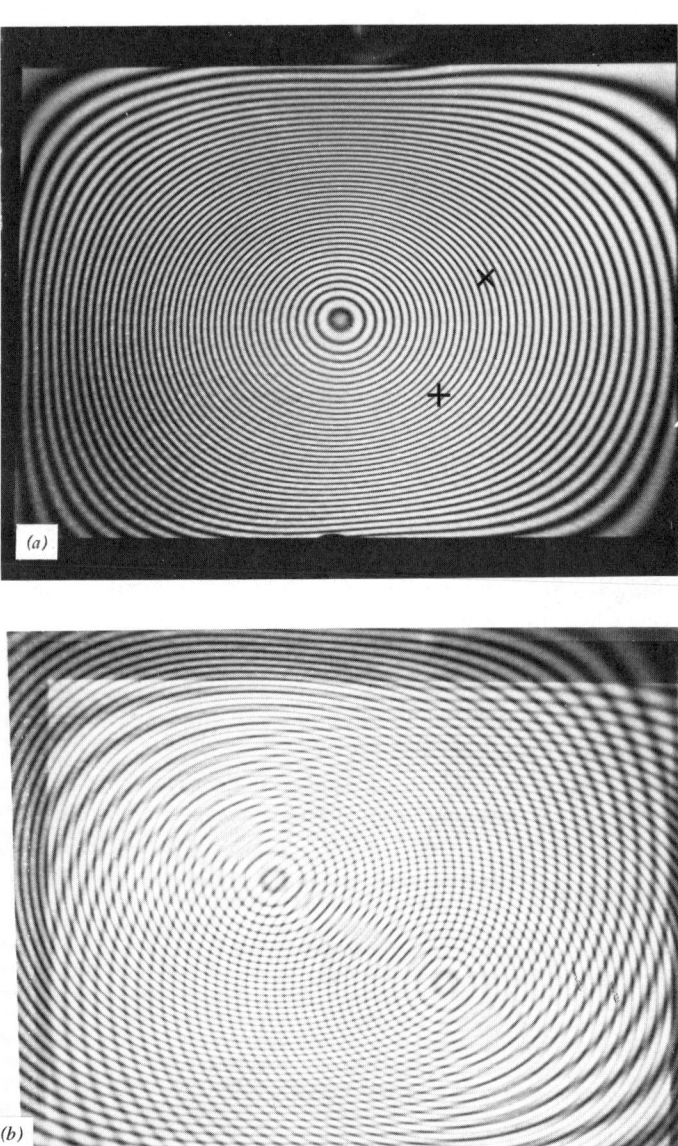

Figure 4.7 Determination of principal strains in a deformed plate by moiré differentiation. (a) Two-exposure hologram. (b) Moiré pattern formed by 180° relative rotation about the point marked by + of two identical photographic transparencies of interferogram (a). (c) Moiré pattern formed at a different location marked x in the same manner.

4.2 MEASUREMENT OF STRAIN, STRESS, AND BENDING MOMENTS

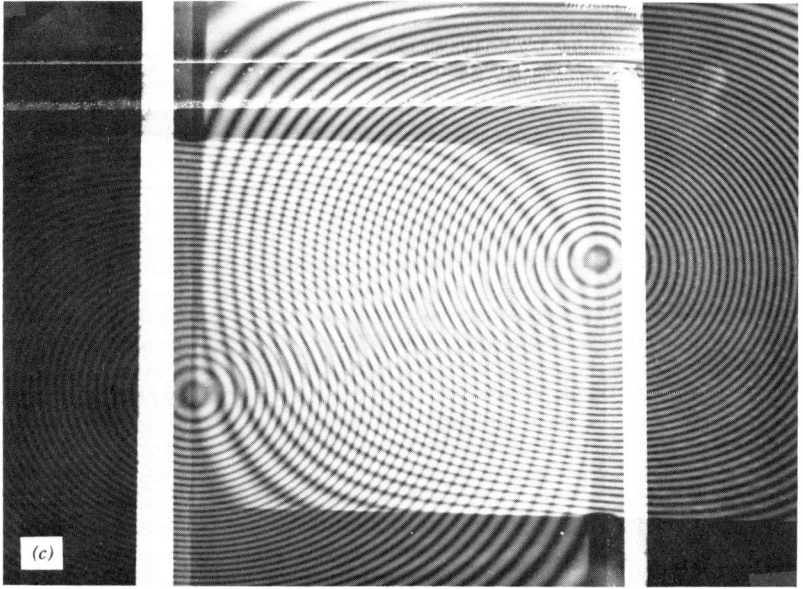

Figure 4.7 Cont.

two-dimensional Taylor series about the point $x=0, y=0$:

$$\Delta\phi(x,y) = \Delta\phi(0,0) + \left(\frac{\partial \Delta\phi}{\partial x}\right)x + \left(\frac{\partial \Delta\phi}{\partial y}\right)x$$

$$+ \frac{1}{2}\left(\frac{\partial^2 \Delta\phi}{\partial x^2}\right)x^2 + \frac{1}{2}\left(\frac{\partial^2 \Delta\phi}{\partial y^2}\right)y^2$$

$$+ \left(\frac{\partial^2 \Delta\phi}{\partial x \partial y}\right)xy, \qquad (4.39)$$

where all derivatives are evaluated at $x=0, y=0$. Assuming that terms of order higher than 2 can be neglected in equation 4.39, we find that the following approximations result:

$$\Delta\phi(x,y) + \Delta\phi(-x,-y) = 2\Delta\phi(0,0) + \left(\frac{\partial^2 \Delta\phi}{\partial x^2}\right)x^2$$

$$+ \left(\frac{\partial^2 \Delta\phi}{\partial y^2}\right)y^2 + 2\left(\frac{\partial^2 \Delta\phi}{\partial x \partial y}\right)xy,$$

$$\Delta\phi(x,y) - \Delta\phi(-x,-y) = 2\left(\frac{\partial \Delta\phi}{\partial x}\right)x + 2\left(\frac{\partial \Delta\phi}{\partial y}\right)y.$$

When these expressions are substituted into the last term of equation 4.38, the moiré pattern is described by

$$\tfrac{1}{2}\cos\left[\Delta\phi(0,0) + \frac{1}{2}\left(\frac{\partial^2\Delta\phi}{\partial x^2}\right)x^2 + \frac{1}{2}\left(\frac{\partial^2\Delta\phi}{\partial y^2}\right)y^2 \right.$$
$$\left. + \left(\frac{\partial^2\Delta\phi}{\partial x\,\partial y}\right)xy\right]\cdot\cos\left[\left(\frac{\partial\Delta\phi}{\partial x}\right)x + \left(\frac{\partial\Delta\phi}{\partial y}\right)y\right].$$

The first cosine term is the slowly varying envelope function that defines the visibility of the fringe system. The moiré fringes are contours of minimum visibility; their equation is

$$\Delta\phi(0,0) + \frac{1}{2}\left(\frac{\partial^2\Delta\phi}{\partial x^2}\right)x^2 + \frac{1}{2}\left(\frac{\partial^2\Delta\phi}{\partial y^2}\right)y^2$$
$$+ \left(\frac{\partial^2\Delta\phi}{\partial x\,\partial y}\right)xy = \pm\frac{(2N_m+1)\pi}{2},$$
$$N_m = 0, 1, \ldots. \qquad (4.40)$$

Now, if we refer displacements to that of the point under study, and orient the axes so that either x or y is parallel to the interference fringes at the origin, then $\partial^2\Delta\phi/\partial x\,\partial y = 0$ and $w(0,0)=0$, and equations 4.40 and 4.30 can be combined to give the following equations for the first moiré fringe:

$$\left(\frac{\partial^2 w}{\partial x^2}\right)x^2 + \left(\frac{\partial^2 w}{\partial y^2}\right)y^2 = \frac{\lambda}{4\cos\theta\cos\psi}. \qquad (4.41)$$

Equation 4.41 represents an ellipse if both partial derivatives have the same sign; if the two partial derivatives have opposite signs, the $\pm$ signs in equation 4.40 can be used to define two conjugate hyperbolas (see Figure 4.7c). If x_d and y_d are the principal diameters of the ellipse or of the conjugate hyperbolas, then

$$\frac{\partial^2 w}{\partial x^2} = \frac{\lambda}{x_d^2 \cos\theta\cos\psi},$$
$$\frac{\partial^2 w}{\partial y^2} = \frac{\lambda}{y_d^2 \cos\theta\cos\psi}. \qquad (4.42)$$

If the deformation of the plate being studied is purely, or predominantly, due to bending, the maximum and minimum surface strains and bending moments can be calculated with equations 4.10 and 4.12, using the experi-

4.2 MEASUREMENT OF STRAIN, STRESS, AND BENDING MOMENTS

mentally determined values of the second derivatives of w along the two principal directions.

When the transparencies are being aligned to produce the moiré pattern, they should be translated by whatever small amount is necessary to form a small, dark moiré fringe centered on the point of interest. This will assure that all measurements are made relative to the same base.

The accuracy of this moiré method of strain measurement should not be expected to exceed 20 percent,[19] and experimental results indicate that errors with a mean absolute value of 50 percent may be more typical.[20] Principal strain directions are measured rather accurately. The primary value of the method is that it provides a visualization of the surface strain field at any point, including the principal directions and relative signs of the maximum and minimum strains or bending moments. Very simple calculations based on the principal diameters of the elliptical or hyperbolic fringes yield estimates of the values of these maximum and minimum strains.

The conditions for *fringe localization*, discussed in Chapter 3, indicate that the position in space where fringes localize depends on the derivatives of displacement, that is, on the rotation and strain of the object. In the early papers of Haines and Hildebrand[21] and Aleksandrov and Bonch-Bruevich[2] it indeed was noted that fringe localization measurements could be used to determine rigid-body rotation. Several workers have noted that the localization condition can be expressed explicitly in terms of strain and rotation. Schumann[22] suggested that determination of strain by fringe localization measurements might be useful. Stetson[23] has proposed a particularly simple procedure for accomplishing this, which is discussed below.

Equation 3.27 relates displacement, fringe spacing, and position of localization when the interferogram is viewed through a narrow slit aperture. If the surface under study is flat and lies in the x-y plane, and if it is observed in the normal direction through a slit aperture that is parallel to the x direction, equation 3.27 can be written as

$$u = \frac{\lambda z_{l_x}}{d_x}, \tag{4.43}$$

where u is the x component of displacement, z_{l_x} is the distance from the object surface at which fringes are localized when viewed through a slit aperture parallel to the x direction, and d_x is the observed fringe spacing in the x direction. Because a slit aperture is used, fringes appear to localize in a plane. Differentiation of equation 4.43 yields

$$\frac{\partial u}{\partial x} = \frac{\lambda}{d_x}\left(\frac{\partial z_{l_x}}{\partial x}\right), \quad \frac{\partial u}{\partial y} = \frac{\lambda}{d_x}\left(\frac{\partial z_{l_x}}{\partial y}\right). \tag{4.44}$$

The derivatives $\partial z_{l_x}/\partial x$ and $\partial z_{l_x}/\partial y$ are the slopes of the plane of localization. Similarly, if observations are made through a slit aperture parallel to the y direction, the derivatives $\partial v/\partial x$ and $\partial v/\partial y$ can be obtained:

$$\frac{\partial v}{\partial x} = \frac{\lambda}{d_y}\left(\frac{\partial z_{l_y}}{\partial x}\right), \qquad \frac{\partial v}{\partial y} = \frac{\lambda}{d_y}\left(\frac{\partial z_{l_y}}{\partial y}\right), \qquad (4.45)$$

where d_y is the fringe spacing in the y direction observed through this aperture. Equations 4.44 and 4.45 indicate that two measurements of fringe spacing and four measurements of slopes of fringe localization planes provide sufficient information to calculate the in-plane strains ϵ_x, ϵ_y, and γ_{xy}, and the in-plane rotation ω_z at each point on the surface, using equations 4.1a, 4.1b, 4.2a, and 4.3c.

Since fringe localization is difficult to measure with precision, this method will be most satisfactory when strain is homogeneous, or at least varies slowly with x and y. The method can be illustrated for the case of homogeneous strain by the experiment depicted in Figure 4.8. A thin rubber membrane was subjected to a plane, uniaxial stress ϵ_x by slightly stretching it as indicated in Figure 4.8a. It was then illuminated as indicated in Figure 4.8b and a two-exposure hologram was recorded, with a plane reference wave, on a film plate that was parallel to the object surface. All but a thin strip of the hologram, parallel to the x axis, was masked off, and a real image was formed by appropriately orienting the hologram and illuminating it with a plane wave of laser light. This is equivalent to viewing through a slit aperture. As indicated in Figures 4.9a and 4.9b, the fringes are localized in a plane having the slopes

$$\frac{\partial z_{l_x}}{\partial x} \cong 1.5, \qquad \frac{\partial z_{l_x}}{\partial y} = 0.$$

The fringe spacing is $d_x = 1$ mm, so the strain is

$$\epsilon_x = \frac{\partial u}{\partial x} = \frac{\lambda}{d_x}\left(\frac{\partial z_{l_x}}{\partial x}\right)$$

$$\cong \frac{0.6328 \times 10^{-6}}{1 \times 10^{-3}}(1.5)$$

$$\epsilon_x \cong 0.9 \times 10^{-3}.$$

When this technique is used to determine rotation, the two slit apertures must be oriented so that neither is parallel to the fringes. If this condition is violated, the localization distance and the fringe spacing are both

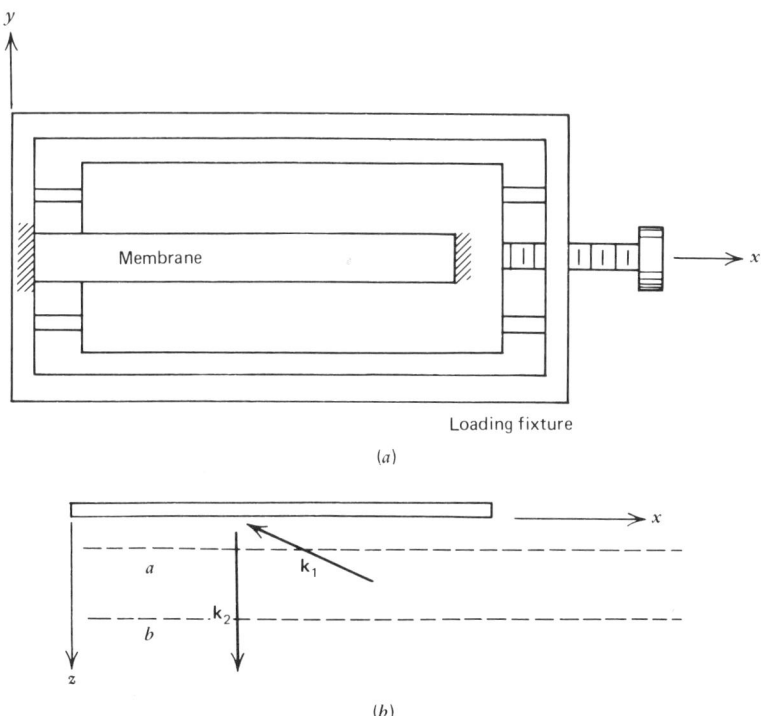

Figure 4.8 Determination of homogeneous strain ϵ_x by observation of fringe localization with a split aperture. (*a*) The membrane was stretched slightly in the *x*-direction. (*b*) Illumination and observation directions, and observation planes. Here *a* and *b* are the planes in which fringes were observed.

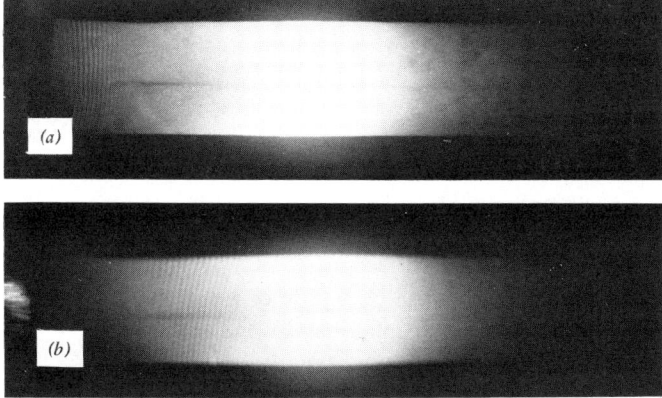

Figure 4.9 Localized fringes observed with a slit aperture parallel to the *x* direction for the configuration depicted in Figure 4.8. (*a*) Fringes in plane *a*. (*b*) Fringes in plane *b*.

infinite; although the corresponding expressions for $\partial u/\partial y$ and $\partial v/\partial x$ are still mathematically correct, they are experimentally indeterminant.

It is helpful to introduce further notation before describing methods for measuring strain and rotation under more general circumstances. Equations 4.1 to 4.3 express strains and rotations in terms of derivatives of displacement. It is convenient to arrange these derivatives into a matrix $[f_{ij}]$ which we term the *deformation gradient matrix*:

$$[f_{ij}] = \begin{bmatrix} \dfrac{\partial u}{\partial x} & \dfrac{\partial v}{\partial x} & \dfrac{\partial w}{\partial x} \\ \dfrac{\partial u}{\partial y} & \dfrac{\partial v}{\partial y} & \dfrac{\partial w}{\partial y} \\ \dfrac{\partial u}{\partial z} & \dfrac{\partial v}{\partial z} & \dfrac{\partial w}{\partial z} \end{bmatrix}. \qquad (4.46)$$

Similarly, the strains and rotations can be arranged into matrices $[e_{ij}]$ and $[\theta_{ij}]$, respectively:

$$[e_{ij}] = \begin{bmatrix} \epsilon_x & \tfrac{1}{2}\gamma_{xy} & \tfrac{1}{2}\gamma_{xz} \\ \tfrac{1}{2}\gamma_{xy} & \epsilon_y & \tfrac{1}{2}\gamma_{yz} \\ \tfrac{1}{2}\gamma_{xz} & \tfrac{1}{2}\gamma_{yz} & \epsilon_z \end{bmatrix}, \qquad (4.47)$$

$$[\theta_{ij}] = \begin{bmatrix} 0 & \omega_z & -\omega_y \\ -\omega_z & 0 & \omega_x \\ \omega_y & -\omega_x & 0 \end{bmatrix}. \qquad (4.48)$$

By examining equations 4.1 to 4.3 and 4.46 to 4.48, it can be seen that, once $[f_{ij}]$ is determined, the strain matrix $[e_{ij}]$ and the rotation matrix $[\theta_{ij}]$ can be obtained by elementary numerical operations:

$$[e_{ij}] = \tfrac{1}{2}\{[f_{ij}] + [f_{ij}]^T\}, \qquad (4.49)$$

$$[\theta_{ij}] = \tfrac{1}{2}\{[f_{ij}] - [f_{ij}]^T\}. \qquad (4.50)$$

The superscript T denotes the transpose of a matrix, which is obtained by interchanging its rows and columns. In the rest of this section we will discuss methods by which the elements of $[f_{ij}]$ can be determined by holographic interferometry. Once this has been done, the principal strains and axes can be determined by a well-known eigenvalue computation.[24]

4.2 MEASUREMENT OF STRAIN, STRESS, AND BENDING MOMENTS

In Chapter 3 it was shown that ρ, the distance along observation direction $\hat{\mathbf{k}}_2$ from the object surface, $z=0$, to a region of localization is

$$\rho = \frac{L_x - k_{2x}\hat{\mathbf{k}}_2 \cdot \mathbf{L}}{(\lambda/2\pi)\hat{\mathbf{i}} \cdot \nabla_n \delta} \quad (4.51)$$

or

$$\rho = \frac{L_y - k_{2y}\hat{\mathbf{k}}_2 \cdot \mathbf{L}}{(\lambda/2\pi)\hat{\mathbf{j}} \cdot \nabla_n \delta} \quad (4.52)$$

Equation 3.63 could also be used, but it is not independent of these two equations. Equations 4.51 and 4.52 are the basis of a scheme for determining the elements of the deformation gradient matrix $[f_{ij}]$. First the object point of interest is viewed from three or more different directions with a small-aperture viewing system, and the displacements L_x, L_y, and L_z are determined by the methods discussed in Chapter 2. Next, the object point is viewed with a large-aperture system along some direction $\hat{\mathbf{k}}_2$ for which fringes are observed to localize, and the distance ρ is measured. Using the measured values of $\mathbf{L}$ and ρ, we can write equations 4.51 and 4.52 as

$$k_x\left(\frac{\partial u}{\partial x}\right) + k_y\left(\frac{\partial v}{\partial x}\right) + k_z\left(\frac{\partial w}{\partial x}\right) = \frac{L_x - k_{2x}\hat{\mathbf{k}}_2 \cdot \mathbf{L}}{\rho} \quad (4.53)$$

$$k_x\left(\frac{\partial u}{\partial y}\right) + k_y\left(\frac{\partial v}{\partial y}\right) + k_z\left(\frac{\partial w}{\partial y}\right) = \frac{L_y - k_{2y}\hat{\mathbf{k}}_2 \cdot \mathbf{L}}{\rho}. \quad (4.54)$$

Since the right-hand sides of equations 4.53 and 4.54 are known, they constitute two linear equations in six unknowns. By repeating the procedure for two more sensitivity vectors, six linear equations can be formulated and solved for the six derivatives of displacement which appear in the first two rows of $[f_{ij}]$. Then the in-plane strains, which comprise the first two rows of $[e_{ij}]$, can be determined, as can be the surface rotations ω_x, ω_y, and ω_z. As noted in Section 4.2, the out-of-plane rotations of the *object surface*, $z=0$, reduce to $\omega_x = \partial w/\partial y$ and $\omega_y = -(\partial w/\partial x)$.

To implement this scheme, it must be noted that the viewing direction $\hat{\mathbf{k}}_2$ cannot be chosen arbitrarily. *When an object point is viewed along some direction, localization will occur only if the apparent displacement $\mathbf{L}_{ob}$ is normal to the fringes observed from that direction.* This result, which was derived in Section 3.5, is illustrated in Figure 3.11. This figure shows a rigid object that rotated about the z axis, which is normal to its surface ($\omega_x = 0, \omega_y = 0, \omega_z = \theta$). It was also viewed along a direction normal to the surface ($\hat{\mathbf{k}}_2 = \hat{\mathbf{k}}$). With this viewing direction, localized fringes can be observed only for object points which lie on the x axis. If one is interested in points above or below the x axis, where **L** has a different orientation, no localization can be observed from this direction. In general, when ρ is to be measured in order to determine strain and rotation at a point, the direction $\hat{\mathbf{k}}_2$ from which the point is viewed must be varied in three dimensions until a direction in which localization occurs is found. One measurement of ρ can then be made. Further measurements of ρ can be made only after searching for other viewing directions for which localization occurs.

Dubas and Schumann[25, 26] note that the locus of localization points traced out while viewing a specific object point from a variety of directions is a curve in space, which they call the *curve of complete localization*. (This should not be confused with the localization curve defined in Chapter 3 as the locus of points of localization when the entire object surface is viewed from a single direction.) The curve of complete localization is tangent to the object displacement **L** at the point where it intersects the object surface. In ref. 26 a viewing device is described which can be used to make localization measurements in a systematic manner, and experimental determinations of strain and rotation are reported. Although rather complex, the method yields good accuracy.

When an object deforms homogeneously, perhaps in combination with a rigid rotation and translation, the deformation gradient matrix $[f_{ij}]$ can be determined conveniently by a method based on the use of the *fringe vector*, a concept introduced by Stetson.[27-32] As noted in Chapter 3, it is sometimes useful to think of the fringes of holographic interferometry on the surface of an object observed through a small-aperture imaging system as being the intersections of the object surface with fringe surfaces, or laminae, extending outward from the object toward the observer. Indeed, we can think of a *fringe locus function* $\delta(\mathbf{r}; \mathbf{K})$ in space which is constant along surfaces in space which are parallel to the fringe laminae. The magnitude of the fringe locus function is the phase difference δ defined in equation 2.10. The position vector **r** of any point in space is measured from an origin on the object surface. The senstivity vector **K** is a parameter because all considerations of fringe formation depend upon the illumination and viewing directions. Let $\mathbf{r}_o$ denote the position vector of any point on the object surface. The fringe vector vector $\mathbf{K}_f$ is defined by the relation

$$\delta(\mathbf{r}_o; \mathbf{K}_f) = \mathbf{K}_f \cdot \mathbf{r}_o + \mathbf{K} \cdot \mathbf{L}_T, \qquad (4.55)$$

where $\mathbf{L}_T$ is the displacement of the object point chosen as the coordinate origin. The fringe vector $\mathbf{K}_f$ is perpendicular to the fringe laminae. Since $\mathbf{K} \cdot \mathbf{L}_T$ is a constant, equation 4.55 implies that the fringe laminae will be equidistant parallel planes.

4.2 MEASUREMENT OF STRAIN, STRESS, AND BENDING MOMENTS

A *homogeneous deformation* is one in which the deformation of each object element is identical. Homogeneous deformations cause spherical solids to deform into elliptical solids, cubic elements to deform into rectangular or trapezoidal parallelepipeds, and so on. Examples include prismatic bars subjected to simple tension or compression, and the expansion of solids due to uniform heating. The utility of the results derived below is enhanced by the fact that more general deformations are often approximately homogeneous over a finite region that can be studied by holographic interferometry. If an object undergoes a homogeneous deformation plus a rigid-body motion, the displacement of any object point is

$$[L_i] = [f_{ij}][r_{o_i}] + [L_{T_i}]. \tag{4.56}$$

It is convenient to introduce matrix notation, so that $[L_i]$, $i = 1, 2, 3$, is the column vector with elements L_x, L_y, and L_z, which represents the displacement $\mathbf{L}$. Also, $[L_{T_i}]$ is the rigid-body translation of the object, and $[f_{ij}]$ is the deformation gradient matrix defined by equation 4.46.

If the fringe vectors $\mathbf{K}_{f_n}$, $n = 1, 2, 3$, can be determined for three different sensitivity vectors $\mathbf{K}_n$, $n = 1, 2, 3$, they will provide sufficient information to compute the nine elements of $[f_{ij}]$. From equations 4.55 and 2.1, we have, for any sensitivity vector $\mathbf{K}$,

$$\mathbf{K}_f \cdot \mathbf{r}_o + \mathbf{K} \cdot \mathbf{L}_T = \mathbf{K} \cdot \mathbf{L}. \tag{4.57}$$

Since $\mathbf{L}_T$ is a constant, it will not enter into the determination of $[f_{ij}]$. If we combine equations 4.56 and 4.57 for each of the three sensitivity vectors, the results can be expressed in matrix form as

$$\begin{bmatrix} K_{f_{11}} & K_{f_{12}} & K_{f_{13}} \\ K_{f_{21}} & K_{f_{22}} & K_{f_{23}} \\ K_{f_{31}} & K_{f_{32}} & K_{f_{33}} \end{bmatrix} \begin{bmatrix} r_{o_1} \\ r_{o_2} \\ r_{o_3} \end{bmatrix} = \begin{bmatrix} K_{11} & K_{12} & K_{13} \\ K_{21} & K_{22} & K_{23} \\ K_{31} & K_{32} & K_{33} \end{bmatrix} \begin{bmatrix} f_{11} & f_{12} & f_{13} \\ f_{21} & f_{22} & f_{23} \\ f_{31} & f_{32} & f_{33} \end{bmatrix} \begin{bmatrix} r_{o_1} \\ r_{o_2} \\ r_{o_3} \end{bmatrix}$$

Clearly, the matrix of fringe vectors is equal to the product of the two square matrices on the right-hand side. In double-subscript notation,

$$[K_{f_{nj}}] = [K_{ni}][f_{ij}],$$

where the subscript n is the number of the observation, and $j = 1, 2, 3$ denotes the x, y, and z components, respectively. Solving for $[f_{ij}]$, we obtain

$$[f_{ij}] = [K_{ni}]^{-1}[K_{f_{nj}}]. \tag{4.58}$$

As with all other fringe interpretation schemes, it is possible to make four or more observations and replace equation 4.55 with a least-mean-square solution in order to suppress the amplification of experimental errors.[29]

The fringe vectors necessary to compute $[f_{ij}]$ by using equation 4.55 can be determined when the shape of the three-dimensional object which is homogeneously deformed is known. Consider the fringes as they appear on the typical object surface depicted in Figure 4.10. Here $\hat{\mathbf{k}}_f$, the unit vector in the direction of the fringe vector $\mathbf{K}_f$, must be normal to the plane defined by points a, b, and c, which lie on one fringe. Assuming that the fringe order increases in the direction from a to d, we have

$$\hat{\mathbf{k}}_f = \frac{\mathbf{r}_{ab} \times \mathbf{r}_{ac}}{|\mathbf{r}_{ab} \times \mathbf{r}_{ac}|}. \tag{4.59}$$

If the fringes are cosinusoidal, δ must increase by 2π from one fringe surface to the next, so the magnitude of the fringe vector is

$$|\mathbf{K}_f| = \frac{2\pi}{\hat{\mathbf{k}}_f \cdot \mathbf{r}_{ad}}. \tag{4.60}$$

Evaluation of equations 4.59 and 4.60 based on experimental observation determines the fringe vector for one viewing direction. After repeating the process for two additional views, $[f_{ij}]$ can be calculated by using equation 4.58, and then the strain $[e_{ij}]$ and rotation $[\theta_{ij}]$ can be separated as indicated in equations 4.49 and 4.50.

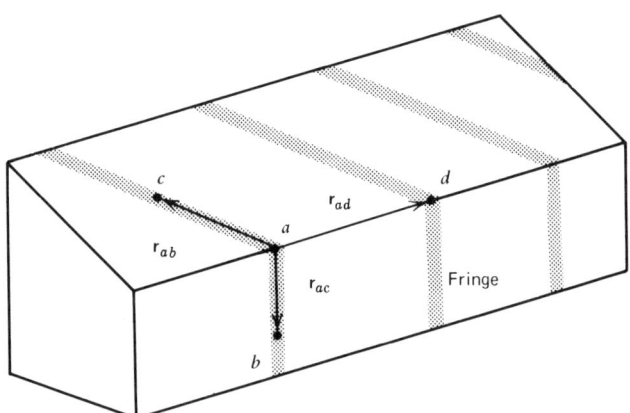

Figure 4.10 Fringes on a three-dimensional object which has deformed homogeneously (After ref. 29). Points a, b, and c define the plane of a fringe surface to which fringe vectors $\mathbf{K}_f$ are normal; r_{ad} is the fringe spacing along the surface.

This formulation has assumed that the sensitivity vector **K** is constant over the region of the object which is studied to determine $\mathbf{K}_f$. In practice, however, this condition may not be satisfied, so corrections to $\mathbf{K}_f$ to account for perspective must be introduced.[30]

Pryputniewicz[31] has used fringe vector analysis to measure carefully controlled rotations and homogeneous strains. In one set of experiments an object was given rigid rotations θ_y in the range 175 to 360 μrad. Analysis of holographic interferograms by the fringe vector method agreed with the actual θ_y to within ± 2 percent. Computed rotations θ_y and θ_z, which should be zero, were indeed very small; they were in the range 2 to 15 μrad. The various components of strain, which should also be zero, were computed as $3-122 \times 10^{-6}$. In a second set of experiments the test object was homogeneously strained in the range $60-210 \times 10^{-6}$ by uniform heating. The normal strains calculated by fringe vector analysis were within 10 to 32 percent of the accepted values, which were calculated on the basis of measured temperature differences. The shear strains and rotations, which should be zero, were calculated to be in the ranges $15-64 \times 10^{-6}$ and 7 to 59 μrad, respectively. These results indicate generally satisfactory measurements by holographic interferometry.

In this section several approaches to the evaluation of strain, rotation, and bending moments have been introduced. The experimenter must attempt to deduce which method seems to be most appropriate for a given application. Factors to be considered are size and shape of the object, homogeneity of the deformation, general nature and magnitude of strains, experimental complexity, and required accuracy. Unfortunately very little guidance in the form of comparative experiments is available in the literature. Most of the methods involve three or more experimental measurements followed by solution of a linear system of equations. Experimental restraints almost invariably dictate that the system of equations will not be well conditioned. This means that small experimental errors may lead to large errors in computed displacements and strains. It is therefore recommended that redundant measurements be made and the system be solved in the least-mean-square sense. In deciding on a method to be used, one can set up the system of equations to be solved for each method and subject them to an error analysis of the type discussed in Section 2.3.

4.3 MEASUREMENT OF MECHANICAL VIBRATIONS

One of the most interesting and useful applications of holographic interferometry is the visualization and measurement of the mechanical vibration of opaque objects. The development of holographic methods for studying vibration was initiated by Powell and Stetson, whose experimental investigation and analysis were published in 1965.[32, 33] They showed that a holographic time exposure of a vibrating surface produces an image of the surface modulated by a system of interference fringes. The brightest of these fringes coincides with the nodal region, that is, the portion of the surface that remains stationary during the vibration. The nodal pattern indicated by this bright fringe is similar to that produced by the Chladni method of sprinkling sand over the surface of a horizontal, vibrating plate. In addition to the bright nodal fringe, several other fringes may be observed in the holographic image; each is a contour of constant vibrational amplitude. If the vibrational motion has a simple time dependence, for example, sinusoidal, the vibrational amplitude of each point on the surface can be determined by straightforward analysis of these fringes. At the expense of additional experimental complexity the relative phase of the vibration of each point can also be determined.

The technique of producing an interferogram by exposing a hologram for a period of time during which the object executes a motion is termed *time-average holographic interferometry*. The hologram records essentially the time-averaged complex amplitude of light scattered by the object to the hologram plane. When an object vibrates sinusoidally, it spends most of the time near its two positions of maximum displacement, where its velocity is zero. Qualitatively, a time-average holographic interferogram of an object vibrating in this manner is like a two-exposure holographic interferogram that displays contours of the object displacement between these two extreme positions. Quantitative interpretation requires a more precise analysis. Such an analysis is presented in this section, followed by a discussion of special experimental techniques for extending or decreasing the sensitivity of vibration measurements, and for determining the relative phase of vibration across an object surface.

4.3.1 Sinusoidal Vibrations

We begin by considering a simple example related to the one which was used in Section 2.2 to introduce fringe interpretation. Consider a cantilever beam with one end fixed to a rigid block. The cantilever beam is illuminated with a plane wave traveling in the $-z$ direction, normal to its front surface (see Figure 4.11a). The beam is vibrating sinusoidally about its equilibrium position, so the displacement of each point is

$$z(x,t) = Z(x)\sin\omega t, \qquad (4.61)$$

where $Z(x)$ is the amplitude of the mechanical vibration at location x, and ω is the circular frequency of the vibration. Let $\mathbf{U}_{os}(x,y)$ represent the complex amplitude of light in the hologram plane which is scattered by the cantilever beam when it is stationary in its equilibrium position:

$$\mathbf{U}_{os}(x,y) = a(x,y)\exp[i\phi(x,y)]. \qquad (4.62)$$

4.3 MEASUREMENT OF MECHANICAL VIBRATIONS

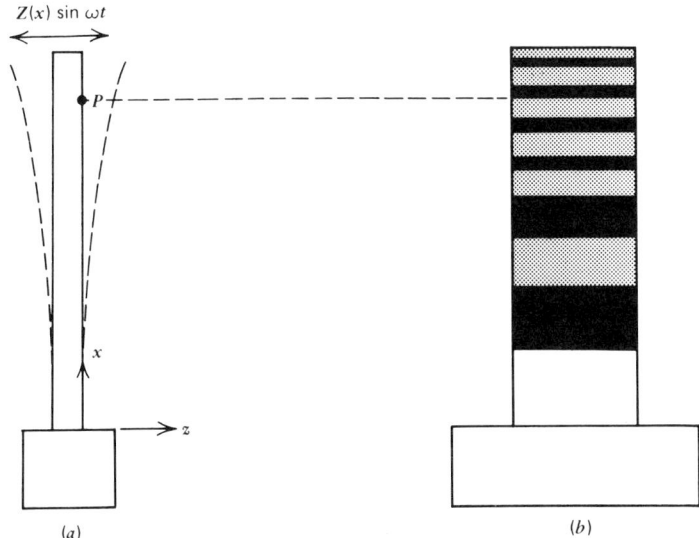

Figure 4.11 Vibrating cantilever beam. (*a*) Side view of the beam. Displacement is in the z direction, and illumination is in the $-z$ direction. (*b*) Fringe pattern observed on the front of the beam.

On the assumption of a large object-to-hologram distance, light that travels a distance l_o from the source to a point on the static object and back to the hologram will travel a distance $l_o - 2Z(x)\sin\omega t$ while the object is vibrating. The corresponding optical phase change is $\Delta\phi(x, y, t) = (2\pi/\lambda)2Z(x)\sin\omega t$; hence at any instant of time the complex amplitude of light in the hologram plane will be

$$\mathbf{U}_o(x,y,t) = a(x,y)\exp\left\{i\left[\phi(x,y) + \left(\frac{4\pi}{\lambda}\right)Z(x)\sin\omega t\right]\right\}. \quad (4.63)$$

A time-average hologram is recorded by simultaneously exposing a film plate to $\mathbf{U}_o(x,y,t)$ and to an off-axis reference wave for a period of time T. When this hologram is developed and illuminated by the reference wave, the reconstructed wave will have a complex amplitude that is proportional to the time average of $\mathbf{U}_o(x,y,t)$ over the exposure interval T:

$$\frac{1}{T}\int_0^T a(x,y)\exp\left\{i\left[\phi(x,y) + \left(\frac{4\pi}{\lambda}\right)Z(x)\sin\omega t\right]\right\}$$
$$= \mathbf{U}_{os}(x,y)\frac{1}{T}\int_0^T \exp\left[i\left(\frac{4\pi}{\lambda}\right)Z(x)\sin\omega t\right]dt.$$

(4.64)

The time-average integral in equation 4.64 is called the *characteristic function* for sinusoidal vibration and is denoted by $\mathbf{M}_T$. Thus the reconstructed complex amplitude is proportional to $\mathbf{U}_0(x,y) \cdot \mathbf{M}_T$, and the irradiance is proportional to

$$I(x,y) = |\mathbf{U}_0|^2 |\mathbf{M}_T|^2 = a^2(x,y)|\mathbf{M}_T|^2. \tag{4.65}$$

The characteristic function for the sinusoidal motion considered here can be evaluated easily. If the exposure time is long compared to the vibrational period, $T \gg 1/\omega$, we have

$$\begin{aligned}\mathbf{M}_T &\equiv \lim_{T \to \infty} \frac{1}{T} \int_0^T \exp\left[i\left(\frac{4\pi}{\lambda}\right) Z(x) \sin \omega t\right] dt \\ &= J_0\left[\left(\frac{4\pi}{\lambda}\right) Z(x)\right],\end{aligned} \tag{4.66}$$

where J_0 is the Bessel function of the first kind of order zero. The corresponding irradiance is proportional to

$$I(x,y) = a^2(x,y) J_0^2\left[\left(\frac{4\pi}{\lambda}\right) Z(x)\right]. \tag{4.67}$$

Equation 4.67 indicates that the virtual image is modulated by a system of fringes described by the square of the zero-order Bessel function, which is shown in Figure 4.12. (Fringes of similar form can be observed by classical interferometry when the vibrating surface is flat and highly polished.[34]) Dark fringes will be centered at each point on the object surface where the amplitude of vibration $Z(x)$ is such that the Bessel function in equations 4.66 and 4.67 is zero. Table 4.1 includes the first 20 zeros of J_0. Higher order zeros are nearly equally spaced and are given by the following asymptotic formula:

$$\xi_n = \left(n - \tfrac{1}{4}\right)\pi + \tfrac{1}{8}\left[\left(n - \tfrac{1}{4}\right)\pi\right]^{-1}. \tag{4.68}$$

Figure 4.11*b* is a sketch of the fringe pattern obtained by time-average holographic interferometry of a cantilever beam. Point P in Figure 4.11*a* lies at the center of the fifth dark fringe. The vibrational amplitude at P can be determined with the aid of Table 4.1:

$$\left(\frac{4\pi}{\lambda}\right) Z(P) = 14.9309,$$

so $Z(P) = 14.9309(\lambda/4\pi) = 1.188\lambda$. The amplitude at any other point can be determined in the same manner.

4.3 MEASUREMENT OF MECHANICAL VIBRATIONS

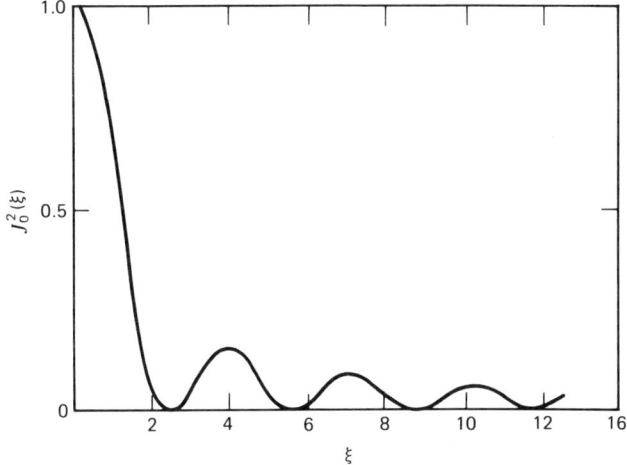

Figure 4.12 Plot of $J_0^2(\xi)$, which is the fringe function for time-average holographic interferometry of sinusoidal vibrations.

The fringes formed in time-average holography of sinusoidal motions have an irradiance described by J_0^2. These fringes differ from the cosinusoidal fringes characteristic of two-exposure holographic interferometry in that their brightness and, to some extent, their spacing decrease with increasing order. *The nodal, or zero-order, fringe is easily identified because it is much brighter than the other fringes.* This is illustrated in Figure 4.13, which is a photograph of an interferogram of a rectangular plate with clamped edges which is vibrating sinusoidally. The nodal area around the periphery of the plate is apparent by its brightness. The decreasing brightness of higher order fringes is also apparent in this figure. This may be contrasted with the cosinusoidal fringes of uniform brightness

TABLE 4.1 ZEROS OF THE BESSEL FUNCTION J_0

n	ξ_n	n	ξ_n
1	2.4048	11	33.7758
2	5.5200	12	36.9170
3	8.6537	13	40.0584
4	11.7915	14	43.1997
5	14.9309	15	46.3411
6	18.0710	16	49.4826
7	21.2116	17	52.6240
8	24.3524	18	55.7655
9	27.4934	19	58.9069
10	30.6346	20	62.0484

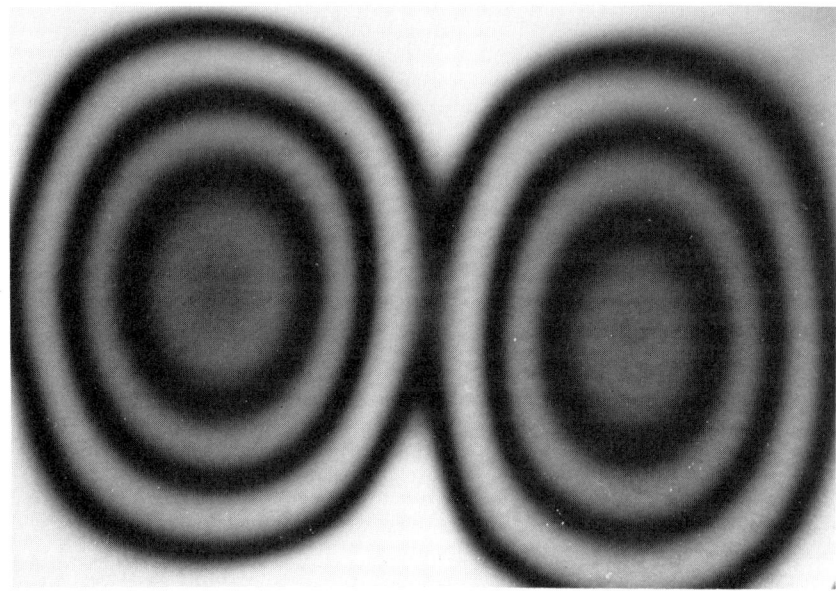

Figure 4.13 Photograph of a time-average holographic interferogram. The object is a rectangular plate made of opaque glass, 1 mm thick, with dimensions 10×13 cm. It was excited to resonant vibration by a sinusoidal acoustic source at a frequency of 1800 Hz.

in Figure 4.7a, which is a two-exposure interferogram of a statically deformed rectangular plate.

The decreasing brightness of J_0^2 fringes makes them difficult to view and photograph when vibrational amplitudes are relatively large. This problem can be minimized, however, by proper care in recording the hologram and in photographing the interferogram. Care should be taken to produce a clean, efficient hologram. Generally, bleaching should not be used to increase efficiency because bleached holograms tend to scatter light and therefore degrade the signal/noise ratio in the reconstructed image. Good results have been reported, however, when a ferricyanide bleach was used.[35] The clarity of higher order fringes appears to be maximized if a ratio of reference beam to object beam irradiance of approximately 1:1 is used when the hologram is recorded.[35] Dramatic increases in the relative brightness of higher order fringes can be attained when the fringe pattern is photographed. To accomplish this, the photograph should be significantly over exposed. Watrasiewicz[36] has demonstrated experimentally that the brightness of fringes in such a photograph declines much less rapidly than J_0^2. Wall[37] has demonstrated that with care approximately 100 fringes can be discerned over an object length of 15 cm in a time-average interferogram. With appropriate beam ratios, exposures, and photographic procedures, it appears that the upper limit to measurable vibrational amplitudes is reached when fringe spacings become comparable to characteristic speckle size or when optical correlation of the scattered wavefronts is lost. These conditions are the same as those which apply in two-exposure interferometry.

4.3.2 Separable Motions

Having introduced the topic of time-average holographic interferometry in terms of simple sinusoidal vibrations, we now proceed to a more general definition of the characteristic function $\mathbf{M}_T$ and to a discussion of its evaluation, interpretation, and application to more complicated motions. The first motions to be considered are those for which the vector displacement of each point on the object surface can be expressed as the product of a displacement vector $\mathbf{L}$ with a function of time $f(t)$.

If all points on the object move according to a single time function $f(t)$, the characteristic function can be written as

$$\mathbf{M}_T(\Delta\phi_A) = \frac{1}{T}\int_0^T \exp\{i[\Delta\phi_A(x,y)\cdot f(t)]\}\, dt, \qquad (4.69)$$

where $\Delta\phi_A$ is the optical phase shift at the hologram plane corresponding to the amplitude of the object displacement. The fringe pattern observed when any point on the object surface is viewed in the holographic virtual image correspondingly is given by

$$I = |\mathbf{M}_T(\mathbf{K}\cdot\mathbf{A})|^2 = \left|\frac{1}{T}\int_0^T \exp[i\mathbf{K}\cdot\mathbf{A}f(t)]\, dt\right|^2, \qquad (4.70)$$

where $\mathbf{K}$ is the sensitivity vector defined in equation 2.9, and $\mathbf{A}$ is the *vector amplitude* of the motion of the observed object point.

Equation 4.70 is quite general and can be applied to any time function, whether or not it is oscillatory. For example, two-exposure holographic interferometry of a static displacement is described by $\mathbf{M}_T$ evaluated for the time function $f(t)$, defined by

$$f(t) = \begin{cases} 0, & 0 \leqslant t < \dfrac{T}{2}, \\ 1, & \dfrac{T}{2} \leqslant t \leqslant T. \end{cases} \qquad (4.71)$$

For this case,

$$\mathbf{M}_T = \frac{1}{T}\int_0^{T/2} dt + \frac{1}{T}\int_{T/2}^T \exp(i\mathbf{K}\cdot\mathbf{A})\,dt$$
$$= \tfrac{1}{2}[1+\exp(i\mathbf{K}\cdot\mathbf{A})],$$

so the interference fringes seen by an observer are the familiar cosinusoidal fringes:

$$I = |\mathbf{M}_T|^2 = \tfrac{1}{2}[1+\cos(\mathbf{K}\cdot\mathbf{A})]$$
$$= \cos^2(\tfrac{1}{2}\mathbf{K}\cdot\mathbf{A}). \qquad (4.72)$$

Characteristic functions can be evaluated for a variety of motions. Table 4.2 indicates several time functions of physical interest which have been investigated, and indicates references where further details can be found. Some of the characteristic functions tabulated in Table 4.2 have quite complicated forms and would rarely be used for direct quantitative evaluation of interferograms; however, a knowledge of them is useful for qualita-

TABLE 4.2 CHARACTERISTIC FUNCTIONS FOR VARIOUS SEPARABLE, SINGLE TIME FUNCTIONS

Two-exposure (static displacement)	$f(t) = \begin{cases} 0, & 0 \leq t < \frac{T}{2} \\ 1, & \frac{T}{2} \leq t \leq T \end{cases}$	$\|\mathbf{M}_T\|^2 = \cos^2(\tfrac{1}{2}\mathbf{K}\cdot\mathbf{A})$	
Constant velocity	$f(t) = \dfrac{t}{T}$	$\|\mathbf{M}_T\|^2 = \dfrac{\sin^2(\tfrac{1}{2}\mathbf{K}\cdot\mathbf{A})}{(\tfrac{1}{2}\mathbf{K}\cdot\mathbf{A})^2}$	Refs. 38, 39
Constant acceleration from rest[a]	$f(t) = \left(\dfrac{t}{T}\right)^2$	$\|\mathbf{M}_T\|^2 = \dfrac{C^2(\sqrt{\mathbf{K}\cdot\mathbf{A}}) + S^2(\sqrt{\mathbf{K}\cdot\mathbf{A}})}{(2/\pi)\mathbf{K}\cdot\mathbf{A}}$	Ref. 40
Harmonic vibration	$f(t) = \sin\omega t$	$\|\mathbf{M}_T\|^2 = J_0^2(\mathbf{K}\cdot\mathbf{A})$	Ref. 33
Exponentially damped harmonic vibration	$f(t) = e^{-\beta t}\sin\omega t$	Graphically presented	Ref. 41
Nonlinear vibration[b]	$f(t) = c_n\left(\dfrac{\omega t}{m}\right)$	Graphically presented	Ref. 42
Nonlinear vibration[b]	$f(t) = s_n\left(\dfrac{\omega t}{m}\right)$	Graphically presented	Ref. 43
Nonlinear vibration[b]	$f(t) = s_n^2\left(\dfrac{\omega t}{m}\right)$	Graphically presented	Ref. 45

[a] C, S are the Fresnel integrals.
[b] c_n, s_n are the Jacobian elliptic functions.

4.3 MEASUREMENT OF MECHANICAL VIBRATIONS

tive assessment of interferograms. For example, consider a surface which is deformed so that each point moves with a constant velocity. This velocity may differ from point to point. From Table 4.2 we see that the resulting fringe pattern is described by $\sin^2(\frac{1}{2}\mathbf{K}\cdot\mathbf{A})/(\frac{1}{2}\mathbf{K}\cdot\mathbf{A})^2$. These fringes are similar to the J_0^2 fringes in that they have a decreasing brightness as the amplitude of the motion increases. On the other hand, if each point on a surface responds to a force by moving with a constant acceleration, the irradiance declines rapidly with increasing amplitude and is modulated only by barely discernible fringes. Their contrast is quite low everywhere, that is, no black fringes occur.

Figures 4.14 a and b show plots of irradiance as a function of vibrational amplitude for two nonsinusoidal vibrations. These vibrations are described by the Jacobian elliptic functions $c_n(\omega t/m)$ and $s_n(\omega t/m)$, which are solutions of Duffing's equation,[44] a common analytical model for nonlinear vibrations. Here c_n describes the vibration of a *stiff spring*, that is, one which responds to a displacement z with a restoring force $P(z)$ that increases with displacement at a rate higher than linear; in particular, $P(z) = -(\alpha z + \gamma z^3)$, where α and γ are positive constants. On the other hand, s_n describes the vibration of a *soft spring*, that is, one with a restoring force that increases at a rate lower than linear; in particular, $P(z) = -(\alpha z - \gamma z^3)$. In the analysis used to generate the curves in Figure 4.14[42, 43] these restoring forces were expressed as

$$P(z) = -\omega^2 \left[(1-2m)z + \left(\frac{2m}{A^2}\right)z^3 \right] \quad (4.73)$$

and

$$P(z) = -\omega^2 \left[(1+m)z - \left(\frac{2m}{A^2}\right)z^3 \right]. \quad (4.74)$$

Here m is a parameter which describes the degree of nonlinearity; $m=0$ corresponds to a linear system executing a sinusoidal vibration. From the curves in Figures 4.14a and b, it is seen that the brightness of higher order fringes is decreased relative to that for a sinusoidal vibration if the spring is stiff. On the other hand, the brightness is increased relative to that for a sinusoidal vibration if the spring is soft. Stetson[46] reached this general conclusion by a different analysis. It should be noted that this effect is a rather subtle one.

The variation of fringe brightness with spring stiffness has a simple physical explanation. When the spring is soft, the object dwells a relatively long time near its positions of maximum displacement. It spends a greater fraction of its vibrational period close to these two positions than does a

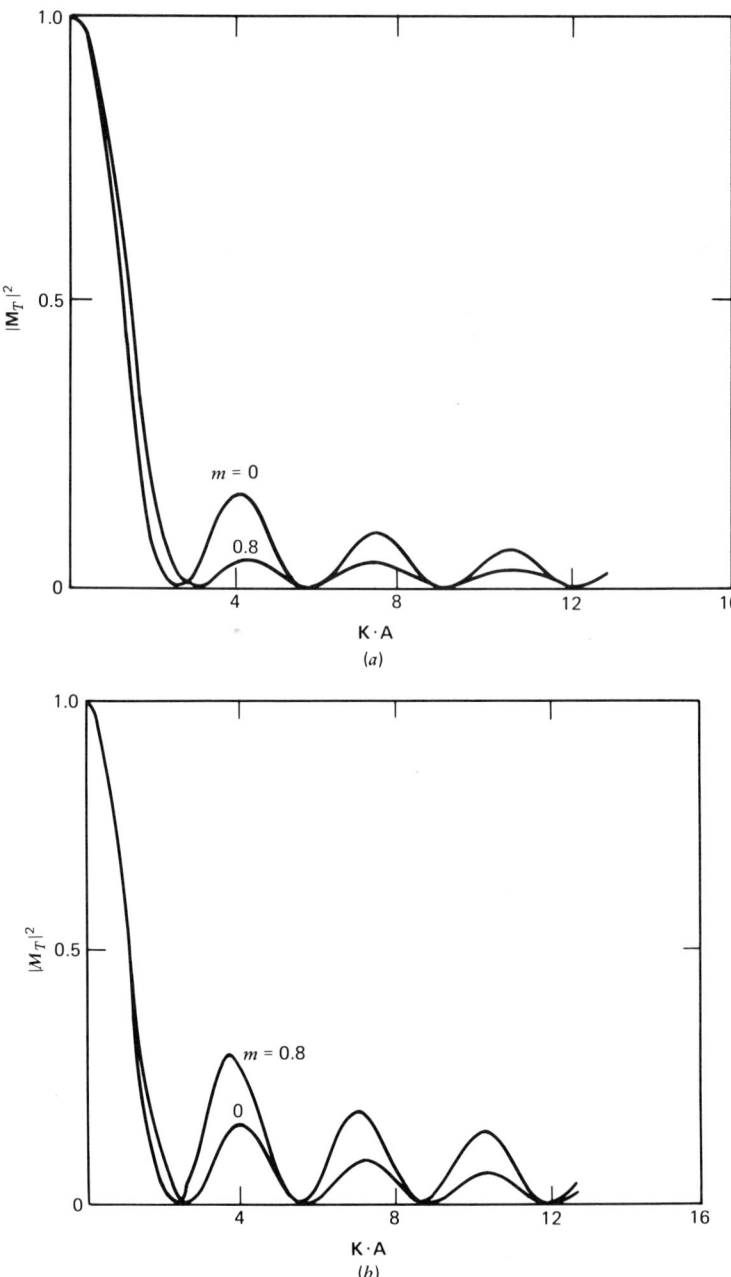

Figure 4.14 Irradiance of fringes due to nonlinear vibrations (Based on refs. 42 and 43). (a) $f(t) = c_n(\omega t/m)$, which is a model of a hard spring. (b) $f(t) = s_n(\omega t/m)$, which is a model of a soft spring.

4.3 MEASUREMENT OF MECHANICAL VIBRATIONS

sinusoidally vibrating object. Constructive interference of light scattered by the surface in these two positions therefore contributes heavily to fringe brightness. When the spring is stiff, the object is decelerated more uniformly during its motion and dwells near the positions of maximum displacement for a smaller fraction of its period than it would if the vibration were sinusoidal. The resulting fringe brightness is low. Clearly, what is important in determining the relative brightness of the fringes in each order is the fraction of the exposure time that the object spends near each location in its displacement interval. Thus the $\cos^2$ fringes of two-exposure interferometry have a uniform maximum brightness in all orders. All other motions yield decreasing fringe brightness with increasing amplitude. This decrease is most rapid for sinc^2 fringes, which occur when each object point moves with constant velocity, because the object spends equal fractions of the exposure time near each location in its displacement interval.

The discussion in the preceding paragraph suggests an alternative interpretation of $\mathbf{M}_T$.[33, 47] The optical wave reconstructed with a time-average hologram is the ensemble average of the optical waves scattered by the object in each position it occupied during exposure. If $f(t)$ is stationary and ergodic, we can introduce a *probability density function* $p(f)$ and rewrite the integral in equation 4.70 as

$$\mathbf{M}_T = \int_{-\infty}^{\infty} p(f)\exp(i\Omega f)\,df, \tag{4.75}$$

where $\Omega \equiv \mathbf{K} \cdot \mathbf{A}$. The integral in equation 4.75 is the Fourier transform of $p(f)$ with respect to Ω:

$$\mathbf{M}_T = \mathcal{F}_\Omega\{p(f)\}; \tag{4.76}$$

this quantity is referred to as the characteristic function of $f(t)$ in probability theory.[48] As an example, if $f(t)$ is random and obeys Gaussian statistics, $p(f) = \exp(-\pi\sigma^2 f^2)$, the characteristic function is $\sigma^{-2}\exp(-\pi\Omega^2/\sigma^2)$. This same interpretation is useful for deterministic time functions $f(t)$ in that tables of Fourier transforms can be used to evaluate $\mathbf{M}_T$. In addition to its computational usefulness, the interpretation of $\mathbf{M}_T$ as a characteristic function in the statistical sense gives physical insight into the process of holographic interferometry. In particular, the simultaneous reconstruction of sequentially recorded wavefronts is illustrative of the ergodic theorem.

4.3.3 Multiple Modes and Nonseparable Motions

In 1967 Barnett[49] presented a sequence of time-average interferograms of a vibrating circular plate with clamped edges. He noted that some of these interferograms appeared to display nodal patterns that were not in agreement with the classical theory of vibrating plates. Molin and Stetson[50] demonstrated experimentally that such patterns actually represented combinations of classical modes. Subsequent analysis and experimentation by Stetson[51-53] and by Wilson[54-56] and their co-workers have produced a detailed description of the formation of fringes in time-average holographic interferometry of objects executing complex motions, particularly multiple vibrational modes. Some of the conclusions they have reached are quite useful in the evaluation of interferograms and will be summarized here.

Let us consider an object surface which executes a combination of N individual, separable motions, so that the characteristic function is

$$M_T = \frac{1}{T} \int_0^T \exp\left[i \sum_{n=1}^{N} \Omega_n f_n(t)\right] dt, \qquad (4.77)$$

where $\Omega_n = \mathbf{K} \cdot \mathbf{A}_n$. An example of such a motion might be that of a plate vibrating simultaneously in two different modes:

$$A_1 \cos(\omega_1 t + \theta_1) + A_2 \cos(\omega_2 t + \theta_2). \qquad (4.78)$$

If the two motions are *temporally independent*, the characteristic function is simply the product of the characteristic functions of the individual motions:

$$M_T = \prod_{n=1}^{N} M_{T_n}. \qquad (4.79)$$

This result was derived by Molin and Stetson[50] and by Wilson for the particular case of multiple sinusoidal vibrational modes.[54] In fact, the result expressed by equation 4.79 is true, in general, for any combination of statistically independent, separable motions.[51] This is apparent if we interpret M_T as the characteristic function of the probability density function of the motion, as in equation 4.75 or 4.76. Then M_T is the characteristic function of the joint probability density function of the N component motions. Equation 4.79 then follows from the general result in probability theory that the characteristic function of the joint probability density function of statistically independent random variables is the product of their individual characteristic functions.[57]

Since M_T is a product for independent motions, a dark fringe will appear everywhere that a dark fringe would result from any one of the individual motions. Dark fringes will therefore be continuous, but bright

4.3 MEASUREMENT OF MECHANICAL VIBRATIONS

fringes will tend to be restricted to rather small regions. As an example of an independent motion, consider an object which vibrates simultaneously in two *irrationally related modes*, for example, the motion described by equation 4.78 when ω_1/ω_2 is *not* a ratio of two integers. The characteristic function is then

$$M_T = J_0(\Omega_1)J_0(\Omega_2), \tag{4.80}$$

where $\Omega_1 = \mathbf{K} \cdot \mathbf{A}_1$ and $\Omega_2 = \mathbf{K} \cdot \mathbf{A}_2$. Such a fringe pattern is shown in Figure 4.15. Figures 4.15a and 4.15b represent two individual, irrationally related modes. Figure 4.15c is the fringe pattern that results when the object vibrates in both of these modes simultaneously. Note the similarity to the pattern that would result if a transparency of Figure 4.15a was laid on top of a transparency of Figure 4.15b. Similar results have been obtained experimentally by Wilson and Strope[55] and via computer analysis by Wilson.[56]

A further useful result is that the fringe system corresponding to each independent motion is localized independently. This has been demonstrated by Molin and Stetson[52] in experiments involving rigid-body rotations about two non intersecting axes. Independent localization aids in qualitative analysis of motions and can sometimes be used to separate fringes due to extraneous motions.

If an object executes two or more *temporally dependent motions*, the characteristic function is more complicated and analysis of fringes may be quite difficult. As an example of a temporally dependent motion, consider an object which vibrates simultaneously in two *rationally related modes*, for example, the motion described by equation 4.78 when ω_1/ω_2 is the ratio of two integers.

Before proceeding to a general analysis, let us consider a special case of practical importance, namely, two simultaneous modes at a single frequency ω. Although the two modes have the same frequency, they can differ in geometrical structure, amplitude, and phase:

$$A_1 \cos \omega t + A_2 \cos(\omega t + \theta). \tag{4.81}$$

Here we let $\theta_1 = 0$ for convenience, and $\theta_2 = \theta$ is the relative temporal phase between the two modes. The characteristic function for this motion is

$$M_T = \frac{1}{T}\int_0^T \exp\left\{i[\Omega_1 \cos \omega t + \Omega_2 \cos(\omega t + \theta)]\right\} dt, \tag{4.82}$$

where $\Omega_1 = \mathbf{K} \cdot \mathbf{A}_1$ and $\Omega_2 = \mathbf{K} \cdot \mathbf{A}_2$. The quantity in the square brackets in

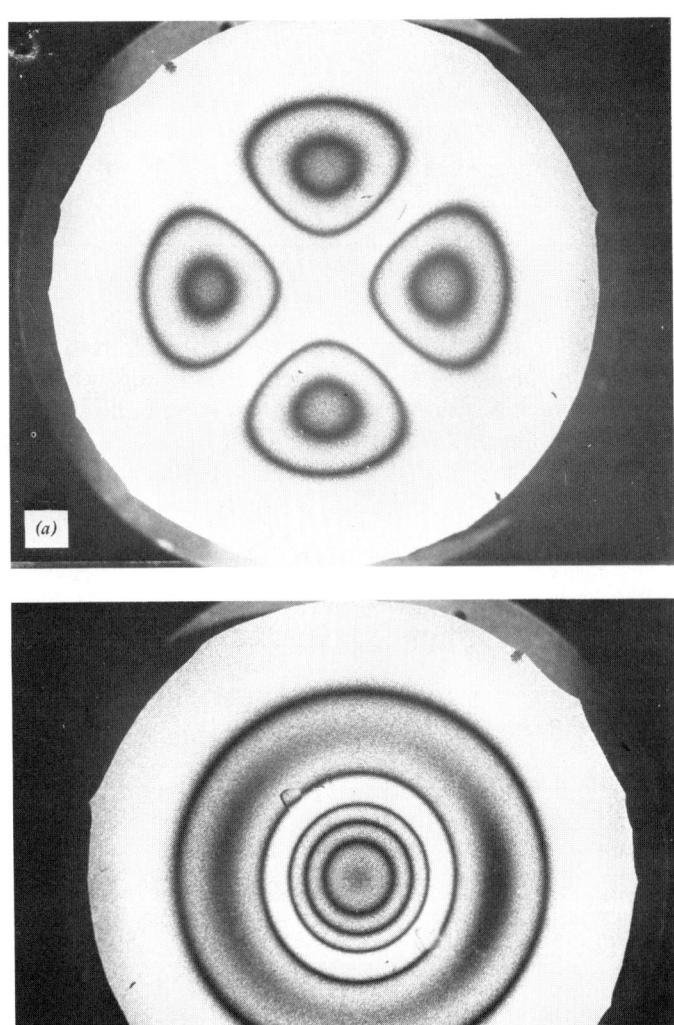

Figure 4.15 Interferograms of irrationally related modes of a circular plate (ref. 50). (*a*) Interferogram of plate excited at 2040 Hz. (*b*) Interferogram of plate excited at 2380 Hz, (*c*) Interferogram of plate excited simultaneously at 2040 Hz and 2380 Hz. (Photographs copyrighted by the Institute of Physics.)

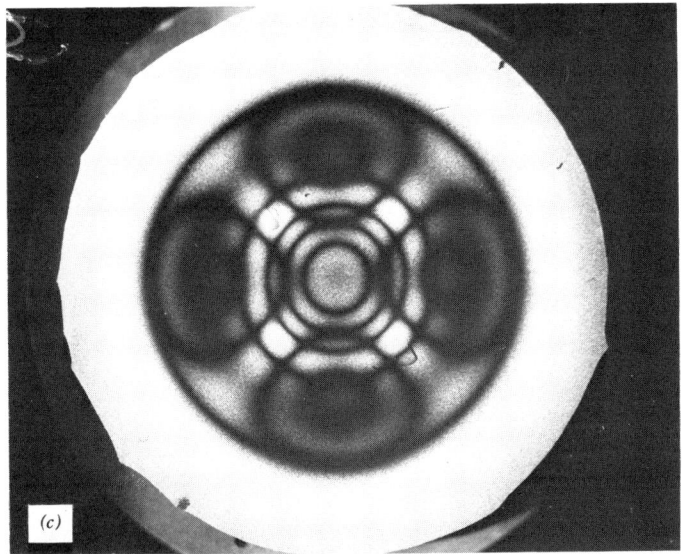

Figure 4.15 Cont.

equation 4.82 can be written as

$$\mathrm{Re}\{\Omega_1 \exp(i\omega t) + \Omega_2 \exp[i(\omega t + \theta)]\}$$
$$= \mathrm{Re}\{[\Omega_1 + \Omega_2 \exp(i\theta)]\exp(i\omega t)\}$$
$$= \mathrm{Re}\{[(\Omega_1 + \Omega_2 \cos\theta)^2 + (\Omega_2 \sin\theta)^2]^{1/2} \exp(i\Delta)\exp(i\omega t)\}$$
$$= (\Omega_1^2 + \Omega_2^2 + 2\Omega_1\Omega_2 \cos\theta)^{1/2} \cos(\omega t + \Delta), \tag{4.83}$$

where $\Delta \equiv \tan^{-1}[(\Omega_2 \sin\theta)/(\Omega_1 + \Omega_2 \cos\theta)]$. Hence the characteristic function is

$$M_T = \frac{1}{T}\int_0^T \exp\left\{i[(\Omega_1 + \Omega_2 \cos\theta)^2 + (\Omega_2 \sin\theta)^2]^{1/2} \cos(\omega t + \Delta)\right\},$$

$$M_T = J_0\left\{[(\Omega_1 + \Omega_2 \cos\theta)^2 + (\Omega_2 \sin\theta)^2]^{1/2}\right\}. \tag{4.84}$$

For this type of vibration, nodal regions, which are indicated by fringes of maximum brightness, occur when both Ω_1 and Ω_2 are zero. Nodal regions other than those near clamped edges tend to shrink to points which are the intersections of nodal curves of the component modes. If each nodal region of the individual modes is an approximately straight line, the fringe contours, which correspond to constant values of the argument in equation 4.84, will be concentric ellipses. Fringes of this type are shown in Figure 4.16. If the amplitudes of the individual modes and the orientation of their

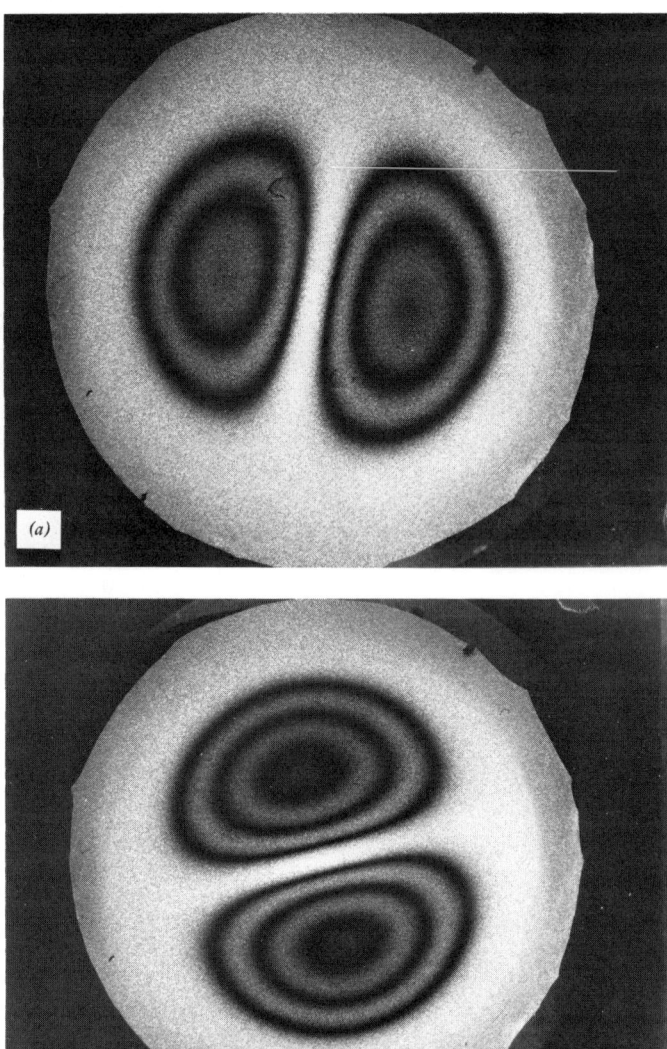

Figure 4.16 Interference patterns illustrating simultaneous vibration in two different modes at the same frequency (ref. 50). (a) and (b) are two modes of a clamped circular plate vibrating with a frequency of 1240 Hz. (c) The plate vibrated simultaneously in the two modes shown in (a) and (b), separated in phase by approximately 90°. (Photographs copyrighted by the Institute of Physics.)

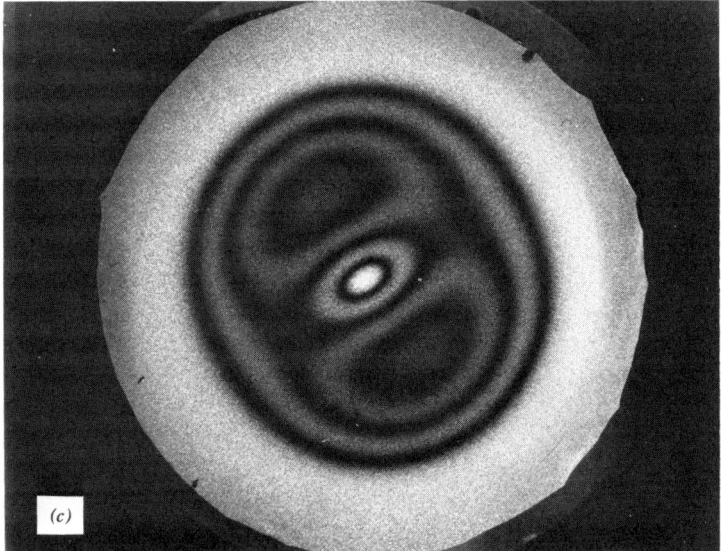

Figure 4.16 Cont.

nodal lines are known a priori, their relative phase θ can be computed from the orientation of the elliptical fringes near the node.[56]

As suggested by this example, the fringes due to simultaneous vibration in any number N of sinusoidal modes at a single frequency are described by the characteristic function

$$\mathbf{M}_T = J_0\left(\left|\sum_{n=1}^{N} \Omega_n\right|\right). \tag{4.85}$$

Here $\Sigma \Omega_n$ represents the *phasor sum* of the component vibrations; that is, the Ω's must be represented as vectors with angular orientations equal to their relative temporal phase angles θ_n. They must be added vectorially.

Now we return to a more general analysis. If an object vibrates simultaneously in N sinusoidal modes of different rationally related frequencies, the characteristic function is

$$\mathbf{M}_T = \frac{1}{T}\int_0^T \exp\left[i\sum_{n=1}^{N} \Omega_n \cos(\omega_n t + \theta_n)\right]. \tag{4.86}$$

Evaluation of $\mathbf{M}_T$ can be simplified by using the following mathematical identity:[50]

$$\exp(iz\cos\phi) = \sum_{n=-\infty}^{\infty} (i)^n J_n(z)\exp(in\phi), \quad (4.87)$$

so that the integrand in equation 4.86 becomes

$$\exp\left[i\sum_{n=1}^{N}\Omega_n\cos(\omega_n t+\theta_n)\right]$$

$$= \prod_{n=1}^{N}\sum_{m=-\infty}^{\infty} (i)^m J_m(\Omega_n)\exp[im(\omega_n t+\theta_n)].$$

When this expression is used, the integral in equation 4.86 can be evaluated. Many terms will integrate to zero; those which remain yield the characteristic function

$$M_T = \sum_{m=1}^{M}\prod_{n=1}^{N} (i)^{l_{nm}} J_{l_{nm}}(\Omega_n)\exp(il_{nm}\theta_n), \quad (4.88)$$

where l_{nm} denotes the $M \times N$ integer roots of the M sets of algebraic equations[43,46]

$$\sum_{n=1}^{N} l_{nm}\omega_n = 0. \quad (4.89)$$

For the general case of N dependent, separable motions described by the time functions $f_n(t)$, the characteristic function is

$$M_T = \sum_{m_1=-\infty}^{\infty}\sum_{m_2=-\infty}^{\infty}\cdots\sum_{m_N=-\infty}^{\infty} A_{m_1 m_2\cdots m_N}(i)^{\sum_{n=1}^{N} m_n}\prod_{n=1}^{N} J_{m_n}(\Omega_n), \quad (4.90)$$

where the coefficients $A_{m_1 m_2\cdots m_N}$ are given by

$$A_{m_1 m_2\cdots m_N} = \frac{1}{T}\int_0^T \prod_{n=1}^{N}\left[f_n t(f_n^2-1)^{1/2}\right]^{m_n} dt. \quad (4.91)$$

Equations 4.88 and 4.90 indicate that the characteristic function for temporally dependent modes, such as vibration at rationally related frequencies, is expressed as an infinite series each term of which is a product of Bessel functions. Although rather complicated, characteristic

4.3 MEASUREMENT OF MECHANICAL VIBRATIONS

functions of this type can be evaluated using a digital computer.[56] The results of such computations have been verified experimentally.[55]

The physical significance of equations 4.88 and 4.90 is not readily apparent; however, Stetson[53] has worked out an approximate characteristic function for temporally dependent motions in which the physics of the fringe formation is more obvious. This approximation is based on an asymptotic analysis using the *method of stationary phase*. To introduce this technique, we consider the approximate computation of M_T for a simple sinusoidal vibration when Ω is large:

$$M_T = \frac{1}{T}\int_0^T \exp(i\Omega\cos\omega t)\,dt. \tag{4.92}$$

Since Ω is large, the integrand in equation 4.92 oscillates rapidly, as shown in Figure 4.17a. Over most of the integration interval, positive and negative contributions tend to cancel each other. This is not the case, however, near $\omega t = 0$, $\omega t = \pi$, $\omega t = 2\pi$, and so on. At these times the derivative of $\cos\omega t$ goes to zero and the oscillation of the integrand slows down. These values of ωt are called *points of stationary phase*. Since only values of ωt near these points of stationary phase contribute appreciably to the integral, we can replace $\cos\omega t$ by an approximation that is valid near these points; for example, near $\omega t = 0$,

$$\cos\omega t \cong 1 - \tfrac{1}{2}(\omega t)^2.$$

The corresponding contribution to the integral in equation 4.92 can be denoted by I_{T_0} and is

$$I_{T_0} = \int_0^\infty \exp\left\{i\Omega\left[1 - \tfrac{1}{2}(\omega t)^2\right]\right\}dt,$$

where the limit has been extended to infinity since only very small values of t contribute appreciably. Evaluation of this integral yields

$$I_{T_0} = \tfrac{1}{2}\sqrt{\pi/(\Omega\omega^2)}\left\{\exp(i\Omega) - \exp\left[i\left(\Omega + \frac{\pi}{2}\right)\right]\right\}.$$

Similarly, near $\omega t = \pi$,

$$\cos\omega t \cong -1 + \tfrac{1}{2}(\omega t - \pi)^2,$$

and the corresponding contribution to the integral in equation 4.92 is I_{T_1}:

$$I_{T_1} = \tfrac{1}{2}\sqrt{\pi/(\Omega\omega^2)}\left\{\exp(-i\Omega) + \exp\left[-i\left(\Omega - \frac{\pi}{2}\right)\right]\right\}.$$

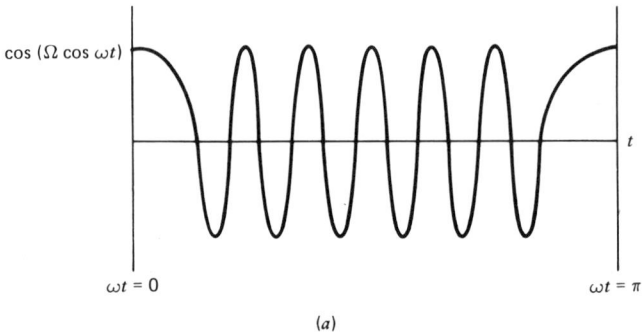

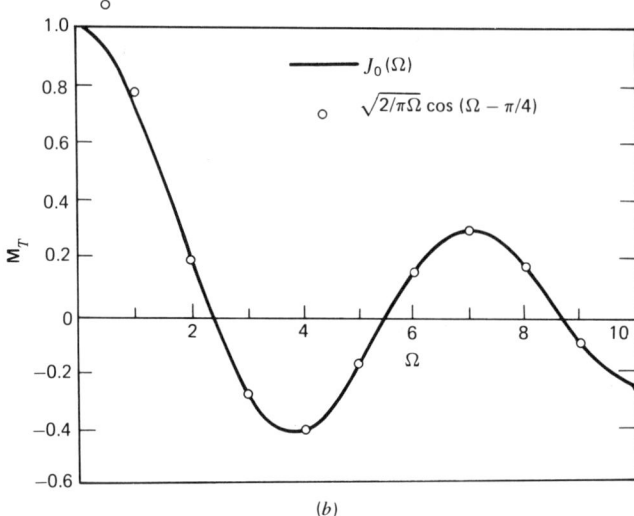

Figure 4.17 Approximation of M_T for a cosinusoidal vibration by the method of stationary phase. (a) The real part of the integrand of equation 4.92. The imaginary part would exhibit similar behavior. (b) Comparison of exact M_t with its approximation by the method of stationary phase.

Since $\mathbf{M}_T$ is an *average* value of the integral, it is sufficient to average it over half the vibrational period, that is, over the time interval $0 \leqslant \omega t \leqslant \pi$; hence

$$M_T \cong \frac{I_{T_0} + I_{T_1}}{\pi/\omega}$$
$$\cong \sqrt{2/(\pi\Omega)} \, \cos\left(\Omega - \frac{\pi}{4}\right). \tag{4.93}$$

4.3 MEASUREMENT OF MECHANICAL VIBRATIONS

This approximation to the characteristic function is plotted in Figure 4.17b, along with the exact characteristic function $M_T = J_0(\Omega)$. The approximation is quite accurate except at very small Ω.

The method of stationary phase is equivalent to the physical statement that time-average fringes are formed primarily by light scattered from the object when it is in positions where it dwells a relatively long time, such as the maximum deformations in a simple harmonic motion. A more general analysis of this type results in the following approximate form of the characteristic function:[53]

$$M_T \cong \frac{1}{T} \sum_{n=1}^{N} \frac{2\Gamma(1/p)\exp\left[i(\Omega_{t_n} \pm q\pi/2p)\right]}{p|(1/p!)d^p\Omega_{t_n}/dt^p|^{1/p}}, \qquad (4.94)$$

where $\Omega_t = \Sigma \Omega_i f_i(t)$, and Γ is the gamma function. Also, $q = \frac{1}{2}[(-1)^p + 1]$ and its sign is that of $d^p\Omega_t/dt^p$ when p is even, p is the order of the first nonvanishing derivative of Ω_t at the stationary phase point t_n, and N is the number of stationary phase points occurring during the exposure time interval T. If the motion is periodic, N can be taken as the number of stationary phase points in one period. In the case of dependent object motions, equation 4.94 represents the sum of contributions from the object in all stationary phase points, each contribution essentially weighted by the amount of time the object spends near that point. This can result in rather complicated fringe patterns. In ref. 53 several examples of the use of this formula are given, and experimental studies of simultaneous harmonic vibrations with frequency ratio $\omega_1/\omega_2 = 2$ are presented.

4.3.4 Real-Time and Stroboscopic Interferometry

In some applications it is desirable to observe the response of a test object to changing excitation in real time. This is especially true in the observation of mechanical vibration, where resonances must be determined by sweeping through a range of excitation frequencies while monitoring the response of the object. This can be accomplished by *real-time holographic interferometry*, in which light scattered by an object interferes with a holographically reconstructed light wave. The terms "concomitant" and "live-fringe" are also used to denote this type of interferometry.

Real-time interferometry can be done with a typical off-axis holographic system such as that shown in Figure 1.27. A single holographic exposure is made of the test object in its initial static configuration. The film plate is removed from the apparatus, processed, dried, and then returned to its original position in the holographic system. If the laser is turned on, the developed hologram will be illuminated simultaneously by the reference wave and by the wave of light scattered by the object. An observer looking through the hologram will receive light from the holographic virtual image of the object and also light which is scattered by the object and transmitted through the hologram. These two optical waves will form an interference pattern if the object is displaced or deformed. Furthermore, if the object vibrates while being observed, a fringe pattern indicative of vibrational amplitude will be seen.[33]

Let us first analyze real-time holographic interferometry subject to three idealizations: the hologram is thin; the irradiances of the object wave and reconstructed wave are equal; and both waves have the same polarization. If $-\mathbf{U}_o(x,y)$ denotes the complex amplitude of the holographically reconstructed object wave, the instantaneous object wave will have complex amplitude $\mathbf{U}_o(x,y) \cdot \exp[i\Delta\phi(x,y,t)]$, where $\Delta\phi$ is the phase change due to displacement or deformation of the object at any time t. The sign of the complex amplitude of the reconstructed object wave is negative because the hologram itself is a photographic negative. The negative sign represents a uniform phase shift of π rad, that is, the amplitude could be written as $\exp(i\pi)\mathbf{U}_o(x,y)$. (This phase shift has been ignored heretofore in this book because it does not effect the irradiance in two-exposure holographic interferometry.) The instantaneous irradiance during a real-time experiment will be

$$I(x,y,t) = |U_o|^2 \cdot |-1 + \exp[i\Delta\phi(x,y,t)]|^2$$
$$= |U_o|^2 \cdot 2\{1 - \cos[\Delta\phi(x,y,t)]\}. \tag{4.95}$$

The fringe pattern observed when any point on the object is viewed will be proportional to

$$I = |M_T|^2 = \tfrac{1}{2}\{1 - \cos[\boldsymbol{K}\cdot\boldsymbol{L}(x,y,t)]\}, \tag{4.96}$$

where $\boldsymbol{K}$ is the sensitivity vector, $\boldsymbol{L}$ is the instantaneous vector displacement of the observed object point, and M_T is the characteristic function for real-time interferometry. Equation 4.96 represents a dark-field interferogram, that is, the fringe corresponding to zero motion will be dark.

If, during observation, the object vibrates sinusoidally with amplitude $\mathbf{A}(x,y)$ and frequency ω, and if it was static in its equilibrium position when the hologram was recorded, the instantaneous irradiance pattern will be proportional to

$$I(x,y,t) = \tfrac{1}{2}[1 - \cos(\boldsymbol{K}\cdot\boldsymbol{A}\sin\omega t)].$$

4.3 MEASUREMENT OF MECHANICAL VIBRATIONS

The human eye retains images on the retina for approximately 0.04 s; Therefore, if the vibrational frequency is above 25 Hz, a *time-averaged irradiance* will be seen. Similar time averaging of irradiance occurs if the pattern is photographed. The characteristic function that describes this *real-time, time-average* technique is

$$|M_T|^2 = \frac{1}{T}\int_0^T [1 - \cos(K \cdot A \sin \omega t)]\, dt, \qquad (4.97)$$

where T is an appropriate averaging time; for example, one vibrational period $T = 2\pi/\omega$. Evaluation of equation 4.97 yields

$$|M_T|^2 = 1 - J_0(K \cdot A), \qquad (4.98)$$

which is plotted in Figure 4.18. This represents a dark-field interferogram with a much lower fringe visibility than occurs in time-average interferometry, for which $|M_T|^2 = J_0^2(K \cdot A)$. Also, the real-time method has half the sensitivity, that is, for a given amplitude **A**, the real-time interferogram will have half as many fringes as a time-average interferogram.

The fringes of real-time interferometry encountered in practice can be described by

$$|M_T|^2 = 1 + C^2 + 2C\cos(K \cdot L) \qquad (4.99)$$

and

$$|M_T|^2 = 1 + C^2 + 2CJ_0(K \cdot A) \qquad (4.100)$$

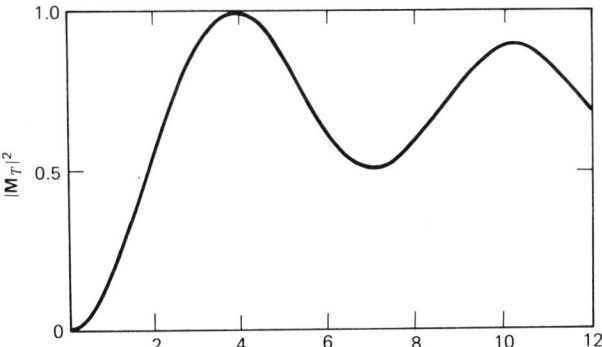

Figure 4.18 Irradiance of fringes for sinusoidal vibration observed by real-time holographic interferometry. This function assumes an ideal thin hologram and equal irradiances and polarizations of the interfering optical waves.

for the cases of deformation and sinusoidal vibration, respectively. The constant C accounts for the relative irradiance of the two interfering waves, polarization components in the object wave which are normal to the laser polarization, and phase shifts due to variation in emulsion thickness caused by processing and drying the holographic plate. These experimental parameters can be adjusted so that $C = \pm 1$, a value which yields fringes of maximum visibility; however, the visibility of fringes described by equation 4.100 will always be significantly less than the J_0^2 time-average fringes.

Precise repositioning of the hologram after development is required if the occurrence of extraneous fringes is to be minimized. Appropriate plate holders are available commercially or can be constructed in-house. In most designs, two edges of the film plate are located against two round stainless steel pins and held in place either by gravity[59] or by a spring. The mount can also be made to pivot about two orthogonal axes so that it can be finely adjusted by turning micrometer screws. Repositioning can be eliminated if the plate is developed without removing it from the holographic system. This can be done, for example, by using a simple holder that suspends the plate vertically so that beakers of photographic chemicals can be lifted from below to submerge it.[60,61]

Swelling, shrinking, and distortion of the emulsion occur during the processing and drying of photographic plates. These cause three effects that may be detrimental in real-time interferometry. First, uniform changes in thickness give rise to a uniform phase shift between the reconstructed and instantaneous object waves. This affects the constant C in equation 4.99 or 4.100. To compensate for this, a phase shifter can be incorporated in either the reference or the object beam, so that either a bright or a dark-field hologram can be attained. Suitable phase shifters are discussed in Section 2.3 in connection with holographic subtraction and addition.

A second effect of uniform swelling or shrinking is a loss of diffraction efficiency so that a dim reconstruction results. The reason for this is that real emulsions have finite thickness, so the fine fringes which form the hologram are in reality nearly parallel fringe *surfaces* extending at some angle through the thickness of the emulsion. Maximum diffraction efficiency occurs when these fringe planes are oriented with respect to the reconstruction wave so that the Bragg condition is met;[62] otherwise, the reconstructed object wave will be attenuated or even extinguished. If the developed emulsion has a thickness identical to the emulsion thickness at the time of exposure, the Bragg condition will be met when the plate is replaced in its original position. If the emulsion shrinks or swells, however, the orientation of the fringe planes will change and poor reconstruction will result. This problem can be minimized by using a relatively small mean angle between the object and reference waves so that the spatial frequency of the hologram is low. Of course, if the object is large, this is impractical and the special processing techniques discussed below should be used.

The third and most important effect, due to nonuniform distortion of the emulsion, is the introduction of extraneous fringes into the interferogram. This problem can also be alleviated by appropriate processing. Nonuniform distortions of photographic emulsions are caused primarily by residual stresses in the emulsion due to strains introduced during manufacture. When the emulsion becomes wet during processing, these residual stresses are relaxed to some extent and deformation results. If this deformation occurs after the plate has been exposed, optical distortions result. This type of deformation can be reduced by stress-relieving the plates before using them. This is done by soaking the plates in water[63] for about 15 min, preferably followed by 5 min in isopropyl or ethyl alcohol,[61,64] and then air-drying them. Presoaking the emulsion in water has the agreeable side effect of sensitizing the plates so that exposure times are decreased by a factor of 3 to 4.[64]

Biedermann and Molin[66] have measured the variation of emulsion thickness with time during a typical processing sequence of a holographic plate. Very rapid swelling occurs during development, and very rapid shrinkage occurs during fixing. It is therefore desirable to minimize development time and to minimize or even eliminate the fixing of the plate.[64] A common technique for reducing emulsion problems is to severely overexpose the plate and then underdevelop and underfix it. Kellie and Stevenson[67] have demonstrated that in this manner development can be restricted to a thin outer layer of the emulsion. For example, AGFA 10E70, which normally would be exposed with He-Ne laser light at 2 $\mu J/cm^2$ and developed in Kodak HRP for 5 min, can be exposed at a level of 100 $\mu J/cm^2$ and developed for 20 s. This creates an active emulsion thickness of less than 2 μm, rather than the full thickness of 7 μm. In effect, a thin hologram is formed which is relatively insensitive to problems of shrinkage and swelling; furthermore, the absolute level of shrinking and swelling is low because of the rapid processing. If the plates are underfixed or unfixed, their useful life is on the order of a few hours, which is sufficient for most real-time experiments. If the plate is unfixed, it is best to use a safelight during the experiment. Appropriate safelights are specified by film manufacturers.

An alternative approach to real-time holographic interferometry is to expose and process the plate *in situ* while it is submerged in a narrow vessel, often referred to as a *liquid gate*. The liquid gate is a rectangular vessel whose front and back surfaces are flat, parallel glass or Plexiglas. A means of clamping the holographic plate in a fixed position inside the liquid gate must be provided. Water and photographic chemicals are circulated into and out of the vessel by gravity feed, a pump, or a vacuum line. Plate holders of the liquid gate type are available commercially, or can be manufactured on the basis of designs or discussions in the literature.[66,68,69] The use of a liquid gate plate holder has four useful features in addition to eliminating the need for precise repositioning of the plate. First, the emulsion can be preswollen to a relatively stable thickness by soaking the plate in water[66] or developer;[64,70] this again decreases exposure times. Second, the emulsion is relatively well matched to the index of refraction of the water or fixer, so the optical effects of emulsion distortions which may occur are minimized. Third, because drying is eliminated, total time between exposure and observation can be reduced to around 100 to 200 s.[66] Fourth, exposure can be visually monitored while the hologram is being recorded by observing the increasing photographic density of the plate. In fact, a small neutral density filter of the correct density can be attached to the plate holder for comparison. Chemicals for the monobath developing and fixing of plates have also been used,[71] and a system using an experimental film which is developed by exposure to a dry gas has been described by Smith et al.[72] A particularly thorough discussion of *in situ* processing for real-time interferometry is presented by Biedermann and Molin.[66]

Polarization effects are important in real-time interferometry. Laser light used for holography is usually linearly polarized in the vertical direction. When this light is scattered by a diffuse object surface, up to 50 percent of it may be shifted into horizontal polarization. Although both components contribute to the total exposure of the plate, only the vertical component, which matches the reference wave, contributes to the formation of the hologram. More importantly, the real-time interference will be between an object wave of mixed polarization and a linearly polarized reconstructed wave. This results in a decrease in fringe visibility of as much as 30 percent —an effect especially detrimental in vibration measurement, where visibility is low to begin with. For this reason, real-time interferograms should be viewed or photographed through a polarizing filter. Another approach is to attach a stable sheet of polarizing material to the front of the liquid gate or dry plate holder. Interestingly, Kellie and Stevenson[73] have found experimentally that depolarization can be minimized by painting the object black. If light economy is not critical, this approach can be used. Retroreflective paints can also be used;[61] these provide excellent light economy and do not alter the polarization of the illumination.

Figure 4.19*a* is a real-time interferogram of a vibrating plate. It was recorded by photographing the vibrating plate through a liquid gate. The system had been carefully adjusted to form a fringe-free image of the object in its static position. Figure 4.19*b* is a time-average interferogram of the same plate vibrating in the same mode. As predicted by equations 4.99 and 4.100, the real-time interferogram shows reduced fringe visibility and

4.3 MEASUREMENT OF MECHANICAL VIBRATIONS

half the number of fringes. Figure 4.19c is also a real-time interferogram of the same vibrating plate, but vertical reference fringes have been introduced by slightly rotating the object about a vertical axis lying in its surface. The characteristic function for real-time interferometry with reference fringes is

$$|M_T|^2 = 1 + C^2 + 2C\cos(\boldsymbol{K} \cdot \boldsymbol{L}_r) J_0(\boldsymbol{K} \cdot \boldsymbol{A}), \qquad (4.101)$$

where $\boldsymbol{L}_r$ is the motion used to introduce the reference fringes. Equation 4.101 describes a set of cosinusoidal fringes modulated by J_0. The sign of J_0 oscillates as its argument increases, so contours of constant vibrational amplitude are the loci of points where the reference fringes exhibit contrast reversal and decreasing visibility. It is apparent that such contours in Figure 4.19c coincide with the dark fringes in Figure 4.19b. This is a common method for searching for resonance and observing mode changes by holographic interferometry.

Finally, we note that vibration measurements are most satisfactorily made using a system in which both real-time and time-average interferograms can be recorded. The real-time and time-average hologram plates are placed next to each other. With the real-time method, modes of interest can be located by varying excitation frequency and amplitude; then a time-average interferogram can be recorded and used for quantitative evaluation since it provides a permanent interferogram with good fringe visibility.

There are two characteristics of time-average and real-time holographic interferometry which might be considered deficiencies: the variation in relative vibrational phase across the object surface is not determined, and the fringe visibility decreases with increasing vibrational amplitude.

Stroboscopic holographic interferometry is a technique which eliminates these deficiencies at the expense of some increase in experimental complexity. This technique was developed and reported by several workers in 1968.[74-79] Stroboscopic holographic interferometry consists of recording a hologram by using a sequence of brief pulses of laser light that are synchronized with the vibrating object. Most commonly, the hologram is formed by illuminating the plate with a brief pulse of light when the object is at its maximum positive displacement and again when it is at its maximum negative displacement. For example, if the motion is $\boldsymbol{A}\sin\omega t$, pulses will be at $\omega t = \pi/2$ and $\omega t = 3\pi/2$. This sequence is then repeated until the total exposure of the plate reaches the desired level. If the pulses are very short, the characteristic function of the interferogram formed in this manner is

$$|M_T|^2 = \cos^2(\boldsymbol{K} \cdot \boldsymbol{A}). \qquad (4.102)$$

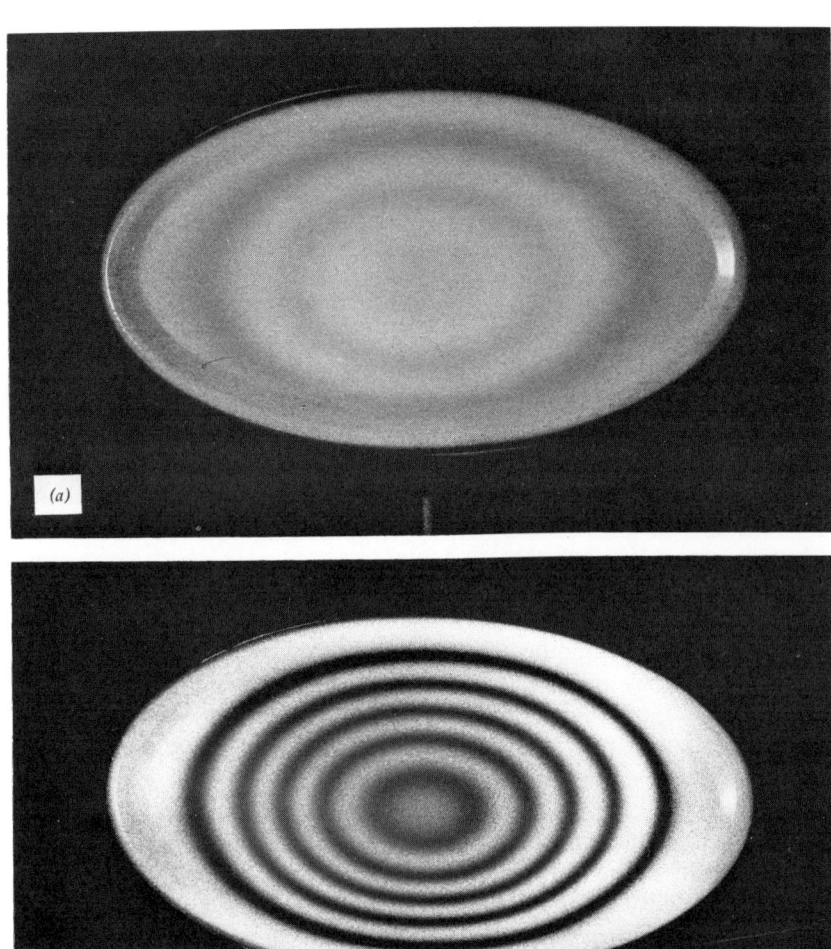

Figure 4.19 Holographic interferograms of a vibrating elliptical plate (ref. 66). (*a*) Real-time interferogram; (*b*) Time-average interferogram; (*c*) Real-time interferogram with reference fringes introduced by rotating the object about a vertical axis. The object vibration is the same in all three cases. The fringes are described by equations 4.100, 4.67, and 4.101, respectively. (Photographs copyrighted by the Institute of Physics.)

4.3 MEASUREMENT OF MECHANICAL VIBRATIONS

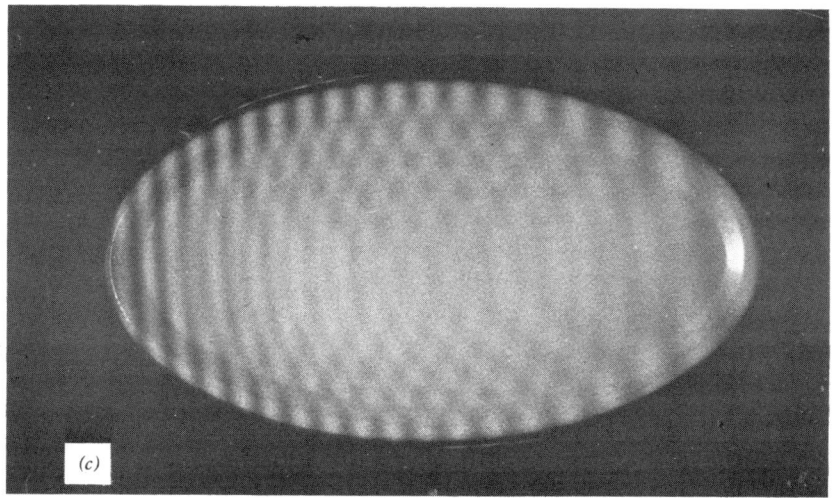

Figure 4.19 Cont.

The fringes described by equation 4.102 have unit visibility for all amplitudes. Figure 4.20 is a stroboscopic interferogram of this type. The object is a 10×15 cm clamped rectangular steel plate, 0.76 mm thick, vibrating at 4110 Hz.

To study the phase distribution or time function of a vibratory motion, or to reduce the sensitivity of the measurement, the light pulses can be triggered at any two times, t_{p_1} and t_{p_2}, during the vibration cycle. The corresponding characteristic function for a sinusoidal motion $\mathbf{A}\sin\omega t$ is

$$M_T = \exp(i\mathbf{K}\cdot\mathbf{A}\sin\omega t_{p_1}) + \exp(i\mathbf{K}\cdot\mathbf{A}\sin\omega t_{p_2}),$$
$$|M_T|^2 = \cos^2\left[\tfrac{1}{2}\mathbf{K}\cdot\mathbf{A}(\sin\omega t_{p_1} - \sin\omega t_{p_2})\right]. \quad (4.103)$$

The form of equation 4.103 emphasizes that a stroboscopic interferogram is essentially a two-exposure interferogram formed by light scattered by the object while it is in two different states of deformation. If $t_{p_1}=0$ and $t_{p_2}=t_p$, the characteristic function for any periodic, separable motion $\mathbf{A}\cdot f(t)$ will be

$$|M_T|^2 = \cos^2\left[\tfrac{1}{2}\mathbf{K}\cdot\mathbf{A}f(t_p)\right]. \quad (4.104)$$

A sequence of interferograms for various t_p will provide a rather complete description of the object motion. Fryer[80] has described a stroboscopic system in which the reference beam scans across the hologram while the object executes the motion under study. The plate was exposed previously

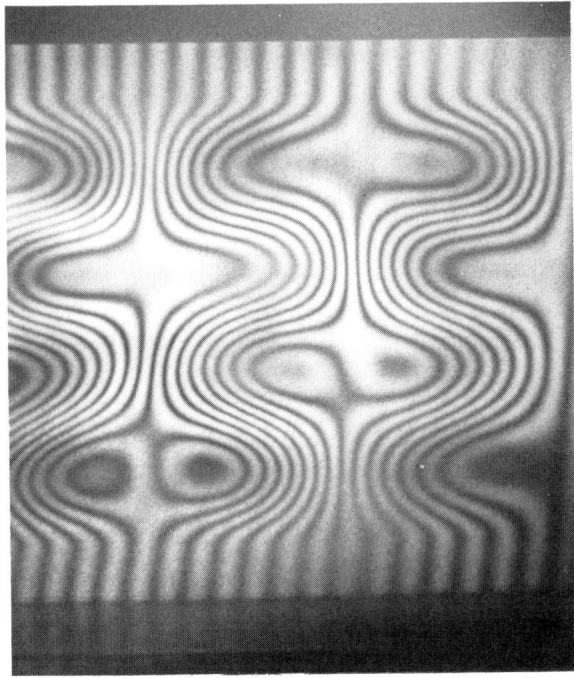

Figure 4.20 Stroboscopic holographic interferogram of a vibrating rectangular plate with clamped edges (ref. 79). The object was rotated between two stroboscopic exposures in order to introduce reference fringes.

while the object was in an initial reference position. Because of the scanning each region of the hologram contains information about the object deformation at a different time. An observer can move his or her eye across the hologram to follow the time variation of deformation. If the motion is periodic, the relative phase of the vibration at each point on the surface can be studied in this manner.

In the above analyses, extremely short, light pulses have been assumed. Real pulses are of finite duration, or "width." Indeed, increasing the pulse width is desirable because it decreases the total exposure time, thereby reducing the requirements for stability of the system and stationarity of the observed vibration. To determine the effect of finite pulse width, let us consider the case in which the object vibrates sinusoidally, $\mathbf{A} \sin \omega t$. Let the pulses have duration Δ and be centered at $t_{p_1} = -t_p$ and $t_{p_2} = +t_p$. The characteristic function can be determined by evaluating equation 4.70, taking T to be the period of the vibration. The result will be proportional

4.3 MEASUREMENT OF MECHANICAL VIBRATIONS

to

$$M_T = \frac{1}{T} \int_{-t_p-\Delta/2}^{-t_p+\Delta/2} \exp(i\mathbf{K}\cdot\mathbf{A}\sin\omega t)\,dt$$
$$+ \frac{1}{T} \int_{t_p-\Delta/2}^{t_p+\Delta/2} \exp(i\mathbf{K}\cdot\mathbf{A}\sin\omega t)\,dt. \quad (4.105)$$

Equation 4.105 can be evaluated to a good approximation by expanding $\sin\omega t$ about the center of each pulse:

$$\sin\omega t \cong \sin\omega t_p + (\omega\cos\omega t_p)(t-t_p).$$

This results in the following characteristic function:

$$M_T = \frac{2\Delta}{T}\mathrm{sinc}\left[\tfrac{1}{2}\mathbf{K}\cdot\mathbf{A}(\omega\cos\omega t_p)\Delta\right]\cos(\mathbf{K}\cdot\mathbf{A}\sin\omega t_p) \quad (4.106)$$

where $\mathrm{sinc}\,\xi = \sin\xi/\xi$. The fringe system in an interferogram of this type can be described by

$$|M_T|^2 = \mathrm{sinc}^2\left[\tfrac{1}{2}\mathbf{K}\cdot\mathbf{A}(\omega\cos\omega t_p)\Delta\right]\cos^2(\mathbf{K}\cdot\mathbf{A}\sin\omega t_p). \quad (4.107)$$

The function $\mathrm{sinc}^2\xi$ is displayed in Figure 3.6. As the pulse width shrinks to zero, the fringes described by equation 4.107 approach full visibility, as described by equation 4.103 for the case of $\omega t_p = \pi/2$. As the pulse width is increased, fringe visibility decreases. This effect is illustrated in Figure 4.21.

Equation 4.107 is an approximate expression valid for small pulse widths Δ. The exact characteristic function has been derived by Listovets and Ostrovskii:[81]

$$|M_T|^2 = \left|\sum_{n=-\infty}^{\infty} J_n(\mathbf{K}\cdot\mathbf{A})\mathrm{sinc}\left(\frac{n\pi\Delta}{T}\right)\exp(i2n\pi t_p)\right|^2 \quad (4.108)$$

and is presented graphically for several values of t_p and Δ in ref. 70. Miler[82] has evaluated M_T for arbitrary t_{p_1}, t_{p_2}, Δ_1, and Δ_2 in an approximation equivalent to that leading to equation 4.107.

Stroboscopic holograms can be recorded by using two pulses from a pulsed laser.[75] More commonly, a sequence of several pulses is generated by chopping a continuous laser beam by mechanical or electro-optical means. The simplest mechanical technique is to chop the beam with a perforated disk driven by a synchronous motor.[77,78] Systems of this type

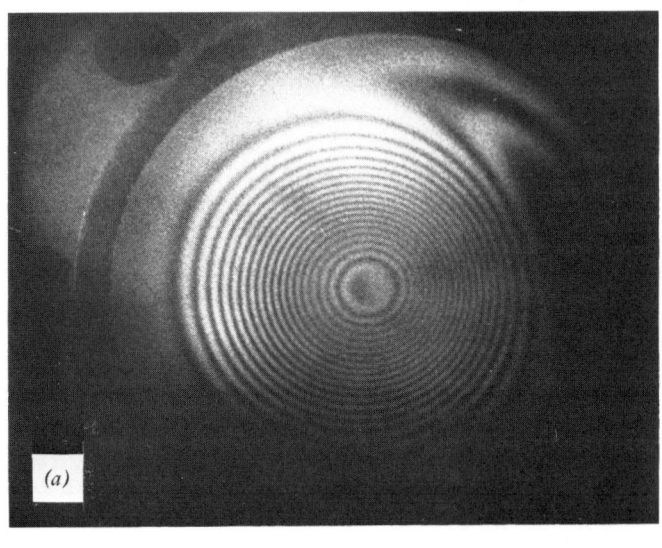

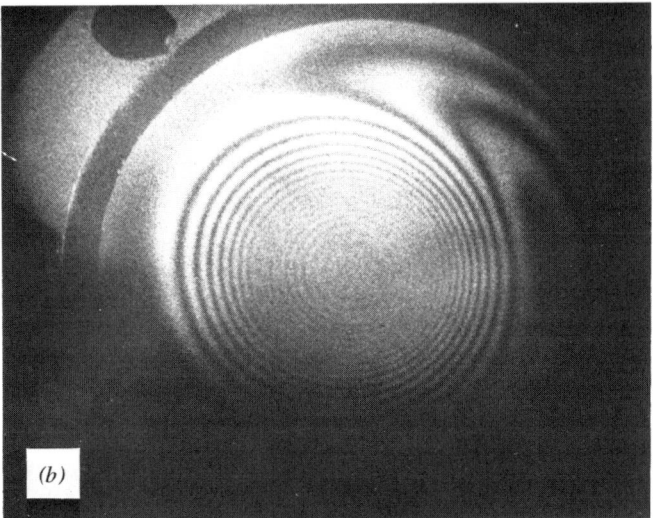

Figure 4.21 Stroboscopic holographic interferogram of a vibrating circular plate (ref. 76). (a) Pulse duration is 30 μs; (b) Pulse duration is 100 μs.

4.3 MEASUREMENT OF MECHANICAL VIBRATIONS

are limited to chopping frequencies of about 2 kHz.[83] The motor can be supplied with a signal whose frequency is derived from the source that excites the vibration in order to maintain precise synchronization. Archbold and Ennos[74,84] have designed a chopper in which the laser beam is focused through a small hole drilled through a spindle driven by an air turbine. Strobe frequencies up to 3.7 kHz were possible with their system, and the range could be extended to about 13 kHz by using a dental drill turbine. Electro-optical devices such as Pockles cells can also be used to modulate a laser beam for stroboscopic holography.[76,79] These are discussed in Section 4.3.4.

To decrease the total exposure time required for stroboscopic interferometry, an initial reference exposure can be used. For example, if the vibration is sinusoidal, $A \sin \omega t$, an initial holographic exposure of the object in its equilibrium position can be recorded, followed by a sequence of short pulses at $t = t_p + 2n\pi$, $n = 0, 1, 2, \ldots$ while the object is vibrating. If ρ denotes the fraction of the total exposure time during which the object is in a static position, the fringes will be described by

$$|M_T|^2 = \rho + (1-\rho)\cos(\mathbf{K} \cdot \mathbf{A}). \tag{4.109}$$

If $\rho = 0.5$, fringes of unit visibility will result:

$$|M_T|^2 = \cos^2\left(\tfrac{1}{2}\mathbf{K} \cdot \mathbf{A}\right). \tag{4.110}$$

If ρ is not 0.5, fringe visibility[74] will be

$$\nu = \frac{2\rho(1-\rho)}{\rho^2 + (1-\rho)^2}, \tag{4.111}$$

where ν decreases rather slowly as the fraction of exposure time used for the initial exposure is increased; for example, if $\rho = \tfrac{2}{3}$, $\nu = 0.8$. Further decreases in total exposure time can be achieved by combining an initial exposure of the object with two[85,86] or four[87] symmetrically located pulses during each vibration period. Characteristic functions have also been computed for stroboscopic interferometry when the reference is a time-average exposure of the vibrating object.[88] This method is suggested for use when the mean position of the object during vibration may differ from the initial static position. Vikram[89] has also suggested the use of stroboscopic holography to study simultaneous vibration in rationally related modes. If the illumination is strobed at the frequency of one mode, the resulting interferogram will contain information about the second mode.

In this discussion of stroboscopic holographic interferometry it has been assumed that a permanent recording of the vibrational amplitude pattern is desired. The method can also be adapted to *real-time* observation.[90] A hologram of the object in its initial static configuration is recorded, developed, and returned to the holographic system. Any of the processing methods used for real-time interferometry can be used. The vibrating object is then stroboscopically illuminated and viewed through the hologram, which is simultaneously illuminated with the reference wave. This technique has the advantage of real-time observation and retains the desirable features of stroboscopic interferometry, such as good fringe visibility.

4.3.5 Temporally Modulated Holography

Time-average and stroboscopic holographic interferometry are particular cases of what is termed *temporally modulated holography*, that is, holography in which the object wave and/or the reference wave are time dependent. A very complete development of the theory and practice of temporally modulated holography has been presented by Aleksoff.[91,92] He demonstrated that temporal modulation of the reference wave introduces several effects which are useful in holographic interferometry. These include vibration measurement with variable sensitivity, determination of relative phase across a vibrating surface, and compensation for extraneous object motion. These applications of temporal modulation are discussed in this section.

Consider an off-axis hologram recorded over an exposure interval T. As usual, we assume that the transmittance of the hologram is proportional to the time-integrated irradiance of the photographic plate. This irradiance is composed of four terms. The term of interest here is the one which yields the virtual image:

$$\int_0^T f_o(t) U_o(x,y,t) f_R^*(t) U_R^*(x,y) \, dt.$$

Here U_o is the complex amplitude at the hologram plane due to illumination of the object with a cw laser, and U_R is the corresponding complex amplitude of the reference wave. Also, $\mathbf{f}_o(t)$ and $\mathbf{f}_R(t)$ describe the temporal modulation of the object wave and reference wave, respectively. These are complex quantities since either the real amplitude or the phase can be modulated. Ordinary time-average holography corresponds to $\mathbf{f}_o(t) = \mathbf{f}_R(t) = 1$. In stroboscopic holography, $\mathbf{f}_o(t)$ and $\mathbf{f}_R(t)$ are equal and represent a sequence of pulses. When the hologram is illuminated by a reconstruction wave proportional to $U_R(x,y)$, the complex amplitude of the reconstructed object wave evaluated at the hologram plane is proportional to

$$U_c(x,y) = \frac{1}{T} \int_o^T f_o(t) f_R(t) U_o(x,y,t) \, dt. \qquad (4.112)$$

4.3 MEASUREMENT OF MECHANICAL VIBRATIONS

One of the most useful applications of temporal modulation to the measurement of mechanical vibration is *frequency-translated holography*. In this technique the reference wave is modulated so that its frequency is translated, that is, increased, by an amount $n\omega$, where n is an integer and ω is the frequency of the vibrating object under study. If the object vibrates sinusoidally, an observer viewing a point in the virtual image described by equation 4.98 will see the object modulated by interference fringes whose irradiance is proportional to

$$I = |M_T|^2 = \left| \frac{1}{T} \int_0^T f_o(t) f_R^*(t) \exp(i\mathbf{K}\cdot\mathbf{A} \sin\omega t) \, dt \right|^2.$$

In this case, $\mathbf{f}_o(t) = 1$, and $\mathbf{f}_R(t) = \exp(in\omega t)$, so the characteristic function is

$$M_T = \frac{1}{T} \int_0^T \exp(-in\omega t)\exp(i\mathbf{K}\cdot\mathbf{A} \sin\omega t) \, dt. \quad (4.113)$$

Evaluation of M_T can be simplified by using the following mathematical identity:[58]

$$\exp(iz\sin\phi) = \sum_{m=-\infty}^{\infty} J_m(z)\exp(im\phi), \quad (4.114)$$

so that the integrand in equation 4.113 becomes

$$\sum_{m=-\infty}^{\infty} \exp(-in\omega t) J_m(\mathbf{K}\cdot\mathbf{A})\exp(im\omega t).$$

Reversing the order of integration and summation yields

$$M_T = \sum_{m=-\infty}^{\infty} J_m(\mathbf{K}\cdot\mathbf{A}) \cdot \frac{1}{T} \int_0^T \exp[i(m-n)\omega t] \, dt. \quad (4.115)$$

If the exposure interval T is long compared with the vibrational period, $T \gg 2\pi/\omega$, the integral in equation 4.115 vanishes except when $m = n$, that is,

$$\frac{1}{T} \int_0^T \exp[i(m-n)\omega t] \, dt = \delta(m-n),$$

so

$$M_T = J_n(\mathbf{K}\cdot\mathbf{A}). \quad (4.116)$$

The fringes observed through the hologram will have an irradiance proportional to

$$I = |M_T|^2 = J_m^2(\mathbf{K} \cdot \mathbf{A}). \tag{4.117}$$

Before discussing the applications of frequency-translated holography, we consider the interpretation of equation 4.117. Using equation 4.114, we expand the periodic object wave $\mathbf{U}_{oS}(x,y) \exp(i\mathbf{K} \cdot \mathbf{A} \sin \omega t) \cdot \exp(i\omega_0 t)$ in a Fourier series, each term of which has a frequency $\omega_0 + m\omega$, where $m = 0, \pm 1, \pm 2,\ldots$ and $\omega_0 \sim 10^{15}$ Hz is the frequency of the laser light. A hologram is recorded by adding to this a reference wave whose frequency is $\omega_0 + n\omega$. This reference wave, in a sense, is coherent with only the one object wave component which has frequency $\omega_0 + n\omega$, so a time-average hologram of only this component is formed. The recording of a frequency-translated hologram therefore is a *temporal filtering* operation in which a single frequency component is selected from a periodic function.[91-93] Ordinary time-average holographic interferometry, $n = 0$, selects the fundamental frequency, and frequency-translated holography can be used to observe any of the higher harmonics. Further physical insight can be gained by considering the interference fringes that form the hologram.[91] If the reference wave has frequency $\omega_0 + n\omega$, the fringes it forms by interfering with an object wave component with frequency $\omega_0 + m\omega$ travel continuously across the hologram plane at a speed such that a stationary photodetector would record a signal at the *beat frequency* $(m - n)\omega$. The fringes are traveling, so their irradiance is smeared, or averaged, during the exposure interval, and does not contribute to formation of the hologram. On the other hand, the component whose frequency equals that of the reference wave yields stationary fringes which form a time-average hologram in the usual manner.

The ability to vary the order of the Bessel function whose square describes the fringes in holographic interferometry is useful for increasing or decreasing the sensitivity of vibration measurements. If very small amplitudes, $\mathbf{K} \cdot \mathbf{A} < 1$, are to be measured, J_0^2 is not a desirable characteristic function because its slope near the origin is zero; $dJ_0^2(\mathbf{K} \cdot \mathbf{A})/d(\mathbf{K} \cdot \mathbf{A}) = 0$ as $\mathbf{K} \cdot \mathbf{A} \to 0$. A time-average interferogram therefore is a bright-field interferogram whose irradiance varies slowly for small vibrational amplitudes. By frequency translation, however, we can form fringes described by $J_1^2(\mathbf{K} \cdot \mathbf{A})$. This represents a dark-field interferogram whose irradiance is $J_1^2(\mathbf{K} \cdot \mathbf{A}) = (\mathbf{K} \cdot \mathbf{A})^2/4$ for $\mathbf{K} \cdot \mathbf{A} \to 0$. Small-amplitude vibrations down to the order of $\lambda/100$ can be detected with this type of hologram.[91,94]

Frequency translation also is useful for reducing the sensitivity of holographic interferometry so that rather large vibrational amplitudes can be measured. The location of the zeros of Bessel functions J_n increases nearly linearly with order n. This is shown in Figure 4.22. Since $|\mathbf{M}_T|^2 =$

4.3 MEASUREMENT OF MECHANICAL VIBRATIONS

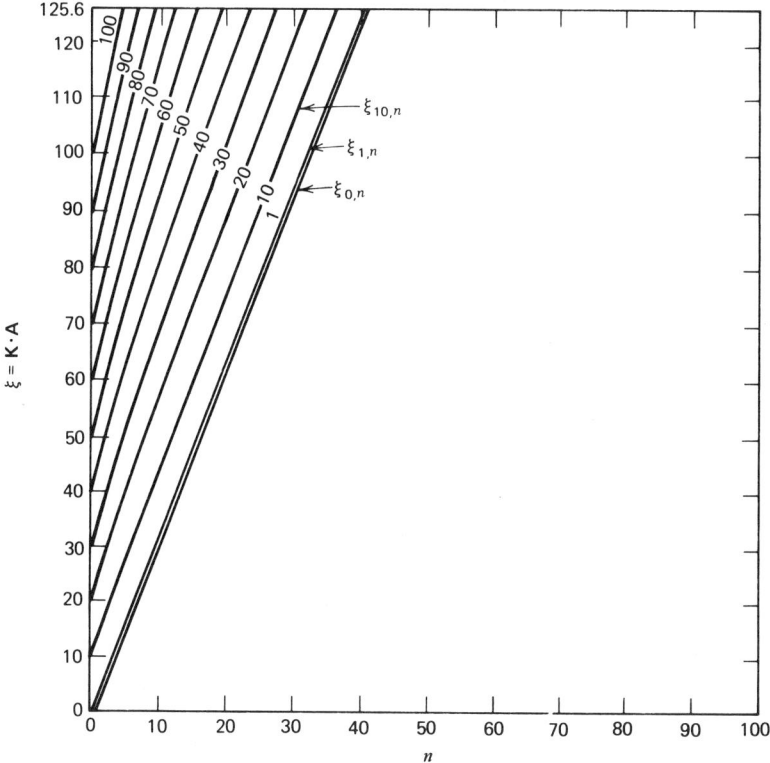

Figure 4.22 Zeros of Bessel functions of the first kind (based on ref. 91). Here $J_n(\xi_{k,n})=0$; $\xi_{k,n}$ is the kth zero of the Bessel function of order n.

$J_n^2(\mathbf{K}\cdot\mathbf{A})$, the number of fringes formed due to a given vibrational amplitude will decrease as the order n is increased. This reduction in sensitivity is illustrated in Figure 4.23, which includes frequency-translated interferograms of a vibrating loudspeaker formed with fringes $J_0^2(\mathbf{K}\cdot\mathbf{A})$, $J_1^2(\mathbf{K}\cdot\mathbf{A})$, $J_9^2(\mathbf{K}\cdot\mathbf{A})$, and $J_{28}^2(\mathbf{K}\cdot\mathbf{A})$.

Amplitude-modulated reference wave holography can be used to map the relative phase as well as the amplitude of a vibrating surface.[95] Suppose that the reference wave is modulated at the same frequency as the vibrating object and with a known phase Δ relative to some portion of the object surface; $\mathbf{f}_o(t)=1$, $\mathbf{f}_R(t)=\exp(i\omega t-\Delta)$. If the analysis leading to equation 4.117 is repeated for this case, the fringes observed in the virtual image reconstructed with a continuous reconstruction wave are found to have an irradiance proportional to

$$I = |M_T|^2 = J_m^2(\mathbf{K}\cdot\mathbf{A})\cos^2\Delta. \quad (4.118)$$

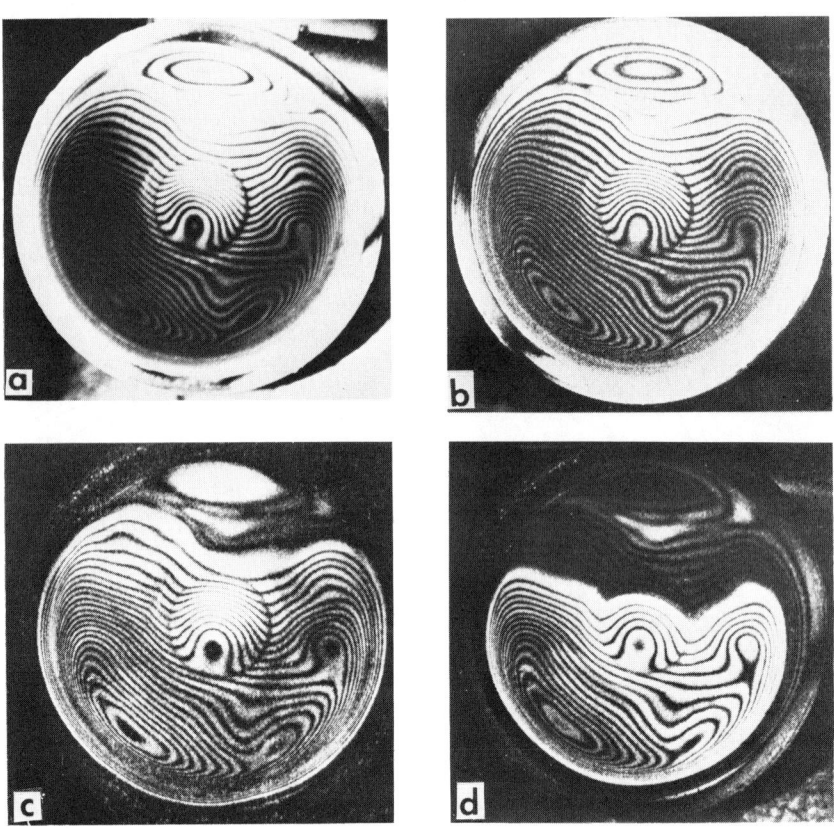

Figure 4.23 Frequency-translated interferograms of a vibrating loudspeaker (ref. 91). The characteristic function for each interferogram is $J_n^2(\mathbf{K}\cdot\mathbf{A})$. (a) $n = 0$; (b) $n = 1$; (c) $n = 9$, (d) $n = 28$.

By recording a sequence of holograms of this type, contours of constant relative phase can be constructed.

Phase modulation of the reference wave is also useful in holographic interferometry. Let the phase of the reference wave be modulated sinusoidally at the frequency of vibration of the object under study. The characteristic function for this situation can be calculated using the modulation functions $\mathbf{f}_o(t) = 1$ and

$$f_R(t) = \exp(iM_R \sin \omega t). \quad (4.119)$$

In equation 4.119, M_R is the *modulation depth*, or amplitude of the cyclic phase shift, and ω is the frequency of the phase modulation and of the

4.3 MEASUREMENT OF MECHANICAL VIBRATIONS

vibrating object. Let us rewrite the expression for the characteristic function as

$$M_T = \frac{1}{T} \int_0^T \exp(iM_R \sin \omega t) \exp[i\mathbf{K}\cdot\mathbf{A} \sin(\omega t - \phi_o)], \quad (4.120)$$

where ϕ_o is the phase of the vibration of the observed object point relative to the reference wave. Applying arguments similar to those used in deriving equation 4.84, we can show that the integral in equation 4.120 gives

$$M_T = J_0\left\{\left[(\mathbf{K}\cdot\mathbf{A})^2 + M_R^2 - 2(\mathbf{K}\cdot\mathbf{A})M_R \cos\phi_R\right]^{1/2}\right\}. \quad (4.121)$$

The relative phase of vibration at each point of the object surface is encoded in the fringe pattern. Neumann et al.[96] have developed an experimental method for measuring relative phase based on this result. Aleksoff[97] also applied this concept to measure the phase and amplitude of resonating ADP crystals. In this application modulation of both the object and reference waves was accomplished by passing laser light through the vibrating transparent crystal.

For simplicity let $\phi_R = 0$; then the fringes formed by phase-modulated holographic interferometry will have an irradiance proportional to

$$I = |M_T|^2 = J_0^2(\mathbf{K}\cdot\mathbf{A} - M_R). \quad (4.122)$$

The fringe of maximum brightness represents a region where $\mathbf{K}\cdot\mathbf{A} = M_R$, so amplitude measurements can be made relative to a controllable datum or "bias." Suppose that we wish to form an interferogram which shows the vibrational amplitude of an object relative to a particular point on its surface. This can be done by attaching a small mirror to that point and reflecting the reference beam from it to the hologram plate. The reference beam is then phase modulated in the same manner as the object wave at the same point, so the fringes have an irradiance proportional to

$$I = J_0^2(\mathbf{K}\cdot\mathbf{A} - \mathbf{K}\cdot\mathbf{A}_R), \quad (4.123)$$

where the subscript R denotes the chosen reference point on the object.

Equation 4.123 is suggestive of a general method for *object motion compensation* in holography and holographic interferometry. Phase modulation of the reference wave by reflecting it from a point on the object is useful in recording holograms of moving objects,[98-100] and for reducing the effect of extraneous rigid-body motions when a holographic interferogram of a vibrating or strained object is to be recorded. Figure 4.24 illustrates this technique; it is an interferogram of a vibrating disk brake which was

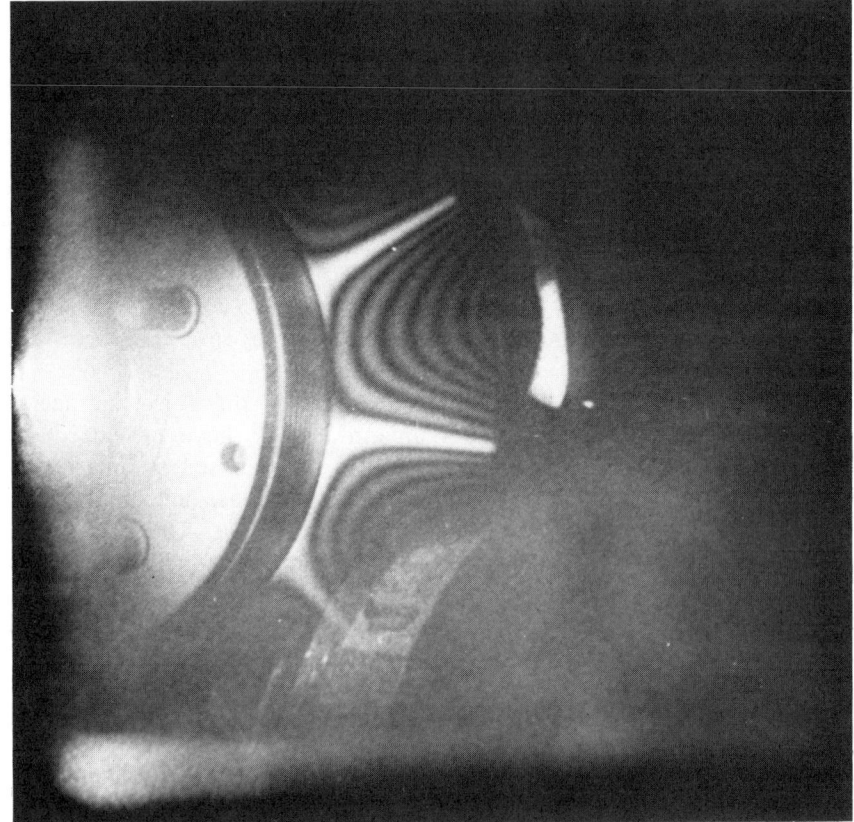

Figure 4.24 Interferogram of a vibrating disk brake recorded with object motion compensation by speckle reference beam scattered by the brake hub.

mounted on a rather soft suspension. The reference wave was derived using a method developed by Waters,[101] by focusing a cw laser beam to a small spot on the object, in this case near the hub of the brake unit. This method produces a *speckle reference beam* which generates a perfectly useful, though slightly noisy, hologram. Without compensation for extraneous motion it was not possible to record holographic interferograms of this object with the available laser power.

Levitt and Stetson[102] have described a technique for obtaining contours of constant phase. Two holograms are recorded with reference waves that are phase modulated at the frequency of object vibration, but with two different phases, ϕ_1 and ϕ_2. A photographic transparency of one of the resulting interferograms is laid on a transparency of the other. In the composite interferogram it is usually easy to identify regions where dark fringes of equal order intersect. In ref. 102 it is shown that such lines are contours of constant phase of the object vibration, ϕ_o. In particular, along the intersection of Nth-order fringes,

$$\phi_o = \tfrac{1}{2}(\phi_1 + \phi_2) \pm N\pi. \tag{4.124}$$

This technique is similar to the method described by Takai et al.[95] for phase mapping by amplitude-modulated holography, but is simpler in that a visual inspection produces the desired phase contour.

If the laser light is modulated before entering a holographic system, $\mathbf{f}_o(t) = \mathbf{f}_R(t) = \mathbf{f}_I(t)$. Stroboscopic holographic interferometry is one example of this type of modulation. Stetson[51,103] has given a general analysis of laser modulation of this type. He has shown that, if a time-average holographic interferogram of an object vibrating with frequency ω is recorded with laser light modulated at 2ω, fringes can be formed whose brightness is nearly uniform in all orders. These fringes are similar to those formed stroboscopically, but do not require the large loss of light efficiency associated with stroboscopic techniques. Other discussions of generalized stroboscopic techniques and of pulse shaping in the holography of moving objects are found in the articles by Mottier[104] and by Redman and Pearce[105] and Decker,[106] respectively.

4.4 APPLICATIONS

In this section we consider the use of holographic interferometry for nondestructive testing, medical and dental research, and experimental mechanics. For each of these, useful characteristics of holographic methods and relevant special techniques will be described, and examples of applications and references to the literature will be given.

4.4.1 Nondestructive Testing

Nondestructive testing is the detection of material or manufacturing imperfections by procedures which do not require destruction or significant alteration of the component being tested. Usually the objective of such testing is to determine the location and size of cracks, voids, disbonds, delaminations, inhomogeneous material properties, residual stresses, imperfect fits, or incorrect dimensions. Nondestructive testing employs a variety of techniques, including transmission imaging using X-rays, gamma rays, or ultrasound, field mapping using eddy current or magnaflux probes, stress detection by acoustic or exoelectron emission, and visual observation enhanced by microscopy or the application of dye penetrants. Holographic interferometry can be used for nondestructive testing if the presence of a flaw results in anomalous deformation of the surface of the component being tested when it is slightly stressed. This anomalous deformation is indicated by a change in the pattern of interference fringes relative to the pattern formed when a flaw-free component is stressed in the same manner. For most applications only a qualitative interpretation of interference fringes is required. Useful characteristics of holographic nondestructive testing techniques include simple, whole-field visual display, applicability to inspection of components having fairly complicated shapes, and lack of special requirements for surface preparation of the test object. Limitations of the technique include stringent requirements for mechanical stability and a restricted range of sensitivity. In this section we discuss holographic nondestructive testing techniques, primarily through examples of applications, and describe practical means for minimizing the limitations noted above.

The basic concept of holographic nondestructive testing is illustrated by comparison of the two interferograms in Figure 4.25. Figure 4.25a is a two-exposure holographic interferogram of the region surrounding a bolt hole in the rib of a steel channel. The fringes are indicative of surface deformation caused by drawing a bolt with a slightly tapered shank into the hole. The fringes are finely spaced in the region above the hole and more sparse below the hole, where the rib meets the heavy channel base. All of the fringes are smooth and continuous. The interferogram in Figure 4.25b was formed in the same manner; however, in this case a thin crack extended radially from the bolt hole. The presence of the crack is indicated by a discontinuity in the curvature of the fringes. Several other stressing techniques were tried and found to be ineffective for this application.[107,108] This example suggests that the design of a holographic inspection technique includes development of an appropriate method of stressing or exciting the object and production of a reference interferogram of a flaw-free component.

Four basic types of stressing or excitation are used in holographic nondestructive testing:

1. Direct mechanical stressing.
2. Pressure or vacuum stressing.
3. Thermal stressing.
4. Vibrational excitation.

Direct mechanical stressing refers to loading the test objects in simple tension, compression, or bending, or applying point loads. The preceding example of crack detection illustrates direct mechanical stressing. Luxmoore[109] has demonstrated the feasibility of holographic methods for detection of cracks and observation of their growth in concrete. Hologra-

4.4 APPLICATIONS

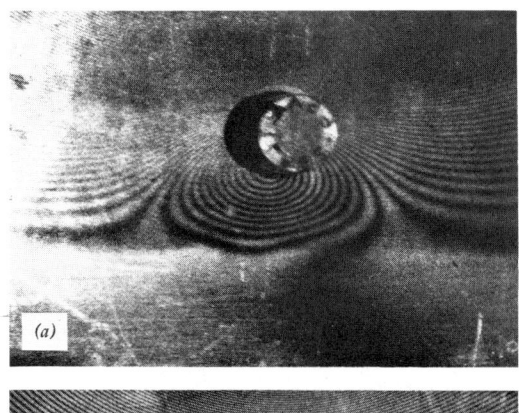

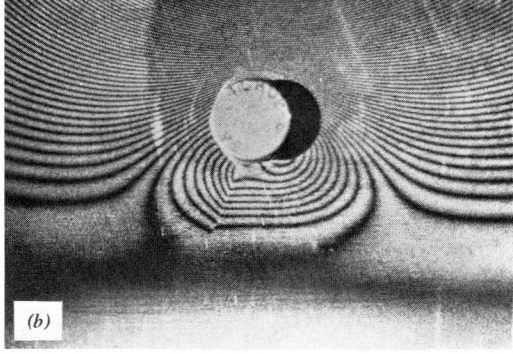

Figure 4.25 Detection of a radial crack near a 79-m-diameter bolt hole in a steel channel. A tapered-shank bolt was drawn into the hole between exposures of a two-exposure hologram (ref. 107). (a) A flaw-free specimen. (b) A specimen with a crack extending radially from the hole.

phy is particularly useful in this application because the coarse, heterogeneous structure of concrete precludes the use of many other testing methods. Crack growth was followed from its early stages to eventual catastrophic failure in cubic specimens of reinforced concrete. The cubes were loaded in compression, and a sequence of two-exposure holographic interferograms was made, with a stress increase of about 4.5 MN/m^2 between each two exposures. Splitting tests were also conducted in which 150-mm-diameter concrete cylinders were loaded in diametral compression. Early detection of fatigue damage in various composite materials by similar holographic methods has been shown to be effective in comparison with other nondestructive testing techniques.[110, 111]

Tension and bending loading has also been used to locate fabric cuts, delaminations, and overlaps in flat sheets of fiberglass-reinforced plastics.

Grünewald et al.[112] have demonstrated that the sensitivity of flaw detection in this type of application depends on the relative orientation of the flaw and the applied stress, and also on the depth of the flaw below the observed surface. In anisotropic materials the orientation of reinforcing fibers is also important. Cuts and delaminations in fiber-resin composite materials may be easier to detect if a significant period of time, perhaps of the order of 10 min, elapses between holographic exposures, because damage in the laminate grows by creeping of the resin.[112]

The most common application of holographic nondestructive testing is the detection of debonds and delaminations in composite materials and structures. Pressure or vacuum stressing is often appropriate for this application. An example is shown in Figure 4.26. The test object is a honeycomb construction panel which is formed by bonding a light-gage aluminum skin to a water-resistant paper honeycomb core. The specimen was tested by placing a simple vacuum chamber, consisting of an aluminum ring with a thick glass cover plate, against the front face of the honeycomb panel. A small vacuum could then be applied to cause the panel to bow out slightly. The panel was illuminated and viewed through

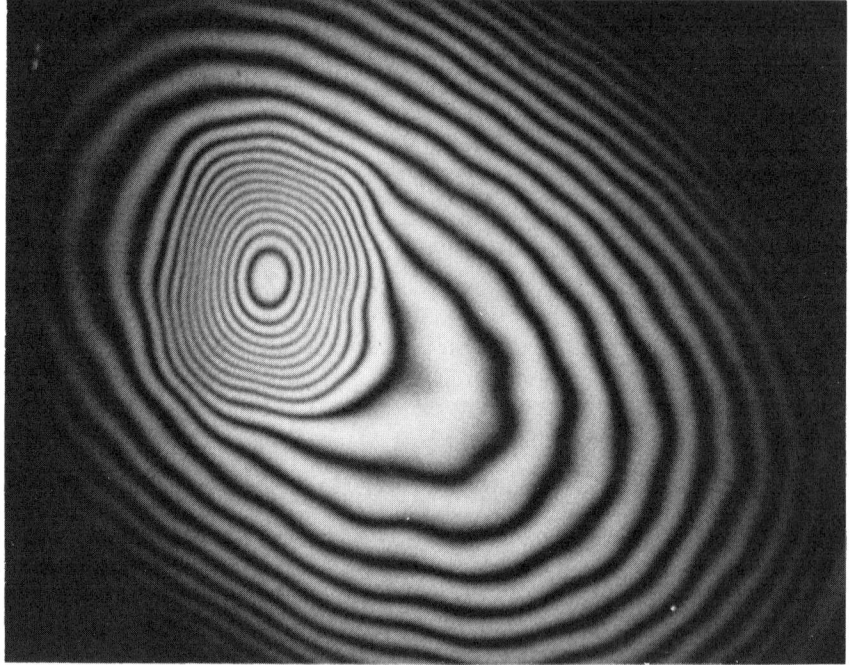

Figure 4.26 Detection of debonded region of a honeycomb construction panel (ref. 115).

4.4 APPLICATIONS

the glass cover plate. The interferogram was formed by changing the vacuum by approximately 7 kPa between exposures of a two-exposure interferogram. The presence of a rectangular debonded region between the core and the skin is clearly indicated by the fringe pattern in Figure 4.26. The skin is not restrained in this debonded region, and therefore it bulges out under the influence of the vacuum more than it does in the surrounding regions, where a good bond exists. Holographic detection of debonds and delamination by vacuum stressing has been used in many commercial and laboratory investigations.[113–115] For thin, flat specimens pressurization from behind is also useful.[112]

Internal pressurization is a convenient way to load test objects which are basically cylindrical in shape. Burchett[116] has studied the applicability of holographic interferometry to nondestructive testing of carbon composite cylinders. The cylinders were subjected to low-level pressurization to generate an interference pattern. Interferograms were quantitatively evaluated to produce plots of circumferential strain as a function of angular position on the cylinder. Burchett demonstrated good correlation between excessive variation of strain at low pressures and failure by fracture at high pressures. Further investigations of this type have been reported by Mayer and Katayanagi.[117] Application of this method to holographic nondestructive testing of solid-fuel propellant grains for rocket motors has been reported by Waters[118] and by others. The grain is a hollow cylinder of fuel of rubberlike consistency surrounded by a thin polymer liner and a thin fiberglass outer case. It is important to detect cracks in the propellant wall and debonds between the fuel and liner because these flaws cause uneven burning. Figure 4.27 is an interferogram of a solid-fuel grain approximately 37 cm long by 4 cm in diameter. Deformation was introduced by a 3.5 kPa pressure differential. The fringe contours clearly indicate the presence of an artificially introduced debond between propellant and liner.

Some commercial systems for nondestructive testing of automotive and aircraft tires utilize pressure and vacuum stressing. The objective of this testing is to locate a variety of flaws such as separations, broken belts, debonds, and voids that can occur in tires. The exterior surface of the tire can be inspected interferometrically by changing the tire pressure slightly between holographic exposures, or during observation by real-time holographic interferometry.[113] Alternatively, the inside surface of the tire can be inspected by placing the tire in a vacuum chamber.[119] A differential vacuum between holographic exposures of around 14 kPa induces a slight change in the overall tire shape and also causes a bulging out of debond regions, which contain entrapped air. This appears to be the most satisfactory method for holographic inspection of tires.[120]

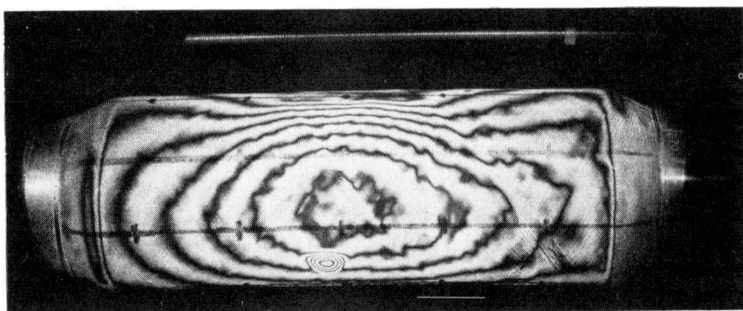

Figure 4.27 Interferogram indicating debonds between a solid propellant fuel grain and its liner (ref. 118).

Thermal stressing has been used for various applications of holographic nondestructive testing. Holographic interferometry is used to observe mechanical deformation which occurs in response to changes in surface temperature. The test object can be heated by a hot air gun, or by radiation from sources such as quartz heaters or infrared lamps, or it can be cooled by evaporation after a volatile substance like alcohol has been sprayed on it. This technique is usually most satisfactory when real-time interferometry is used, so that the dynamically changing fringe patterns can be observed. Debonds in honeycomb or laminate structures result in large local deformation when bending stresses are induced by temperature gradients. Also, air spaces created by debonds conduct less heat than do regions where a good bond exists, so the temperature field itself is distorted. Similar thermal distortions can be caused by crushed honeycomb cores or by voids, delaminations, or other nonhomogeneities in composite materials.

An unusual application of holographic nondestructive testing is the inspection of ancient Italian panel paintings, reported by Amadesi and colleagues.[121] Figure 4.28 is a holographic interferogram they recorded of "Santa Caterina," a fifteenth century painting by Pier Francesco Fiorentino. The painting is on a poplar wood panel coated with priming layers of gesso and glue. Early disclosure of detachments within such laminate structures is valuable to art conservation efforts. Detached regions are readily apparent in this interferogram. Deformation was caused by heating the surface with a stream of moderately warm air. In a related application Vlasov et al. used holographic interferometry to determine the optimum relative humidity for storage of fresco materials.[122]

Typical engineering applications of holographic nondestructing testing by thermal stressing include inspection of carbon composite cylinders,[116] tires,[113] and solid propellant rocket motor casings.[123] A somewhat different

4.4 APPLICATIONS

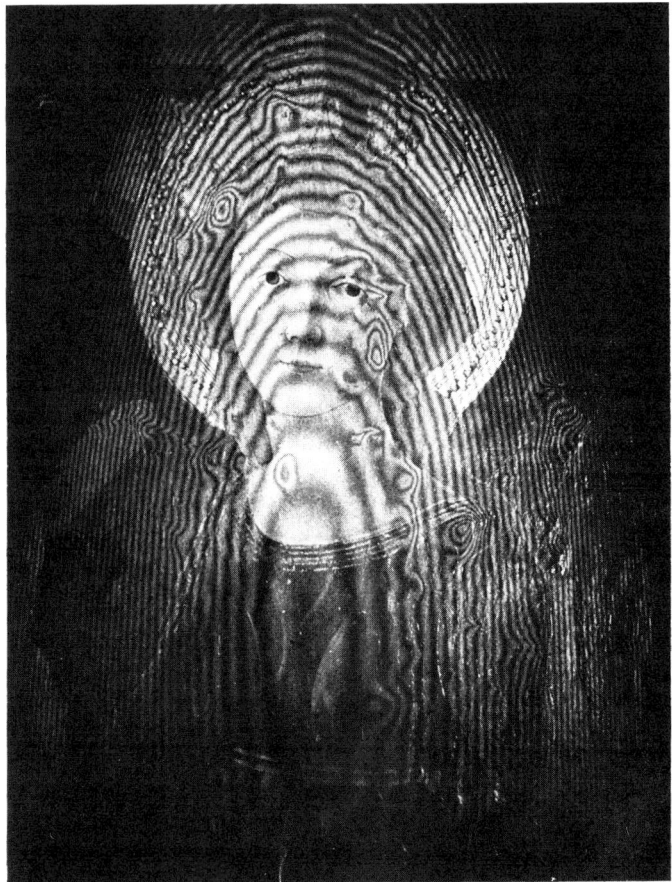

Figure 4.28 Holographic interferogram indicating delaminations in a fifteenth century panel painting (ref. 121).

application is inspection of electronic components and circuit boards.[123,125] In this case thermal stresses due to the heating of circuit elements cause small deformations of the circuit board and components, which can be observed by using holographic interferometry. Once the fringe pattern of a properly functioning circuit is known, defective elements such as overheated resistors or transistors can be located because they create perturbations in the normal pattern. This technique can be used to optimize the placement of components on a board to minimize heating.[124] Thermal stressing has also been applied for holographic detection of interference fits in metal fasteners.[126]

Vibrational excitation is useful for detecting debonds in composite materials and for disclosing inhomogeneities in materials or structures. The technique can be used in two ways. First, the entire structure being analyzed can be vibrated in a resonant mode. This creates bending stresses which may cause anomalous deformation near flaws such as voids or buildups in hollow structures. This approach has been used to test turbine blades[127] and to detect flaws in fiberglass-reinforced plastics.[112] Second, debonds near the surface, for example, between the skin and the core of a honeycomb panel, create a locally flexible structure which can be caused to resonate. Real-time holographic interferometry can be used to observe the vibration of the surface of a structure while the frequency of excitation is varied, to locate such local resonances.

Vibrational techniques, particularly for excitation of local resonances, may require high frequencies so that vibrational amplitudes are quite low. Temporally modulated holographic interferometry (see Section 4.3.5) therefore may be quite useful in these applications. A result of an investigation by Kersch[127] which illustrates this is shown in Figure 4.29. Figure 4.29a is a time-average holographic interferogram of a laminated structure vibrating at 250 kHz. The amplitude is so low that no fringe pattern is observable. A second interferogram of the same object excited in the same

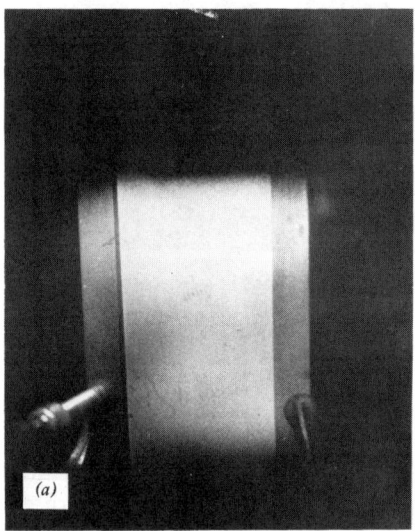 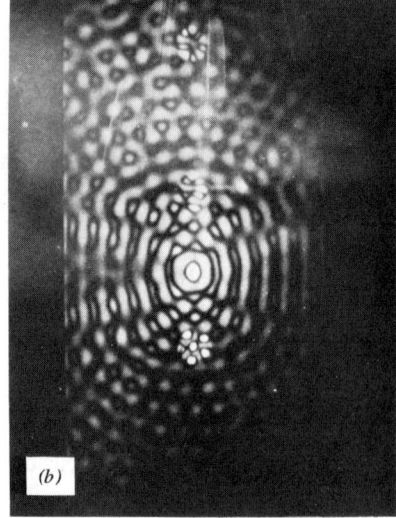

Figure 4.29 Interferograms of a laminated structure vibrating at 250 kHz (ref. 127). (*a*) Time-average interferogram with no visible fringes. (*b*) Interferogram recorded with a phase-shifted reference wave. (Reprinted from *Materials Evaluation*, **29**, 126 (1971) by permission of the American Society for Nondestructive Testing, Inc.)

way is shown in Figure 4.29b; however, a phase-shifted reference wave was used to increase the sensitivity. A rather symmetrical pattern can be seen radiating from a bright central spot located opposite the transducer. Three debonds are detectable in this interferogram; they lie along the vertical centerline and cause local disturbances in the fringe pattern.

It is clear from the examples cited above that holographic interferometry is useful for nondestructive testing if a stressing technique can be devised such that flaws induce detectable perturbations in the deformation of the object surface. Usually the stressing technique is chosen empirically with guidance provided by simple analysis of the anticipated deformation and by previous results published in the literature.

The high sensitivity of holographic interferometry, which is the basis for its usefulness, also leads to certain difficulties in testing, particularly in a nonlaboratory environment. Interferometric mechanical stability is required; this is difficult to achieve when noise and building or machine vibrations are present. Also, a stressing technique may generate such a fine pattern of fringes, due to gross deformation of the object, that perturbations caused by flaws may not be distinct or obvious to an observer.

Considerable ingenuity has been displayed in alleviating problems associated with lack of mechanical stability. Holographic interferograms can be recorded in an industrial environment if care is taken with the design of the optical configuration and of the mechanical fixtures. This is well illustrated in the articles by Hockley and Butters[128] and by Abramson and Bjelkhagen.[129] If extraneous object motion is primarily an out-of-plane translation or vibration, *object motion compensation* can be used to eliminate unwanted fringes. This is accomplished by reflecting or scattering the holographic reference wave from a point on the object surface. This technique is discussed in Section 4.3.5. Often the holographic system can be oriented so that its sensitivity is low for displacements in the direction of anticipated extraneous motion. This can be done with the aid of the holodiagram described in Section 2.4. Stability requirements can also be reduced by the use of image plane holography,[124] in which an optical system is employed to form an image of the test object near the hologram.

The *sandwich hologram* devised by Abramson[59,130] can serve to remove interference fringes generated by certain rigid-body motions executed by the test object between holographic exposures. Two film plates, P_1 and P_2 (without antihalation backing), are positioned in a plate holder with their emulsions facing the object, as shown in Figure 4.30a. A single holographic exposure is then made with a spherical reference wave while the object is in its initial unstressed configuration. Plates P_1 and P_2 are removed, and a second pair of plates, P_3 and P_4, are placed in the holder and positioned in the same manner (Figure 4.30b). The object is then

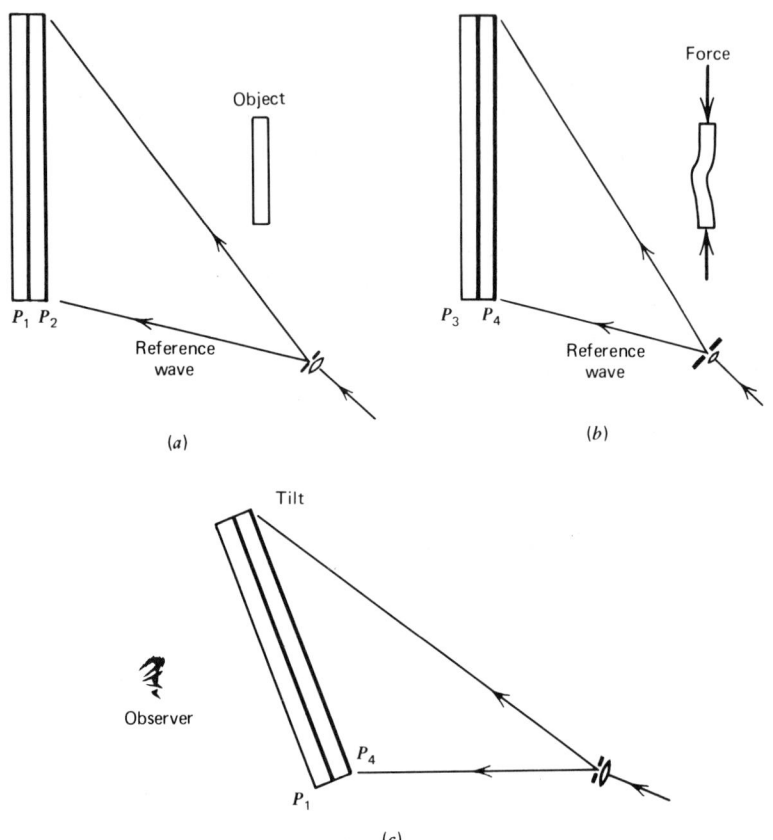

Figure 4.30 Sandwich hologram. (a) Recording the first hologram pair; (b) Recording the second hologram pair; (c) Reconstruction of a sandwich hologram suitable for object motion compensation.

stressed, and a single exposure is made on the pair of plates P_3 and P_4. After the four plates have been developed in the usual manner, a sandwich hologram is formed by combining P_1 and P_4, as shown in Figure 4.30c. Plates P_2 and P_3 can also be used to form a second sandwich hologram if desired. The holograms can be combined in a plate holder, or they can be bonded together with epoxy along their edges. The sandwich hologram is then reconstructed by illuminating it in the usual manner. Tilting the sandwich with respect to its initial position produces fringes equivalent to those caused by a tilt of the object. By tilting the sandwich in the same direction as the actual object, fringes due to rigid-body tilting of the object can be removed. A modified form of sandwich holography in which only two plates are used has been devised by Hariharan and Hegedus.[131]

4.4 APPLICATIONS

Abramson and Bjelkhagen have demonstrated industrially oriented applications of sandwich holography. For example, Abramson has used cw holographic interferometry to measure the deformation of large machine tools on a factory floor.[132] The technique has been adapted to pulsed-laser holographic interferometry by Bjelkhagen.[133,134] The two photographic plates used to form the sandwich are mounted in a plate holder which rotates at a high rate of speed (of the order of 90 revolutions per second) about an axis normal to the plates. The holograms formed by two consecutive laser pulses are angularly separated. When the sandwich hologram is constructed, the two plates are simply rotated with respect to each other until the two holograms are brought into registration. This is the only modification of the basic method which is required.

If extraneous object motion is too large to permit the recording of an ordinary off-axis hologram or to be compensated for by the techniques described above, the film plate can be rigidly attached to the object and *single-wave holography*[135] used. Neumann and Penn[136] have adapted this method to nondestructive testing and mechanical measurements. As shown in Figure 4.31, a film plate is rigidly attached to the surface of the test object, using a fixture which separates the emulsion from the object surface by a distance of 1 cm or less. The object surface must be painted with a retroflective material such as Codit®. For each exposure of a two-exposure holographic interferogram the plate and object are illuminated by a single spherical laser wave of large radius of curvature. In such a hologram the retroreflected light forms the object wave, and the laser light transmitted through the hologram in the opposite direction, serves as the reference wave. The phase difference between these two waves is nearly independent of the distance between the point source and the hologram, so the interferogram is quite insensitive to rigid-body translations of the object. Furthermore, for a 5 mm spacing between emulsion and object it was found that object rotations of up to 17 mrad were permissible. If fringes localize near the object surface, holographic interferograms of this type can be viewed in white light.

Compensation for gross deformation, as well as rigid-body motion, can be accomplished by using the *fringe control* methods developed by Champagne.[137] Consider, as an example, inspection of a honeycomb panel for debonds. If vacuum stressing is used, a relatively large region of the skin will deform and generate a pattern of concentric fringes similar to those in Figure 4.6a. The fringe spacing may be very small, especially if the test area is large, because fringe spacing is inversely proportional to the fourth power of the diameter of the stressed region. The fringes may be too fine to detect with the unaided eye, and even if they can be seen, perturbations due to flaws may not be discernible. If *real-time* interferometry is used, the frequency and form of the interference can be controlled to a large extent

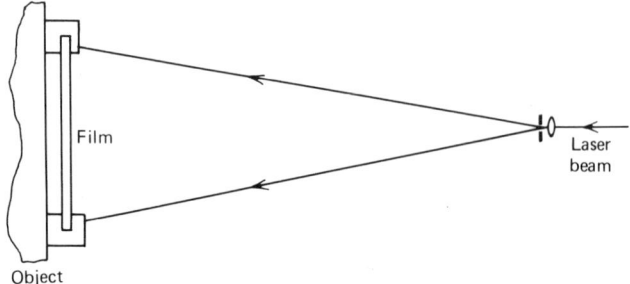

Figure 4.31 Single-wave holographic interferometry for nondestructive testing (after ref. 136). The hologram is attached to the object, which is painted with a retroreflective coating.

by using the system shown schematically in Figure 4.32. A first exposure is made in the usual manner, and the hologram is either developed in place or removed, developed, and replaced in the system. An observer looking through the hologram at the illuminated object sees a fringe pattern due to the interference of the instantaneous and reconstructed wavefronts. The variation in relative phase of these two wavefronts can be adjusted by changing the apparent location of the point source of object illumination. The apparent angular position of the source can be varied by tilting the mirror slightly about a vertical or horizontal axis, and the apparent distance from the point source to the object can be varied by translating the lens along its optical axis. By these adjustments one can compensate to first approximation for object deformation over relatively large regions of the object. Mean fringe spacing is reduced, and fringe perturbations which indicate the location of flaws become more apparent. This is illustrated in Figure 4.33.

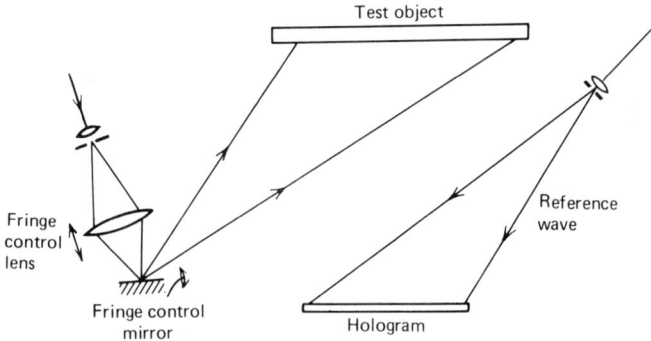

Figure 4.32 Real-time fringe control.

4.4 APPLICATIONS

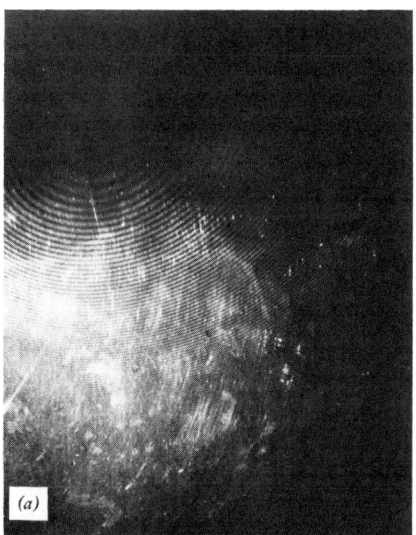 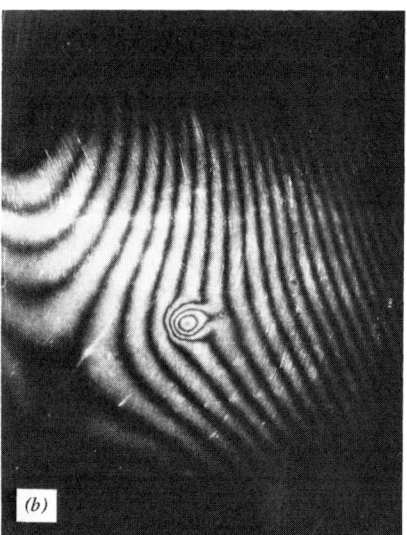

Figure 4.33 Application of fringe control to nondestructive testing of an aluminum honeycomb structure (ref. 127). (*a*) The flaw is not apparent because the fringe frequency is too high. (*b*) The same structure as in (*a*); the presence of a flaw is apparent when fringe control is used. (Reprinted from *Materials Evaluation*, **29**, 129 (1971) by permission of the American Society for Nondestructive Testing, Inc.)

Fringe control in *two-exposure* holographic interferometry must be accomplished by manipulation of the reconstruction wave. Two different, angularly separated reference waves are used as in the electronic phase detection scheme discussed in Section 2.3. Good fringe control requires the use of image plane holography, that is, there must be a one-to-one correspondence between individual points on the object and points on the hologram. An appropriate recording system is shown schematically in Figure 4.34*a*. A holographic exposure of the undeformed object is recorded using reference wave R_1. After the object has been deformed, a second hologram is recorded using reference wave R_2. The interferogram is produced by illuminating the hologram simultaneously with two reconstruction waves similar to R_1 and R_2. With a system like that shown in Figure 4.34*b*, one reconstruction wave can be varied relative to the other. As described by the holographic imaging equations (Section 1.5), this causes an apparent translation, rotation, or deformation of one reconstructed object wave with respect to the other. As in the real-time technique, it is then possible to reduce the spacing or perhaps reorient the fringes over a relatively large object area. The theory of fringe control, as

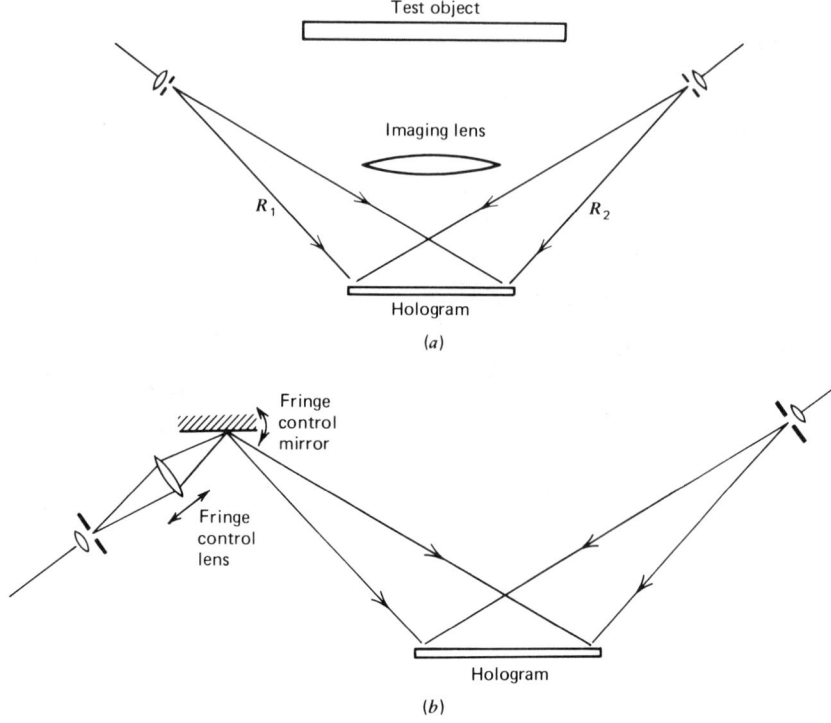

Figure 4.34 Two-exposure fringe control using separate reference waves. (*a*) Recording system. (*b*) Reconstruction system.

well as its use in quantitative analysis of fringes, was presented by Champagne[137] and has been developed in greater detail by Schumann and Dubas.[138, 139] The application of fringe control to nondestructive testing is described and illustrated by Klappert[114] and by Kersch.[127] A related method is described by de Larminant and Wei.[140]

One final application of holographic interferometry to nondestructive testing, broadly defined, is the precise comparison of the shape of machined parts with a standard or master component. In general this cannot be done holographically because the microstructures of two different parts will differ greatly on the scale of the wavelength of light. Wavefronts of laser light scattered by the two parts will not be correlated, so they cannot be compared interferometrically. However, when a typical precision-ground surface is illuminated with light, the *specular component* of the reflected light, which leaves the surface at an angle of reflection equal to the angle of incidence, is nearly independent of the surface microstructure. In some applications it is possible to compare two different objects by holographic interferometry using only specularly reflected object waves. Archbold et al.[141] developed a system based on this principle to compare cylinder bores with a master with respect to roundness, straightness, and dimension. Larinov et al.[142] have used spatial filtering to isolate the specular component of light for interferometric comparison of ground optical elements. Schaefer and Blodgett[143] have applied specular reflection to study the removal of metal by electropolishing. Archbold et al. also suggested the use of longer wavelength radiation, for example, 10.6 μm light from a CO_2 laser, for comparison of complicated shapes. Since most surfaces are smooth on this scale, direct comparison of the wavefronts of laser light scattered by two different objects would be possible and could easily be done holographically. Current efforts in the development of recording media for use at 10.6 μm may make this technique feasible in the near future.

4.4.2 Medical and Dental Research

Since holographic interferometry can be used to make whole-field measurements of the deformation of objects of complex shape without mechanical contact, it is a potentially helpful tool for medical and dental research.[144] The amount of work in this area to date is small, but the examples cited in this section suggest a potential value of holographic interferometry in medical and dental research.

Deformation of the human chest during inhalation has been measured by Zivi and Humberstone.[145] Holographic interferograms were formed, using a Q-switched ruby laser which emitted two consecutive pulses 150 μs apart. The object wave was diffused by ground glass and irradiated the subject, who wore very dark goggles because the irradiance exceeded the nominal safety limit of 0.07 J/cm^2. An assessment of the motion of the entire chest can be made when the interferogram is viewed. Bulging or shrinking areas of the chest are easily identified by closed-loop fringes. Over most regions of the chest, fringes localized very near the surface, indicating that the motion is largely rotational. Since the time interval between laser pulses was known, the velocity of chest expansion could be determined; a typical value was about 0.2 cm/s. Bjelkhagen et al.[146] have also used double-pulse holographic interferometry to determine the velocity of chest motion in order to demonstrate a potential application to diagnostic cardiology. The laser pulses were triggered by the patient's electrocardiogram, and the resulting velocity contours were qualitatively different for a patient with heart disease and for a healthy person. Another illustration of the usefulness of holographic interferometry for studying the motion of entire physiological structures is given by Fuchs and Schott.[147] They recorded interferograms of a mazerized skull when force was applied through the teeth as in chewing. The resulting distribution of deformation over the entire facial structure was evident in the interference pattern.

Holographic interferometry has been used in dental research to study the motion of teeth and the deformation of the associated facial structure and also to evaluate and nondestructively test prosthodontic devices. The deformation of gold dental bridges with soldered joints was measured by Wictorin et al.[148] Defective solder joints could be detected by excessive deformation under small loads. Quantitative measurements of the variation in elastic deformation with distance from a joint to the point of load application were also made. A detailed study of models of prosthodontic appliances and fixed bridgework was made by Wendendal and Bjelkhagen.[149] The specimens were subjected to 1.5 N differential loads in simple three-point bending, and 170 different two-exposure interferograms, recorded with an He-Ne laser, were analyzed to determine the distribution and maximum value of deformation. Comparisons were then made among bar elements which differed in type and method of manufacture. Some correlation of deformation with casting defects such as voids was also found when the results of the interferometric study were compared with X-ray shadowgraphs of the elements.

Wendendal and Bjelkhagen[150] also demonstrated the feasibility of using holographic interferometry to measure the motion of teeth and the deformation of related structures *in vivo*. The patients' lips and cheeks were retracted by an acrylic hook so that the teeth and gums were visible. Salivation was inhibited, and the exposed regions of jaw and teeth were coated with a gold paint. This increased the reflectance of the surface and overcame the difficulty in forming interferograms of biological materials, due to their translucence. The patient was asked to bite, and a subminiature force sensor between two teeth triggered a pulse of the ruby laser when a prespecified force of about 0.5 N was reached. A second pulse was triggered 450 μs later. The resulting interferograms were evaluated to determine the relative displacement of teeth and the deformation of the surrounding structure.

Time-average holographic interferometry has been used to study tympanic membrane vibrations in both laboratory animals and human beings.[151, 152] The quantitative results from laboratory animals and the qualitative results from human cadavers have led to increased understanding of the mechanics of hearing. Khanna and Tonndorf[151] recorded over 1000 time-average holograms of tympanic membranes in both living cats and fresh cat cadavers. After surgery to increase optical access to the tympanic membrane the latter was coated with a thin layer of submicron-size bronze particles to increase reflectance and eliminate translucence. Holograms were then recorded while the membrane responded to sound in the range 600 to 6000 Hz at various sound pressure levels. Figure 4.35

4.4 APPLICATIONS

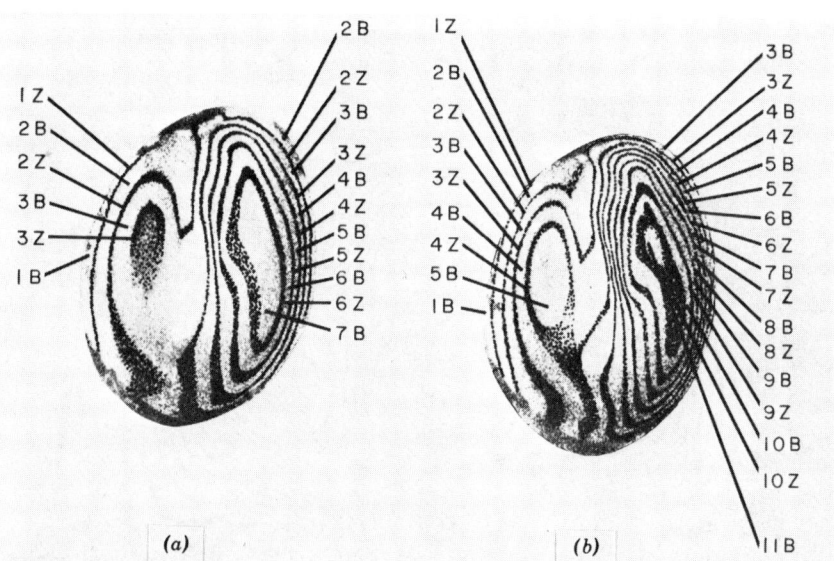

Figure 4.35 Vibrational amplitude patterns of the tympanic membrane of a cat responding to a 600 Hz sound at two different sound pressure levels (ref. 151). Fringe order numbers are identified.

shows two typical interferograms from this study. The modal patterns observed with holographic interferometry support the concept of a curved membrane developed by Helmholtz, rather than the more recent rigid-plate model based on measurements made with capacitive probes. Extension of this study to include measurements in fresh human cadaver ears supported the same conclusions.[152]

Pulsed-laser holographic interferometry also has been used to study deformation of the tympanic membrane in response to acoustic impulses.[153, 154] The objective of this study was to increase understanding of the mechanism of ear damage due to explosions and weapon discharges. Two-exposure holograms of the response of guinea pig tympanic membranes to noise generated by bursting electrical fuses were recorded using 2 ns pulses of a Q-switched ruby laser triggered by a microphone. The observed location of maximum deformation was in agreement with clinical observations of the common location of lesions caused by acoustic impulses.

4.4.3 Solid Mechanics

The early work of Haines and Hildebrand[21] demonstrated the potential of holographic interferometry for measuring the deformation of complex,

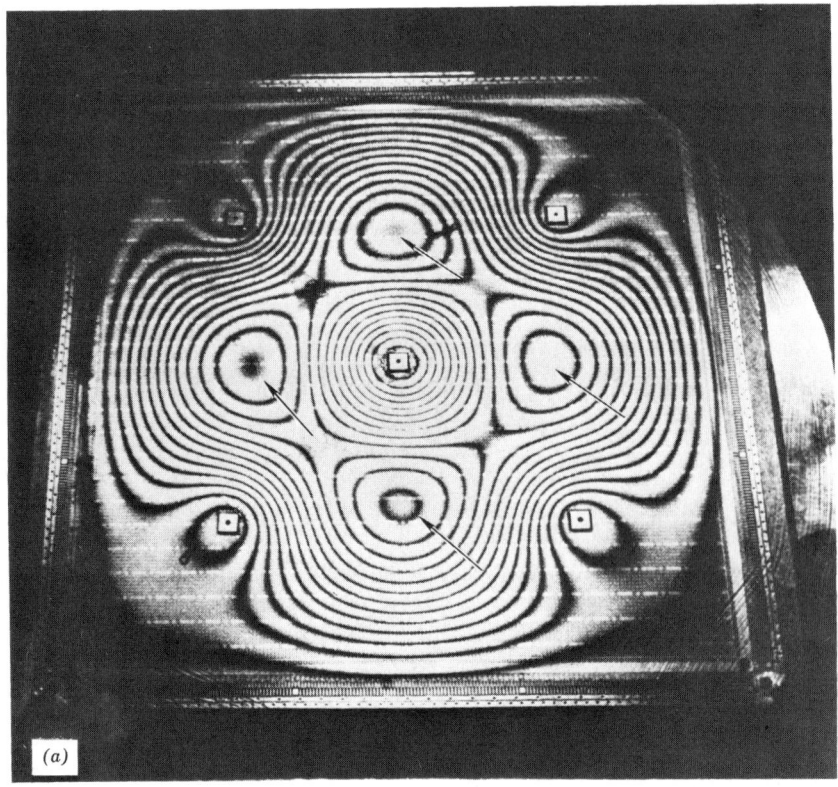

Figure 4.36 Deflection of a rectangular plate fastened with five struts and subjected to a uniform pressure (ref. 159). (a) Holographic interferogram. (b) Prediction of computer model.

opaque objects in response to a mechanical load. Subsequently, Gottenberg[155] made precise measurements of the bending of a prismatic bar and demonstrated their agreement with calculations based on three-dimensional elasticity theory. Sampson[156] also demonstrated the applicability of holographic interferometry to solid mechanics by measuring the normal deflection of cantilever beams and of clamped circular plates subjected to uniform pressure. These measurements also were in excellent agreement with theory. Both normal and in-plane deformation of a circular disk was measured by Sciammarella and Gilbert,[157] whose experiments are discussed in Section 2.3. The experiments of Wilson,[4] which are described in Section 4.2, showed that membrane strains could be measured by holographic interferometry. Small rotation and torsion of cylindrical shafts were also measured by the same author.[158] Rotation of a cylinder about its

4.4 APPLICATIONS

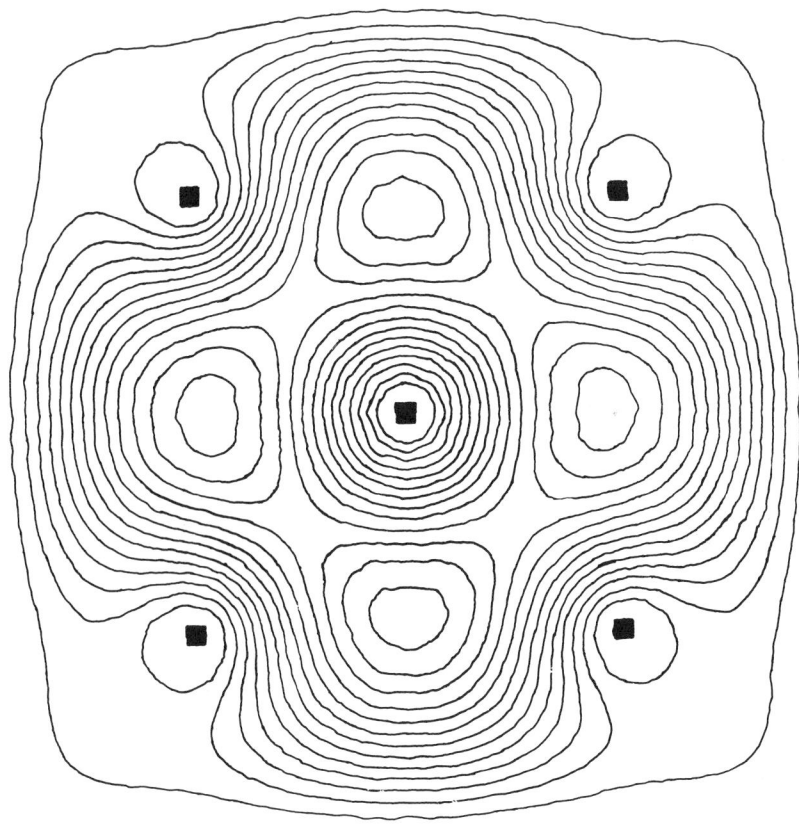

(b)

Figure 4.36 Cont.

shaft gives rise to a set of uniformly spaced fringes parallel to the axis of rotation; their spacing is a direct measure of the rotation. If a cylinder is subjected to torsional strain, the fringe spacing varies in the axial direction and provides an accurate measurement of the local rotation.

Interaction of holographic interferometry and computer modeling in a real design situation is illustrated in a study reported by Wilson et al.[159] They dealt with a rectangular epoxy-laminated plate which formed one side of a mildly pressurized enclosure. The plate was connected to the opposite face of the enclosure by five struts used to stiffen the structure. Figure 4.36a is a two-exposure interferogram of the stiffened plate. The fringes are contours of constant normal deflection in response to a pressure change of 0.196 kPa between holographic exposures. Figure 4.36b is a plot of deflection contours determined analytically. The interferograms were useful for verifying the general validity of the model, for developing appropriate boundary conditions, and for determining numerical values of the stiffness factor.

Full-field fringe patterns are very useful for demonstrating anisotropic response to mechanical loads. Quantitative deformation analysis can serve to determine the numerical values of coefficients used in analytical models of anisotropic materials, and to determine parameters such as critical buckling loads.[160] Bending of isotropic plates has been studied by Boone and Verbiest[17] (see Section 4.2) and Saito et al.[18] Deformation and strain of cylindrical shells[161, 162] and hemispherical shells[163] have also been measured by holographic interferometry.

The sensitivity of holographic interferometry to very small deflections was used to advantage by Hunter and Morton[164] in the development of torsional relaxation tests. The objective of their research program was to devise short-term (~ 100 to 1000 h) tests to predict long-term (~ 8 years) energy release due to relaxation of torsion bars. A torsion bar was subjected to a torsional strain. Real-time holographic interferometry was used to monitor the rate of relaxation by observing the changing fringe pattern on a uniform stress beam attached to the bar. Periodically during the test a hologram was recorded, and the fringe pattern was observed as a function of time. In this manner relaxation rates on the order of 10^{-8} N·m/s could be measured. This represents an increase in accuracy by a factor of about 400 over strain gage measurements. In addition, creep in the strain-gage adhesive was far greater than the relaxation effects, so a complicated sequence involving periodic unloading of the torsion bar was required if the optical method was not used. A model of relaxation with time was developed based on the interferometric measurements. Measurement periods as short as 100 h can serve as a basis for reasonably accurate prediction of long-time relaxation.

The high sensitivity of holographic interferometry has also been used to advantage in a study of the deformation of concrete due to absorption of water. Goldberg[165] developed a holographic interferometer for this purpose. Four separate object waves are used. One pair of object waves has propagation vectors lying in a horizontal x-y plane, and another pair has propagation vectors lying in a vertical y-z plane; the z axis is normal to the object surface. A single holographic exposure is made while the object is illuminated by all four waves. This hologram is then used to form real-time interferograms while the object is illuminated by each object wave, one at a time. All interferograms are formed by photographing the object from a single viewing direction along the z-axis. The four interferograms enable one to calculate the vector displacement of any point on the surface and

4.4 APPLICATIONS

provide one redundant measurement, which is used to check the consistency of the calculated displacement. Goldberg et al.[166] used this interferometer to study the swelling of rectangular bars of concrete due to absorption of water from above. Sequences of interferograms were recorded over intervals ranging from 1.83 to 5.75 h, and were analyzed to produce maps of both normal and in-plane deformation of the concrete specimens. This study required interferometric sensitivity, but the rough surfaces precluded the use of nonholographic techniques.

The noncontacting nature of interferometric methods enables one to make measurements in certain hostile or corrosive environments. This is illustrated by deformation measurements made by Hsu and Moyer[167] at high temperatures. Their test specimen was a Poco-graphite cantilever beam which was mechanically deformed inside a furnace at temperatures as high as 800°C. Holographic interferograms were formed, with the object being viewed through a quartz window. Hsu and Lewak[168] have also studied temperature-induced deformations of composite plates over a wide range of temperatures. Another example of measurement in a hostile environment is the real-time interferometric observation of stress-corrosion cracking of metal specimens submerged in methanol.[169]

The applications cited thus far in this section involve static or very slowly varying deformations. Holography also has application to the study of the vibration and impact response of mechanical components and systems. Aprahamian and Evensen have reported extensive experiments in which vibrational modes of beams[170] and plates[171] were identified. A detailed study of triangular plates has been made by Williams et al.[172] Holographic interferometry is particularly useful for the study of high-frequency vibrations for which amplitudes are often too small to measure by noninterferometric methods. When searching for resonant modes, real-time holography (Section 4.3.4) or speckle interferometry (see Section 7.3) is used to observe changes in vibration pattern while the exciting frequency is varied. It is important to distinguish single resonant modes from combinations of modes. As one tunes through resonance, node lines for combination modes will move about the vibrating surface, whereas node lines for a single mode will remain stationary but become quite narrow.[173] Once the mode of interest has been established, a time-average hologram can be recorded in order to provide a permanent record and an interferogram with better fringe visibility. Figure 4.37 includes typical mode patterns of a rectangular plate studied by Aprahamian and Evensen.[171] They identified a large number of modes in the frequency range 162 to 20,000 Hz and also studied selected modes at frequencies as high as 100 kHz.

Vibration measurements extending into ultrasonic frequency ranges also have been reported by Wilson et al.[174] and by Tuschak and Allaire.[175] The

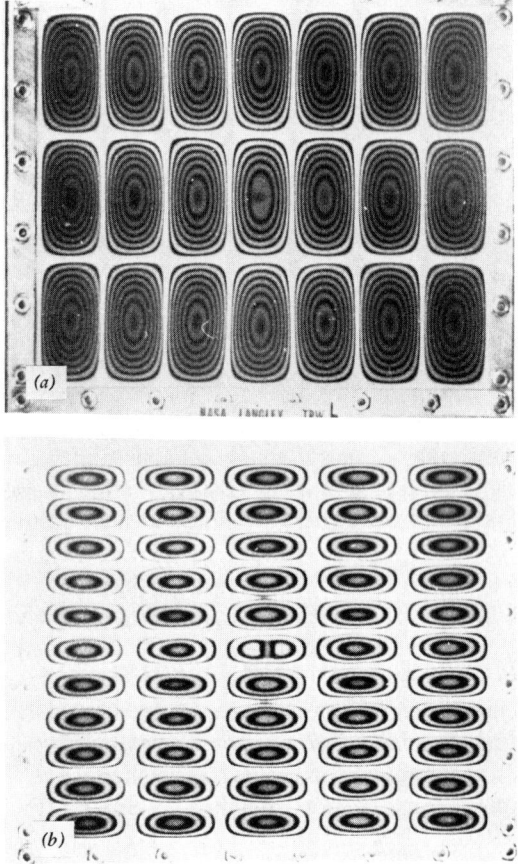

Figure 4.37 High-order vibration modes of a rectangular plate of dimensions 203 × 254 × 1.58 mm (ref. 171). (a) 5,132 Hz. (b) 12,670 Hz. (c) 18,158 Hz. (d) 20,892 Hz.

former authors studied the vibration of the horn and tool of an ultrasonic bonder at its resonant frequency of 60 kHz. Time-average holography was used to locate nodal regions and measure amplitudes. These measurements provided information which cannot be obtained by other methods. However, a detailed explanation of the mode structures was difficult because combined modes were clearly present, but the detailed phase information required for a complete analysis of the interferogram was not available (see Section 4.3.3). Tuschak and Allaire[175] studied an ultrasonic resonator in the form of a solid circular cylinder of steel driven by piezoelectric disks at 25.1 kHz. They measured vector amplitudes of the surface of the resonator with an estimated accuracy of ±9 percent of the maximum amplitude. The

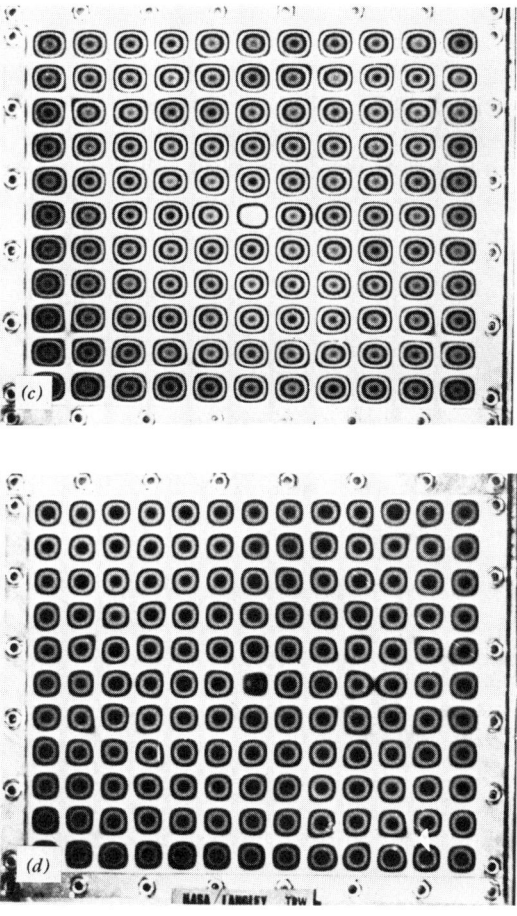

Figure 4.37 Cont.

interferometric measurements agreed well with theoretical results normalized to the same maximum displacement. Borissov et al.[176] have studied the vibration of piezoelectric quartz resonators.

Chomat and Miler[177] have demonstrated amplitude measurements of the vibration of various electromechanical components such as piezoelectric resonators, tuning fork filters, telephone receiver diaphragms, and circular and elliptical loudspeakers. Lashkari and Weingarten[178] recorded holographic interferograms of a segmented shell consisting of a circular cylinder topped with a coaxial conical shell. Their measurements demonstrated that the two segments essentially act independently and may not even vibrate at the same circumferential harmonic at a given frequency. This knowledge aided the development of finite-element analysis of segmented shells. Wakashima et al.[179] have used time-average holographic interferometry to determine the viscoelastic parameters of plasticized epoxy resins.

The potential usefulness of the interaction of holographic vibration measurement with design of complicated systems was illustrated in an appealing manner by Ågren and Stetson's[180] work on resonances of the treble viol. This musical instrument is considered to be weak and uneven in both tone and volume, and seldom balances well with other members of the viol family. Holographic interferometry was used to study the resonances of the bottom plate of a classical treble viol. This technique was found to be more accurate and informative than the traditional sound-level-meter measurements. By using a combination of the results of the holographic analysis and intuition, a new design for a "magnum treble viol" was developed with the aim of approaching an optimum spacing of resonant peaks. The instrument was manufactured and found to be stronger and more even as tested holographically and by musical performance.

Techniques for the interaction of holographic vibration measurements with analysis and design of vibrating or statically deflecting structures have been developed by Stetson and Taylor[173, 181–183] in a more conventional engineering context. Their approach is summarized here. From the theory of structural mechanics it is known that the normal deflection of a thin, nearly flat plate or membrane can be expressed as a series of products of temporal and spatial functions:

$$w(x,y,t) = \sum_{j=1}^{\infty} \tau_j(t) \Phi_j(x,y). \qquad (4.125)$$

The $\Phi_j(x,y)$ are termed *normal mode functions*. The time-varying amplitude coefficients $\tau_j(t)$ can be determined by substituting equation 4.125 into the equations of motion and utilizing the fact that the Φ_j are orthogonal with respect to the distributed mass of the plate. For example, if a concentrated normal force which varies sinusoidally with frequency ω and amplitude F is applied at the point (x_o, y_o), the resulting normal displacement is

$$w(x,y,t) = \text{Re}\left\{ F \exp(i\omega t) \sum_{j=1}^{n} \frac{\Phi_j(x,y)\Phi_j(x_o,y_o)/M_j}{2\beta_j i\omega + (\omega_j^2 - \omega^2)} \right\}, \qquad (4.126)$$

where M_j is the generalized distributed mass for the jth mode:

$$M_j = \iint_A \rho(x,y) \Phi_j^2(x,y) \, dx \, dy, \qquad (4.127)$$

4.4 APPLICATIONS

β_j is a coefficient which describes the damping of the jth mode, ω_j is the resonant frequency of the jth mode, $\rho(x,y)$ is the mass per unit area of the plate, and the series has been truncated to n terms. Equation 4.126 can be used to describe accurately the response of a plate, including combination modes, if the damping is low enough that mode coupling is negligible.[152] The normal mode functions $\Phi_j(x,y)$ can be determined by recording time-average holographic interferograms of the plate being studied at resonant frequencies ω_j. Stetson and Taylor[173] have demonstrated this for the case of asymmetric vibrations of circular disks. The amplitudes of two component modes were measured to determine Φ_j. Equation 4.126 was then used to predict a combined mode, and the predictions agreed well with holographic measurements.

It is interesting that normal mode theory coupled with vibration measurements can be used to predict *static deflections*.[181] If we let $\omega \to 0$ in equation 4.126, we obtain an expression for deflection in response to a concentrated static load F applied at x_0, y_0:

$$w(x,y) = F \sum_{j=1}^{n} \frac{\Phi_j(x,y)\Phi_j(x_0,y_0)}{M_j \omega_j^2}. \tag{4.128}$$

Equation 4.128 can be interpreted as expressing the response to a concentrated force at x_0, y_0 in terms of a set of *influence coefficients*. The point is that direct experimental determination of static influence coefficients is very difficult at best; however, the normal mode functions can be determined rather easily by time-average holography. To demonstrate this approach, Stetson and Taylor[181] determined Φ_j for the first five modes of a rectangular steel plate and then used equation 4.128 to compute the response to concentrated forces at several different locations. The agreement of these computations with experimental results was good, with an error of 11 percent in the worst case.

It has been shown that perturbation theory can be used with holographically determined normal mode functions as a starting point to predict the effect of small changes in parameters such as mass distribution and stiffness moduli.[182] This could be applied for correction of undesirable responses in existing structures, or in evolutionary design procedures starting with a prototype and proceeding toward design goals.

Holographic interferometry has been used to measure the amplitude of vibrating objects which rotate at high speeds, such as magnetic disks for computers, propellers, and turbine blade assemblies. One approach is to rotate the hologram on the same axis as the object and at the same speed.[184] Alternatively, two-exposure interferograms can be recorded with an appropriately synchronized Q-switched pulsed laser. Sikora and Mendenhall[185] have recorded interferograms of a rotating marine propellor in this manner. Synchronization was achieved by triggering the pulsed laser with an He-Ne laser beam which was intercepted by the propellor blades. Chen[186] triggered a pulsed laser by using a timing mark on the hub of a high-speed (105 rad/s) rotating magnetic disk, whose dynamic behavior was being studied. An electronic record of the time interval between actuation of a detector by the timing mark and pulsing of the laser permitted the disk to be repositioned accurately after rotation ceased. A second holographic exposure was then made to produce an interferogram.

A different approach to the study of rotating objects has been reported by Stetson,[187] who used an *optical derotator* in the form of a folded Abbé inverting prism mounted in the hollow shaft of an electric motor. When the prism is axially aligned with the rotating object and rotates at half of its speed, an image transmitted through it remains stationary. The derotation is sufficiently accurate to permit the recording of double-pulse holographic interferograms. Figure 4.38 is a dramatic example of the ability of this system to operate under nonlaboratory conditions. This figure is a two-exposure interferogram indicative of the vibration of a turbine fan rotating at 4460 rpm while being viewed through the intake of an operating jet aircraft engine.

Very rapid phenomena such as stress waves in solids can be studied using holographic interferometry with a pulsed-laser source. Experiments of this type are inherently more costly and sophisticated than those involving cw lasers, but they provide measurements of the spatial distribution of highly transient deformations which cannot be obtained by other techniques. Gottenberg[155] demonstrated the use of a Q-switched ruby laser to measure the displacement of an aluminum bar in response to compressional impact loading when a 22-caliber magnum bullet was fired axially at one end. A similar study was made by Robertson and King.[188] In experiments of this type the first holographic exposure is made while the object is stationary, and the second exposure is made at a predetermined time after impact. The time delay is usually achieved by electronic means after electrical, optical, or acoustical triggering associated with the impact of a projectile or pendulum on the test object.

Flexural waves in beams[189] and plates[190] due to concentrated impact loads have been measured by Aprahamian and his co-workers. These investigations show that high-quality, two-exposure interferograms can be recorded using ruby lasers Q-switched with Pockels cells, and that accurate quantitative evaluation is possible. Figure 4.39 shows two interferograms of an aluminum beam which was struck from behind by a ballistic pendulum. The interferograms were recorded at 12.5 μs and 25 μs after impact and show the evolution of the stress wave from a radially symmetric form to a nearly one-dimensional wave. The development of stress

4.4 APPLICATIONS

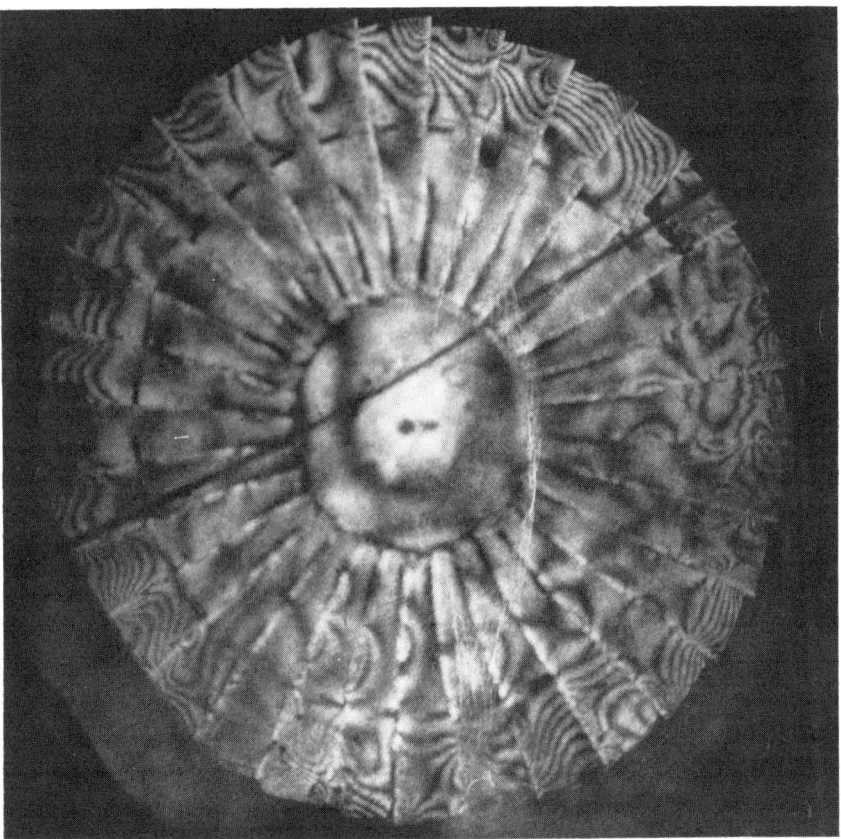

Figure 4.38 Two-exposure interferogram of a vibrating turbine fan rotating at 4460 rpm in an operating jet aircraft engine. (Photograph courtesy of Dr. K. A. Stetson, United Technology Laboratories.)

waves in a rectangular plate was studied[190] by recording interferograms at eight different time delays varying from 10 to 150 μs after impact. Fringe interpretation in experiments of this type requires some physical insight because normal deformation changes sign along the wave. Interpretation is quite difficult after the stress wave has been reflected at the boundaries because no zero-order fringe can be located. The measurements of Aprahamian et al.[190] deviated less than 10 percent from analytical stress wave solutions. Similar techniques have been used to observe shock waves due to the impact of high-energy electron beam pulses with solids[191] and the impact of projectiles with ceramic materials,[192] and to assess the nature of

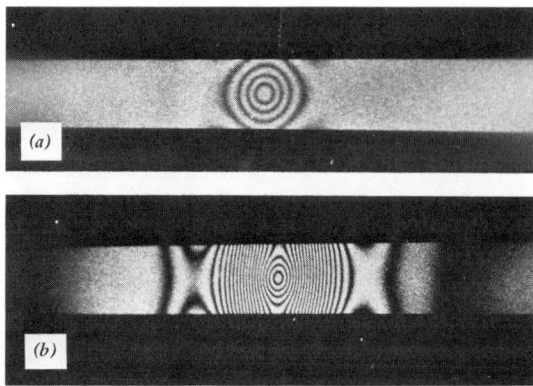

Figure 4.39 Propagating stress wave in an aluminum beam after impact by a ballistic pendulum: (a) 12.5 μs after impact; (b) 25 μs after impact (ref. 189).

mass ejection from solids.[193] Impact loading has been used as an excitation technique in holographic nondestructive testing to locate debonds between a boron-epoxy composite and a titanium substrate.[194]

When transient deformations are large, it is not possible to make measurements with respect to the undeformed state; however, *differential interferometry* can be applied to determine the change of deformation during some interval of time, usually of the order of microseconds, during the transient.[195] This requires the use of a laser which produces a double pulse with the desired temporal spacing.

Double-pulse differential holographic interferometry was applied by Wilson and Strope[196] to study the deformation of a printer type-piece after impact by the hammer of the printer mechanism. Their holographic interferograms were recorded using two 20 ns laser pulses separated by about 10 μs. The first pulse was triggered just before the hammer struck the type-piece; the second pulse occurred after deformation of the type-piece. The normal deformation data were fitted with cubic splines and differentiated numerically to determine in-plane strain (see Section 4.2). Agreement of these strains with values obtained by finite-element analysis was generally good. An unanticipated twisting motion of the type-piece was revealed by the interferograms.

Two limitations of the measurement of transient deformation by pulsed-laser holographic interferometry should be noted. First, the object motion must be only a fraction of a wavelength during each pulse. For normal deformation and a 20 ns pulse this limits the rate of deformation to about 4 m/s. Second, the temporal development of a phenomenon such as propagation of a stress wave must be studied by performing multiple experiments, each with a different time delay.

REFERENCES

1. I. H. Shames, *Mechanics of Deformable Solids*, Prentice-Hall, Englewood Cliffs, NJ, 1964.
2. E. B. Aleksandrov and A. M. Bonch-Bruevich, Investigation of surface strains by the hologram technique, *Sov. Phys. Tech. Phys.*, **12**, 258–265 (1967).
3. A. E. Ennos, Measurements of in-plane surface strain by hologram interferometry, *J. Sci. Instrum.*, Ser. II, **1**, 731–746 (1968).
4. A. D. Wilson, Inplane displacement of a stressed membrane with a hole measured by holographic interferometry, *Appl. Opt.*, **10**, 908–912 (1971).
5. J. Wallach, J. M. Holeman, and F. A. Passanti, Holographic strain measurement on a tensile specimen, *Proceedings of Symposium on Engineering Applications of Holography* (February 16–17, 1972, Los Angeles, CA), Society of Photo-Optical Instrumentation Engineers, Redondo Beach, CA, 1972, pp. 167–175.
6. Reference 1, p. 99.
7. S. P. Timoshenko and S. Woinowsky-Krieger, *Theory of Plates and Shells*, McGraw-Hill, New York, 1959.
8. L. H. Taylor and G. B. Brandt, An error analysis of holographic strains determined by cubic splines, *Exp. Mech.*, **12**, 543–548 (1972).
9. G. B. Brandt and L. H. Taylor, Holographic strain analysis using spline functions, *Proceedings of Symposium on Engineering Applications of Holography* (February 16–17, 1972, Los Angeles, CA), Society of Photo-Optical Instrumentation Engineers, Redondo Beach, CA, 1972, pp. 123–131.
10. B. Carnahan and J. O. Wilkes, *Digital Computing and Numerical Methods*, Wiley, New York, 1973, pp. 307–311.
11. J. H. Ahlberg, E. N. Nilson, and J. L. Walsh, *The Theory of Splines and Their Applications*, Academic Press, New York, 1967.
12. J. E. Sollid, Holography applied to structural components, *Opt. Eng.*, **14**, 460–469 (1975).
13. P. Bézier, *Numerical Control, Mathematics and Applications*, Wiley, New York, 1972.
14. D. Post, The moiré grid-analyzer method for strain analysis, *Exp. Mech.*, **5**, 368–377 (1965).
15. P. S. Theocaris, Moiré fringes: a powerful measuring device, *Appl. Mech. Rev.*, **15**, 333–339 (1962).
16. J. P. Duncan and D. G. Sabin, An experimental method for recording curvature contours in flexed elastic plates, *Exp. Mech.*, **15**, 333–339 (1962).
17. P. Boone and R. Verbiest, Application of hologram interferometry to plate deformation and translation measurements, *Opt. Acta*, **16**, 555–567 (1969).
18. H. Saito, I. Yamaguchi, and T. Nakajima, Applications of holographic interferometry to mechanical experiments, in *Applications of Holography*, E. S. Barrekette, W. E. Kock, T. Ose, J. Tsujiuchi, and G. W. Stroke (Eds.), Plenum Press, New York-London, 1971, pp. 105–126.
19. K. A. Stetson, Moiré method for determining bending moments from hologram interferometry, *Opt. Laser Technol.*, **2**, 80–84 (1970).
20. R. K. Erf, J. P. Waters, R. M. Gagosz, K. A. Stetson, G. Whitney, and H. Aas, *Nondestructive Holographic Techniques for Structure Inspection*, Annual Technical Report M991208-24, United Aircraft Research Laboratories, East Hartford, CT, July 1973, pp. 27–54.

21. K. A. Haines and B. P. Hildebrand, Surface-deformation measurements using the wavefront reconstruction technique, *Appl. Opt.*, **5**, 595–602 (1966).
22. W. Schumann, Some aspects of the optical techniques for strain determination, *Exp. Mech.*, **13**, 225–231 (1973).
23. K. A. Stetson, Holographic strain analysis by fringe localization planes, *J. Opt. Soc. Am.*, **66**, 627 (1976).
24. W. Prager, *Introduction to Mechanics of Continua*, Ginn, Boston, 1961, pp. 21–28.
25. M. Dubas and W. Schumann, The determination of strain at a nontransparent body by holographic interferometry, *Opt. Acta*, **21**, 547–562 (1974).
26. M. Dubas and W. Schumann, On direct measurement of strain and rotation in holographic interferometry using the line of complete localization, *Opt. Acta*, **22**, 807–819 (1975).
27. K. A. Stetson, Fringe interpretation for hologram interferometry of rigid-body motions and homogeneous deformations, *J. Opt. Soc. Am.*, **64**, 1–10 (1974).
28. K. A. Stetson, Fringe vectors and observed-fringe vectors in hologram interferometry, *Appl. Opt.*, **14**, 272–273 (1975).
29. K. A. Stetson, Homogeneous deformations: determination by fringe vectors in hologram interferometry, *Appl. Opt.*, **14**, 2256–2259 (1975).
30. R. Pryputniewicz and K. A. Stetson, Holographic strain analysis: extension of fringe-vector method to include perspective, *Appl. Opt.*, **15**, 725–728 (1976).
31. R. J. Pryputniewicz, Holographic strain analysis: an experimental implementation of the fringe vector theory, *Appl. Opt.*, **17**, 3613–3618 (1978).
32. R. L. Powell and K. A. Stetson, Interferometric analysis by wavefront reconstruction, *J. Opt. Soc. Am.*, **55**, 1593–1598 (1965).
33. K. A. Stetson and R. L. Powell, Interferometric hologram evaluation and real-time vibration analysis of diffuse objects, *J. Opt. Soc. Am.*, **55**, 1694–1695 (1965).
34. H. Osterberg, An interferometer method of studying the vibrations of an oscillating quartz plate, *J. Opt. Soc. Am.*, **22**, 19–35 (1932).
35. K. A. Stetson and K. Singh, Measurement of signal-to-noise ratio in hologram reconstructions, by vibration interferograms, *Opt. Laser Technol.*, **3**, 104–108 (1971).
36. B. M. Watrasiewicz, Mechanical vibration analysis by holographic methods, *Opt. Technol.*, **1**, 20–23 (1968).
37. M. R. Wall, Recording high-order holographic vibration fringes, *Opt. Technol.*, **1**, 266–270 (1969).
38. J. D. Redman, Holographic velocity measurement, *J. Sci. Instrum.*, **44**, 1032–1033 (1967).
39. M. Zambuto and M. Lurie, Holographic measurement of general forms of motion, *Appl. Opt.*, **9**, 2066–2072 (1970).
40. C. S. Vikram and R. S. Sirohi, Holographic images of objects moving with constant acceleration, *Appl. Opt.*, **10**, 672–673 (1971).
41. J. Janta and M. Miler, Time-average holographic interferometry of damped oscillations, *Optik*, **36**, 185–195 (1972).
42. P. C. Gupta and K. Singh, Time-average hologram interferometry of periodic, noncosinusoidal vibrations, *Appl. Phys.*, **6**, 233–240 (1975).
43. P. C. Gupta and K. Singh, Characteristic fringe function for time-average holography of periodic nonsinusoidal vibrations, *Appl. Opt.*, **14**, 129–133 (1975).
44. W. T. Thomson, *Vibration Theory and Applications*, Prentice-Hall, Englewood Cliffs, NJ, 1965, pp. 140–142.

REFERENCES

45. P. C. Gupta and K. Singh, Hologram interferometry of vibrations represented by the square of a Jacobian elliptic function, *Nouv. Rev. Opt.*, **7**, 95–100 (1976).
46. K. A. Stetson, Finges of hologram interferometry for simple nonlinear oscillations, *J. Opt. Soc. Am.*, **62**, 297–298 (1972).
47. K. A. Stetson, Hologram interferometry of nonsinusoidal vibrations analyzed by density functions, *J. Opt. Soc. Am.*, **61**, 1359–1362 (1971).
48. W. B. Davenport, Jr., and W. L. Root, *An Introduction to the Theory of Random Signals and Noise*, McGraw-Hill, New York, 1958, p. 67.
49. N. E. Barnett, Vibration analysis by holography, *J. Opt. Soc. Am.*, **57**, 1406(A) (1967).
50. N.-E. Molin and K. A. Stetson, Measuring combination mode vibration patterns by hologram interferometry, *J. Phys. E: Sci. Instrum.*, **2**, 609–612 (1969).
51. K. A. Stetson, Effects of beam modulation on fringe loci and localization in time-average hologram interferometry, *J. Opt. Soc. Am.*, **60**, 1378–1388 (1970).
52. N.-E. Molin and K. A. Stetson, Finge localization in hologram interferometry of mutually independent and dependent rotations around orthogonal, non-intersecting axes, *Optik*, **33**, 399–422 (1971).
53. K. A. Stetson, Method of stationary phase for analysis of fringe functions in hologram interferometry, *Appl. Opt.*, **11**, 1725–1731 (1972).
54. A. D. Wilson, Characteristic functions for time-average holography, *J. Opt. Soc. Am.*, **60**, 1068–1071 (1970).
55. A. D. Wilson and D. H. Strope, Time-average holographic interferometry of a circular plate vibrating simultaneously in two rationally related modes, *J. Opt. Soc. Am.*, **60**, 1162–1165 (1970).
56. A. D. Wilson, Computed time-average holographic interferometric fringes of a circular plate vibrating simultaneously in two rationally or irrationally related modes, *J. Opt. Soc. Am.*, **61**, 924–929 (1971).
57. Reference 48, p. 54.
58. I. S. Gradshteyn and I. W. Ryzhik, *Table of Integrals, Series, and Products*, Academic Press, New York, 1965, p. 973.
59. N. Abramson, Sandwich hologram interferometry: a new dimension in holographic comparison, *Appl. Opt.*, **13**, 2019–2025 (1974).
60. J. O. Bolstad, Holograms and spatial filters processed and copied in place, *Appl. Opt.*, **6**, 170 (1967).
61. W. F. Fagan, P. Waddell, and W. McCracken, The study of vibration patterns using real-time hologram interferometry, *Opt. Laser Technol.*, **4**, 167–172 (1972).
62. E. N. Leith, A. Kozma, J. Upatnieks, J. Marks, and N. Massey, Holographic data storage in 3-D media, *Appl. Opt.*, **5**, 1303–1311 (1966).
63. A. A. Friesem and J. L. Walker, Experimental investigation of some anomalies in photographic plates, *Appl. Opt.*, **8**, 1504–1506 (1969).
64. D. E. Duffy, Reducing photographic emulsion shrinkage for real-time holographic interferometry, *J. Phys. E: Sci. Instrum.*, **3**, 561–562 (1970).
65. J. N. Butters, D. Denby, and J. A. Leendertz, A method for reducing movement in holographic emulsions, *J. Phys. E.: Sci. Instrum.*, **2**, 116–117 (1969).
66. K. Biedermann and N.-E. Molin, Combining hypersensitization and *in situ* processing for time-average observation in real-time hologram interferometry, *J. Phys. E.: Sci. Instrum.*, **3**, 669–680 (1970).
67. T. F. Kellie and W. H. Stevenson, Experimental techniques in real-time holographic interferometry. 1: Reduction of emulsion shrinkage effects, *Opt. Eng.*, **12**, 47–49 (1973).

68. W. van Deelen and P. Nisenson, Mirror blank testing by real-time holographic interferometry, *Appl. Opt.*, **8**, 951–955 (1969).
69. B. U. Achia, A simple plate holder for on-site wet processing of holograms in real-time holographic interferometry, *J. Phys. E.: Sci. Instrum.*, **5**, 128–129 (1972).
70. D. H. Casler and H. D. Pruett, Simultaneous exposure-development of holograms on 649-F film, *Appl. Phys. Lett.*, **10**, 341–342 (1967).
71. H. F. Dietrich, R. J. Raine, and R. N. O'Brien, A 5-minute monobath for Kodak 649-F plates used in holography and holographic interferometry, *J. Photogr. Sci.*, **24**, 120–123 (1976).
72. H. M. Smith, M. H. Sewell, and J. R. King, Real-time holographic interferometry: a system, *Appl. Opt.*, **15**, 729–733 (1976).
73. T. F. Kellie and W. H. Stevenson, Experimental techniques in real time holographic interferometry. 2: Minimization of depolarization effects, *Opt. Eng.*, **12**, 131–132 (1973).
74. E. Archbold and A. E. Ennos, Observation of surface vibration modes by stroboscopic hologram interferometry, *Nature*, **217**, 942–943 (1968).
75. J. T. La Macchia, Stroboscopic holography with a mode-locked laser, *J. Appl. Phys.*, **39**, 5340–5341 (1968).
76. P. Sajenko and C. D. Johnson, Stroboscopic holographic interferometry, *Appl. Phys. Lett.*, **13**, 22–24 (1968).
77. B. M. Watrasiewicz and P. Spicer, Vibration analysis by stroboscopic holography, *Nature*, **217**, 1142–1143 (1968).
78. A. N. Zaidel', L. G. Malkhasyan, G. V. Markova, and Yu. I. Ostrovskii, Holographic strobe method for studying vibrations, *Sov. Phys. Tech. Phys.*, **13**, 1470–1473 (1969).
79. G. M. Mayer, Vibration phase measurement by rotation-strobe holography, *J. Appl. Phys.*, **40**, 2863–2866 (1969).
80. P. A. Fryer, A scanning technique for allowing whole vibration cycles to be stored on one hologram, *Appl. Opt.*, **9**, 1216 (1970).
81. V. S. Listovets and Yu. I. Ostrovskii, Hologram interferometry for vibration analysis (review), *Sov. Phys. Tech. Phys.*, **19**, 847–860 (1975).
82. M. Miler, Stroboscopic holography in ramp approximation, *Opt. Commun.*, **14**, 406–408 (1975).
83. P. A. Fryer, Vibration analysis by holography, *Rep. Prog. Phys.*, **33**, 489–531 (1970).
84. A. E. Ennos and E. Archbold, Vibrating surface viewed in real time by interference holography, *Laser Focus*, **4**, 58–59 (1968).
85. C. S. Vikram, Direct approach to reduce recording time in stroboscopic holographic interferometry, *Nouv. Rev. Opt. Appl.*, **4**, 147–149 (1973).
86. C. S. Vikram, Further reduction of recording time in stroboscopic holographic interferometry, *Opt. Commun.*, **7**, 347–348 (1973).
87. K. N. Chopra and G. S. Bhatnagar, Quadruple-exposure technique in stroboscopic holographic interferometry, *Appl. Opt.*, **13**, 2468–2470 (1974).
88. C. S. Vikram, C. Bose, and J. N. Maggo, Combination of time-average and stroboscopic techniques in holographic interferometry of sinusoidal vibration, *Nouv. Rev. Opt. Appl.*, **6**, 55–59 (1975).
89. C. S. Vikram, Stroboscopic holographic interferometry of vibration simultaneously in two modes, *Opt. Commun.*, **11**, 360–364 (1974).

90. D. S. Elinevskii, R. S. Bekbulatov, Yu. N. Shaposhnikov, V. A. Eryshev, A. M. Burenkin, and Yu. G. Yurtaev, Stroboholographic method used for studying vibrations, *Strength Mater.*, **8**, 95–99 (1976).
91. C. C. Aleksoff, Temporally modulated holography, *Appl. Opt.*, **10**, 1329–1341 (1971).
92. C. C. Aleksoff, Temporal modulation techniques, in *Holographic Nondestructive Testing*, R. K. Erf (Ed.), Academic Press, New York, 1974, pp. 247–263.
93. J. W. Goodman, Temporal filtering properties of holograms, *Appl. Opt.*, **6**, 857–859 (1967).
94. M. H. Zambuto and W. K. Fischer, Shifted reference holographic interferometry, *Appl. Opt.*, **12**, 1651–1655 (1973).
95. N. Takai, M. Yamada, and T. Idogawa, Holographic interferometry using a reference wave with a sinusoidally modulated amplitude, *Opt. Laser Technol.*, **8**, 21–23 (1976).
96. D. B. Neumann, C. F. Jacobson, and G. M. Brown, Holographic technique for determining the phase of vibrating objects, *Appl. Opt.*, **9**, 1357–1362 (1970).
97. C. C. Aleksoff, Time average holography extended, *Appl. Phys. Lett.*, **14**, 23–24 (1969).
98. V. J. Corcoran, R. W. Herron, Jr., and J. G. Jaramillo, Generation of a hologram from a moving target, *Appl. Opt.*, **5**, 668–669 (1966).
99. F. M. Mottier, Holography of randomly moving objects, *Appl. Phys. Lett.*, **15**, 44–45 (1969).
100. H. J. Caulfield, Holography of randomly moving objects, *Appl. Phys. Lett.*, **16**, 234–235 (1970).
101. J. P. Waters, Object motion compensation by speckle reference beam interferometry, *Appl. Opt.*, **11**, 630–636 (1972).
102. J. A. Levitt and K. A. Stetson, Mechanical vibrations: mapping their phase with hologram interferometry, *Appl. Opt.*, **15**, 195–199 (1976).
103. K. A. Stetson, Envelope factors due to laser modulation in time-average, holographic, vibration analysis, *J. Opt. Soc. Am.*, **62**, 698–700 (1972).
104. F. M. Mottier, Vibration analysis by interferometric fringe modulation, *Nouv. Rev. Opt. Appl.*, **1** (Suppl.), 12 (1970).
105. J. D. Redman and I. K. Pearce, Effect of laser pulse shape on the form of holographic velocity fringes, *Appl. Opt.*, **12**, 947–949 (1973).
106. A. J. Decker, Pulse modulation effect on velocity fringes, *Appl. Opt.*, **14**, 1061–1063 (1975).
107. C. M. Vest, E. L. McKague, and A. A. Friesem, Holographic detection of microcracks, *Trans. ASME*, Ser. D-J. Basic Engineering, **93**, 237–246 (1971).
108. A. A. Friesem and C. M. Vest, Detection of microfracture by holographic interferometry, *Appl. Opt.*, **8**, 1253–1254 (1969).
109. A. D. Luxmoore, Holographic detection of cracks in concrete, *Nondestr. Test.*, **6**, 258–263 (1973).
110. M. J. Salkind, Early detection of fatigue damage in composite materials, *J. Aircr.*, **13**, 764–769 (1976).
111. J. J. Nevadunsky, J. J. Lucas, and M. J. Salkind, Early fatigue damage detection in composite materials, *J. Compos. Mater.*, **9**, 394–408 (1975).
112. K. Grünewald, W. Fritzsch, A. V. Harnier, and E. Roth, Nondestrictive testing of plastics by means of holographic interferometry, *Polym. Sci. Eng.*, **15**, 16–28 (1975).
113. R. M. Grant and G. M. Brown, Holographic nondestructive testing (HNDT), *Mater. Eval.*, **27**, 79–84 (1969).

114. W. R. Klappert, The setting of double-exposure optical holographic NDT parameters, *Mater. Eval.*, **34**, 160–164 (1976).
115. C. M. Vest and D. W. Sweeney, Applications of holographic interferometry to nondestructive testing, *Int. Adv. Nondestr. Test.*, **5**, 17–21 (1977).
116. O. J. Burchett, Analysis techniques for the inspection of structures by holographic interferometry, *Mater. Eval.*, **30**, 25–31 (1972).
117. M. D. Mayer and T. E. Katayanagi, Holographic examination of a composite pressure vessel, *J. Test. Eval.*, **5**, 47–52 (1977).
118. J. P. Waters, Holographic inspection of solid propellant to linear bonds, *Appl. Opt.*, **10**, 2364–2365 (1971).
119. G. M. Brown, Holography used in nondestructive testing with particular emphasis on pneumatic tires, *SPIE Proc.*, **34**, 75–84 (1973).
120. G. M. Brown, Pneumatic tire inspection, in *Holographic Nondestructive Testing*, R. K. Erf (Ed.), Academic Press, New York, 1974, pp. 355–364.
121. S. Amadesi, F. Gori, R. Grella, and G. Guattari, Holographic methods for painting diagnostics, *Appl. Opt.*, **13**, 2009–2013 (1974).
122. N. G. Vlasov, V. M. Ginzburg, V. G. Novgorodov, V. N. Protsenko, B. M. Stepanov, and F. V. Ushkov, Application of holographic-interference methods to the determination of the optimal temperature-humidity conditions for preserving monumental paintings and works of art, *Sov. Phys. Dokl.*, **20**, 859–861 (1976).
123. W. J. Harris and D. C. Woods, Thermal stress studies using optical holographic interferometry, *Mater. Eval.*, **32**, 50–56 (1974).
124. I. S. Klimenko, E. G. Matinyan, and L. G. Dubitskii, Use of focused image holography for the nondestructive testing of electronic parts, *Sov. J. Nondestr. Test.*, **10**, 696–699 (1975).
125. J. R. Crawford and R. Benson, Holographic interferometry of circuit board component failure, *Appl. Opt.*, **15**, 24–25 (1976).
126. C. S. Bartolotta and B. J. Pernick, Holographic nondestructive evaluation of interference fit fasteners, *Appl. Opt.*, **12**, 885–886 (1973).
127. L. A. Kersch, Advanced concepts of holographic nondestructive testing, *Mater. Eval.*, **29**, 125–140 (1971).
128. B. S. Hockley and J. N. Butters, Holography as a routine method of vibration analysis, *J. Mech. Eng. Sci.*, **12**, 37–48 (1970).
129. N. Abramson and H. Bjelkhagen, Industrial holographic measurements, *Appl. Opt.*, **12**, 2792–2796 (1973).
130. N. Abramson, Sandwich hologram interferometry. 2: Some practical calculations, *Appl. Opt.*, **14**, 981–984 (1975).
131. P. Hariharan and Z. S. Hegedus, Two-hologram interferometry: a simplified sandwich technique, *Appl. Opt.*, **15**, 848–849 (1976).
132. N. Abramson, Sandwich hologram interferometry. 4: Holographic studies of two milling machines, *Appl. Opt.*, **16**, 2521–2531 (1977).
133. H. Bjelkhagen, Pulsed sandwich holography, *Appl. Opt.*, **16**, 1727–1731 (1977).
134. N. Abramson and H. Bjelkhagen, Pulsed sandwich holography. 2: Practical application, *Appl. Opt.*, **17**, 187–191 (1978).
135. T. H. Jeong, *A Study Guide on Holography*, Integraf, Inc., P.O. Box 586, Lake Forest, IL, 1975.
136. D. B. Neumann and R. C. Penn, Off-table holography, *Exp. Mech.*, **15**, 241–244 (1975).

137. E. B. Champagne, Quantitative data reduction with the use of fringe control techniques in conjunction with holographic interferometry, *Proceedings of Symposium on Engineering Applications of Holography* (February 16–17, 1972, Los Angeles, CA), Society of Photo-Optical Instrumental Engineers, Redondo Beach, CA, 1972, pp. 133–145.
138. W. Schumann and M. Dubas, On the motion of holographic images caused by movements of the reconstruction light source, with the aim of application to deformation analysis, *Optik*, **46**, 377–392 (1976).
139. W. Schumann and M. Dubas, On the holographic interferometry used for deformation analysis with one fixed and one movable reconstruction source, *Optik*, **47**, 391–404 (1977).
140. P. M. de Larminant and R. P. Wei, A fringe-compensation technique for stress analysis by reflection holographic interferometry, *Exp. Mech.*, **16**, 241–248 (1976).
141. E. Archbold, J. M. Burch, and A. E. Ennos, The application of holography to the comparison of cylinder bores, *J. Sci. Instrum.*, **44**, 489–494 (1967).
142. N. P. Larinov, A. V. Lukin, and K. S. Mustafin, Holographic inspection of shapes of unpolished surfaces, *Sov. J. Opt. Technol.*, **39**, 154–155 (1972).
143. R. J. Schaefer and J. A. Blodgett, Holographic study of electropolishing, *J. Electrochem. Soc.*, **123**, 1701–1705 (1976).
144. P. Greguss, Holographic interferometry in biomedical sciences, *Opt. Laser Technol.*, **8**, 153–159 (1976).
145. S. M. Zivi and G. N. Humberstone, Chest motion visualized by holographic interferometry, *Med. Res. Eng.*, **9**, 5–7 (1970).
146. H. Bjelkhagen, B. Hök, and K. Nilsson, Imaging of chest motion due to heart action by means of holographic interferometry (to be published, 1978).
147. von P. Fuchs and D. Schott, The application of holography to the measurement of deformation of human facial bones, *Dtsch. Zahnärztl. Z.*, **28**, 90 (1973).
148. L. Wictorin, H. Bjelkhagen, and N. Abramson, Holographic investigation of the elastic deformation of defective gold solder joints, *Acta Odontol. Scand.*, **30**, 659–670 (1972).
149. P. R. Wendendal and H. I. Bjelkhagen, Holographic interferometry on the elastic deformation of prosthodontic appliances as simulated by bar elements, *Acta Odontol. Scand.*, **32**, 189–199 (1974).
150. P. R. Wendendal and H. I. Bjelkhagen, Dynamics of human teeth in function by means of double pulsed holography: an experimental investigation, *Appl. Opt.*, **13**, 2481–2485 (1974).
151. S. M. Khanna and J. Tonndorf, Tympanic membrane vibrations in cats, studied by time-averaged holography, *J. Acoust. Soc. Am.*, **51**, 1904–1920 (1972).
152. J. Tonndorf and S. M. Khanna, Tympanic-membrane vibrations in human cadaver ears studied by time-averaged holography. *J. Acoust. Soc. Am.*, **52**, 1221–1233 (1972).
153. F. Able, A. L. Dancer, H. Fagot, R. B. Franke, and P. Smigielski, Holographic interferometry applied to investigation of tympanic-membrane displacements in guinea-pig ears subjected to acoustic impulses, *J. Acoust. Soc. Am.*, **58**, 223–228 (1975).
154. P. Smigielski, F. Able, H. Fagot, A. Dancer, and R. Franke, Application de l'interferometrie holographique a l'étude des deformations du cobaye sous l'effet de bruits de durée brève, *Nouv. Rev. Opt. Appl.*, **6**, 49–54 (1975).
155. W. G. Gottenberg, Some applications of holographic interferometry, *Exp. Mech.*, **8**, 405–410 (1968).
156. R. C. Sampson, Holographic-interferometry applications in experimental mechanics, *Exp. Mech.*, **10**, 313–320 (1970).

157. C. A. Sciammarella and J. A. Gilbert, Strain analysis of a disk subjected to diametral compression by means of holographic interferometry, *Appl. Opt.*, **12**, 1951–1956 (1973).
158. A. D. Wilson, Holographically observed torsion in a cylindrical shaft, *Appl. Opt.*, **9**, 2093–2097 (1970).
159. A. D. Wilson, C. H. Lee, H. R. Lominac, and D. H. Strope, Holographic and analytic study of a semiclamped rectangular plate supported by struts, *Exp. Mech.*, **11**, 1–6 (1971).
160. R. E. Rowlands and I. M. Daniel, Applications of holography to anisotropic composite plates, *Exp. Mech.*, **12**, 75–82 (1972).
161. C. A. Sciammarella and T. Y. Chang, Holographic interferometry applied to the solution of a shell problem, *Exp. Mech.*, **14**, 217–224 (1974).
162. T. Matsumoto, K. Iwata, and R. Nagata, Measurement of deformation in a cylindrical shell by holographic interferometry, *Appl. Opt.*, **13**, 1080–1084 (1974).
163. M. C. Collins and C. E. Watterson, Surface-strain measurements on a hemispherical shell using holographic interferometry, *Exp. Mech.*, **15**, 128–132 (1975).
164. A. R. Hunter and T. M. Morton, Application of holographic interferometry to predict long-time torsional relaxation, *Exp. Mech.*, **15**, 153–160 (1975).
165. J. L. Goldberg, A holographic interferometer for the measurement of the vector displacement of a slowly deforming rough surface, *Jap. J. Appl. Phys.*, **14** (Suppl. 14-1), 253–258 (1975).
166. J. L. Goldberg, K. M. O'Toole, and H. Roper, Holographic interferometry for measuring swelling of hardened concrete, *J. Test. Eval.*, **3**, 263–270 (1975).
167. T. R. Hsu and R. G. Moyer, Application of holography in high-temperature displacement measurements, *Exp. Mech.*, **12**, 431–432 (1972).
168. T. R. Hsu and R. Lewak, Measurements of thermal distortion of composite plates by holographic interferometry, *Exp. Mech.*, **16**, 182–187 (1976).
169. C. M. Vest, Holographic interferometry in material testing, *Int. J. Nondestr. Test.*, **3**, 351–374 (1972).
170. R. Aprahamian and D. A. Evensen, Applications of holography to dynamics. I: High-frequency vibrations of beams, *Trans. ASME*, Ser. E: *J. Appl. Mech.*, **37**, 287–291 (1970).
171. R. Aprahamian and D. A. Evensen, Applications of holography to dynamics. I: High-frequency vibrations of plates, *Trans. ASME*, Ser. E: *J. Appl. Mech.*, **37**, 1083–1090 (1970).
172. R. Williams, Y. T. Yeow, and H. F. Brinson, An analytical and experimental study of vibrating equilateral triangular plates, *Exp. Mech.*, **15**, 339–346 (1975).
173. K. A. Stetson and P. A. Taylor, The use of normal mode theory in holographic vibration analysis with application to an asymmetrical circular disk, *J. Phys. E (Sci. Instrum.)*, **4**, 1009–1015 (1971).
174. A. D. Wilson, B. D. Martin, and D. H. Strope, Holographic interferometry applied to motion studies of ultrasonic bonders, *IEEE Trans.*, **SU-19**, 453–461 (1972).
175. P. A. Tuschak and R. A. Allaire, Axisymmetric vibrations of a cylindrical resonator measured by holographic interferometry, *Exp. Mech.*, **15**, 81–88 (1975).
176. M. Borissov, K. T. Stoytchev, and M. I. Kovachev, Holographic study of vibrations of piezoelectric AT-cut quartz resonators, *C. R. Acad. Bulg. Sci.*, **29**, 477–480 (1976).
177. M. Chomat and M. Miler, Application of holography to the analysis of mechanical vibration electronic components, *TESLA Electron.*, **3**, 83–93 (1973).

178. M. Lashkari and V. I. Weingarten, Vibrations of segmented shells, *Exp. Mech.*, **13**, 120-125 (1973).
179. H. Wakashima, T. Miyazaki, T. Uyemura, and Y. Yamamoto, Measurement of dynamic viscoelasticity by holography, *J. Jap. Soc. Precis. Eng.*, **41**, 984-989 (1975).
180. C.-H. Ågren and K. A. Stetson, Measuring the resonances of treble viol plates by hologram interferometry and designing an improved instrument, *J. Acoust. Soc. Am.*, **6**, 1971-1983 (1972).
181. K. A. Stetson and P. A. Taylor, Analysis of static deflections by holographically recorded vibration modes, *J. Phys. E.: Sci. Instrum.*, **5**, 923-926 (1972).
182. K. A. Stetson, Perturbation method of structural design relevant to holographic vibration analysis, *AIAA J.*, **13**, 457-459 (1975).
183. K. A. Stetson, Holographic vibration analysis, in *Holographic Nondestructive Testing*, R. K. Erf (Ed.), Academic Press, New York, 1974, pp. 181-220.
184. T. Tsuruta and Y. Itoh, Holographic interferometry for rotating subject, *Appl. Phys. Lett.*, **17**, 85-87 (1970).
185. J. P. Sikora and F. T. Mendenhall, Jr., Holographic vibration study of a rotating propellor blade, *Exp. Mech.*, **14**, 230-232 (1974).
186. Y. Chen, Holographic interferometry applied to rotating disks, *Trans. ASME*, Ser. E: *J. Appl. Mech.*, **42**, 499-501 (1975).
187. K. A. Stetson, The use of an image derotator in hologram interferometry and speckle photography of rotating objects, *Exp. Mech.*, **18**, 67-73 (1978).
188. E. R. Robertson and W. King, The technique of holographic interferometry applied to the study of transient stresses, *J. Strain Anal.*, **9**, 44-49 (1974).
189. R. Aprahamian, D. A. Evensen, J. S. Mixen, and J. E. Wright, Application of pulsed holographic interferometry to the measurement of propagating transverse waves in beams, *Exp. Mech.*, **11**, 309-314 (1971).
190. R. Aprahamian, D. A. Evensen, J. S. Mixen, and J. L. Jacoby, Holographic study of propagating transverse waves in plates, *Exp. Mech.*, **11**, 357-362 (1971).
191. F. C. Perry and L. P. Mix, Application of holographic interferometry to shock waves in solids, *Appl. Phys. Lett.*, **24**, 624-626 (1974).
192. F. Able, P. Smigielski, and H. Fagot, Effective application of holographic interferometry by double exposure to the study of deformation of a ceramic under projectile impact, *Opt. Commun.*, **8**, 369-371 (1973).
193. J. R. Asay, L. P. Mix, and F. C. Perry, Ejection of material from shocked surfaces, *Appl. Phys. Lett.*, **29**, 284-287 (1976).
194. W. P. Chu, D. M. Robinson, and J. H. Goad, Holographic nondestructive testing with impact excitation, *Appl. Opt.*, **11**, 1644-1645 (1972).
195. J. W. C. Gates, R. G. N. Hall, and I. N. Ross, Holographic interferometry of impact loaded objects using a double-pulse laser, *Opt. Laser Technol.*, **4**, 72-75 (1972).
196. A. D. Wilson and D. H. Strope, Holographic interferometry deformation study of a printer type-piece, *IBM J. Res. Dev.*, **16**, 258-268 (1972).

5

Transparent Objects: Fringe Formation and Localization

5.1 INTRODUCTION

In this chapter we consider applications of holographic interferometry for which the "object" is a transparent medium with nonhomogeneous refractive index. This is the foundation of measurement techniques used in aerodynamics, heat transfer, plasma diagnostics, and stress analysis of transparent models. In Chapters 2 to 4 we discussed situations in which the object wave was scattered by an opaque surface, which was displaced or deformed between holographic exposures. In this chapter we consider that the object wave propagates through a transparent medium. Phase shifts giving rise to an interference pattern are due to variations in the speed of light within the medium. The resulting interferogram can be analyzed to determine physical properties of the object such as mass density, temperature, electron number density, or state of strain.

An example of a two-exposure holographic interferogram of a transparent object is shown in Figure 5.1. During the first exposure the object wave passed through air which was heated by a small flame. The temperature distribution in the air near the flame caused a nonhomogeneous distribution of refractive index and therefore introduced phase changes in the object wave. During the second exposure the flame was extinguished, so the air had uniform temperature and refractive index. When the disturbed and undisturbed object waves were simultaneously reconstructed, they interfered to form the fringe pattern shown in the figure.

The optical wave used to probe transparent objects can be a simple plane or spherical wave, or it may be a complicated wave scattered by a diffusing screen. When a plane object wave is used, the holographic interferograms are quite similar to those produced by classical instruments such as Mach-Zehnder and Michelson interferometers.[1] There are differences, however, which are of considerable practical importance. When a diffuse object wave is used, holographic methods are, for practical purposes, unique and provide multidirectional interferometric data. Diffuse-light

5.2 PHASE OBJECTS AND REFRACTION IN TRANSPARENT MEDIA

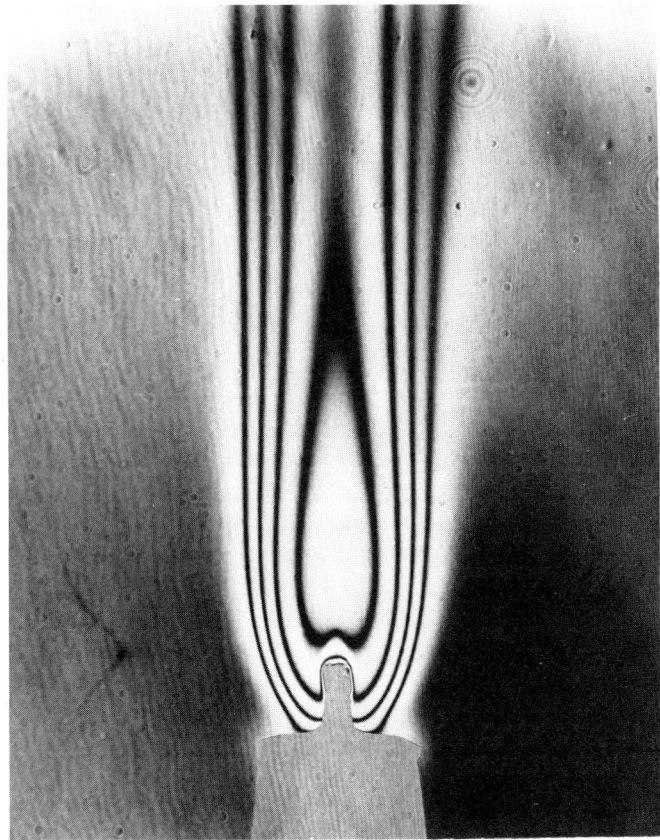

Figure 5.1 Holographic interferogram of a flame.

holographic interferometry of phase objects gives rise to fringe localization effects similar to those discussed in Chapter 3. In the present chapter we will discuss methods of recording holographic interferograms of transparent objects and will analyze the formation and localization of interferometric fringes. The application of nonlinear holography to interferometry is also considered.

5.2 PHASE OBJECTS AND REFRACTION IN TRANSPARENT MEDIA

We begin by presenting some basic information regarding the propagation of monochromatic light in nonhomogeneous media. This will serve as background for the subsequent discussion of holographic interferometry of transparent objects. The speed of light in a nonpolar dielectric medium depends primarily on the density of the medium and the wavelength of the light. In a vacuum the speed of light c_0 is given by electromagnetic theory as

$$c_0 = (\mu_0 \epsilon_0)^{-1/2} = 2.99776 \times 10^8 \text{ m/s},$$

where μ_0 is the magnetic permeability, and ϵ_0 the permittivity of free space. In other media the speed of light c deviates from this value, so it is convenient to introduce a dimensionless ratio, called the *refractive index n*, defined by

$$n = \frac{c_0}{c}. \tag{5.1}$$

An optically nonhomogeneous medium is one in which n varies from point to point. The dependence of n on other physical properties of liquids, gases, plasmas, and transparent solids is discussed in Chapter 6.

When a light wave propagates through a nonhomogeneous medium, its wavefronts, that is, surfaces of constant phase, are distorted. This is obvious because light travels more rapidly through some regions of the medium than through others. For our purposes it is convenient to utilize *geometric optics*, that is, introduce the concept of light rays, to describe the effects of nonhomogeneity on the propagation of light. A ray, which is defined as a curve in space that is everywhere tangent to the normal to the wavefront, describes the path of light through a medium. In homogeneous media rays are straight lines. In nonhomogeneous media, rays are curves.

In Figure 5.2 several rays and some corresponding wavefronts in a nonhomogeneous medium are depicted. The equations of geometric optics are the small wavelength approximations to the electromagnetic theory of light. The equation relating a ray to the refractive index distribution in a medium is[2]

$$\frac{d}{ds}\left(n \frac{d\mathbf{r}}{ds}\right) = \nabla n, \tag{5.2}$$

where $\mathbf{r}$ is the position vector of points on the ray, s is the length of the ray, measured from an arbitrary point on it, and ∇ is the gradient operator, that is,

$$\nabla n = \hat{\mathbf{i}} \frac{\partial n}{\partial x} + \hat{\mathbf{j}} \frac{\partial n}{\partial y} + \hat{\mathbf{k}} \frac{\partial n}{\partial z}. \tag{5.3}$$

5.2 PHASE OBJECTS AND REFRACTION IN TRANSPARENT MEDIA

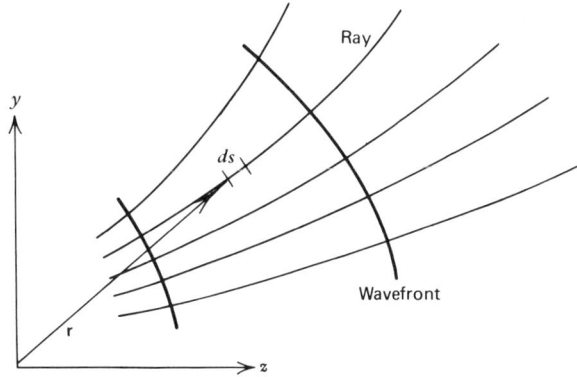

Figure 5.2 Wavefronts and rays in a nonhomogeneous medium.

In a homogeneous medium n is constant, so the ray equation 5.2 becomes

$$\frac{d^2\mathbf{r}}{ds^2} = 0. \tag{5.4}$$

The solution of equation 5.4 is

$$\mathbf{r} = \mathbf{a}s + \mathbf{b},$$

where $\mathbf{a}$ and $\mathbf{b}$ are vector constants. This is in accordance with the previous statement that rays are straight lines within homogeneous media.

There are two types of nonhomogeneities. The first kind is a discontinuity at the interface between two media. This occurs at the surfaces of lenses, prisms, and windows. The second kind, a continuously varying distribution of refractive index, occurs in flames, plasmas, aerodynamic fields, stressed transparent solids, and so on.

In the case of a discontinuity at the interface between two homogeneous media, rays are straight lines within each medium but have discontinuous slopes at the interface. Consideration of the electromagnetic boundary conditions shows that part of the light will be reflected and part will be transmitted through the interface.[3] If the boundary is smooth (on the scale of the wavelength of light), reflection will be governed by *Newton's law of reflection*,

$$\theta_r = \theta_i, \tag{5.5}$$

where θ_r is the angle of reflection and θ_i is the angle of incidence. The direction of the transmitted ray is governed by *Snell's law*,

$$n_1 \sin \theta_1 = n_2 \sin \theta_2, \tag{5.6}$$

where $n_{1,2}$ are the refractive indices in the two media, and $\theta_{1,2}$ are the angles between the rays in each medium and the normal to the interface (Figure 5.3). From equation 5.6 it can be seen that light rays bend into a region of higher refractive index. The bending of rays due to changes in refractive index is referred to as *refraction*.

Let us consider three simple, but important, examples of refraction at an interface. The first example is refraction due to a plane slab of thickness l, as shown in Figure 5.4. We write equation 5.6 at each of the interfaces:

$$n_2 \sin \theta_2 = n_1 \sin \theta_1,$$
$$n_3 \sin \theta_3 = n_2 \sin \theta_2.$$

Eliminating the term $n_2 \sin \theta_2$, we find that

$$\sin \theta_3 = \frac{n_1}{n_3} \sin \theta_1. \tag{5.7}$$

Thus the ray emerges at the angle θ_3, given by equation 5.7. A particularly important case is a glass window of refractive index n. The air on each side of the window has refractive index $n \simeq 1$, so $\theta_3 = \theta_1$. The ray emerges parallel to its original direction, but is displaced by an amount

$$d = l \sin \theta_2 = l \left(\frac{1}{n} \right) \sin \theta_1. \tag{5.8}$$

If the ray is normal to the window, no refraction occurs.

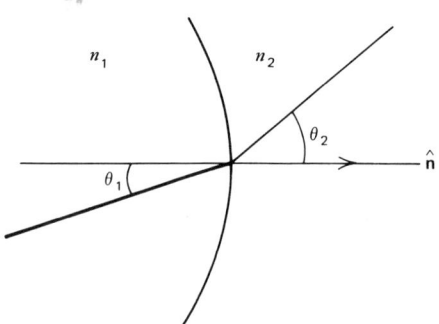

Figure 5.3 Refraction at an interface between two media.

5.2 PHASE OBJECTS AND REFRACTION IN TRANSPARENT MEDIA

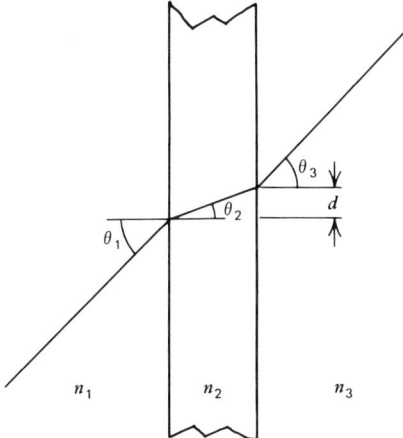

Figure 5.4 Refraction by a plane slab.

As a second example, consider refraction of a ray by the wedge prism shown in Figure 5.5. Let the entering ray be normal to the back face, and let the wedge angle be denoted by γ. Refraction will occur only at the second interface, where the angle of incidence is $\theta_1 = \gamma$, so $n_2 \sin\theta_2 = n_1 \sin\gamma$. If the prism is surrounded by air, with $n \simeq 1$, then $\sin\theta_2 = n\sin\gamma$. The ray will be deviated from its original direction by an angle $\beta = \theta_2 - \theta_1$. For small angles, $\sin\gamma \simeq \gamma$, so

$$\beta = (n-1)\gamma. \tag{5.9}$$

The third example illustrates the effect of boundary curvature. As shown in Figure 5.6, light passes through a cylinder with radius R and refractive

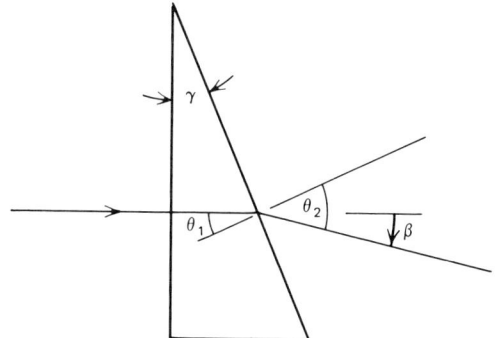

Figure 5.5 Refraction by a wedge prism.

index n, which is surrounded by air with $n \simeq 1$. The entrance angle is

$$\theta_1 = \tan^{-1}\left[\frac{y}{(R^2-y^2)^{1/2}}\right] = \sin^{-1}\left(\frac{y}{R}\right).$$

Tracing the ray through the object, using equation 5.6, shows that $\theta_{\text{exit}} = \theta_{\text{in}}$; however, unlike the case of the plane window, the exiting ray is not parallel to its original orientation. Figure 5.6a shows several rays

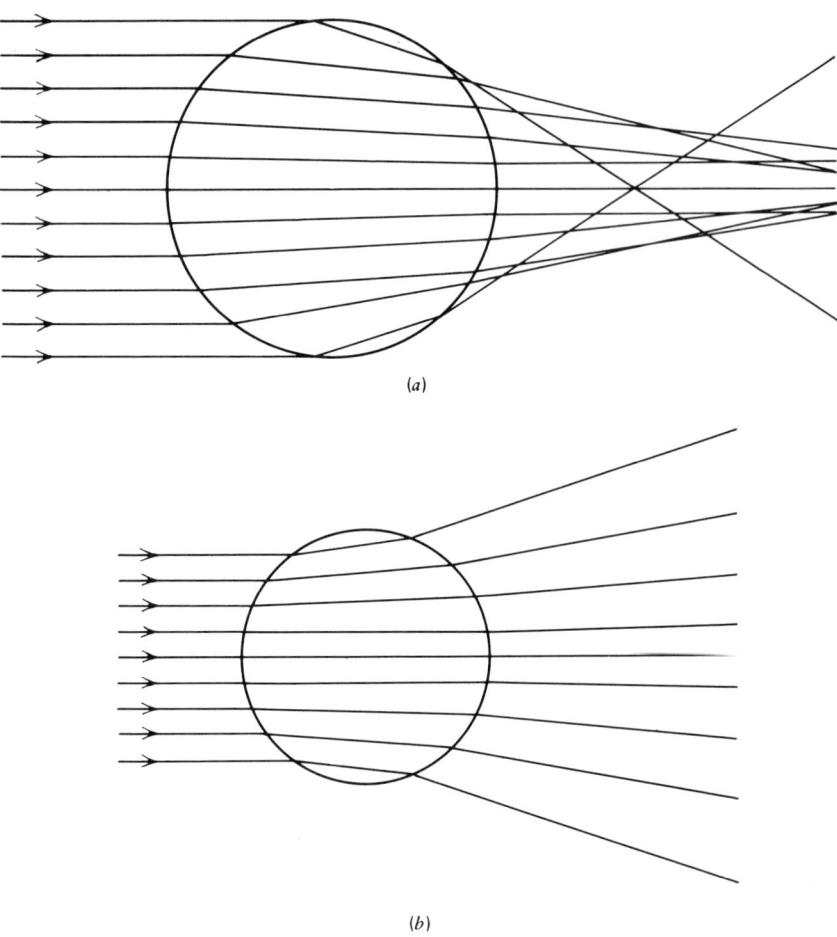

Figure 5.6 Refraction by a homogeneous cylinder. (a) The refractive index of the cylinder is greater than that of the surroundings. (b) The refractive index of the cylinder is less than that of the surroundings.

5.2 PHASE OBJECTS AND REFRACTION IN TRANSPARENT MEDIA

passing through a cylinder with refractive index slightly greater than that of the surrounding medium. Figure 5.6b shows refraction when the refractive index of the cylinder is less than that of the surrounding medium.

In interferometry we are usually concerned with transparent objects with continuous, smoothly varying refractive indices. Determining the path of a ray through such media is usually difficult. There are a few cases in which simple exact or approximate solutions to equation 5.2 can be found. Two examples of practical importance for interferometry are considered here. In the first we assume that n varies in only one direction, which is perpendicular to the entering ray. As shown in Figure 5.7, a ray enters the medium at a location y_0 and is parallel to the z axis. The medium is stratified so that its refractive index is a function only of y. The y- and z-components of the vector equation 5.2 are

$$\frac{d}{ds}\left[n(y)\frac{dy}{ds}\right] = \frac{dn}{dy},$$

$$\frac{d}{ds}\left[n(y)\frac{dz}{ds}\right] = 0.$$

Since

$$ds = \left[1+\left(\frac{dy}{dz}\right)^2\right]^{1/2} dz,$$

the preceding two equations can be combined to give

$$\frac{y''\,dy}{1+(y')^2} = \frac{dn}{n}, \qquad (5.10)$$

where the prime denotes differentiation with respect to z, and the variables

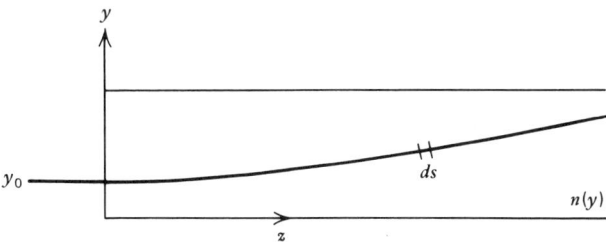

Figure 5.7 Path of an optical ray through a stratified medium.

have been separated. If the ray enters at $z=0$ with

$$y(0) = y_0, \quad y'(0) = 0, \quad n(0) = n_0,$$

equation 5.10 can be integrated[4] to give

$$1 + (y')^2 = \left(\frac{n}{n_0}\right)^2. \tag{5.11}$$

As a specific case, let the variation of n be linear:

$$n(y) = n_0 + n'(y - y_0).$$

For small gradients, we can make the approximation

$$\left(\frac{n}{n_0}\right)^2 \simeq 1 + 2\left(\frac{n'}{n_0}\right)(y - y_0),$$

so equation 5.11 is approximated by

$$(y')^2 = 2\left(\frac{n'}{n_0}\right)(y - y_0),$$

whose solution, with $y'(0) = 0$, is

$$y - y_0 = \tfrac{1}{2}\left(\frac{n'}{n_0}\right)z^2. \tag{5.12}$$

Thus a ray in a linearly stratified medium will travel a path which is *parabolic* to a first approximation.

As a second example, consider a medium in which the refractive index is radially symmetric, $n = n(r)$. In terms of the vector $\mathbf{s}$, defined by

$$\mathbf{s} = \frac{d\mathbf{r}}{ds},$$

which is tangent to the ray, equation 5.2 can be written as

$$\frac{d}{ds}[n(r)\mathbf{s}] = \nabla n. \tag{5.13}$$

Following Born and Wolf,[2] we consider the derivative of the quantity

5.2 PHASE OBJECTS AND REFRACTION IN TRANSPARENT MEDIA

$\mathbf{r} \times n\mathbf{s}$:

$$\frac{d}{ds}(\mathbf{r} \times n\mathbf{s}) = \frac{d\mathbf{r}}{ds} \times n\mathbf{s} + \mathbf{r} \times \frac{d}{ds}(n\mathbf{s}). \tag{5.14}$$

The first term on the right-hand side of equation 5.14 is identically zero since $d\mathbf{r}/ds$ is parallel to $n\mathbf{s}$. Using equation 5.13, we see that the second term also vanishes because the vector ∇n points in the radial direction, so

$$\frac{d}{ds}(\mathbf{r} \times n\mathbf{s}) = 0$$

and

$$\mathbf{r} \times n\mathbf{s} = \text{contant}. \tag{5.15}$$

Using the notation indicated in Figure 5.8, we can express this result as

$$rn(r)\sin\theta = p, \tag{5.16}$$

where p, called the *impact parameter*, has a constant value for each ray. Equation 5.16 is known as *Bouguer's formula*[5] and is one of the few exact solutions to the refraction equation.

These examples have been introduced to serve as background information for the discussion of the holographic interferometry of transparent

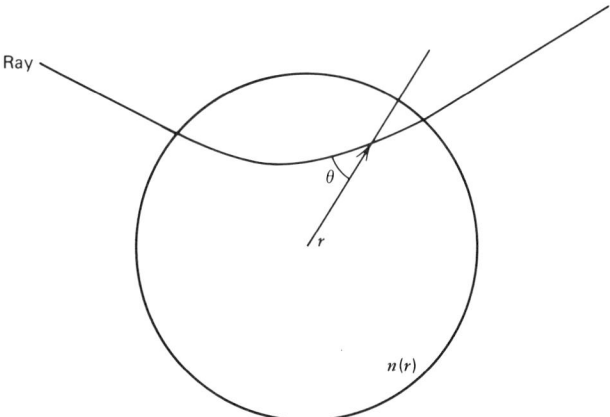

Figure 5.8 Notation for Bouguer's formula, equation 5.16.

objects. Detailed discussions of refraction and ray tracing can be found in monographs such as the one by Stavroudis.[6]

The quantity of primary interest in the interferometry of transparent media is the *optical pathlength* Φ of a ray through the medium. This is defined as

$$\Phi = \int n\,ds, \tag{5.17}$$

which is the path integral of the refracting index along the ray. When refraction is negligible, rays remain straight lines, and the path integral becomes a line integral. If the ray is parallel to the z axis, we simply replace ds by dz, so

$$\Phi(x,y) = \int n(x,y,z)\,dz. \tag{5.18}$$

In this *refractionless limit* the effect of passing through the medium is a change of phase relative to that of light traversing the same straight path in a homogeneous medium. Transparent objects in which refraction is negligible are often referred to as *phase objects*.

5.3 HOLOGRAPHY AND INTERFEROMETRY WITH PHASE OBJECTS

We begin by considering the holographic interferometry of phase objects using a plane object wave. A typical off-axis system is shown schematically in Figure 5.9a. Within the phase object the refractive index distribution is $n_1(x,y,z)$ during the initial holographic exposure and $n_2(x,y,z)$ during the second. When the hologram is developed and reilluminated by the reference wave, two waves are reconstructed simultaneously:

$$\mathbf{U}_{o1} = a_1(x,y)\exp\left[i\frac{2\pi}{\lambda}\Phi_1(x,y)\right]$$

and

$$\mathbf{U}_{o2} = a_2(x,y)\exp\left[i\frac{2\pi}{\lambda}\Phi_2(x,y)\right],$$

where Φ_1 and Φ_2 are given by equation 5.18. The fringe pattern is the irradiance of the sum of $\mathbf{U}_{o1}$ and $\mathbf{U}_{o2}$ in the image plane (Figure 5.9b). Assuming for simplicity that a_1 and a_2 are uniform, unit amplitudes, we

5.3 HOLOGRAPHY AND INTERFEROMETRY WITH PHASE OBJECTS

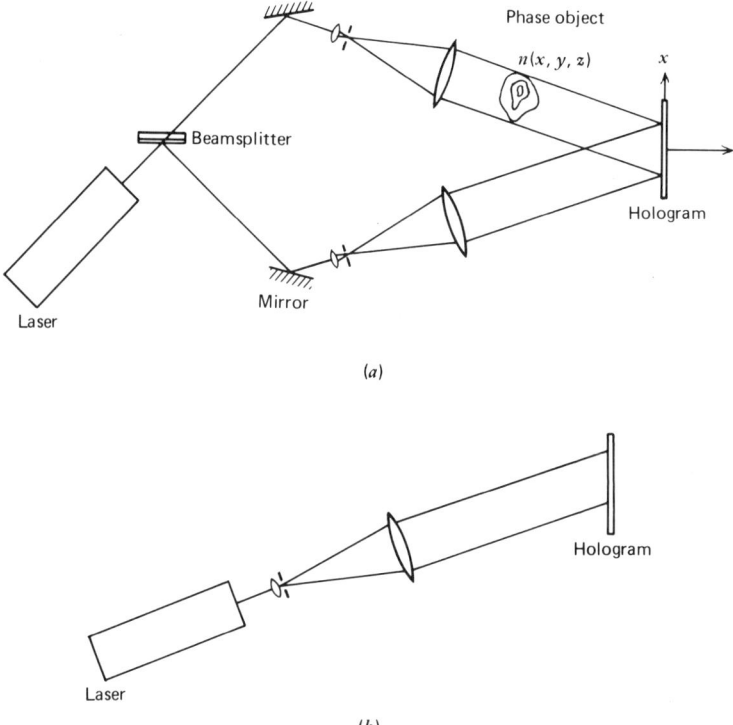

Figure 5.9 Typical off-axis configuration for holographic interferometry. (a) System for recording. (b) System for reconstruction.

can write this irradiance as

$$I(x,y) = 2\left\{1 + \cos\frac{2\pi}{\lambda}\left[\Phi_2(x,y) - \Phi_1(x,y)\right]\right\}. \quad (5.19)$$

In most applications the refractive index during one exposure, say the first, is uniform and can be denoted by n_0. Then the fringe pattern is

$$I(x,y) = 2\left[1 + \cos\left(\frac{2\pi}{\lambda}\Delta\Phi(x,y)\right)\right], \quad (5.20)$$

where

$$\Delta\Phi(x,y) = \int\left[n(x,y,z) - n_0\right]dz \quad (5.21)$$

is the *optical pathlength difference*. The equation of a bright fringe is

$$\Delta\Phi(x,y) = \int [n(x,y,z) - n_0] \, dz = N\lambda. \quad (5.22)$$

If the object has a variation of refractive index only in the y direction, the pathlength difference $\Delta\Phi$ is simply

$$\Delta\Phi(y) = [n(y) - n_0] l,$$

where l is the length of the object. The fringe spacing is determined by the gradient of n. Two examples are shown schematically in Figure 5.10. In Figure 5.10a the refractive index varies linearly, as

$$n(y) = n_0 + n'y.$$

This results in a series of equally spaced parallel fringes,

$$y = \frac{N\lambda}{n'l}, \quad N = 0, 1, 2, \ldots.$$

In Figure 5.10b the refractive index is of the form

$$n(y) = n_0 - n_1 \exp(-ay).$$

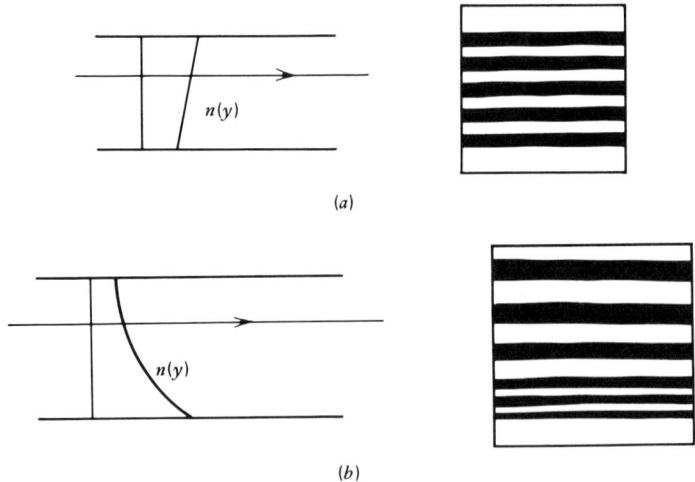

Figure 5.10 Fringe patterns formed by holographic interferometry of stratified media. (*a*) Linear stratification, $n(y) = n_0 + n'y$. (*b*) Exponential variation of refractive index, $n(y) = n_0 - n_1 \exp(-ay)$.

5.3 HOLOGRAPHY AND INTERFEROMETRY WITH PHASE OBJECTS

This distribution is similar to the ones which occur in the measurement of thermal boundary layers. The fringes are parallel straight lines with variable spacing. The spacing is large in regions of small gradient and small in regions of high gradient.

Radially symmetric phase objects are of great practical importance in aerodynamics, heat and mass transfer, and plasma diagnostics. A radially symmetric phase object, $n(r)$, is shown in Figure 5.11a. The probing plane wave is traveling in the z direction, as indicated by the typical ray shown in the figure. The pathlength difference $\Delta\Phi(x)$ for a two-exposure hologram is found by evaluating equation 5.21, with $dz = (r^2 - x^2)^{-1/2} r\, dr$:

$$\Delta\Phi(x) = 2 \int_x^R \frac{[n(r) - n_0] r\, dr}{(r^2 - x^2)^{1/2}}. \tag{5.23}$$

The integral in equation 5.23 is the Abel transform[7,8] of $[n(r) - n_0]$. The interferogram can be said to display contours of the Abel transform of a radially symmetric phase object. Figure 5.11b shows the form of the fringe pattern due to a phase object with constant refractive index n inside the cylindrical region, and Figure 5.11c shows the fringe pattern due to a phase object with a Gaussian distribution of refractive index:

$$n(r) - n_0 = n_1 \exp\left(\frac{-r^2}{a^2}\right).$$

The interferogram of a flame shown in Figure 5.1 is an example of a fringe pattern due to a radially symmetric phase object.

We noted in Chapter 1 that holographic interferometry is differential in time, that is, the two waves which interfere are separated temporally rather than spatially. In a Mach-Zehnder interferometer only one wave passes through the phase object. It then interferes with a plane comparison wave which has traveled a different path through the interferometer. In a holographic interferometer, like that shown in Figure 5.9, both the object and comparison waves travel across the same object space. Holographic interferometers are therefore *single-path interferometers*. This important feature permits the use of test sections with windows of rather poor optical quality. If a test section window is not optically flat and homogeneous, it will introduce a *pathlength error*, or noise, $\Delta\Phi_n(x,y)$ into an optical wave passing through it. Let $\Delta\Phi(x,y)$ represent the pathlength difference due to the phase objects inside the test section. It is contours of $\Delta\Phi(x,y)$ which we desire the interferometer to display as a fringe pattern. In a Mach-Zehnder interferometer the fringe pattern for this test section and phase object

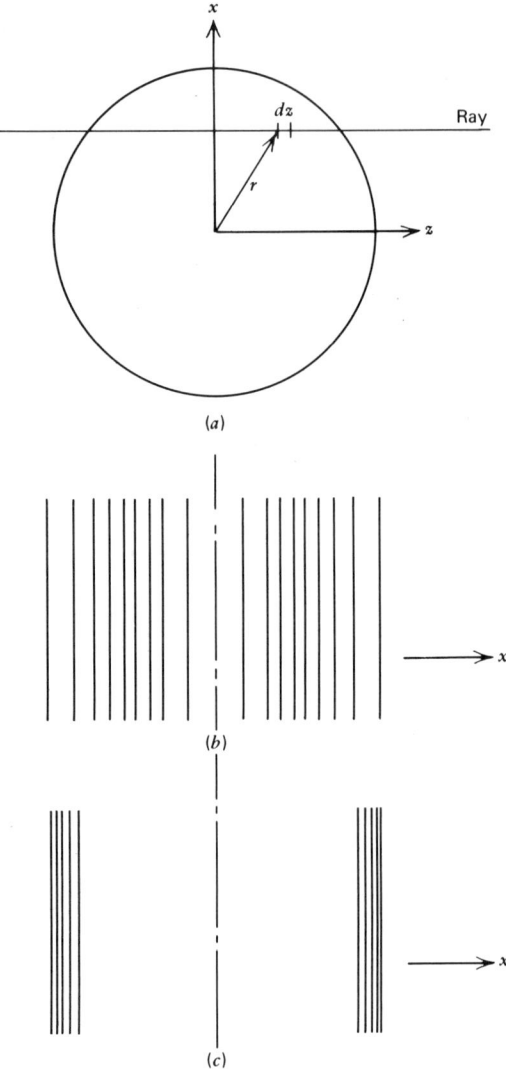

Figure 5.11 Interferometry of a cylindrical phase object. (*a*) Notation. (*b*) Fringe pattern for a constant refractive index. (*c*) Fringe pattern for a Gaussian distribution of refractive index, $n(r) - n_0 = n_1 e^{-r^2/a^2}$.

5.3 HOLOGRAPHY AND INTERFEROMETRY WITH PHASE OBJECTS

would be

$$I(x,y) = 2\left\{1 + \cos\frac{2\pi}{\lambda}\left[\Delta\Phi(x,y) + \Delta\Phi_n(x,y)\right]\right\}. \quad (5.24)$$

Thus pathlength variations in text section windows lead to errors in the fringe pattern which can be eliminated only by using optically flat, homogeneous windows so that $\Delta\Phi_n(x,y)$ = constant. In a holographic interferometer only *changes* in pathlength between exposures are displayed (see equation 5.19). Since the test section windows are present during both exposures, the effect of $\Phi_n(x,y)$ is canceled, and the interference pattern is given by the intensity distribution in equation 5.19.

The cancellation of phase errors due to windows of low optical quality is illustrated in Figure 5.12. The test section is constructed of ordinary Plexiglas and filled with water. The phase object is the thermal plume above a heated horizontal wire submerged in the water. Figure 5.12a shows a Mach-Zehnder interferogram. The desired fringe pattern is obscured completely by the fringes due to pathlength variations in the windows. Figure 5.12b, an interferogram made by two-exposure holographic interferometry with the same test section and phase object, is a relatively clear display of the desired fringe pattern. This somewhat extreme example illustrates the important property of cancellation of pathlength errors in holographic interferometry. It should be noted, however, that errors due to *refraction* by curved test section windows are not canceled and must be accounted for in quantitative measurements.

An interferogram described by equation 5.20 is called an *infinite-fringe interferogram*. This term implies that a field of uniform irradiance, that is, an infinitely wide fringe, results when $\Delta\Phi(x,y)=0$. This type of interferogram displays contours of constant value of $\Delta\Phi(x,y)$. There is a sign ambiguity in such interferograms because $+\Delta\Phi(x,y)$ and $-\Delta\Phi(x,y)$ yield the same fringe pattern. When interpreting the fringes it is not clear whether the optical pathlength increases or decreases from one fringe to the next. In Chapter 6 we will show that this sign ambiguity can be resolved by introducing *reference fringes* into the interferogram. These fringes are most commonly parallel, straight "wedge fringes" with equal spacing corresponding to a constant phase gradient of known sign. The resulting interferograms are called *finite-fringe interferograms* and are commonly used in classical interferometry.

Wedge reference fringes are easily introduced by tilting the object beam by a small angle $\Delta\beta_0$ between the exposures of a two-exposure hologram. The resulting interferogram has the irradiance distribution

$$I(x,y) = 2\left\{1 + \cos\frac{2\pi}{\lambda}\left[\Delta\Phi(x,y) + \Delta\beta_o y\right]\right\}. \quad (5.25)$$

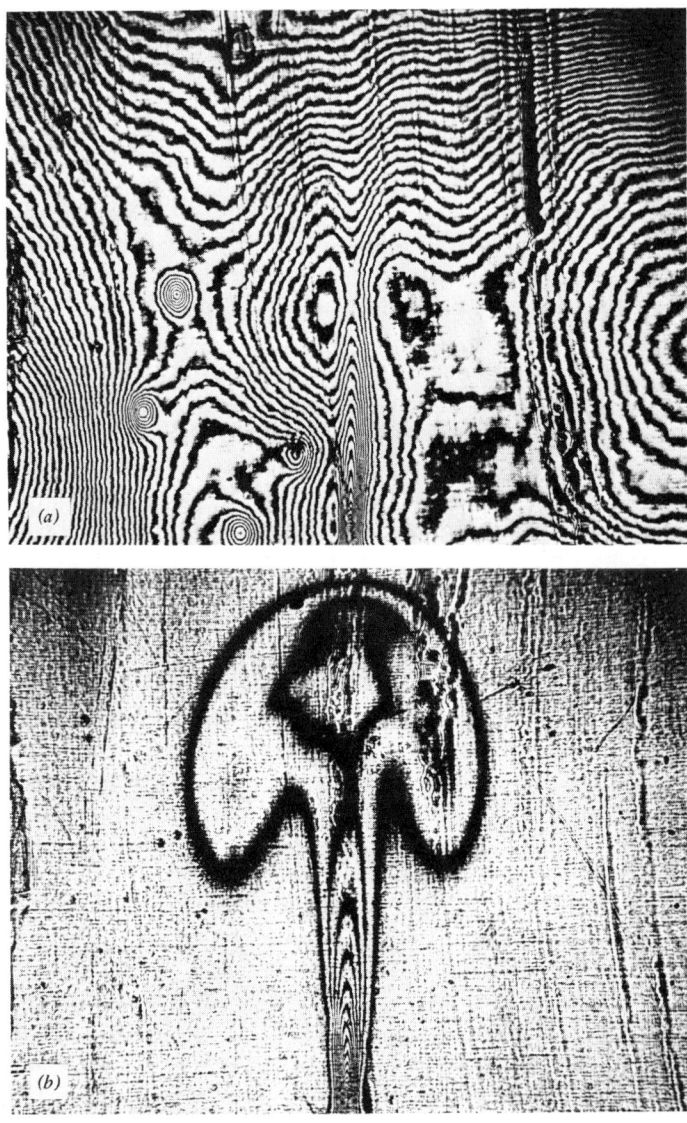

Figure 5.12 Effect of nonuniform test section windows. (*a*) Interferogram formed with a Mach-Zehnder interferometer. (*b*) Interferogram formed by holographic interferometry. The phase object is the same in (*a*) and (*b*), and is contained in a plastic test section of poor optical quality. In (*b*) the pathlength error due to the test section windows is eliminated by two-exposure holographic interferometry.

5.3 HOLOGRAPHY AND INTERFEROMETRY WITH PHASE OBJECTS

An example of an interferogram of this type is shown in Figure 5.13a. The tilt can also be introduced in other orientations, as in Figure 5.13b. Usually reference fringes are oriented so that they are parallel to the gradient of interest, as in Figure 5.13b, although reference fringes of other geometric forms are useful in certain cases.[9,10] Several other techniques can also be used to produce wedge fringes in holographic interferometry. If the reference beam, which is assumed here to be collimated, is tilted by an angle $\Delta\beta_R$ between exposures, an equivalent tilt of the object beam is introduced.[11,12] The magnitude of the equivalent tilt can be found by applying the holographic imaging equations discussed in Section 1.5. For the configurations shown in Figure 5.14, $\alpha_o = \alpha_R = \alpha_c = \alpha_I = 0$, $\lambda_c = \lambda_R$, and all R's in the imaging equations are infinite, so equation 1.96 becomes

$$\sin\beta_I = \sin\beta_c + \sin\beta_o - \sin\beta_R.$$

To fix ideas, we assume that the first exposure is made with beam orientations β_R and β_o, and with the object space undisturbed. The second exposure is made with the reference beam oriented at $\beta_R + \Delta\beta_R$ and the phase object present. Reconstruction is with a beam oriented at $\beta_c = \beta_R + \Delta\beta_R$. The effect of this is to tilt the reconstruction of the initial object wave by a small angle $\Delta\beta_I$ with respect to the second object wave, where

$$\sin(\beta_o + \Delta\beta_I) = \sin(\beta_R + \Delta\beta_R) + \sin\beta_o - \sin\beta_R.$$

For small $\Delta\beta$, $\sin(\beta + \Delta\beta) \cong \sin\beta + (\cos\beta)\Delta\beta$, so this equation yields

$$\Delta\beta_I = \left(\frac{\cos\beta_R}{\cos\beta_o}\right)\Delta\beta_R. \tag{5.26}$$

The irradiance of the interferogram is

$$I(x,y) = 2\left\{1 + \cos\frac{2\pi}{\lambda}\left[\Delta\Phi(x,y) + \left(\frac{\cos\beta_R}{\cos\beta_o}\right)\Delta\beta_R y\right]\right\}. \tag{5.27}$$

Finally, wedge fringes can be introduced by tilting the plate an angle $\Delta\beta$ between exposures. Because the coordinate system for equation 1.96 is attached to the plate, this is equivalent to tilting the reference beam by $\Delta\beta_R = -\Delta\beta$, so the reconstruction of the initial object wave will be tilted by

$$\Delta\beta_I = -\left(\frac{\cos\beta_R}{\cos\beta_o}\right)\Delta\beta.$$

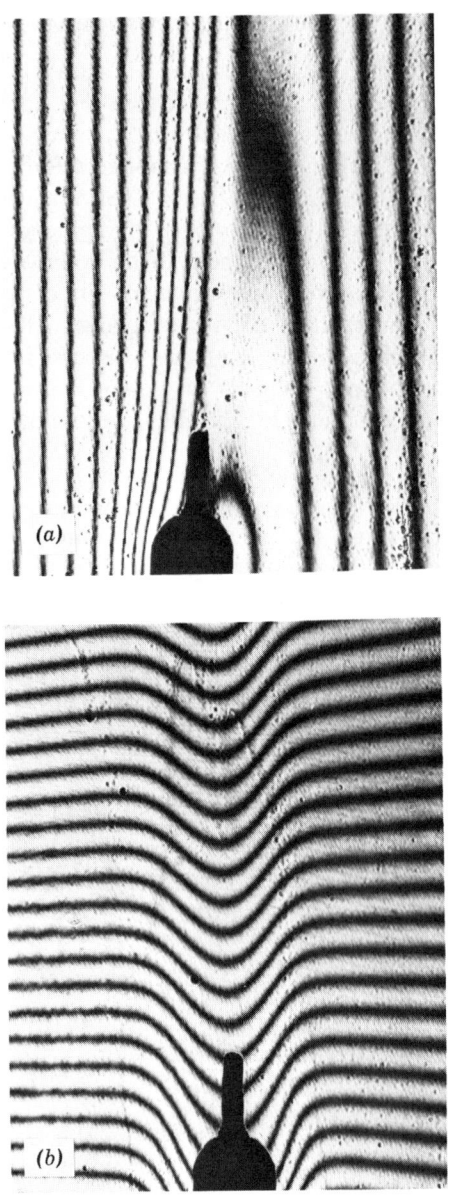

Figure 5.13 Interferograms with wedge reference fringes. (*a*) Reference fringes are vertical. (*b*) Reference fringes are horizontal; the phase object is the same as in (*a*).

5.3 HOLOGRAPHY AND INTERFEROMETRY WITH PHASE OBJECTS

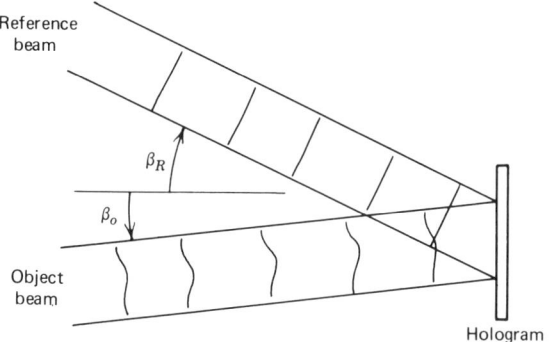

Figure 5.14 Reference fringes can be formed by changing the reference beam angle β_R or the object beam angle β_o, or by tilting the plate.

The orientation of the second object wave with respect to the plate in its final position is $\beta_{I_2} = \beta_o - \Delta\beta$, so the angle between the two reconstructed object waves is

$$\beta_{I_2} - \beta_{I_1} = \left(\frac{\cos\beta_R}{\cos\beta_o} - 1\right)\Delta\beta. \tag{5.28}$$

Interferograms of this type have been discussed by various workers[13,14] and are sometimes referred to as *hologram-moiré interferograms*.

So far we have discussed the introduction of reference fringes in the context of two-exposure holographic interferometry. This yields a permanent interferogram with reference fringes of fixed spatial orientation and period. Alternatively, dual holograms or dual reference waves can be used. These permit one to vary the orientation and period of reference fringes at the time of reconstruction while still maintaining a permanent holographic record. In the *dual-hologram method* one hologram of the original object wave U_o is recorded. A hologram of the disturbed object wave U'_o is recorded on a different photographic plate. Then U_o and U'_o reconstructed by simultaneously illuminating both holograms, as shown in Figure 5.15. The holograms can be tilted or translated independently to produce wedge reference fringes or shear.[15-17] A particularly convenient form of the dual-hologram method is the *sandwich hologram* introduced by Abramson,[18,19] which does not require accurate kinematic control of the plate holder. (See Section 4.4.1.)

The *dual-reference-beam method*[20] permits dynamic control of the reference fringes with a single holographic plate. As shown in Figure 5.16a, an initial off-axis hologram of U_o is recorded with reference wave U_{R_1}. A

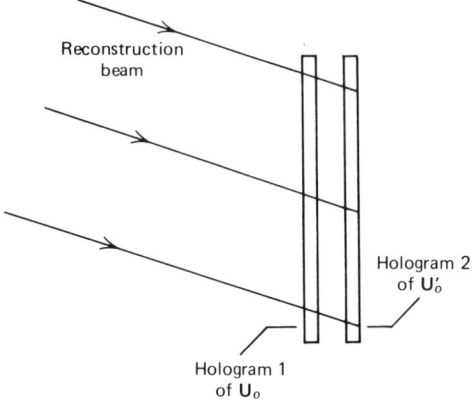

Figure 5.15 The dual-hologram method of forming interferograms with reference fringes.

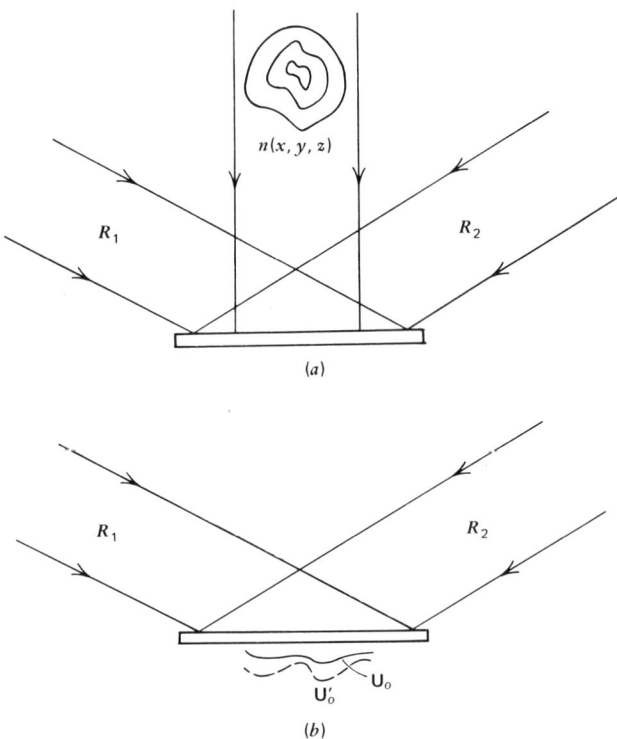

Figure 5.16 The dual-reference-beam method of forming interferograms with reference fringes. (*a*) Each exposure is recorded with a separate reference beam. (*b*) During reconstruction the plate is illuminated by two beams.

5.3 HOLOGRAPHY AND INTERFEROMETRY WITH PHASE OBJECTS

second exposure of the same plate records U'_o with reference wave U_{R_2}. The interferogram is obtained by exposing the developed hologram simultaneously to U_{R_1} and U_{R_2}, as in Figure 5.16b. By tilting the reference beams slightly, the reconstructions of U_o and U'_o can be tilted independently to produce reference fringes with any desired orientation and period.

Thus far in this chapter we have considered holographic interferometers with plane object waves. Holography offers important practical advantages such as cancellation of pathlength errors, permanent holographic records, and flexibility in the use of reference fringes; however, in principle, most applications of plane wave interferometry can be accomplished with either classical or holographic methods. We now turn our attention to *diffuse-illumination interferometry* of phase objects, for which holography is almost uniquely suited. This type of holographic interferometry was first reported in the important 1965 and 1966 papers by Heflinger, Wuerker, and Brooks.[21,22] Their pulsed-laser holographic interferograms of high-speed aerodynamic flow fields clearly illustrated the practical potential of holographic interferometry of phase objects, and served as the basis of much subsequent work in the field.

Several advantages are gained by incorporating a diffusing screen, such as a ground glass plate, into a holographic interferometer, as shown in Figure 5.17. Some of the useful characteristics of diffuse-illumination holography and interferometry are as follows:

1. The average irradiance of light is nearly uniform over the photographic plate.
2. Noise in the form of diffraction rings due to dust particles or small scratches on optical elements is eliminated.
3. The hologram can be viewed with the unaided eye.
4. The hologram can be viewed from many directions.

The first two characteristics were emphasized in the pioneering work of Leith and Upatnieks[23] on off-axis holography. They are important in the holographic recording of any transmission object. Uniformity of average irradiance is desirable because the ratio of reference beam to object beam irradiance must be the same in all regions on the plate. When a plane object wave is used, any obstacle, such as a dust particle, diffracts light into a nearly spherical wave. At the photographic plate this wave interferes with the object wave to form a set of concentric fringes similar to a zone plate pattern. When the hologram is reconstructed, an undesirable pattern of concentric rings is superimposed on the image or interferogram. When diffuse illumination is used, the dust particle no longer gives rise to a distinct zone plate pattern; consequently, the undesirable diffraction

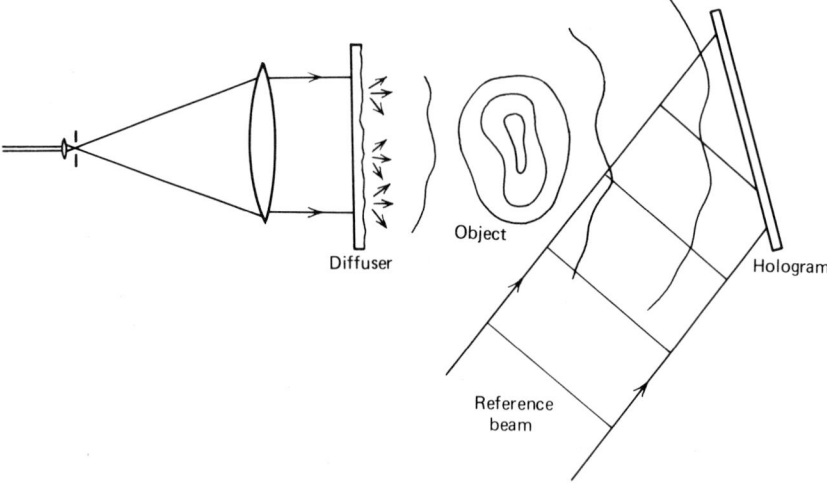

Figure 5.17 Holographic interferometer with diffuse object illumination.

pattern is effectively eliminated from the holographic image or interferogram.

Characteristics 3 and 4 provide the principal motivation for diffuse-illumination interferometry. When one directly views a two-exposure hologram with plane wave illumination, a bright spot is seen at the pinhole of the object beam, but the fringe pattern is not discernible. Furthermore, the fringes usually cannot be photographed with an ordinary camera, since the field of view is limited to an area the size of the iris. Hence fringes can by observed only by projecting them onto a diffuse screen or by photographing them with a large-diameter lens. If the object illumination is diffuse, an observer receives light scattered by all points on the diffuser, so ordinary imaging can be used. The fringe pattern in the virtual image can be observed by looking through the hologram with the unaided eye or can be photographed with a conventional camera. Furthermore, with diffuse illumination the fringes are localized in space, generally near the phase object. Fringe localization enables the viewer to form a qualitative impression of the three-dimensional structure of the phase object. A diffuse-illumination hologram can be viewed from any direction which is compatible with the apertures of the hologram and the illuminated diffuser. As the observer shifts his or her viewing direction, the fringe pattern seen through a two-exposure, or real-time, hologram changes. A single diffuse-illumination holographic interferogram is thus equivalent to a large number of plane wave interferograms, each recorded with a different orientation of the phase object. Quantitative evaluation of such multidirectional inter-

5.3 HOLOGRAPHY AND INTERFEROMETRY WITH PHASE OBJECTS

ferograms makes it possible to measure three-dimensional, asymmetric refractive index distributions. This will be discussed in Chapter 6.

If a plane reference wave is used to record a diffuse-illumination holographic interferogram, as in Figure 5.17, the fringe pattern in the real image can be photographed or examined. The real image is formed by rotating the developed holographic plate so that it is illuminated with the conjugate of the reference wave. The fringe pattern can then be examined by placing a diffuse viewing screen in the real image space, or it can be photographed directly by exposing film to the real image of the fringes. A fine fringe pattern can also be examined in detail in the real image with a short-focal-length lens system such as a low-power microscope objective.

In many cases the interferometric fringes viewed in a given direction are localized in a complicated three-dimensional surface. Since the real image formed by a typical large-aperture hologram has a small depth of focus, it may not be possible to record a clear photograph of the fringe system. This suggests the use of the *thin-beam reconstruction* technique,[24] depicted in Figure 5.18. A standard recording geometry, Figure 5.17, is used. The hologram is then rotated 180° and illuminated with a beam of small, and preferably variable, diameter. The real image (and interferogram) is formed with rays emanating from a small illuminated region of the hologram, as shown in Figure 5.18. If the distance between hologram and object is large, these rays will be nearly parallel. By illuminating different points on the hologram, the orientation of the nearly collimated beam

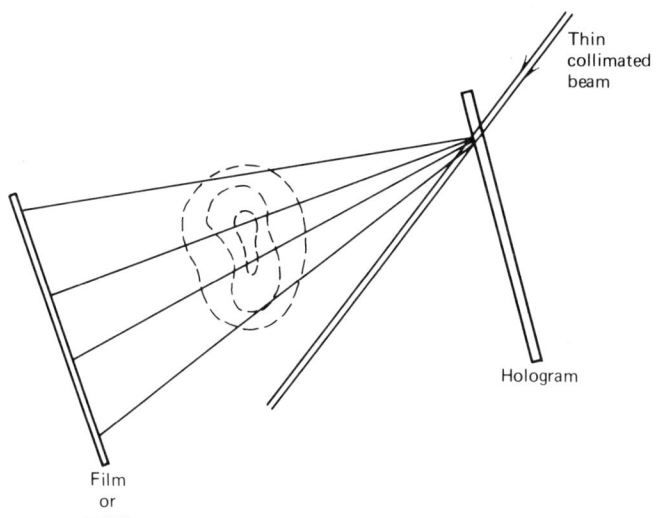

Figure 5.18 Thin-beam reconstruction of a diffuse-illumination holographic interferogram.

which forms the interference pattern can be varied. If this distance is not large, the interference pattern is formed with a *fan beam*, as in the figure. It will be shown in Chapter 6 that such an interferogram is suitable for quantitative evaluation. Since only a small area of the hologram is illuminated, the effective region of localization is extended, and the entire fringe system can be brought into focus. On the other hand, such a small aperture increases the characteristic speckle size. The optimum aperture of illumination at the hologram is determined as a trade-off between these two effects.

A knowledge of diffuser characteristics is useful in designing diffuse-illumination interferometers. Two characteristics of particular relevance are the angular distribution of irradiance and the state of polarization of light leaving the diffuser. There are two broad classes of diffusers. *Surface diffusers*, such as ground glass, affect only the optical pathlength of light transmitted by them and can be characterized accurately by a simple transmittance function:

$$\mathbf{t}(x,y) = \exp\left[-i\phi(x,y)\right], \tag{5.29}$$

where $\phi(x,y)$ is real. An example of a surface diffuser is a plate of glass having a homogeneous refractive index, with one surface microscopically cracked and pitted into an irregular contour by grinding with abrasives having grain sizes typically on the order of 1 to 100 μm. There is strong experimental evidence that ground glass surfaces can be modeled accurately as Gaussian random processes.[25] Surface diffusers do not appreciably alter the polarization of light. *Volume diffusers*, such as opal glass, consist of a three-dimensional collection of very small scattering centers. The light is diffused by multiple scattering. This affects both the amplitude and the phase of transmitted light, so volume diffusers are not as easily characterized as are surface diffusers. Light is randomly polarized by multiple scattering in opal glass.

Goniometric curves indicating the angular variation in irradiance of scattered light for some typical ground and opal glasses are shown in Figure 5.19. Similar data for some other diffusers are given by Gusev and Konstantinov.[26] The outer curve in this figure is a plot of $\cos\theta$, which would be the curve for a perfect Lambertian diffuser. It is seen that the irradiance of light diffused by ground glass is reduced to half its maximum value at an angle of 10° or less. Opal glass provides much greater uniformity of irradiance over larger angles, and in fact closely approximates a Lambertian diffuser. Thus opal glass is preferable when large viewing angles are desired. The uniformity of wide-angle illumination can be enhanced by using two or more diffusers in series. Very uniform irradiance over small viewing angles can be achieved by etching finely

5.3 HOLOGRAPHY AND INTERFEROMETRY WITH PHASE OBJECTS

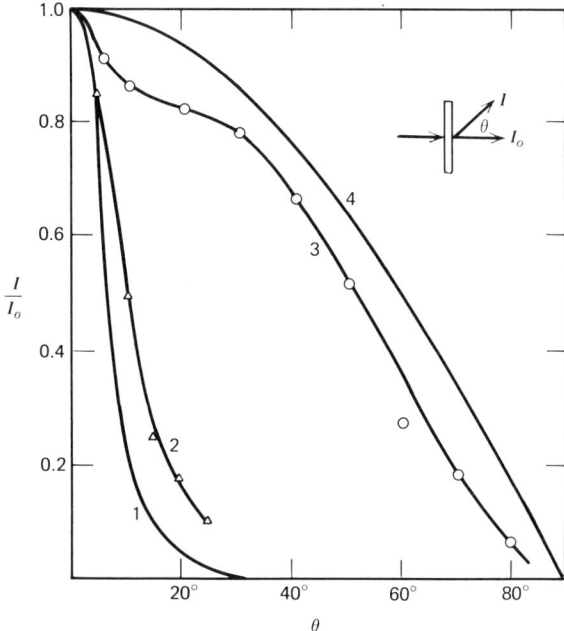

Figure 5.19 Angular distribution of irradiance of light scattered by various diffusers: 1. ground glass, 500 grit polish (ref. 25); 2. ground glass, 80 μm grit (ref. 26); 3. opal glass, 500 μm thick (ref. 31); 4. Lambertian diffuser.

ground glass with dilute hydrofluoric acid.[27] Special-purpose diffusers, some useful in interferometry, can be produced by kinoform or bleached hologram techniques (see, e.g., refs. 28 to 30).

Depolarization by opal glass has been studied by George et al.[31] For a diffuser 100 μm thick only about 5 percent of the irradiance is shifted to a new polarization. For thicknesses of 500 μm or more the light is completely depolarized. Depolarization effects are important in holographic interferometry. Only light with a polarization parallel to that of the reference wave can interfere with it to form the hologram. The other component of polarization, however, does contribute to the total exposure of the plate—a fact that should be recognized when computing exposure times and setting reference beam/object beam ratios. This effect can be eliminated by placing a polarizer in front of the hologram during recordings. It must, of course, be interferometrically stable. Depolarization is especially important in real-time interferometry[32] because the reconstructed wave is linearly polarized, but the instantaneous wave with which it interferes is not. This adds an incoherent background which significantly reduces fringe visibility. Full fringe visibility can be restored by viewing the fringes through a linear polarizer.

Burch et al.[33] and Gates[29] developed a holographic interferometer which utilizes a diffusing element termed a *scatter plate*. A scatter plate can be formed by lightly polishing ground glass with a very fine abrasive. This results in a surface contour like that shown schematically in Figure 5.20c. Similar elements can be produced by lightly smoking glass[34] or by forming appropriate bleached holograms.[29] A portion of light passing through a scatter plate is specularly transmitted, and the rest is diffused. The ratio of diffused to specularly transmitted light can be controlled by the polishing process. Scatter plate holographic interferometers are depicted in Figures 5.20a and 5.20b. The specularly transmitted light serves as the reference wave, and the diffuse component as the object wave. All optical rays which interfere near some point in the film plane to form the hologram originate

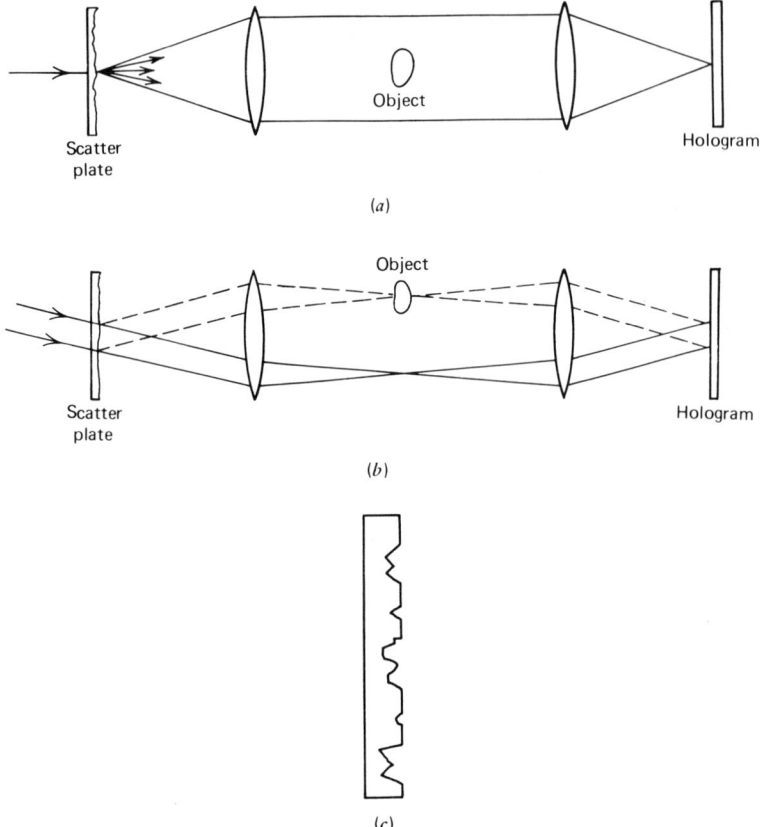

Figure 5.20 The Burch-Gates scatter plate holographic interferometer. (a) On-axis version. (b) Off-axis version. (c) Profile of scatter plate.

5.3 HOLOGRAPHY AND INTERFEROMETRY WITH PHASE OBJECTS

in the neighborhood of a single point of the scatter plate. This minimizes coherence requirements, so the scatter plate interferometer is particularly useful with pulsed lasers or multimode cw lasers and has even been used with filtered mercury and xenon lamps.[29] Another attractive feature of this system is its simplicity. However, its use is restricted, by the lens diameter, to examination of rather small objects.

Unfortunately, fringe clarity and resolution in diffuse-illumination interferometry are limited by laser speckle. In Section 1.4 we found that the characteristic speckle size b_s is:

$$b_s \simeq 1.22\lambda \left(\frac{f}{d}\right), \quad (5.30)$$

where f/d is the f number of the imaging system. If the spacing of interference fringes d_f is of the same order as the speckle width, it may be difficult to resolve the interference fringes. The criterion

$$\frac{b_s}{d_f} < 1 \quad (5.31)$$

should be satisfied. In visible light this criterion is usually satisfied if a lens with low f number, say $f/2$, can be used. Unfortunately, fringe localization often forces the use of high-f-number imaging systems to obtain sufficient depth of localization. If the fringe spacing exceeds a few lines per millimeter, speckle problems are likely to be encountered.

Tanner[35,36] studied fringe clarity in diffuse-illumination interferometry and found that it varies in a complicated manner with b_s/d_f and with the visibility and geometric form of the fringes. He defined fringe visibility as

$$v = \frac{\bar{I}_b - \bar{I}_d}{\bar{I}_b + \bar{I}_d}, \quad (5.32)$$

where $\bar{I}_b$ and $\bar{I}_d$ are the average irradiances along the bright and dark fringes, respectively. If fringes are of a simple geometric form, such as parallel straight lines, b_s/d_f approaching, or even slightly exceeding, unity may be acceptable. If fringes are of a complicated shape, however, good subjective clarity may require b_s/d_f to be 0.1 or less even if v is close to unity. Tanner also showed that in the presence of speckle the clarity of fringes cannot be enhanced significantly by photographing them with high contrast, as would be the case if speckle were not present.

The speckle problem in multidirectional interferometry can be alleviated by using a discrete set of plane waves for object illumination. A set of plane waves can be produced by an array of point sources in the back focal plane of a collimating lens.[35] An alternative procedure is to replace the diffuser in the system shown in Figure 5.17 by a sinusoidal phase grating.[37-39] Such gratings are easily produced by recording the interference pattern of two angularly separated plane waves on a photographic plate. The developed plate can be bleached to form a phase grating which modulates the phase of light transmitted by it (Section 1.3). The amplitude transmittance of a sinusoidal phase grating of infinite extent is

$$\mathbf{t}(x) = \exp\left[i\left(\frac{M}{2}\right)\sin(2\pi f_x x)\right], \quad (5.33)$$

where M is the phase modulation depth, and f_x is the spatial frequency of the grating. By using the identity[40]

$$\exp\left[i\left(\frac{M}{2}\right)\sin x\right] = \sum_{n=-\infty}^{\infty} J_n\left(\frac{M}{2}\right)\exp(inx), \quad (5.34)$$

where J_n is the Bessel function of the first kind of order n, it is seen that, if the phase grating is illuminated by a normally incident plane wave of unit amplitude, the transmitted light will have the amplitude

$$\mathbf{U}(x,y) = \sum_{n=-\infty}^{\infty} J_n\left(\frac{M}{2}\right)\exp(in2\pi f_x x). \quad (5.35)$$

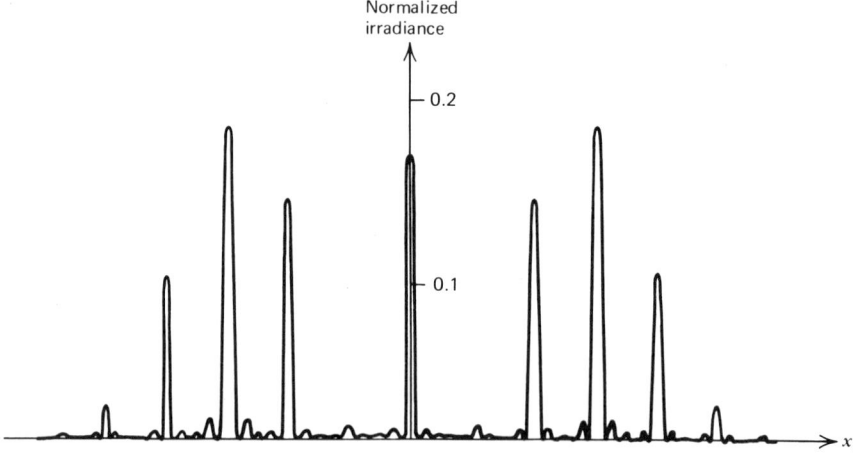

Figure 5.21 Fraunhofer diffraction pattern of a sinusoidal phase grating. This is the irradiance pattern in the back focal plane of a lens which accepts the wavefront of light diffracted by the grating. (Ref. 43)

5.3 HOLOGRAPHY AND INTERFEROMETRY WITH PHASE OBJECTS

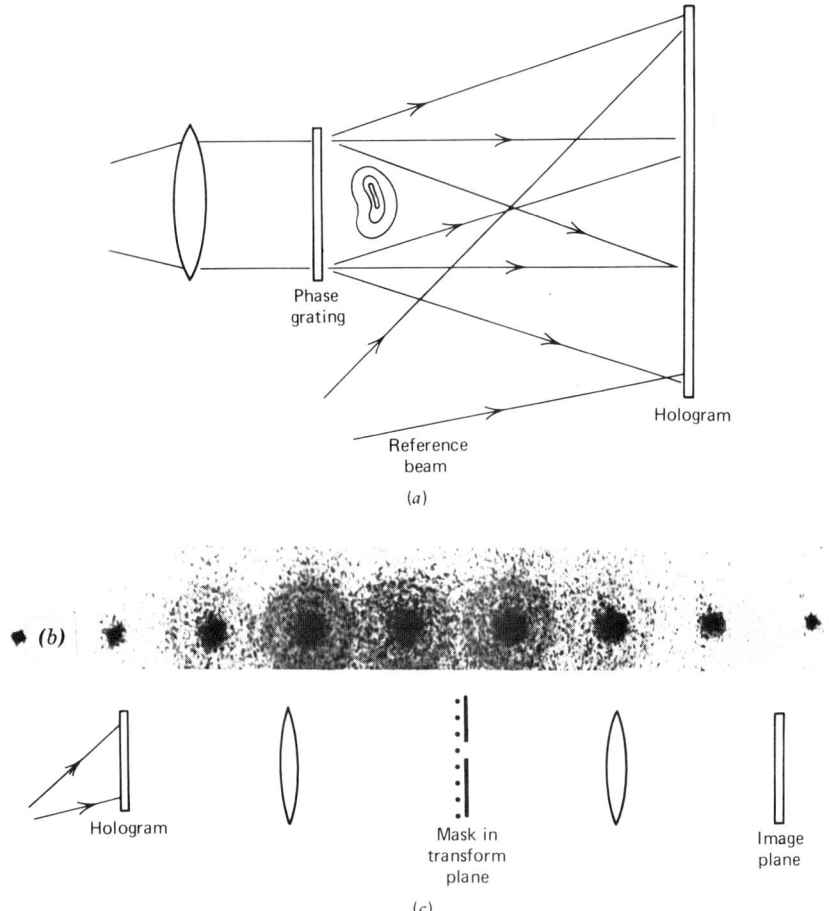

Figure 5.22 Holographic interferometer with object illumination derived from a phase grating. (*a*) System for recording the hologram. (*b*) Irradiance in back focal plane of transforming lens receiving the reconstructed wavefront. (*c*) System for viewing the interferogram.

Thus the phase grating diffracts the incident light into a discrete set of plane waves propagating at angles

$$\theta_n = \sin^{-1}(nf_x\lambda), \qquad n = 0, \pm 1, \pm 2, \ldots. \qquad (5.36)$$

The plane waves have intensities proportional to $J_n^2(M/2)$, and M depends on exposure and bleaching procedure.[41,42] Goodman[43] has calculated the irradiance far from a sinusoidal phase grating which is square with sides of length l,

$$I(x,y) = \left(\frac{l^2}{\lambda z}\right)^2 \sum_{n=-\infty}^{\infty} J_n^2\left(\frac{M}{2}\right)$$
$$\cdot \text{sinc}^2\left[\frac{l}{\lambda z}(x - nf_x\lambda z)\right] \text{sinc}^2\left(\frac{ly}{\lambda z}\right), \quad (5.37)$$

where x and y are coordinates in a plane a distance z from the grating. The cross section of this pattern for a typical grating ($M=8$) is illustrated in Figure 5.21, which shows that the wavefront from a phase grating of finite extent closely approximates a set of plane waves. A holographic interferometer with phase grating illumination is shown schematically in Figure 5.22a. On reconstruction of a holographic interferogram recorded with this interferometer, the interferograms corresponding to the various illumination directions can be separated by elementary spatial filtering, as shown in Figures 5.22b and 5.22c. Interferograms produced in this manner are nearly speckle-free, and fringe resolution can be improved by factors of about 3 to 5 in comparison to diffuse-illumination interferograms.

5.4 FRINGE LOCALIZATION IN DIFFUSE-ILLUMINATION INTERFEROMETRY

The fringes observed in a diffuse-illumination holographic interferogram of a phase object appear to be localized in space. They move and change shape as the viewing direction is varied, and they tend to localize near steep gradients of refractive index. For example, if an interferogram of a flame is recorded, the fringes will localize near the center of the flame. If the phase object has a more complicated configuration, the surface or curve of fringe localization may be very convoluted. In this section we derive equations which relate the surface of fringe localization to the distribution of refractive index change and the geometry of the holographic system.

Fringe localization in diffuse-illumination interferometry of phase objects can be analyzed with an approach and notational system similar to that used in Chapter 3 to relate fringe localization to the displacement of an opaque surface. The analysis presented here is restricted to the refractionless limit, that is, rays are assumed to be straight lines. An illuminated diffuser is depicted in Figure 5.23. In front of the diffuser is a region whose refractive index is n_0 during one holographic exposure, and $n(x,y,z)$ during the second exposure of a two-exposure hologram. It is useful to imagine that a field $f(x,y,z)$ is located in front of the diffuser, where

$$f(x,y,z) = n(x,y,z) - n_0. \quad (5.38)$$

5.4 FRINGE LOCALIZATION

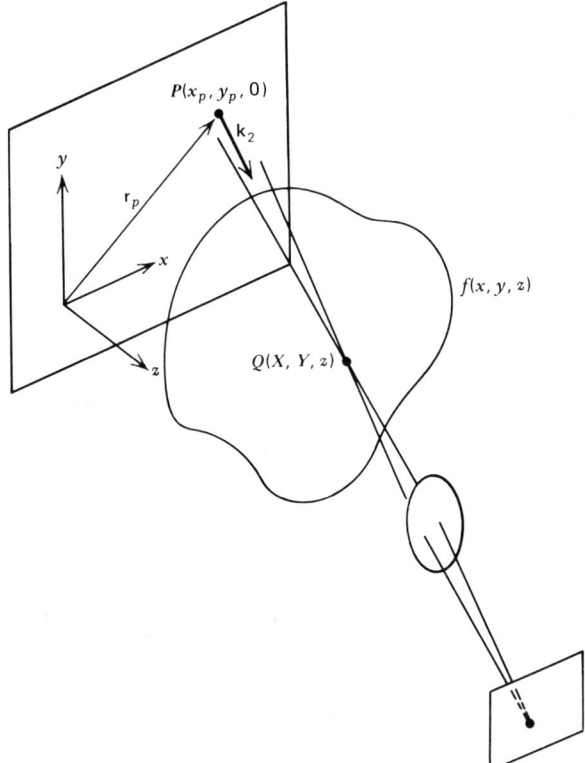

Figure 5.23 Notation for analysis of fringe localization.

This field also can be denoted by $f(\mathbf{r})$, where $\mathbf{r}$ is the position vector of any point in the field. In the analysis which follows, the Cartesian coordinate axes can be located arbitrarily. For convenience we assume that the diffuser lies in the x-y plane; however, the reader should note that the analysis is completely independent of this assumption.

Consider an observer who examines the interferogram through a viewing system which receives light with a mean propagation vector $\mathbf{k}_2$. The system is focused on a plane normal to $\mathbf{k}_2$ that contains a point Q, as indicated in Figure 5.23. As was discussed in Section 3.2, the fringes will appear to be localized, for this viewing direction, near those points Q where the phase difference is nearly constant over the small cone of ray pairs which pass through Q and are collected by the viewing system. It is convenient to introduce the optical pathlength difference $\Delta\Phi$, defined by

$$\Delta\Phi = \int f(\mathbf{r})\,ds. \qquad (5.39)$$

Analytically, we determine localization by seeking the points Q along all rays with a given $\mathbf{k}_2$ for which the variation of $\Delta\Phi$ with viewing direction is zero. A light ray which passes through points $P(x_p, y_p, 0)$ and $Q(X, Y, z)$ and strikes a detector surface is shown in Figure 5.23. Light traveling along this ray has the propagation vector $\mathbf{k}_2$. The condition for fringe localization is

$$d(\Delta\Phi) = \frac{\partial(\Delta\Phi)}{\partial x_p} dx_p + \frac{\partial(\Delta\Phi)}{\partial y_p} dy_p = 0, \tag{5.40}$$

where dx_p and dy_p are differential changes in the intersection of the ray with the diffuser. The optical pathlength difference is

$$\Delta\Phi = \int_{-\infty}^{\infty} f(\mathbf{r}) \, ds = \int_{-\infty}^{\infty} f(\mathbf{r}_p + \hat{\mathbf{k}}_2 s) \, ds, \tag{5.41}$$

where s is distance along the ray. The limits $\pm\infty$ are used in equation 5.41 for convenience. It is implied that the refractive index difference $f(x, y, z)$ is nonzero only in a finite region between the diffuser and the entrance aperture of the viewing system, and has no discontinuities.

Since the limits of the integral in equation 5.41 are fixed, the derivative

$$\frac{\partial(\Delta\Phi)}{\partial x_p} = \int_{-\infty}^{\infty} \nabla f \cdot \left(\frac{\partial \mathbf{r}}{\partial x_p}\right) ds, \tag{5.42}$$

where

$$\nabla f = \hat{\mathbf{i}} f^x + \hat{\mathbf{j}} f^y + \hat{\mathbf{k}} f^z, \tag{5.43}$$

is the gradient of f, and

$$\frac{\partial \mathbf{r}}{\partial x_p} = \hat{\mathbf{i}} + \left(\hat{\mathbf{i}} k_{2x}^{x_p} + \hat{\mathbf{j}} k_{2y}^{x_p} + \hat{\mathbf{k}} k_{2z}^{x_p}\right) s. \tag{5.44}$$

The superscripts in these equations denote parital derivatives. Therefore

$$\frac{\partial(\Delta\Phi)}{\partial x_p} = \int_{-\infty}^{\infty} \left[(1 + k_{2x}^{x_p} s) f^x + k_{2y}^{x_p} f^y s + k_{2z}^{x_p} f^z s \right] ds. \tag{5.45}$$

Similarly, the derivative of $\Delta\Phi$ with respect to y_p is

$$\frac{\partial(\Delta\Phi)}{\partial y_p} = \int_{-\infty}^{\infty} \left[k_{2x}^{y_p} f^x s + (1 + k_{2y}^{y_p} s) f^y + k_{2z}^{y_p} f^z s \right] ds. \tag{5.46}$$

5.4 FRINGE LOCALIZATION

If the viewing aperture is of roughly the same dimension in all directions (e.g., a square or circular aperture), dx_p and dy_p can be varied independently in equation 5.40. Therefore $\partial(\Delta\Phi)/\partial x_p$ and $\partial(\Delta\Phi)/\partial y_p$ must vanish independently. The expressions given in equation 3.7 for derivatives of the components of $\hat{\mathbf{k}}_2$ with respect to coordinates x_p, y_p in the plane $z=0$ can be used in equations 5.45 and 5.46, to give the following criteria for fringe localization:

$$\int_{-\infty}^{\infty} \left\{ \left[1 - \frac{k_{2z}}{z_l}(k_{2y}^2 + k_{2z}^2)s \right] f^x + \frac{k_{2z}}{z_l} k_{2x} k_{2y} f^y s + \frac{k_{2z}}{z_l} k_{2x} k_{2z} f^z s \right\} ds = 0, \quad (5.47)$$

$$\int_{-\infty}^{\infty} \left\{ \frac{k_{2z}}{z_l} k_{2x} k_{2y} f^x s + \left[1 - \frac{k_{2z}}{z_l}(k_{2x}^2 + k_{2z}^2)s \right] f^y + \frac{k_{2z}}{z_l} k_{2y} k_{2z} f^z s \right\} ds = 0. \quad (5.48)$$

Here a subscript l has been introduced to emphasize that z_l is the z coordinate of the point of localization along the ray in question.

If the distance s is measured from the diffuser, that is, the plane $z=0$,

$$s = \frac{z}{k_{2z}}. \quad (5.49)$$

Therefore equations 5.47 and 5.48 can be rewritten as

$$z_l = \frac{\int_{-\infty}^{\infty} \left[(k_{2y}^2 + k_{2z}^2) f^x - k_{2x} k_{2y} f^y - k_{2x} k_{2z} f^z \right] z \, dz}{\int_{-\infty}^{\infty} f^x \, dz}, \quad (5.50)$$

$$z_l = \frac{\int_{-\infty}^{\infty} \left[-k_{2x} k_{2y} f^x + (k_{2x}^2 + k_{2z}^2) f^y - k_{2y} k_{2z} f^z \right] z \, dz}{\int_{-\infty}^{\infty} f^y \, dz}. \quad (5.51)$$

For a particular field of refractive index change, $f(x,y,z)$, evaluation of equations 5.50 and 5.51 yields the equations of two surfaces. Fringes localize along the curve which is the intersection of these surfaces.

If the interferogram is viewed along the z axis, then $\hat{\mathbf{k}}_2 = \hat{\mathbf{k}}$, and the localization condition is

$$z_l = \frac{\int_{-\infty}^{\infty} f^x z \, dz}{\int_{-\infty}^{\infty} f^x \, dz}, \qquad (5.52)$$

$$z_l = \frac{\int_{-\infty}^{\infty} f^y z \, dz}{\int_{-\infty}^{\infty} f^y \, dz}. \qquad (5.53)$$

Equations 5.52 and 5.53 indicate that fringe localization is determined by the gradient of the field normal to the viewing direction. Specifically, *the fringes localize where the first moments of f^x and f^y in the z direction vanish.* The simple form of equations 5.52 and 5.53 suggests that the more general equations 5.50 and 5.51 might be interpreted in terms of components of ∇f normal to the viewing direction. To see that this is indeed the case, let $\hat{\mathbf{k}}_h$ and $\hat{\mathbf{k}}_v$ be unit vectors which lie in the aperture plane; $\hat{\mathbf{k}}_h$ is perpendicular to $\hat{\mathbf{k}}_2$ and coplanar with the x axis. Similarly, $\hat{\mathbf{k}}_v$ is perpendicular to $\hat{\mathbf{k}}_2$ and coplanar with the y axis. Note that $\hat{\mathbf{k}}_h$ and $\hat{\mathbf{k}}_v$ are special cases of the vector $\hat{\mathbf{k}}_{ap}$, which is defined by equations 3.23 and 3.24. To determine $\hat{\mathbf{k}}_h$, let $(dy/dx)_{ap} = 0$ in these two equations. Similarly, $\hat{\mathbf{k}}_v$ is determined by letting $(dy/dx)_{ap} \to \infty$:

$$\hat{\mathbf{k}}_h = \frac{\hat{\mathbf{i}}(k_{2y}^2 + k_{2z}^2) - \hat{\mathbf{j}} k_{2x} k_{2y} - \hat{\mathbf{k}} k_{2x} k_{2z}}{(k_{2y}^2 + k_{2z}^2)^{1/2}}, \qquad (5.54)$$

$$\hat{\mathbf{k}}_v = \frac{-\hat{\mathbf{i}} k_{2x} k_{2y} + \hat{\mathbf{j}}(k_{2x}^2 + k_{2z}^2) - \hat{\mathbf{k}} k_{2y} k_{2z}}{(k_{2x}^2 + k_{2z}^2)^{1/2}}. \qquad (5.55)$$

The general localization conditions, equations 5.50 and 5.51, can therefore be rewritten in more compact form as

$$z_l = \frac{k_{2z} k_{hx} \int_{-\infty}^{\infty} \hat{\mathbf{k}}_h \cdot \nabla f s \, ds}{\int_{-\infty}^{\infty} f^x \, ds}, \qquad (5.56)$$

$$z_l = \frac{k_{2z} k_{vy} \int_{-\infty}^{\infty} \hat{\mathbf{k}}_v \cdot \nabla f s \, ds}{\int_{-\infty}^{\infty} f^y \, ds}. \qquad (5.57)$$

5.4 FRINGE LOCALIZATION

We now consider a few examples of fringe localization. Many refractive index fields of practical importance are radially symmetric. The optical pathlength difference for a radially symmetric field is

$$\Delta\Phi(x) = 2 \int_x^R \frac{f(r)r\,dr}{(r^2 - x^2)^{1/2}}, \tag{5.58}$$

where x is understood to be the coordinate normal both to the axis of symmetry and to the viewing direction. Equations 5.52 and 5.53 indicate that the fringes localize where the moment of the gradient of $f(r)$ normal to the viewing direction vanishes. The fringes therefore will localize in the center plane of the object, that is, in the plane which is normal to the viewing direction and contains the axis of symmetry of the phase object. This is illustrated in the photograph in Figure 5.1, which was made by placing a sheet of film in the center plane of the real image formed by a two-exposure hologram of the flame. If the film were placed in another plane in front of, or behind, the center plane, the fringes would be blurred or unobservable.

Fringe localization is illustrated further in Figures 5.24b and 5.24c, which are interferograms of two identical, radially symmetric phase objects separated by 2.5 cm, as shown in Figure 5.24a. The phase objects were plumes of heated water above small, identical electrical heating elements. When viewed along the direction $\hat{\mathbf{k}}_2$, the optical axis passes through the center of both phase objects. The moments of the gradient normal to the viewing direction vanish in the plane midway between the two objects. Figures 5.24b and 5.24c indicate that fringes indeed localize in that plane. If the same holographic interferogram is viewed along a direction such as $\mathbf{k}_2'$, the fringes due to each phase object localize in their respective center planes.

Surfaces or curves of fringe localization may be quite convoluted. For example, consider two cylindrical phase objects, one in front of the other, with axes of symmetry which are parallel to the y axis. Let each object have a radial distribution of refractive index change which is Gaussian. The phase object, which is centered at $z = z_{o1}$, has a refractive index distribution $f_1(r_1)$, where

$$f_1(r_1) = f_{o1} \exp\left[-\left(\frac{r_1}{r_{o1}}\right)^2\right]. \tag{5.59}$$

Similarly, the second object is centered at $z = z_{o2}$ and is described by

$$f_2(r_2) = f_{o2} \exp\left[-\left(\frac{r_2}{r_{o2}}\right)^2\right], \tag{5.60}$$

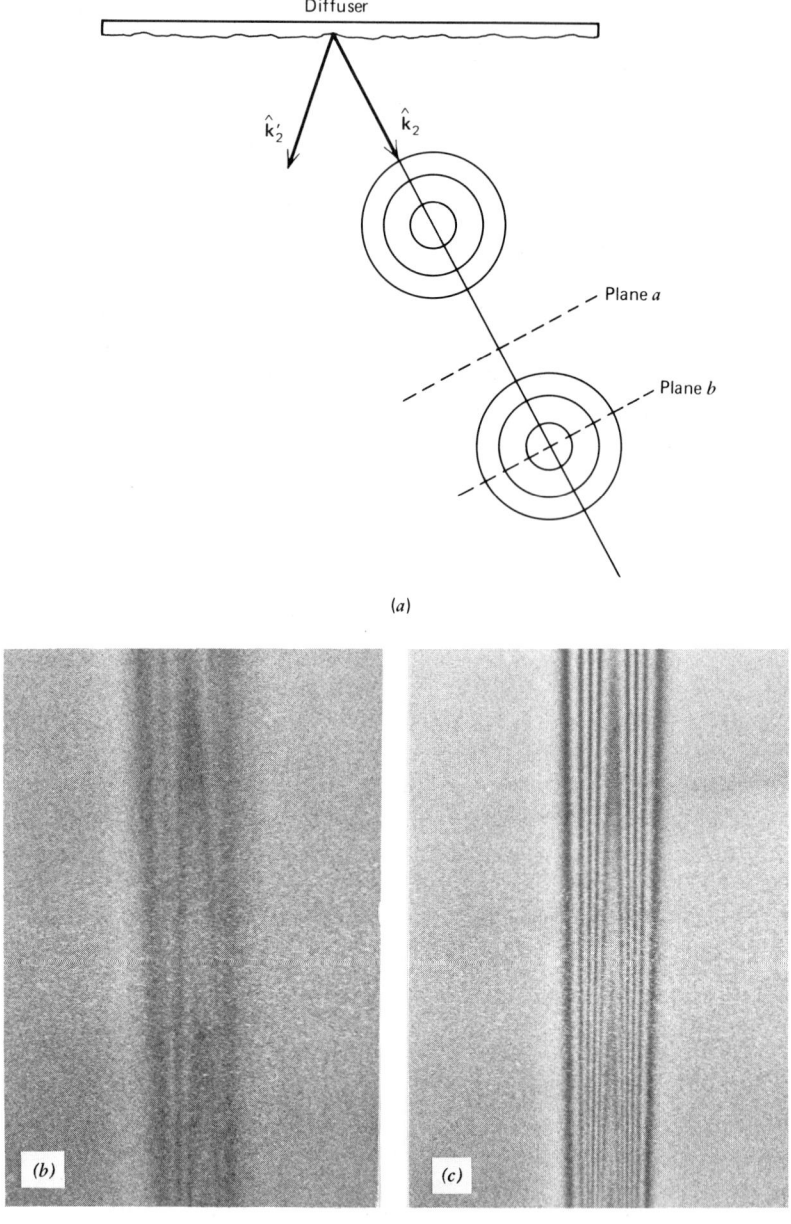

Figure 5.24 An example of fringe localization. (*a*) The viewing direction is parallel to the line connecting the centers of two identical, radially symmetric phase objects. (*b*) The interferogram in plane *b*. (*c*) The image in plane *a*.

5.4 FRINGE LOCALIZATION

where f_{o1}, f_{o2} are the maximum changes in refractive index, and r_{o1}, r_{o2} are characteristic radii. If a holographic interferogram of this object is viewed along the z axis, localization is determined by equation 5.52. The second equation, 5.53 is not used since the basic relation, equation 5.48, is identically satisfied when $f' \equiv 0$. The distribution of change in refractive index is

$$f(x,y) = f_{o1} \exp\left[-\frac{x^2+(z-z_{o1})^2}{r_{o1}^2}\right] + f_{o2} \exp\left[-\frac{x^2+(z-z_{o2}^2)}{r_{o2}^2}\right]. \quad (5.61)$$

Evaluation of the integrals in equation 5.52 yields the equation of the surface of localization:

$$z_l = \frac{z_{o1}(f_{o1}/r_{o1}^2)\exp\left[-(x/r_{o1})^2\right] + z_{o2}(f_{o2}/r_{o2}^2)\exp\left[-(x/r_{o2})^2\right]}{(f_{o1}/f_{o1}^2)\exp\left[-(x/r_{o1})^2\right] + (f_{o2}/r_{o2}^2)\exp\left[-(x/r_{o2})^2\right]}. \quad (5.62)$$

If the two phase objects are identical, $r_{o1} = r_{o2}$ and $f_{o1} = f_{o2}$, so the surface of localization is the plane

$$z = \tfrac{1}{2}(z_{o1} + z_{o2}), \quad (5.63)$$

which is consistent with the preceding discussion. If the two phase objects have different characteristic radii and maximum values, the surface of localization will be curved. As an example, let $z_{o1} = 7.0$, $z_{o2} = 19.7$, $r_{o1} = 0.843$, $r_{o2} = 0.474$, $f_{o1} = 0.847 \times 10^{-4}$, and $f_{o2} = 0.113 \times 10^{-3}$. Figure 5.25 is a plot of the resulting surface of localization. Figure 5.26 includes photographs of the interference patterns in the planes $z = z_{o1}$, $z = (z_{o1} + z_{o2})/2$, and $z = z_{o2}$ in the real image of a holographic interferogram of a phase object with a structure similar to that in this example.

If the interferogram of the phase object described by equation 5.61 is viewed from an angle specified by k_{2x} and k_{2z}, with $k_{2y} = 0$, the surface of localization can be determined by applying equation 5.50. The integrals in this equation are integrals along the line

$$x = x_p + \left(\frac{k_{2x}}{k_{2z}}\right)z. \quad (5.64)$$

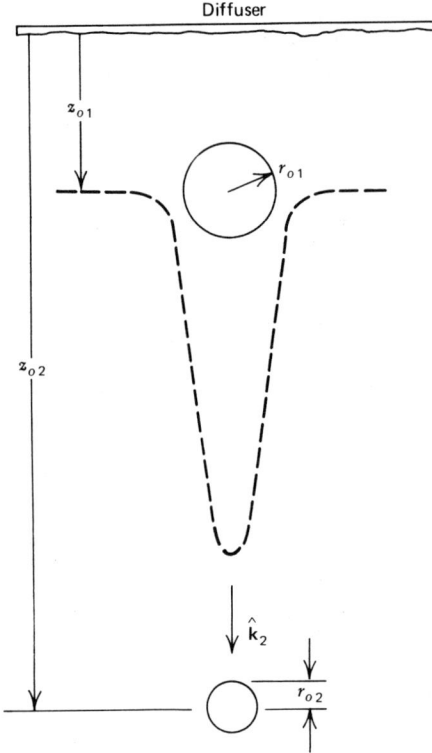

Figure 5.25 Surface of localization for the phase object described by equation 5.61.

The derivatives of the field are

$$f^x = -\left(\frac{2f_{o1}}{r_{o1}^2}\right) x \exp\left\{-\frac{[x^2+(z-z_{o1})^2]}{r_{o1}^2}\right\}$$
$$- (2f_{o2}/r_{o2}) x \exp\left\{-\frac{[x^2+(z-z_{o2})^2]}{r_{o2}^2}\right\}, \quad (5.65)$$

$$f^z = -\left(\frac{2f_{o1}}{r_{o1}^2}\right)(z-z_{o1}) \exp\left\{-\frac{[x^2+(z-z_{o1})^2]}{r_{o1}^2}\right\}$$
$$- \left(\frac{2f_{o2}}{r_{o2}^2}\right)(z-z_{o2}) \exp\left\{-\frac{[x^2+(z-z_{o2})^2]}{r_{o2}^2}\right\}. \quad (5.66)$$

Equations 5.64 to 5.66 can be combined and substituted into equation 5.50 to yield the equation of the surface of localization. This equation is quite

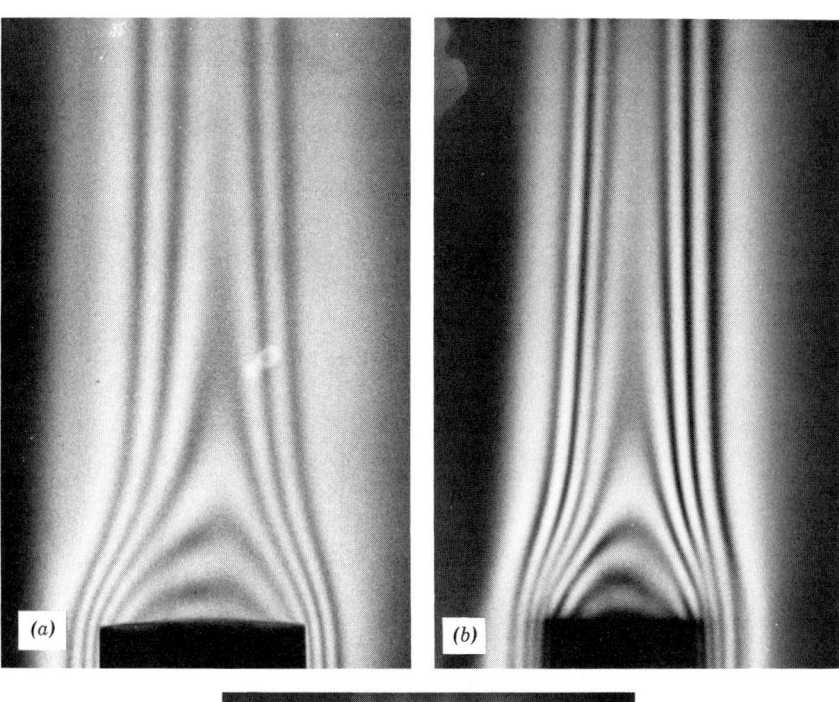

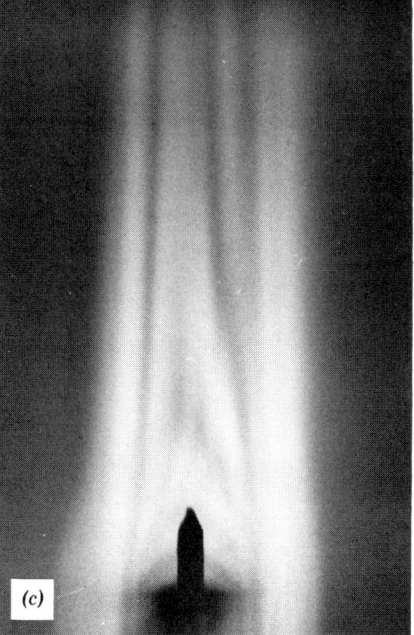

Figure 5.26 Photographs of the interference pattern due to a phase object similar to that depicted in Figure 5.25. (a) Fringe pattern in plane $z = z_{01}$. (b) Fringe pattern in plane $z = (z_{01} + z_{02})/2$. (c) Fringe pattern in plane $z = z_{02}$.

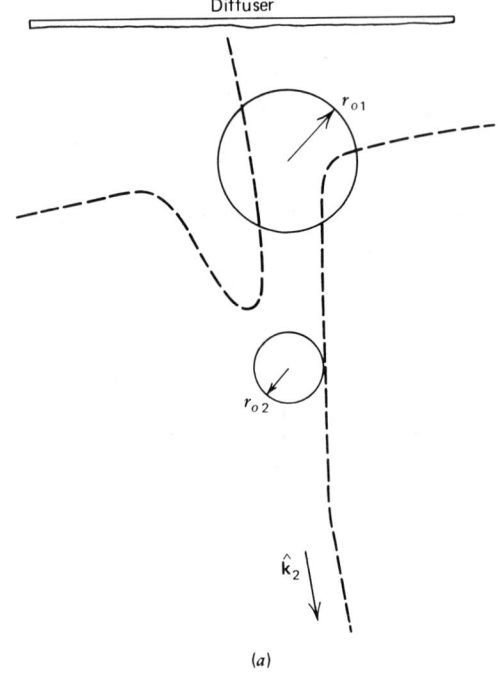

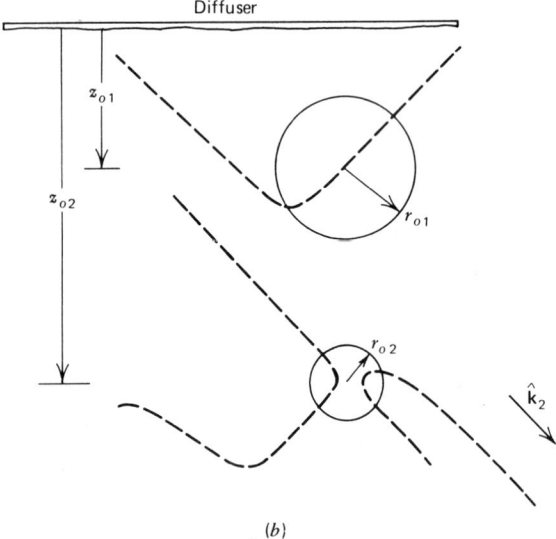

Figure 5.27 Surface of localization for the phase object depicted in Figure 5.25 when the viewing angle is varied. (a) The object is viewed at an angle of 10°. (b) The object is viewed at an angle of 45°.

lengthy, so the results are presented in graphical form in Figure 5.27, where the surfaces of localization are shown for two different viewing directions. It is clear from this example that a qualitative understanding of the structure of a refractive index field can be obtained by visual observation of a diffuse-illumination holographic interferogram. As the viewing direction is varied, fringe localization and spacing give a distinct impression of regions of rapid change in refractive index. Quantitative evaluation of interferograms of phase objects is discussed in Chapter 6.

5.5 NONLINEAR HOLOGRAPHIC INTERFEROMETRY

For most applications it is desirable to record and process holograms so that their amplitude transmittance varies linearly with exposure. This is at best an approximation, however, because the amplitude transmittance of most recording materials is a nonlinear function of exposure. When holography is used for three-dimensional photography or information storage, nonlinear effects are detrimental, giving rise to ghost images and noise.[44-48] In holographic interferometry it is possible to take advantage of nonlinearity. In this section we will consider the nature of nonlinear holograms and discuss their application to phase difference amplification and multiple-beam interferometry.

Consider a simple hologram formed by exposing a thin, plane emulsion to two interfering plane waves of equal amplitude. If the recording material is linear, fringes of sinusoidal profile (Figure 5.28a) are recorded, and the transmittance of the hologram is given by equation 1.34. If this hologram is illuminated by a plane wave for reconstruction, the optical wave leaving the hologram consists of three components: the directly transmitted component and two components referred to as the ± 1-order diffracted waves. If the recording medium is nonlinear, the recorded fringes have a nonsinusoidal profile such as that shown in Figure 5.28b. The amplitude transmittance corresponding to these nonsinusoidal fringes can be expanded in a Fourier series. The higher harmonic terms of these series give rise to additional diffracted orders. If one of the waves used to form a nonlinear hologram has traversed a phase object, the higher order diffracted waves can be combined in various ways to form interferograms with increased sensitivity or increased spatial resolution.

A typical transmittance-exposure $(t-E)$ curve for holographic film is shown in Figure 5.29. The transmittance represented by this curve can be expressed as a Taylor series in exposure E:

$$t = K_0 + K_1 E + K_2 E^2 + K_3 E^3 + \cdots . \tag{5.67}$$

Inclusion of terms through the third order gives a reasonable fit to the t-E curves of photographic emulsions over their useful range.[44] By using an

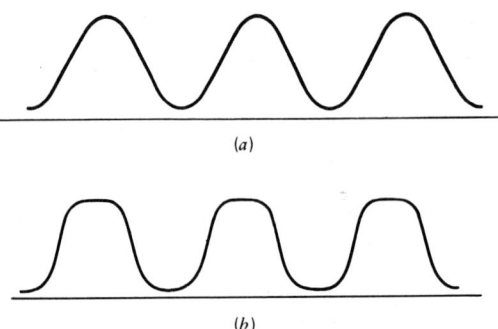

Figure 5.28 Fringe profiles. (*a*) Sinusoidal profile typical of linear interferometry. (*b*) Nonsinusoidal profile typical of nonlinear interferometry.

exposure greater or smaller than that required for linear holography, the importance of the quadratic and cubic terms can be enhanced. Since exposure is proportional to irradiance, a hologram of an object wave with amplitude $U_o(x,y)$ recorded with reference wave $U_R(x,y)$ would have an amplitude transmittance given by

$$t(x,y) = K_o + K_1(|U_o + U_R|^2) + K_2(|U_o + U_R|^2)^2 + \cdots \quad (5.68)$$

or, more explicitly,

$$\begin{aligned}
t(x,y) = K_o &+ K_1\{|U_o|^2 + |U_R|^2 + U_o U_R^* + U_o^* U_R\} \\
&+ K_2\{(|U_o|^2)^2 + (|U_R|^2)^2 + (U_o U_R^*)^2 + (U_o^* U_R)^2 \\
&+ 4|U_o|^2|U_R|^2 + 2|U_o|^2 U_o U_R^* + 2|U_o|^2 U_o^* U_R \\
&+ 2|U_R|^2 U_o U_R^* + 2|U_R|^2 U_o^* U_R\} + \cdots .
\end{aligned} \quad (5.69)$$

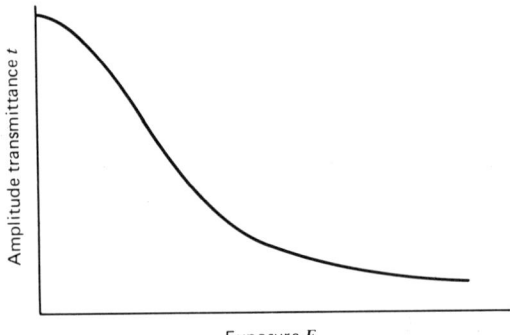

Figure 5.29 Typical transmittance-exposure curve for holographic film.

5.5 NONLINEAR HOLOGRAPHIC INTERFEROMETRY

Collecting similar terms yields a series of the form

$$t(x,y) = k_0 + k_1(U_oU_R^* + U_o^*U_R)$$
$$+ k_2[(U_oU_R^*)^2 + (U_o^*U_R)^2] + \cdots. \quad (5.70)$$

Equation 5.70 is the general expression for the transmittance of a nonlinear hologram.

As a specific example, consider the system depicted in Figure 5.30. The plane object wave passes through a phase object and impinges on the film plate at normal incidence. The reference wave is also plane and is oriented at an angle θ. At the hologram plane

$$U_o(x,y) = a_o\exp[-i\phi(x,y)],$$
$$U_R(x,y) = a_R\exp(i2\pi fy), \quad (5.71)$$

where $f = (\sin\theta)/\lambda$. The amplitude transmittance of a hologram formed by exposing the plate to U_o and U_R in a nonlinear portion of the t-E curve is

$$t(x,y) = k_0 + k_1 a_o a_R\big(\exp\{-i[2\pi fy + \phi(x,y)]\}$$
$$+ \exp\{i[2\pi fy + \phi(x,y)]\}\big)$$
$$+ k_2 a_o^2 a_R^2[\exp\{-i2[2\pi fy + \phi(x,y)]\}$$
$$+ \exp\{i2[2\pi fy + \phi(x,y)]\}]. \quad (5.72)$$

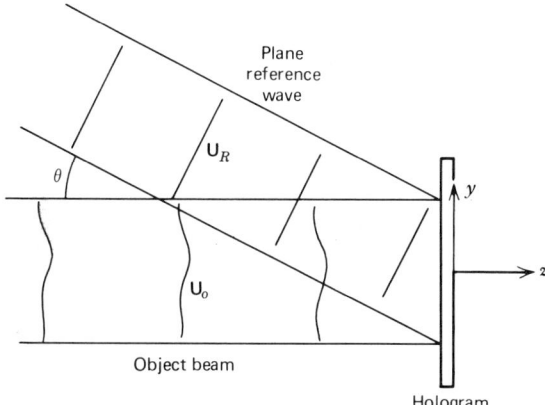

Figure 5.30 Recording an off-axis hologram of a phase object for nonlinear interferometry. A single holographic exposure is made.

This transmittance can also be expressed as

$$t(x,y) = \sum_{n=-\infty}^{\infty} c_n \exp\{-in[2\pi fy - \phi(x,y)]\}, \qquad (5.73)$$

where the c_n are constants. In reality only a finite number of components can be generated because only those with propagation angles between $-\pi/2$ and $+\pi/2$ rad have physical significance. If this hologram is illuminated by U_R for reconstruction, the amplitude of the diffracted wave leaving the hologram is

$$\begin{aligned}U(x,y) &= U_R(x,y)t(x,y) \\ &= k_0 a_R \exp(i2\pi fy) \\ &\quad + k_1 a_o a_R^2 (\exp[-i\phi(x,y)] + \exp\{i[4\pi fy + \phi(x,y)]\}) \\ &\quad + k_2 a_o^2 a_R^3 (\exp\{-i[2\pi fy + 2\phi(x,y)]\} \\ &\quad + \exp\{i[6\pi fy + 2\phi(x,y)]\}).\end{aligned} \qquad (5.74)$$

Equation 5.74 describes five component waves, each propagating in a different direction indicated by the coefficient of $i2\pi fy$ in the exponential. These waves are shown schematically in Figure 5.31a. Additional components may be present and can be accounted for by including higher order terms in the expansion, equation 5.68.

The various wave components described by equation 5.74 have phase distributions which are positive- or negative-integer multiples of $\phi(x,y)$. As first pointed out by Bryngdahl and Lohman,[49] these harmonics can be used for amplifying small phase shifts $\phi(x,y)$. For example, if the wave of the second diffracted order, which has an amplitude proportional to $\exp[-i2\phi(x,y)]$, is combined with a plane wave of equal amplitude and the same propagation direction, the resulting interference pattern will be proportional to

$$I(x,y) = 1 + \cos[2\phi(x,y)]. \qquad (5.75)$$

Equation 5.75 indicates a phase difference amplification by a factor of 2, that is, there are twice as many fringes as in a linear interferogram.

To describe an alternative approach to phase difference amplification,[50] we suppose that the hologram described by equation 5.73 is illuminated with the conjugate of U_R:

$$U_R^*(x,y) = a_R \exp(-i2\pi fy). \qquad (5.76)$$

The resulting diffracted wave components are indicated in Figure 5.31b. Compare the first diffracted orders, which propagate along the z direction,

5.5 NONLINEAR HOLOGRAPHIC INTERFEROMETRY

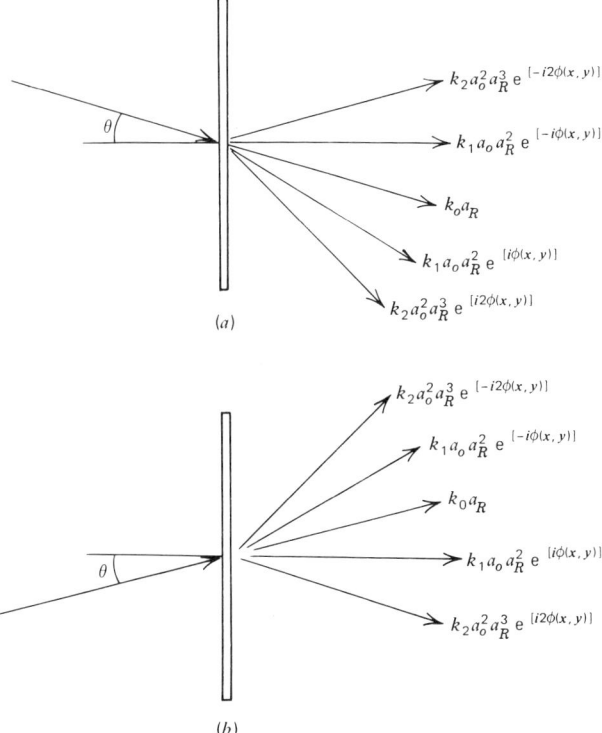

Figure 5.31 Propagation directions of component waves produced by the nonlinear hologram described by equation 5.73. (a) The hologram is illuminated by a duplicate of the original reference wave. (b) The hologram is illuminated by the conjugate of the original reference wave.

in Figures 5.31a and 5.31b. Although these components travel in the same direction, they have phase distributions with opposite signs. If the hologram is illuminated *simultaneously* by U_R and U_R^*, these two components add to form an irradiance pattern:

$$I(x,y) = 2k_1 a_o a_R^2 \{1 + \cos[2\phi(x,y)]\}. \tag{5.77}$$

Again the sensitivity has been doubled. It should be noted that only the linear part of the hologram has been utilized. Greater phase amplification can be obtained by combining higher harmonics. An example is shown in Figures 5.32a and 5.32b, where the reconstruction waves are

$$\begin{aligned} U_R'(x,y) &= a_R \exp(i4\pi f y), \\ U_R'^*(x,y) &= a_R \exp(-i4\pi f y). \end{aligned} \tag{5.78}$$

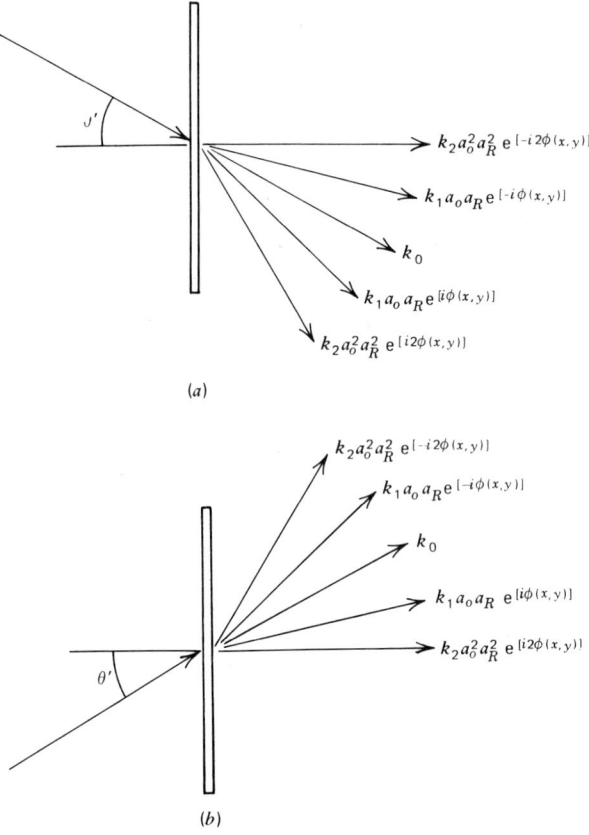

Figure 5.32 Propagation directions of component waves produced by the nonlinear hologram described by equation 5.73 when the illuminating direction is varied. (a) The hologram is illuminated by a plane wave with orientation $\theta' = \sin^{-1} 2\lambda f$. (b) The hologram is illuminated by a plane wave with orientation $-\theta'$.

These waves travel at angles $\pm \theta'$, where $\theta' = \sin^{-1} 2\lambda f$. If the hologram is illuminated by $\mathbf{U}'_R$ and $\mathbf{U}'^*_R$ simultaneously, the two components propagating in the z direction add to form an irradiance pattern

$$I(x,y) = 2(k_2 a_o^2 a_R^3)^2 \{1 + \cos[4\pi\phi(x,y)]\}. \quad (5.79)$$

This irradiance pattern represents an interferogram with fourfold fringe multiplication. Obviously the approach can be generalized by using higher harmonics to attain phase multiplications of 2, 4, 6, ... A reconstruction system for phase multiplication is shown in Figure 5.33. The dual reconstruction waves can be formed with a Mach-Zehnder[51] or Michelson

5.5 NONLINEAR HOLOGRAPHIC INTERFEROMETRY

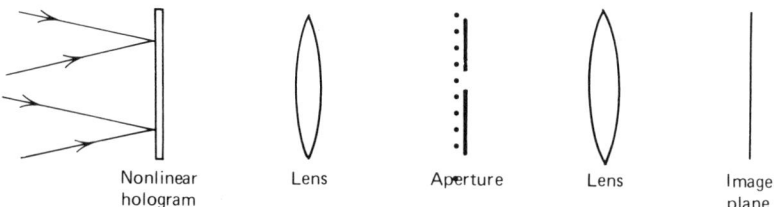

Figure 5.33 A reconstruction system for phase amplification by nonlinear holographic interferometry.

interferometer, or by using a Ronchi grating.[52] Matsumoto and Takashima[51] achieved phase amplification by a factor of 14 with this technique. An alternative approach which requires the recording of two holograms has been suggested by Velzel.[53]

The generation of higher harmonic component waves can also be accomplished with bleached holograms, which convert a recorded irradiance pattern into either a surface relief variation or a refractive index variation. If a linear hologram recorded using U_o and U_R, given by equation 5.71, is bleached, its amplitude transmittance will be of the form

$$t(x,y) = \exp\left\{i\left(\frac{M}{2}\right)\sin\left[2\pi fy - \phi(x,y)\right]\right\}, \qquad (5.80)$$

where M is the depth of phase modulation of the hologram. Using the identity given in equation 5.34, we can express the transmittance as

$$t(x,y) = \sum_{n=-\infty}^{\infty} J_n\left(\frac{M}{2}\right)\exp\left\{-in\left[2\pi fy - \phi(x,y)\right]\right\}, \qquad (5.81)$$

which is of the same form as equation 5.73 for a nonlinear amplitude hologram. Thus phase holograms can be used in place of nonlinear amplitude holograms in the interferometers discussed in this section. When higher harmonics are to be used for accurate phase measurements, it is desirable to follow the careful processing, prehardening, annealing, and drying procedures outlined by Toyooka[54] and by Pennington and Harper[55] in order to reduce emulsion shrinkage, extraneous surface relief, and light scattering.

It should be noted that the phase amplification schemes discussed above involve spatial division of amplitude by a single-exposure hologram. There is no phase error cancellation, as in two-exposure holographic interferometry. Furthermore, any phase errors due to imperfect optical elements are also amplified by the same factor. Shrinkage of the photographic emulsion is another source of significant error. Toyooka[54] has developed a scheme for removal of phase errors. A plane wave of light which has passed through a phase object in its initial state is recorded by off-axis holography. The hologram is then developed and replaced in its original position. The desired change in the phase object is then made, and the light passing through it is diffracted by the hologram and recorded on a second hologram. The second hologram is processed nonlinearly, and phase amplification is achieved by superimposing a pair of higher harmonic component waves. With this system, which is described in detail in ref. 54, all phase errors except those due to emulsion shrinkage during development of the first hologram are eliminated. Although this interferometer is effective in eliminating phase errors, it requires particularly careful experimental procedures and precise alignment.

Another application of nonlinear holography is *multiple-beam interferometry* of phase objects. In this context multiple-beam interferometry is used to produce sharpened, nonsinusoidal fringes whose positions can be located with good precision by visual inspection. Multiple-beam interferometry of transparent objects by nonholographic techniques requires multiple reflections by the surfaces of the object; consequently, these surfaces must have rather high reflectances (see, e.g., Tolansky[56]). Holographic techniques, on the other hand, can be used to study phase objects with negligible surface reflectance.

Reconsider a hologram with amplitude transmittance given by equation 5.73. This is an off-axis, nonlinear hologram of a phase object described by the phase distribution $\phi(x,y)$, recorded with a plane reference wave. The complex amplitudes of the object and reference waves are given in equations 5.71. Each component wave diffracted by this hologram has a phase distribution which is an integral multiple of $\phi(x,y)$. Multiple-beam interferograms can be formed by addition of several of these component waves. For this addition to occur, the various component waves must propagate in the same direction. The configuration depicted in Figure 5.34a can be used to combine several components in an appropriate manner. A grating, such as a Ronchi grating or a holographic phase grating, having a fringe frequency equal to the carrier frequency of the hologram is placed in the illuminating beam. The resulting reconstruction wave at the hologram plane is

$$\mathbf{U}_R(x,y) = \sum_{m=-\infty}^{\infty} a_m \exp(-im2\pi fy), \qquad (5.82)$$

so the complex amplitude of light leaving the hologram is

$$\mathbf{U}(x,y) = \left[\sum_{m=-\infty}^{\infty} a_m \exp(-im2\pi fy) \right] \cdot \left(\sum_{n=-\infty}^{\infty} c_n \exp\{-in[2\pi fy - \phi(x,y)]\} \right). \qquad (5.83)$$

5.5 NONLINEAR HOLOGRAPHIC INTERFEROMETRY

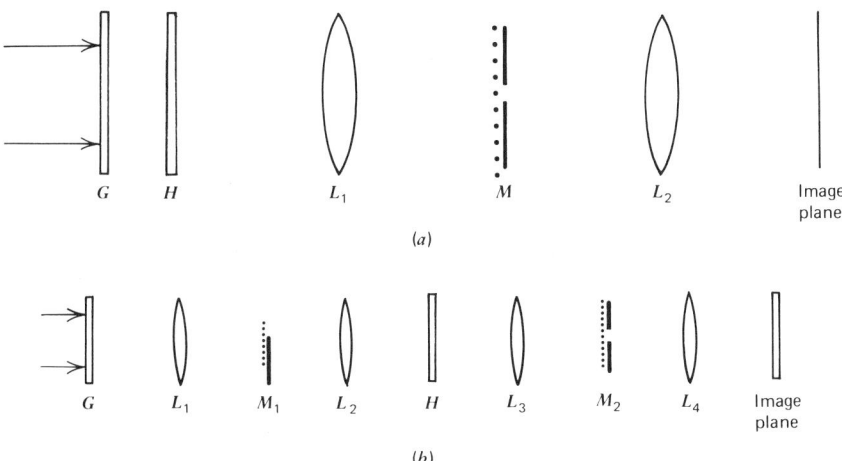

Figure 5.34 Systems for fringe sharpening by multiple-beam nonlinear holographic interferometry (ref. 58). (a) The nonlinear hologram H is illuminated by several orders of light diffracted by grating G; a transform-plane mask M is used to select waves traveling in a single direction. (b) The nonlinear hologram H is illuminated by imaging grating G onto the hologram. Mask M_1 blocks out all negative orders of the grating, and mask M_2' selects waves traveling in a single direction.

Equation 5.83 represents an array of component waves traveling in various directions; however, a simple telecentric imaging system with a small aperture in the Fourier transform plane can be used to select the light propagating in one direction, as shown in Figure 5.34a. For example, the complex amplitude of light propagating in the z direction, which is the direction of the original object wave, is

$$\begin{aligned}\mathbf{U}_o(x,y) &= a_0 c_0 - a_1 c_1 \{\exp[-i\phi(x,y)] - \exp[i\phi(x,y)]\} \\ &\quad - a_2 c_2 \{\exp[-i2\phi(x,y)] + \exp[i2\phi(x,y)]\} \\ &\quad - \cdots \\ &= a_0 c_0 - \sum_{n=0}^{\infty} a_n c_n \cos[n\phi(x,y)],\end{aligned} \quad (5.84)$$

where the relations $a_n = -a_{-n}$, $c_n = -c_{-n}$ have been used. The corresponding interferogram is given by the irradiance pattern

$$I(x,y) = \left\{ a_0 c_0 - \sum_{n=0}^{\infty} a_n c_n \cos[n\phi(x,y)] \right\}. \quad (5.85)$$

If the aperture is used to select the first diffracted order, that is, the component wave propagating in the direction of the reference wave used to record the hologram, a different amplitude distribution results:

$$\mathbf{U}_1(x,y) = \sum_{n=0}^{\infty} a_{n+1} c_n \exp[-in\phi(x,y)]. \quad (5.86)$$

This yields the interferogram

$$I(x,y) = \left| \sum_{n=0}^{\infty} a_{n+1} c_n \exp[-in\phi(x,y)] \right|^2. \quad (5.87)$$

A similar system of fringes can be produced by placing the grating behind, rather than in front of, the hologram.[57] The precise profiles of the fringes described by equation 5.85 or 5.87 depend on the relative values of the coefficients a_n and c_n and on the number of orders actually collected by the optical system.

Bryngdahl[58] introduced a system in which the hologram described by equation 5.73 is illuminated by imaging a grating, with a frequency identical to the carrier frequency of the hologram, onto the hologram. The telecentric system used to image the grating has a mask in the transform plane which blocks out all negative orders, as indicated in Figure 5.34b. The reconstruction wave at the hologram has an amplitude distribution like that described by equation 5.83, except that the summation is from $n=0$ to $n=\infty$, so the amplitude of light leaving the hologram in the z direction is

$$\mathbf{U}_o(x,y) = \sum_{n=0}^{\infty} a_n c_n \exp[-in\phi(x,y)]. \quad (5.88)$$

The corresponding interferogram has the irradiance distribution

$$I(x,y) = \left| \sum_{n=0}^{\infty} a_n c_n \exp[-in\phi(x,y)] \right|^2, \quad (5.89)$$

which is similar to that of equation 5.87. The relative values of the coefficients a_n can be adjusted by placing neutral density filters over the illuminated spots, which correspond to each diffracted order, in the plane of mask M_1, shown in Figure 5.34b.

If the irradiance of the higher orders drops off so that the coefficients in equation 5.87 or 5.89 decay geometrically, the complex amplitude is of the form

$$\begin{aligned}\mathbf{U}_o(x,y) &= \sum_{n=1}^{\infty} A^n \exp[in\phi(x,y)] \\ &= \frac{A \exp[-i\phi(x,y)]}{1 - A \exp[-i\phi(x,y)]},\end{aligned} \quad (5.90)$$

5.5 NONLINEAR HOLOGRAPHIC INTERFEROMETRY

where A is a constant. The corresponding irradiance pattern is

$$I(x,y) = \frac{A^2}{1+A^2-2A\cos\phi}. \tag{5.91}$$

This is the *Airy distribution*, which also describes multiple-reflection fringes in classical interferometry. Figure 5.35 includes several examples of fringe profiles produced by holographic multiple-beam interferometry. Figure 5.35*a* shows the fringe profiles described by equation 5.91 for several values of A. In reality only a finite number of components contribute to the interference pattern. This is illustrated in Figure 5.35*b*, which indicates the fringe profiles corresponding to inclusion of only small numbers of components in the summation in equation 5.90. To indicate the effect of relative irradiances of the different orders, similar fringe profiles are depicted in Figure 5.35*c* for the case in which all coefficients $a_n c_n$ have the same value.

The above analyses of holographic multiple-beam interferometry have neglected relative phase delays in each component wave diffracted by the grating and by the hologram. Such delays occur if the grating and hologram are not pure amplitude structures. These delays cause deviations from the predicted fringe profiles. Matsumoto[59] has analyzed the system depicted in Figure 5.34*a* in detail and included these delays. He also applied Fresnel diffraction theory to calculate additional relative phase delays among the various orders due to propagation over the distance d from the grating to the hologram. By varying the spacing d, the fringe profile can be adjusted to optimize the fringe sharpening effect.

All of the multiple-beam interferometry systems described thus far in this section use single-exposure holograms. It also is possible to utilize nonlinear effects to produce sharpened fringe profiles in two-exposure holographic interferometry.[60] The total exposure of a two-exposure hologram is proportional to

$$E = |U_o + U_R|^2 + |U'_o + U_R|^2, \tag{5.92}$$

where U_o is the complex amplitude of the object wave during the first exposure, U'_o is the complex amplitude of the object wave during the second exposure, and U_R is the complex amplitude of the reference wave. If a nonlinear hologram is formed by under- or overexposure, or by bleaching, its transmittance is determined by substituting equation 5.92 into equation 5.67:

$$t(x,y) = \sum_{n=0}^{\infty} K_n\left(|U_o + U_R|^2 + |U'_o + U_R|^2\right). \tag{5.93}$$

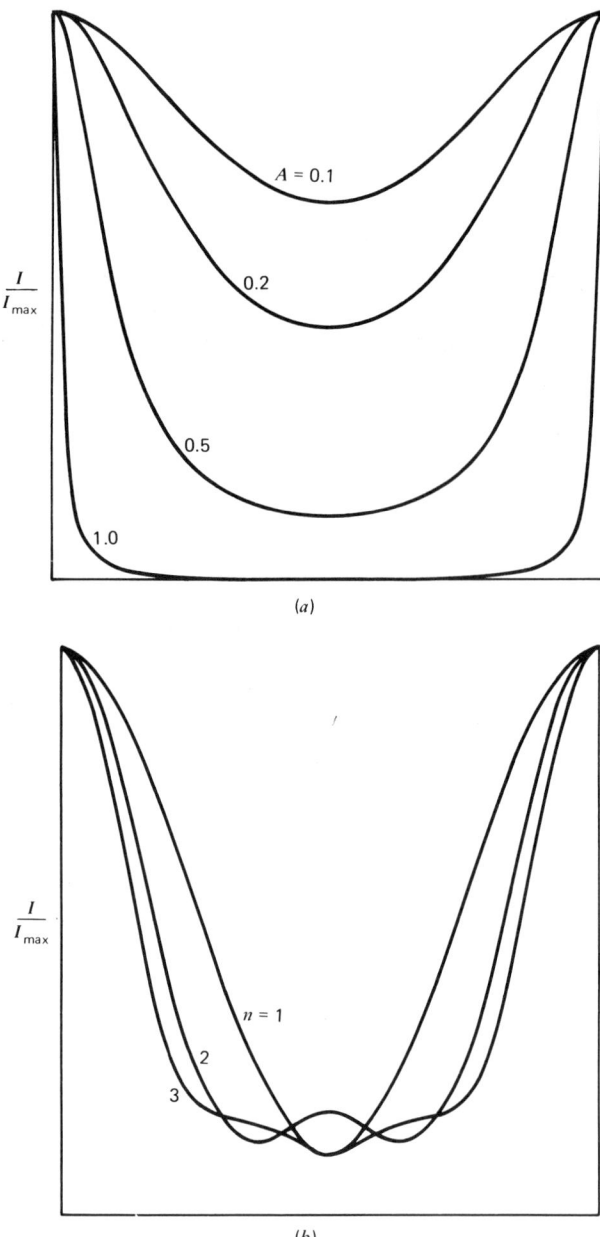

Figure 5.35 Fringe profiles produced by multi-beam interferometry. (a) The Airy distribution (equation 5.91) for various values of A. (b) Fringe profiles when a small number of components are combined as in the summation in equation 5.90. (c) Same as (b) except the coefficients $a_n c_n$ all have the same value (equation 5.89).

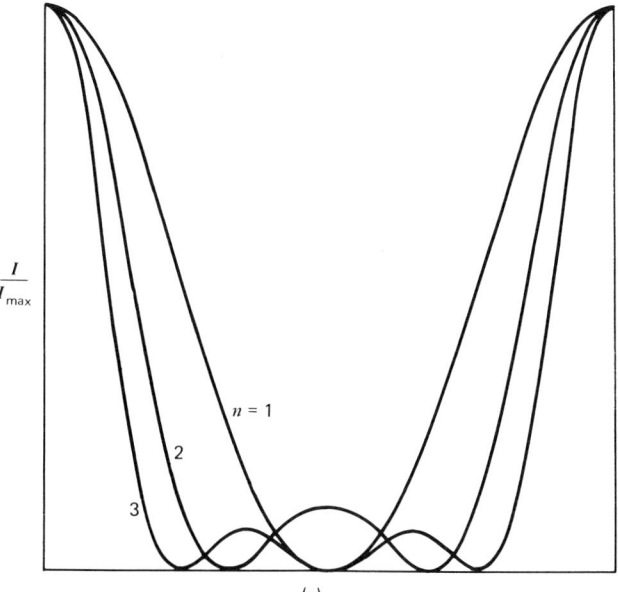

(c)

Figure 5.35 Cont.

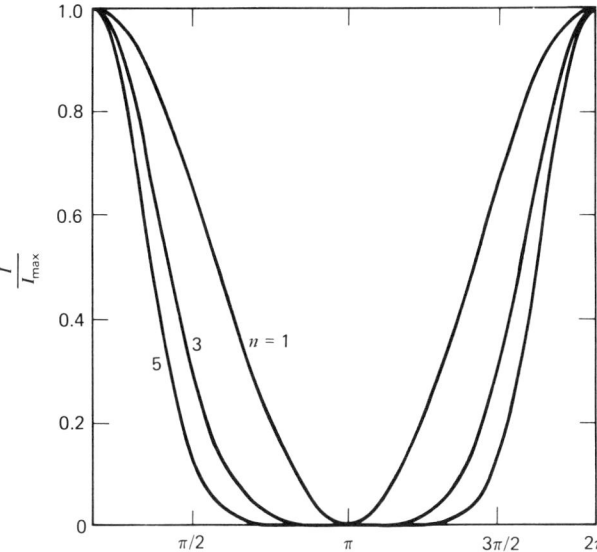

Figure 5.36 Fringe profiles in various orders of a two-exposure nonlinear holographic interferogram (equation 5.95).

If the expressions for the object wave amplitudes

$$U_o(x,y) = a_o \exp[-i\phi_1(x,y)], \qquad U'_o(x,y) = a_o \exp[-i\phi_2(x,y)] \tag{5.94}$$

are substituted into equation 5.93, and if like powers of U_R and U_R^* are collected, the irradiance of the wavefront in the nth diffracted order is found to be

$$I(x,y) = I_0 \left(1 + \cos\left\{\frac{2\pi}{\lambda}[\phi_2(x,y) - \phi_1(x,y)]\right\}\right)^n. \tag{5.95}$$

Thus the fringe profiles in the higher orders are moderately sharpened, as illustrated by Figure 5.36. This scheme requires no extra optics or critical alignment beyond that used for ordinary two-exposure, off-axis holographic interferometry.

REFERENCES

1. W. H. Steel, *Interferometry*, Cambridge University Press, Cambridge, 1967.
2. M. Born and E. Wolf, *Principles of Optics*, 2nd ed., Pergamon Press, New York, 1964, p. 122.
3. Reference 2, pp. 4–7.
4. W. Hauf and U. Grigull, Optical methods in heat transfer, in *Advances in Heat Transfer*, Vol. 6, J. P. Hartnett and T. F. Irvine, Jr. (Eds.), Academic Press, New York, 1970, p. 148.
5. Reference 2, p. 123.
6. O. N. Stavroudis, *The Optics of Rays, Wavefronts, and Caustics*, Academic Press, New York, 1972.
7. R. N. Bracewell, *The Fourier Transform and Its Applications*, McGraw-Hill, New York, 1965, pp. 262–266.
8. I. H. Sneddon, *The Use of Integral Transforms*, McGraw-Hill, New York, 1972, pp. 318–323.
9. F. P. Kupper and C. A. van Dijk, Reference fringes in holographic interferometry, *Opt. Laser Technol.*, **5**, 69–74 (1970).
10. O. Bryngdahl and W.-H. Lee, Shearing interferometry in polar coordinates, *J. Opt. Soc. Am.*, **64**, 1606–1615 (1974).
11. H. H. M. Chau and G. J. Mullany, Holographic moiré patterns; their application to flow visualization in aerodynamics, *Appl. Opt.*, **6**, 1428–1430 (1967).
12. N. Nakatani, K. Kawata, and T. Yamada, Flow visualization by an improved double exposure method in holography, *Opt. Laser Technol.*, **6**, 82–83 (1974).
13. R. C. Spencer and S. A. T. Anthony, Real-time holographic moiré patterns for flow visualization, *Appl. Opt.*, **7**, 561 (1968).
14. B. U. Achia and D. W. Thompson, Real-time hologram-moiré interferometry for liquid flow visualization, *Appl. Opt.*, **11**, 953–954 (1972).

15. J. W. C. Gates, Holographic phase recording by interference between reconstructed wavefronts from separate holograms, *Nature*, **220**, 473–474 (1968).
16. A. G. Havener and R. J. Radley, Dual hologram interferometry, *Opto-Electronics*, **4**, 349–357 (1972).
17. R. J. Radley, Jr., and A. G. Havener, Application of dual hologram interferometry to wind-tunnel testing, *AIAA J.*, **11**, 1332–1333 (1973).
18. N. Abramson, Sandwich hologram interferometry: a new dimension in holographic comparison, *Appl. Opt.*, **13**, 2019–2025 (1974).
19. N. Abramson, Sandwich hologram interferometry. 2: Some practical calculations, *Appl. Opt.*, **14**, 981–984 (1975).
20. G. S. Ballard, Double-exposure holographic interferometry with separate reference beams, *J. Appl. Phys.*, **39**, 4846–4848 (1968).
21. R. E. Brooks, L. O. Heflinger, and R. F. Wuerker, Interferometry with a holographically reconstructed comparison beam, *Appl. Phys. Lett.*, **7**, 248–249 (1965).
22. L. O. Heflinger, R. F. Wuerker, and R. E. Brooks, Holographic interferometry, *J. Appl. Phys.*, **37**, 642–649 (1966).
23. E. N. Leith and J. Upatnieks, Wavefront reconstruction with diffused illumination and three-dimensional objects, *J. Opt. Soc. Am.*, **54**, 1295 (1964).
24. H. H. Chau and O. S. F. Zucker, Holographic thin-beam reconstruction technique for the study of 3-D refractive-index field, *Opt. Commun.*, **8**, 336–339 (1973).
25. C. N. Kurtz, Transmittance characteristics of surface diffusers and the design of nearly band-limited binary diffusers, *J. Opt. Soc. Am.*, **62**, 982–989 (1972).
26. O. B. Gusev and V. B. Konstantinov, Diffusers in holography, *Sov. Phys. Tech. Phys.*, **14**, 258–262 (1969).
27. J. Dyson, Optical diffusing screens of high efficiency, *J. Opt. Soc. Am.*, **50**, 519–520 (1960).
28. C. N. Kurtz, H. O. Hoadley, and J. J. De Palma, Design and synthesis of random phase diffusers, *J. Opt. Soc. Am.*, **63**, 1080–1092 (1973).
29. J. W. C. Gates, Holography with scatter plates, *J. Phys. E: Sci. Instrum.*, Ser. 2, **1**, 989–994 (1968).
30. W. J. Dallas, Deterministic diffusers for holography, *Appl. Opt.*, **12**, 1179–1187 (1973).
31. N. George, A. Jain, and R. D. S. Melville, Jr., Speckle, diffusers, and depolarization, *Appl. Phys.*, **6**, 65–70 (1975).
32. T. F. Kellie and W. H. Stevenson, Experimental techniques in real-time interferometry. 2: Minimization of depolarization effects, *Opt. Eng.*, **12**, 131–132 (1973).
33. J. M. Burch, J. W. C. Gates, R. G. N. Hall, and L. H. Tanner, Holography with a scatter-plate as a beam splitter and a pulsed ruby laser as a light source, *Nature*, **212**, 1347 (1966).
34. S. M. Fraser and K. A. R. Kinloch, Large viewing angle holograms, *J. Phys. E: Sci. Instrum.*, **7**, 774–776 (1974).
35. L. H. Tanner, The scope and limitations of three-dimensional holography of phase objects, *J. Sci. Instrum.*, **44**, 1011–1014 (1967).
36. L. H. Tanner, A study of fringe clarity in laser interferometry and holography, *J. Phys. E: Sci. Instrum.*, **1**, 517–522 (1968).
37. C. M. Vest and D. W. Sweeney, Holographic interferometry of transparent objects with illumination derived from phase gratings, *Appl. Opt.*, **9**, 2321–2325 (1970).
38. M. M. Butusov, Use of a moiré scatterer in holographic interferometry, *Sov. Phys. Tech. Phys.*, **17**, 325–328 (1972).

39. L. O. Heflinger and R. E. Brooks, *Holographic Instrumentation Studies*, Report 12122-6007-R0-00, TRW Systems Group, Redondo Beach, CA, December 1970, pp. 42–50.
40. I. S. Gradshteyn and I. W. Ryzhik, *Table of Integrals Series and Products*, 2nd ed., Academic Press, New York, 1965, p. 973.
41. J. Upatnieks and C. D. Leonard, Characteristics of dielectric holograms, *I.B.M. J. Res. Dev.*, **14**, 527–532 (1970).
42. H. M. Smith, Photographic relief images, *J. Opt. Soc. Am.*, **58**, 533–539 (1968).
43. J. W. Goodman, *Introduction to Fourier Optics*, McGraw-Hill, New York, 1968, pp. 69–70.
44. A. Kozma, Photographic recording of spatially modulated coherent light, *J. Opt. Soc. Am.*, **56**, 428–432 (1966).
45. A. A. Friesem and J. S. Zelenka, Effects of film nonlinearities in holography, *Appl. Opt.*, **6**, 1755–1759 (1967).
46. J. W. Goodman and G. R. Knight, Effects of film nonlinearities on wavefront reconstruction images of diffuse objects, *J. Opt. Soc. Am.*, **58**, 1276–1283 (1968).
47. O. Bryngdahl and A. Lohmann, Nonlinear effects in holography, *J. Opt. Soc. Am.*, **58**, 1325–1334 (1968).
48. A. Kozma, Analysis of the film nonlinearities in hologram recording, *Opt. Acta*, **15**, 527–551 (1968).
49. O. Bryngdahl and A. W. Lohman, Interferograms are image holograms, *J. Opt. Soc. Am.*, **58**, 141–142 (1968).
50. K. S. Mustafin, V. A. Seleznev, and E. I. Shtyrkov, Use of the nonlinear properties of a photoemulsion for enhancing the sensitivity of holographic interferometry, *Opt. Spectrosc.*, **28**, 638–640 (1970).
51. K. Matsumoto and M. Takashima, Phase-difference amplification by nonlinear holography, *J. Opt. Soc. Am.*, **60**, 30–33 (1970).
52. O. Bryngdahl, Longitudinally reversed shearing interferometry, *J. Opt. Soc. Am.*, **59**, 142–146 (1969).
53. C. H. F. Velzel, Small phase differences in holographic interferometry, *Opt. Commun.*, **2**, 289–291 (1970).
54. S. Toyooka, Elimination of wavefront aberration of optical elements used in phase difference amplification, *Appl. Opt.*, **13**, 2014–2018 (1974).
55. K. S. Pennington and J. S. Harper, Techniques for producing low-noise improved efficiency holograms, *Appl. Opt.*, **9**, 1643–1650 (1970).
56. S. Tolansky, *Multiple-Beam Interferometry of Surfaces and Films*, Oxford University Press, London, 1949.
57. K. Matsumoto, Holographic multiple-beam interferometry, *J. Opt. Soc. Am.*, **59**, 777–778 (1969).
58. O. Bryngdahl, Multiple-beam interferometry by wavefront reconstruction, *J. Opt. Soc. Am.*, **59**, 1171–1175 (1969).
59. K. Matsumoto, Analysis of holographic multiple-beam interferometry, *J. Opt. Soc.*, **61**, 176–181 (1971).
60. D. Dameron and C. M. Vest, Fringe sharpening and propagation effects in nonlinear two-exposure holography, *J. Opt. Soc. Am.*, **66**, 1418–1421 (1976).

6
Transparent Objects: Fringe Interpretation, Special Techniques, and Applications

6.1 INTRODUCTION

In this chapter we consider the application of holographic interferometry to the measurement or visualization of changes in the physical properties of transparent objects. Such properties include mass density of fluids, electron density of plasmas, temperature of fluids, chemical species concentrations in reacting gases, and state of stress in solids. Each of these properties can be related to refractive index, which is measured by interferometry.

In Chapter 5 the formation of holographic interferograms of transparent objects was discussed. This chapter deals with the inverse situation: given a holographic interferogram, how can we determine the refractive index distribution? In addition, a few specialized techniques of holographic interferometry of transparent objects will be considered. Applications of holographic interferometry to several types of measurement or testing problems are discussed. Emphasis is placed on quantitative techniques and on applications for which unique characteristics of holographic interferometry are utilized. The initial discussions of fringe interpretation are quite general, with refractive index being the quantity to be measured. In Section 6.5, which deals with specific applications, the relationships between refractive index and other physical properties are discussed.

6.2 FRINGE INTERPRETATION

This section deals with the quantitative evaluation of holographic interferograms to determine distributions of change of refractive index. The analysis applies to changes between the exposures of two-exposure holographic interferograms and to real-time interferometry. The transparent media under consideration are treated as phase objects, that is, the probing optical rays are assumed to be straight lines. In this refractionless limit the optical pathlength difference of a ray traveling in the z direction through a phase object is given by equation 5.22:

$$\Delta\Phi(x,y) = \int \left[n(x,y,z) - n_0 \right] dz = N\lambda. \quad (5.22)$$

Here n_0 is the refractive index of the phase object at the time of the initial holographic exposure and is usually a constant, and $n(x,y,z)$ is the refractive index distribution at the time of the second exposure, or at the instant a real-time interferogram is photographed. Evaluation of an interferogram consists of determining $\Delta\Phi$ from the fringe pattern and then *inverting* equation 5.22 to calculate $n(x,y,z) - n_0$. A third step, relating $n(x,y,z)$ to other physical properties, is discussed for various applications in Section 6.5.

It is possible to measure $\Delta\Phi$ directly in units of wavelength λ by assigning order numbers to the fringes. Figure 6.1a is a sketch of a fringe pattern due to a high-speed flow of air past a wedge-shaped airfoil. The number $N=0$ is assigned to the large, bright fringe in the undisturbed ambient gas. The centers of all subsequent bright fringes are assigned the numbers $N=1,2,3,\ldots$ consecutively. The centers of all dark fringes are assigned numbers $N=0.5$, 1.5, 2.5, etc. At any location the optical pathlength difference is

$$\Delta\Phi = N\lambda. \quad (6.1)$$

There is a sign ambiguity in fringe order numbers which arises because $+\Delta\Phi$ and $-\Delta\Phi$ yield identical fringe patterns. The numbers $N=0, -1, -2, -3,\ldots$ could have been assigned to the fringe pattern in this example. Often the experimenter has sufficient knowledge of the field being examined to infer the appropriate sign. If this is not the case, reference fringes are advantageous.

When reference fringes are used, the sign ambiguity can be removed. Figure 6.1b is a sketch of a finite-fringe interferogram of the same phase object as in Figure 6.1a. The system of parallel reference fringes could have been introduced by any of the methods discussed in Section 5.3. The experimenter governs whether the reference optical pathlength difference increases from right to left, or from left to right. Assume that it increases from right to left. As a fringe is traced from the ambient toward the airfoil, it bends to the left if $\Delta\Phi$ is increasing, and toward the right if $\Delta\Phi$ is decreasing. To assign fringe order numbers to locations in a finite-fringe interferogram a straight line is drawn tangent to the ambient reference

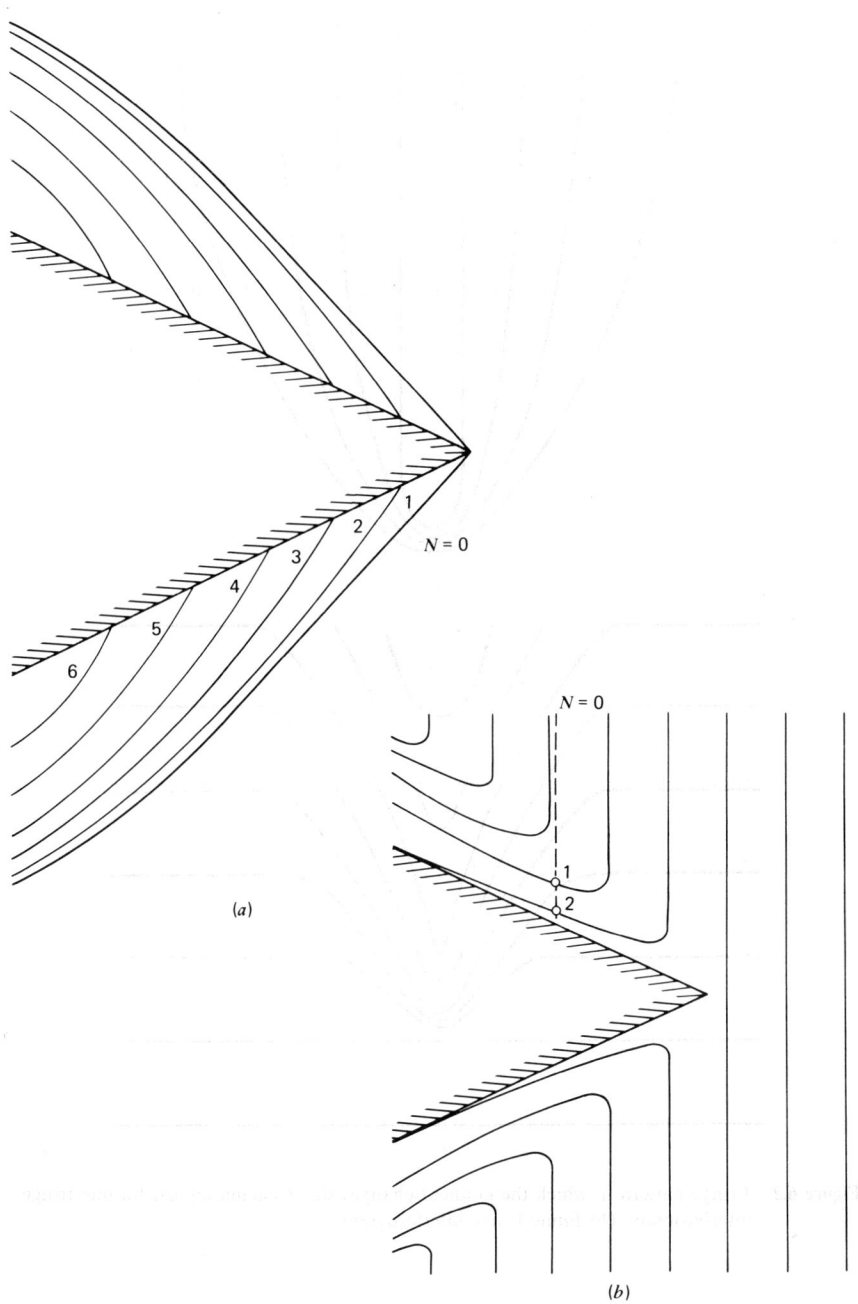

Figure 6.1 Fringe pattern due to high-speed flow past an airfoil (schematic). (*a*) Infinite fringe interferogram. (*b*) Finite fringe interferogram.

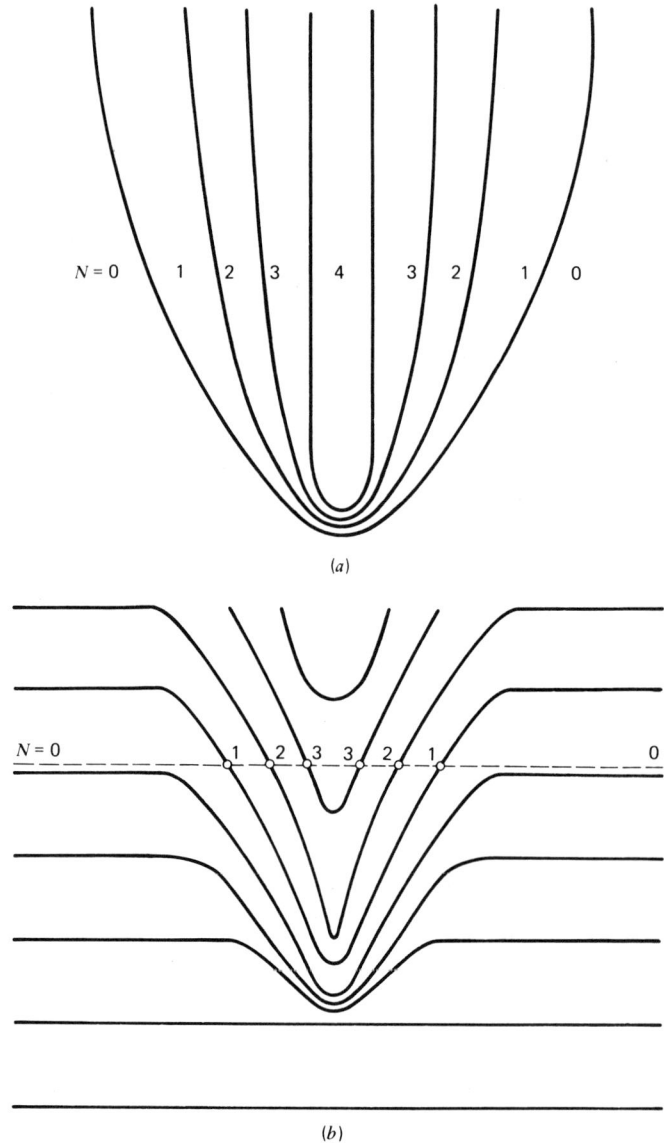

Figure 6.2 Fringe pattern in which the gradient changes sign (schematic). (*a*) Infinite fringe interferogram. (*b*) Finite fringe interferogram.

6.2 FRINGE INTERPRETATION

fringes, for example, the dotted line in Figure 6.1b. Then $N=0$ is assigned to all points where the fringe does not deviate from the dotted line. Order numbers $N=1, 2, 3, \ldots$ are assigned to each consecutive point where the center of a new dark fringe crosses the dotted line, and $N=0.5, 1.5, 2.5, \ldots$ to each point where the center of a bright fringe crosses the dotted line. If fringes whose slopes have the opposite sign are encountered, the fringe order numbers will decrease rather than increase. This is illustrated in Figure 6.2. When a fringe order number is assigned to any point in a finite-fringe interferogram, the optical pathlength difference is again given by equation 6.1.

6.2.1 Inversion Techniques

Once $\Delta\Phi$ has been evaluated for an interferogram, equation 5.22 must be inverted to determine $n(x,y,z) - n_0$. The ease with which this inversion can be accomplished depends on the structure of the phase object. Three cases arise in practice:

1. Two-dimensional phase objects, with no variation of refractive index in the z direction.
2. Radially symmetric phase objects.
3. Asymmetric phase objects.

Inversion of equation 5.22 for a *two-dimensional phase object* is straightforward. Let the phase object have length L in the z direction, which is the propagation direction of the object wave. The unknown change of refractive index is a function only of x and y, so equation 5.22 becomes

$$\lambda N(x,y) = \int_0^L [n(x,y) - n_0] \, dz$$
$$= [n(x,y) - n_0] \cdot L.$$

Therefore

$$n(x,y) - n_0 = N(x,y) \cdot \frac{\lambda}{L}. \tag{6.2}$$

This simple relation is often useful because flow around airfoils, temperature distributions near long, heated cylinders, and many other fields encountered in practice are approximately two-dimensional.

Radially symmetric phase objects are commonly encountered in investigations of flow around cones, jets, thermal plumes, flames, plasma arcs, and so on. The object may be either spherical or cylindrical. In either case its refractive index is a function of radius only. For convenience let

$$f(r) = n(r) - n_0.$$

Figure 5.11a depicts an optical ray traveling in the z direction through a radially symmetric phase object. Since $dz = d(r^2 - x^2)^{1/2} = (r^2 - x^2)^{-1/2} r \, dr$, equation 5.22 becomes

$$N(x) \cdot \lambda = 2 \int_x^R \frac{f(r) r \, dr}{(r^2 - x^2)^{1/2}}. \tag{6.3}$$

Many phase objects encountered in practice decay smoothly to zero at large radius and have no discontinuities; it is then convenient to rewrite equation 6.3 as

$$N(x) \cdot \lambda = 2 \int_x^\infty \frac{f(r) r \, dr}{(r^2 - x^2)^{1/2}}. \tag{6.4}$$

The right-hand side of equation 6.4 is the *Abel transform* of $f(r)$. An interferogram of a radially symmetric phase object therefore displays contours of constant value of the Abel transform of $n(r) - n_0$. The inverse of equation 6.4 is

$$f(r) = -\frac{\lambda}{\pi} \int_r^\infty \frac{(dN/dx) \, dx}{(x^2 - r^2)^{1/2}}. \tag{6.5}$$

This inversion formula is well known and can be derived by classical methods[1] or, more conveniently, by transform methods.[2]

In the analysis of interferometric data the fringe order number $N(x)$ is known only at a finite number of discrete locations and must be inverted numerically. Inversion schemes can be based on numerical approximations of either equation 6.4 or 6.5. In either case we consider that the phase object is divided into I discrete annular elements of constant width Δr, as shown in Figure 6.3. The objective is to determine I discrete values of $f(r)$ from data consisting of I values of $N(x)$.

Methods based on approximating equation 6.4 lead to a system of algebraic equations which must be solved. The simplest approach is to consider that each annular element has a uniform refractive index. Let f_i denote the value of $n - n_0$ for the element $r_i \leq r \leq r_{i+1}$, where $r_i = i \cdot \Delta r$. With this approximation, equation 6.4 becomes

$$N_i \cdot \lambda = 2 \sum_{k=i}^{I-1} f_k \int_{r_k}^{r_{k+1}} \frac{r \, dr}{(r^2 - r_i^2)^{1/2}}. \tag{6.6}$$

6.2 FRINGE INTERPRETATION

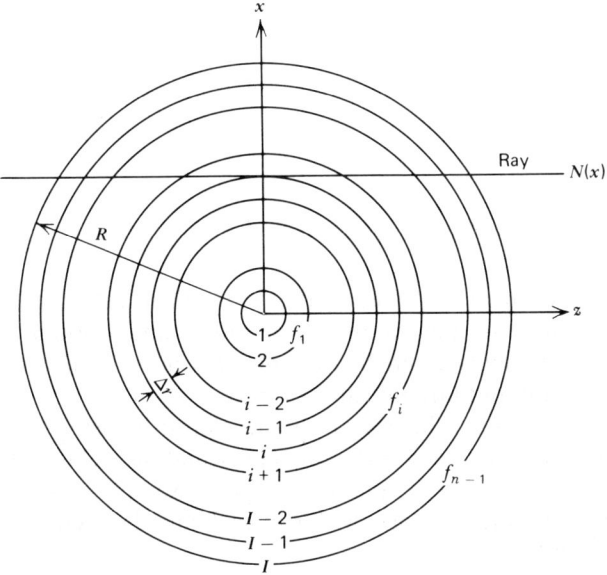

Figure 6.3 Cross section of a radially symmetric phase object divided into discrete annular elements of width Δr.

The integral in equation 6.6 has the value

$$(r_{k+1}^2 - r_i^2)^{1/2} - (r_k^2 - r_i^2)^{1/2} = \Delta r \{[(k+1)^2 - i^2]^{1/2} - (k^2 - i^2)^{1/2}\},$$

so

$$\sum_{k=i}^{I-1} A_{ki} f_k = \left(\frac{\lambda}{2\Delta r}\right) N_i, \qquad (6.7)$$

where the coefficients A_{ki} are

$$A_{ki} = \{[(k+1)^2 - i^2]^{1/2} - (k^2 - i^2)^{1/2}\}. \qquad (6.8)$$

Equation 6.7 represents a set of simultaneous linear algebraic equations which must be solved for the unknown values f_k. This solution is quite simple because the coefficients A_{ki} form a matrix of lower triangular form, so that

$$A_{(I-1),(I-1)} f_{(I-1)} = \frac{\lambda}{2\Delta r} N_{(I-1)},$$

$$A_{(I-2),(I-2)} f_{(I-2)} + A_{(I-1),(I-2)} f_{(I-1)} = \frac{\lambda}{2\Delta r} N_{(I-2)},$$

$$A_{(I-3),(I-3)} f_{(I-3)} + A_{(I-2),(I-3)} f_{(I-2)} + A_{(I-1),(I-3)} f_{(I-1)} = \frac{\lambda}{2\Delta r} N_{(I-3)}.$$

The outermost element is located so that $f_I = 0$. Each value of f can thus be calculated in turn, starting at the outer radius and working toward the center. The coefficients given by equation 6.8 have been computed and tabulated by Hauf and Grigull[3] up to $I = 25$.

Alternative inversion schemes can be developed by using more complex representations of $f(r)$; each different representation yields a different set of coefficients A_{ki} for use in equation 6.7. For example, Ladenburg et al.[4] assumed that $f(r)$ varies linearly with r in each annular element. This leads to the coefficients

$$A_{ki} = (k+i)[(k+1)^2 - i^2]^{1/2} - k(k^2 - i^2)^{1/2}$$
$$- i^2 \ln \frac{k+1+[(k+1)^2 - i^2]^{1/2}}{k^2 + (k^2 - i^2)^{1/2}}. \tag{6.9}$$

Barakat[5] and Sweeney[6] have developed methods based on the representation of $f(r)$ by sampling series. The method proposed by Sweeney permits acceptance of fringe data at arbitrary locations, not just at equally spaced intervals, and can be used with redundant data to help suppress errors.

Inversion schemes based on approximating equation 6.5 lead to formulas for directly computing f without solving a system of equations. The simplest approach is to interpolate the fringe data by considering that $N(x)$ varies linearly with x^2 between each data point; then in the zone $x_i \leq x \leq x_{i+1}$

$$\frac{dN}{dx} = 2 \frac{N_{i+1} - N_i}{x_{i+1}^2 - x_i^2} x.$$

We can then approximate equation 6.2 by

$$f_i = -\frac{\lambda}{\pi} \sum_{k=i}^{I-1} 2 \frac{N_{k+1} - N_k}{x_{k+1}^2 - x_k^2} \int_{r_k}^{r_{k+1}} \frac{x \, dx}{(x^2 - r_i^2)^{1/2}}. \tag{6.10}$$

6.2 FRINGE INTERPRETATION

The integral in equation 6.10 is identical to that in equation 6.6 and has the value $\Delta r\{[(k+1)^2 - i^2]^{1/2} - (k^2 - i^2)^{1/2}\}$. Also, $x_{k+1}^2 - x_k^2 = \Delta r[(k+1)^2 - k^2] = \Delta r(2k+1)$, so

$$f_i = -\frac{2\lambda}{\pi \Delta r} \sum_{k=i}^{I-1} B_{ki}(N_{k+1} - N_k), \quad (6.11)$$

where

$$B_{ki} = \frac{[(k+1)^2 - i^2]^{1/2} - (k^2 - i^2)^{1/2}}{2k+1}. \quad (6.12)$$

These coefficients have been tabulated by Gooderum and Wood.[7] A similar method was also described by Bennett et al.[1]

Several methods have been proposed which avoid the difference operation in equation 6.11, so that the inversion formula is a simple summation:

$$f_i = \frac{-2}{\pi \Delta r} \sum_{k=i}^{I-1} \beta_{ki} N_k. \quad (6.13)$$

A set of coefficients β_{ki} can be generated by noting that equation 6.5 can be rewritten[8] as

$$f(r) = -\frac{\lambda}{2\pi r} \frac{dF}{dr}, \quad (6.14)$$

where

$$F(r) = 2 \int_r^R \frac{N(x) x \, dx}{(x^2 - r^2)^{1/2}} \quad (6.15)$$

if $N(x)$ is a well-behaved continuous function. If we again consider $N(x)$ to vary linearly with x^2 between each data point, then, in the zone $x_i \leq x \leq x_{i+1}$,

$$N = N(x_i) - \frac{[N(x_{i+1}) - N(x_i)](x - x_i)^2}{x_{i+1}^2 - x_i^2}. \quad (6.16)$$

The coefficients β_{ij} obtained by substituting equation 6.16 into equations 6.15 and 6.14 are

$$\begin{aligned} \beta_{kk} &= -B_{kk}, \\ \beta_{ki} &= B_{k,i-1} - B_{ki}, \quad i \geq k+1, \end{aligned} \quad (6.17)$$

where B_{ki} are the coefficients given by equation 6.12. This set of coefficients β_{ki} has been tabulated by Nestor and Olsen.[9] Barr[10] modified this procedure by least-mean-square fitting a quadratic function in each interval to the $f(r)$ determined by the Nestor-Olsen method. This results in a somewhat different set of coefficients β_{ki} for use in equation 6.13. Barr's method is intended to smooth the solution slightly and thus minimize the effect of small errors in the data. His coefficients are tabulated in ref. 10.

Clearly a set of coefficients β_{ki} can be generated for each possible analytical function which can be fitted to the data $N(x)$. Several possibilities are described in the literature. Bockasten[11] exactly fitted a cubic polynomial in x to the four data points closest to each interval. This yields

$$\beta_{ki} = -\left(\frac{1}{24\pi}\right)[g(k,i+1) - 4g(k,i+1) \\ + 6g(k,i) - 4g(k,i-1) + g(k,i-2)] \quad (6.18)$$

for $k+1 < i < I-2$ and $i > 2$, where

$$g(k,i) = (3k^2 + 6i^2 - 2)\ln\left[i + (i^2 - k^2)^{1/2}\right] \\ - 9i(i^2 - k^2)^{1/2},$$

$$k \leq i \leq I-1, i > 0, \text{ and } i < I-1. \quad (6.19)$$

Other coefficients are influenced by the integration limits; all are tabulated in ref. 11 for $I = 20$ and $I = 40$. Other approaches include that of Bradley,[12] who fitted a power series in $(1 - x^2/R^2)$ to $N(x)$ by a least-mean-square method, and that of South,[13] who fitted cubic splines to $N(x)$ in each interval. The latter method is intended for use when $N(x)$ is a relatively complicated function which cannot be represented accurately by a low-order polynomial. Herlitz[14] expanded $N(x)$ in a series of Chebyshev functions and derived an expression for $f(r)$ as a series of Zernike polynomials. Bracewell[15] gives a simple numerical deconvolution technique for inverting data of the type considered here.

A special problem occurring in aerodynamics and plasmas in discontinuous data due to a step change of N across a shock. Such data are of the form shown in Figure 6.4. This problem can be handled easily if ΔN can be estimated accurately. The data in the region $r_0 < x < R$ can be inverted by most of the methods discussed above since $f(r)$ is always given in terms of $r \leq x \leq R$. The data in the region $0 < x < r_0$ can also be inverted by adapting a formula given by Bracewell:[2, 15]

$$f(r) = -\frac{\lambda}{\pi}\int_r^{r_0} \frac{(dN/dx)\,dx}{(x^2 - r^2)^{1/2}} + \frac{\Delta N}{\pi(r_0^2 - r^2)^{1/2}}. \quad (6.20)$$

6.2 FRINGE INTERPRETATION

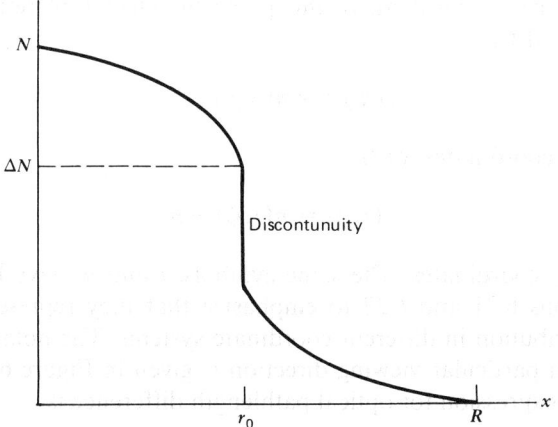

Figure 6.4 Discontinuous distribution of fringe number. Data of this type commonly occur when shock waves are present.

Any of the methods which use equation 6.11 can be applied simply by appending the second term in equation 6.20 to the result and using r_0 as the upper limit of integration. Methods which use equation 6.13 can also be used if N_k is replaced by $(N_k - \Delta N)$ and the second term in equation 6.20 is appended to the result. This problem is discussed in greater detail by Bennett et al.[1]

To summarize, several methods are available for the analysis of interferograms of radially symmetric phase objects. All such methods are numerical approximations of equation 6.4 or 6.5. The best method for a particular application depends on the structure and accuracy of the data and on the number of data points available. A study by Kulagin et al.[16] suggests that the method of Nestor and Olsen[9] is the best general-purpose method and can be expected to give reasonable accuracy and to have low sensitivity to errors in the data. The safest procedure is to try several methods with known functions $f(r)$ similar to the one expected to be measured, and to compare the methods with regard to accuracy and sensitivity to random errors in the data.

Asymmetric phase objects have been investigated by interferometry only recently. Determination of an asymmetric refractive index distribution requires analysis of a large number of interferograms, each recorded from a different viewing direction. Holographic interferometry is ideally suited for recording such data, which we term *multidirectional interferometric data*. Holographic interferometers which can be used for this purpose were discussed in Section 5.3.

Let the change of refractive index of an asymmetric phase object be $n(x,y,z) - n_0$. Consider the problem of determining this distribution in a particular plane $z =$ constant. In this plane the change of refractive index can be denoted by

$$f(x,y) = n(x,y) - n_0 \tag{6.21}$$

in Cartesian coordinates, or by

$$f(r,\phi) = n(r,\phi) - n_0 \tag{6.22}$$

in cylindrical coordinates. The same symbols, f and n, have been used in both equations 6.21 and 6.22 to emphasize that they represent the same *physical* distribution in different coordinate systems. The notation for data recorded in a particular viewing direction is given in Figure 6.5. With this notation the expression for optical pathlength difference is

$$N(p,\theta) \cdot \lambda = \int_{-\infty}^{\infty} \int_{-\infty}^{\infty} f(r,\phi) \delta[p - r\sin(\phi - \theta)] \, dx \, dy. \tag{6.23}$$

For a given observation direction, specified by θ, $N(p,\theta)$ is the fringe order data read from the interferogram in the manner described at the beginning of this section. Also, δ is the Dirac delta function, so the integral on the

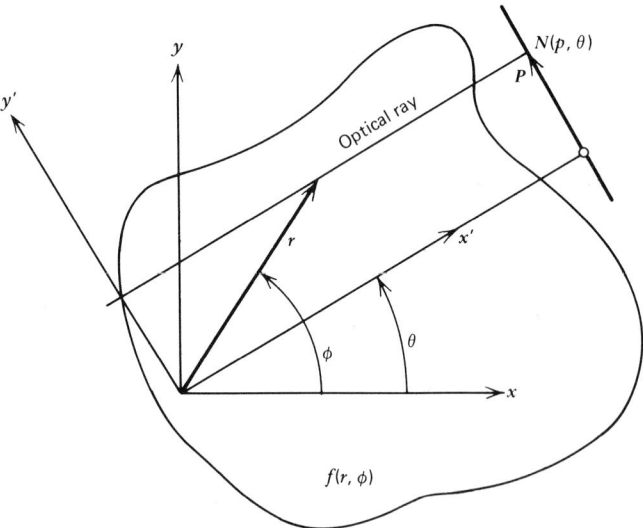

Figure 6.5 Notation for equation 6.23 and 6.24. A typical optical ray is shown passing through an asymmetric phase object $f(r,\phi)$ and impinging on a particular observation plane where an interferogram is recorded. Here θ defines the observation direction, and p locates a probing ray oriented in this direction.

6.2 FRINGE INTERPRETATION

right-hand side represents the line integral of $f(r,\phi)$ along a straight ray passing through the phase object. The limits $\pm\infty$ can be used for convenience since it is assumed that the phase object decays smoothly to zero at the edges of the field. The right-hand side of equation 6.23 is the two-dimensional *Radon transform*[17] of $f(r,\phi)$. An interferogram of an asymmetric phase object therefore displays contours of constant value of the Radon transform of $n(r,\phi) - n_0$. The inverse of equation 6.23 is

$$f(r,\phi) = \frac{\lambda}{2\pi^2} \int_{-\pi/2}^{\pi/2} d\theta \int_{-\infty}^{\infty} \frac{(\partial N/\partial p)\,dp}{r\sin(\phi-\theta)-p}. \qquad (6.24)$$

This inversion formula was given by Berry and Gibbs[18] and is a special case of the inverse Radon transform discussed in refs. 17 to 22.

Interferometric data consisting of a discrete set of measured values of fringe order number $N(p,\theta)$ must be inverted numerically. Inversion schemes can be based on numerical approximations of either equation 6.23 or 6.24. Methods based on approximating equation 6.23 lead to a system of algebraic equations which must be solved. The simplest approach is to consider the region occupied by the phase object to be divided into discrete rectangular elements of dimension Δx by Δy, where

$$\Delta x = \frac{L_x}{M+1}, \qquad \Delta y = \frac{L_y}{N+1},$$

as shown in Figure 6.6. Each element is considered to have a uniform refractive index. Let f_k denote the value of refractive index change of the element centered on the point $x = m\Delta x$, $y = n\Delta y$, where m and n are integers. The subscript k is related to m,n by

$$k = n(M+1) + m + 1. \qquad (6.25)$$

Let the fringe order number N_i be associated with the *i*th ray traversing the phase object. This ray is specified by the coordinates p,θ. If A_{ki} denotes the length of the segment of the *i*th ray which lies in the *k*th element, the total optical pathlength difference for the *i*th ray is given by

$$\sum_{k=1}^{K} A_{ki} f_k = \lambda N_i, \qquad (6.26)$$

where $K = (M+1)(N+1)$ is the total number of elements. Equation 6.26 is a finite-sum approximation to equation 6.23. The coefficients A_{ki} can be

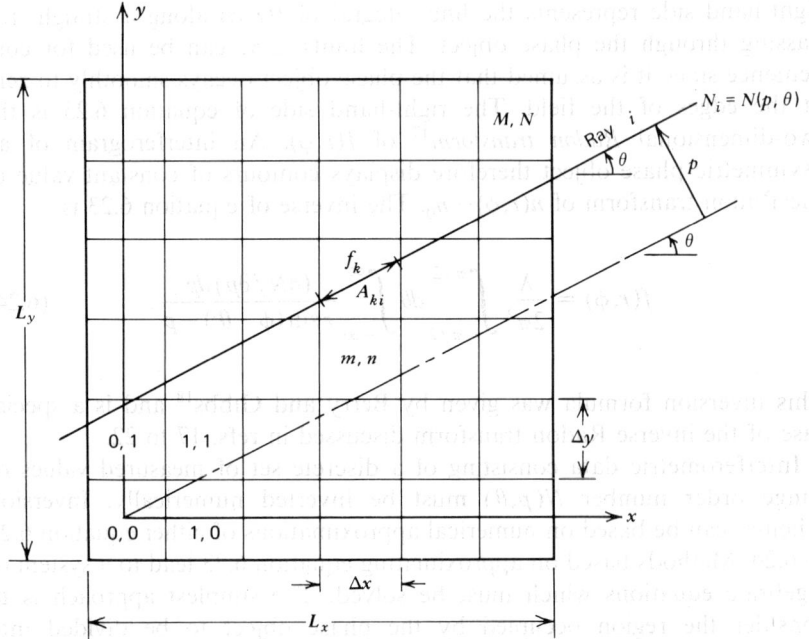

Figure 6.6 Cross section of an asymmetric phase object divided into discrete rectangular elements of dimension $\Delta x \times \Delta y$. A typical probing ray, denoted by the index i, is shown traversing the object.

determined geometrically;[23] their values are as follows:

$$A_{ki} = \begin{cases} \Delta x \sec\theta & \text{for } |b| \leq \dfrac{|\Delta y - \Delta x|\tan\theta|}{2} \\ & \text{and } |\tan\theta| \leq \dfrac{\Delta y}{\Delta x}, \\[2pt] \dfrac{\Delta y \sec\theta}{|\tan\theta|} & \text{for } |b| \leq \dfrac{\Delta x|\tan\theta| - \Delta y}{2} \\ & \text{and } |\tan\theta| > \dfrac{\Delta y}{\Delta x}, \\[2pt] \dfrac{\sec\theta}{|\tan\theta|}\left(\dfrac{\Delta x|\tan\theta| + \Delta y}{2} - |b|\right) & \text{for } \dfrac{|\Delta y - \Delta x|\tan\theta\|}{2} < |b| \\ & \leq \dfrac{|\Delta y + \Delta x|\tan\theta\|}{2}, \\[2pt] \Delta y & \text{for } |c| < \dfrac{\Delta x}{2} \text{ and } |\tan\theta| = \infty, \\[2pt] 0 & \text{for } |b| > \dfrac{\Delta x|\tan\theta| + \Delta y}{2}, \\[2pt] 0 & \text{for } |c| > \dfrac{\Delta x}{2} \text{ and } |\tan\theta| = \infty, \quad (6.27a) \end{cases}$$

6.2 FRINGE INTERPRETATION

where

$$b = p\sec\theta + m\Delta x \tan\theta - n\Delta y \tag{6.27b}$$

and

$$c = p + m\Delta x. \tag{6.27c}$$

If the number I of pathlength measurements is equal to the number of elements K, a system of K linear algebraic equations of form 6.26 can be generated and solved for the K unknown values f_k. This approach to the analysis of holographic interferograms was introduced by Alwang et al.[24]

Alternative inversion schemes can be developed by using more complex representations of $f(x,y)$. Each different representation yields a different set of coefficients for use in equation 6.26. For example, Sweeney and Vest[23] used a representation based on sampling theory:

$$f(x,y) = \sum_{m=0}^{M-1} \sum_{n=0}^{N-1} f(m\Delta x, n\Delta y) \cdot \frac{\sin[\pi(x-m\Delta x)/\Delta x]}{\pi(x-m\Delta x)/\Delta x} \cdot \frac{\sin[\pi(y-n\Delta y)/\Delta y]}{\pi(y-n\Delta y)/\Delta y}. \tag{6.28}$$

This leads to the following coefficients for use in equation 6.26:

$$A_{ki} = \begin{cases} \Delta x \sec\theta \operatorname{sinc}\left[\dfrac{(p\sec\theta + m\Delta x\tan\theta - n\Delta y)}{l_y}\right] \\ \qquad \text{for } 0 \leqslant |\tan\theta| \leqslant \dfrac{\Delta x}{\Delta y}, \\[6pt] \left(\dfrac{\Delta y \sec\theta}{|\tan\theta|}\right) \operatorname{sinc}\left[\dfrac{(p\sec\theta + m\Delta x\tan\theta - n\Delta y)}{\Delta x \tan\theta}\right] \\ \qquad \text{for } \dfrac{\Delta y}{\Delta x} \leqslant |\tan\theta| \leqslant \infty, \\[6pt] \Delta y \operatorname{sinc}\left(\dfrac{p+m\Delta x}{\Delta x}\right) \quad \text{for } |\tan\theta|=\infty, \end{cases} \tag{6.29}$$

where

$$\operatorname{sinc} x \equiv \frac{\sin\pi x}{\pi x}. \tag{6.30}$$

A similar representation using an expansion in terms of

$$\frac{J_0\{\pi[(x-m\Delta)^{1/2}+(y-n\Delta)^{1/2}]/\Delta\}}{\pi[(x-m\Delta)^{1/2}+(y-n\Delta)^{1/2}]},$$

where $\Delta x = \Delta y = \Delta$, was used in ref. 21. Expansions like equation 6.28 give a smoothed representation of $f(x,y)$ and are convenient for interpolating between the values f_k at the *sample points* $x = m\Delta x, y = n\Delta y$. It has been found that good accuracy can be attained using equation 6.26 only if redundant measurements are made, that is, the number of measurements I must exceed the number of values K of f_k to be determined. This is particularly true when the range of viewing angles θ is significantly less than 180°, which is often the case in applications.[23] Equations 6.26 can then be solved in the least-mean-square sense. Efficient computer techniques for lms solutions are available (e.g., ref. 25).

Inversion schemes based on approximating equation 6.24 lead to formulas for directly computing $f(r,\phi)$ without solving a system of equations. A detailed discussion of these schemes would exceed the scope of this book; however, we shall briefly outline one useful approach. Let the interferometric data be recorded for N equally spaced viewing angles, separated by $\Delta\theta$. For each viewing direction the fringe order number is measured at M equally spaced points separated by Δp (see Figure 6.7). An interpolating series based on sampling theory can be fitted to these data:

$$N(p,\theta) = \sum_{m=-\infty}^{\infty} N(m\Delta p, \theta) \operatorname{sinc}\left(\frac{p - m\Delta p}{\Delta p}\right). \qquad (6.31)$$

If we denote the integral over p in equation 6.24 by I_1, and substitute equation 6.31 into it, the resulting expression can be algebraically arranged

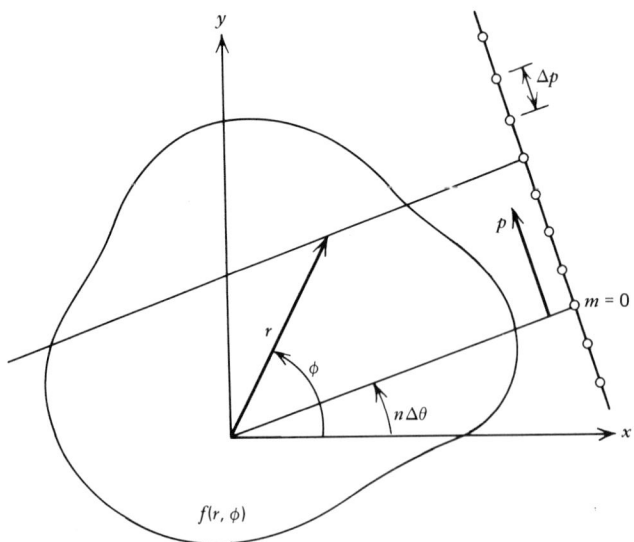

Figure 6.7 Notation for equations 6.31 and 6.37. The asymmetric phase object, probing ray, and interferogram plane correspond to those depicted in Fig. 6.5.

6.2 FRINGE INTERPRETATION

to yield

$$I_1 = -\sum_{m=-\infty}^{\infty} \frac{1}{\Delta p} N(m\Delta p, \theta) \int_{-\infty}^{\infty} \frac{(d/d\xi)(\operatorname{sinc}\xi)\,d\xi}{\xi - x}, \quad (6.32)$$

where $x = (r/\Delta p)\sin(\phi - \theta) - m$. The integral in equation 6.32 is the Hilbert transform of $d(\operatorname{sinc}\xi)/d\xi$, which has the value

$$\pi^2\left(-\operatorname{sinc} x + \tfrac{1}{2}\operatorname{sinc}^2 \tfrac{1}{2}x\right).$$

Now, if we let $\Delta\theta = \pi/M$ and $\Delta\phi = \pi/K$, where K and M are integers, and let

$$j = \frac{r}{\Delta p}\sin(k\Delta\phi - n\Delta\theta) \quad (6.33)$$

also be an integer, then

$$I_1 = \frac{\pi^2}{\Delta p}\sum_{m=-\infty}^{\infty} N(m\Delta p, n\Delta\theta)\left\{\operatorname{sinc}(j-m) - \tfrac{1}{2}\operatorname{sinc}^2\left[\tfrac{1}{2}(j-m)\right]\right\}. \quad (6.34)$$

Note that

$$\tfrac{1}{2}\operatorname{sinc}^2\left[\tfrac{1}{2}(j-m)\right] = \begin{cases} \tfrac{1}{2} & \text{for } j-m=0, \\ 0 & \text{for } j-m \text{ even}, \\ \dfrac{2}{\pi^2(j-m)^2} & \text{for } j-m \text{ odd}. \end{cases} \quad (6.35)$$

The use of equations 6.35 and 6.34 plus algebraic rearrangement yields

$$I_1 = \frac{\pi^2}{\Delta p}\left\{\tfrac{1}{2}N(j\Delta p, n\Delta\theta) - \frac{2}{\pi^2}\sum_{m \text{ odd}} \frac{N[(j+m), n\Delta\theta]}{i^2}\right\}. \quad (6.36)$$

where $i = j - m$.

If the integral over θ in equation 6.24 is replaced by a simple summation, and equation 6.36 is used, the following inversion formula is obtained:

$$f(tr_0, k\Delta\phi) = -\frac{\lambda\Delta\theta}{4\Delta p}\sum_{n=1}^{M}\left\{N(j\Delta p, n\Delta\theta) - \frac{4}{\pi^2}\right.$$
$$\left.\cdot \sum_{i \text{ odd}} \frac{N[(j-i), n\Delta\theta]}{i^2}\right\}. \quad (6.37)$$

In this formula all variables have been discretized, with k and t being integers and r_0 denoting a finite interval in r. Since $r = tr_0$ may not be compatible with j being an integer in equation 6.33, interpolation will be required to carry out the summation in equation 6.37. Equation 6.37 is equivalent to a formula first given by Ramachandran and Lakshminarayanan.[26] A more detailed discussion of this inversion scheme can be found in ref. 26.

Multidirectional interferometric data can also be inverted by direct application of fast Fourier transform (FFT) codes. To see how this can be done, let $F(u,v)$ denote the two-dimensional Fourier transform of $f(x,y)$, the unknown refractive index distribution. It is possible to determine $F(u,v)$ rather directly from the interferometric data. The fringe order number for a given viewing direction (see Figure 6.5) is

$$\lambda \cdot N(y') = \int_{-\infty}^{\infty} f(x',y')\,dx'. \tag{6.38}$$

The one-dimensional Fourier transform of $N(y')$ is

$$\mathcal{F}_1\{\lambda \cdot N(y')\} = \int_{-\infty}^{\infty} N(y')\exp(-i2\pi v'y')\,dy'$$
$$= \int_{-\infty}^{\infty}\int_{-\infty}^{\infty} f(x',y')\exp(-i2\pi v'y')\,dx'\,dy'. \tag{6.39}$$

When equation 6.39 is compared with the definition of the two-dimensional Fourier transform,

$$F(u,v) = \int_{-\infty}^{\infty}\int_{-\infty}^{\infty} f(x,y)\exp[-i2\pi(ux+vy)]\,dx\,dy, \tag{6.40}$$

it is clear that the right-hand side of equation 6.39 is the two-dimensional Fourier transform of $f(x,y)$ evaluated along the line $u' = 0$. This result is true for any viewing direction, so we can state the following general results, due to Bracewell,[15, 27] which is referred to as the *central section theorem*:

The one-dimensional Fourier transform of the integral of $f(x,y)$ along all lines parallel to $y = x \cdot \tan\theta$ is equal to the two-dimensional transform of $f(x,y)$ evaluated along the radial line $v = u \cdot \tan(\theta + \pi/2)$.

This theorem can be applied to the analysis of interferograms. Fringe order data for each individual viewing direction are numerically transformed by the FFT method.[28] In this manner values of the Fourier transform of $f(x,y)$ are computed at equally spaced points along radial lines in the transform plane, as indicated in Figure 6.8. By interpolation, values of the transform on a square grid of points can be determined. This numerical representation of $F(u,v)$ is then inverted by the use of the

6.2 FRINGE INTERPRETATION

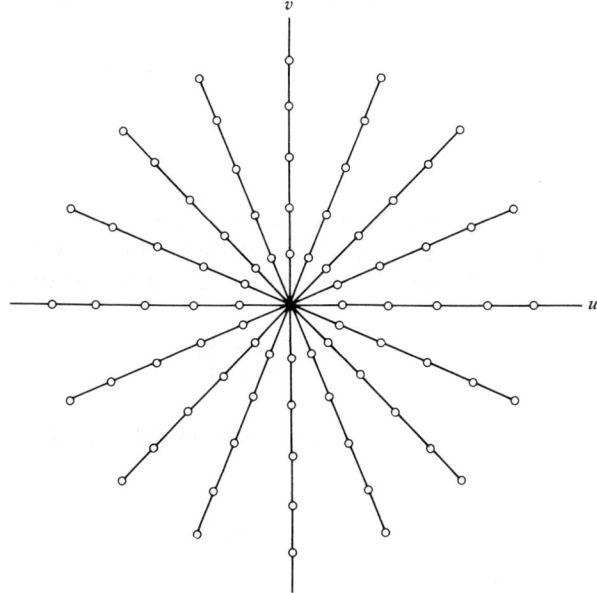

Figure 6.8 Fourier transform plane of the function representing an asymmetric phase object. Values of this transform can be computed at points on radial lines in this plane.

two-dimensional FFT to obtain $f(x,y)$. This scheme has been applied to holographic interferometry by Zien et al.[29]

Other schemes for reducing multidirectional interferometric data have been presented by Matulka and Collins,[30] Junginger and van Haeringer,[31] and Presnyakov.[32] The problem of reconstructing objects or fields from their line integrals arises in a variety of applications such as radio astronomy, electron microscopy, X-ray imaging, and plasma diagnostics. Solution techniques developed for these applications have been reviewed by Gordon;[33] many of them could be adapted for use in holographic interferometry.

6.2.2 Errors Due to Refraction

In the preceding discussion of fringe interpretation it was assumed that all probing optical rays are straight lines, that is, the effect of ray curvature due to refraction was neglected. When the length of the test section or the refractive index gradients are large, however, refraction is not negligible, and the various equations for fringe interpretation given above will be in error. In this section we briefly discuss errors caused by refraction and indicate how they can be minimized in some important cases.

Consider an experiment in which the refractive index of the medium being studied varies only in the y direction. The medium is confined to a test section of length L, as shown in Figure 6.9. A typical probing ray enters the test section normally at a position y_0. The path $y(z)$ of this ray is governed by equation 5.10. We restrict the analysis to the case of a linear variation

$$n(y) = n_0 + n'(y - y_0). \tag{6.41}$$

Equation 6.41 is a first-order approximation to a general variation $n(y)$ if $n_0 = n(y_0)$ and n' is the refractive index gradient at y_0. The equation of the probing ray is then given to a first approximation by equation 5.12:

$$y - y_0 = \frac{1}{2}\left(\frac{n'}{n_0}\right) z^2. \tag{6.42}$$

At the exit of the test section the ray has been deflected by an amount δ, where

$$\delta = \frac{1}{2}\left(\frac{n'}{n_0}\right) L^2 \tag{6.43}$$

and has a slope given by

$$y'_L = \left(\frac{n'}{n_0}\right) L. \tag{6.44}$$

The optical pathlength of this ray through the test section is given by equation 5.17, which can be written as

$$\Phi = \int_0^L n(y)\left[1 + \left(\frac{dy}{dz}\right)^2\right]^{1/2} dz. \tag{6.45}$$

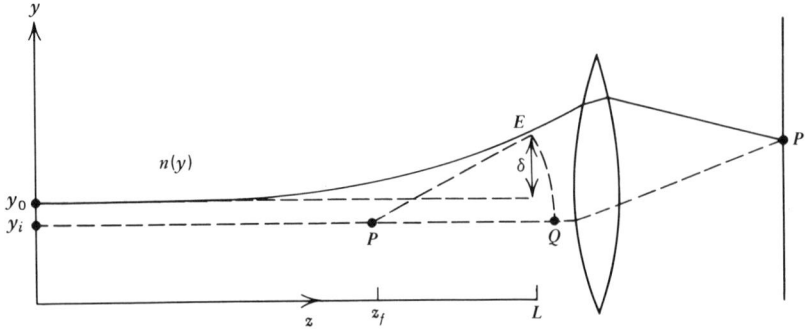

Figure 6.9 Formation of an interferogram when refraction of the probing rays is significant.

6.2 FRINGE INTERPRETATION

Combining equations 6.41, 6.42, and 6.45 and evaluating the integral gives the optical pathlength

$$\Phi = n_0 L \left[1 + \frac{1}{3} \left(\frac{n'}{n_0} \right)^2 L^2 \right], \tag{6.46}$$

where terms smaller than $(n'/n_0)^2$ have been neglected.

Equation 6.46 can be used to develop a criterion for estimating the importance of refraction effects. The first term of this equation is the optical pathlength of a ficticious straight horizontal ray through y_0, and the second term is the deviation due to refraction. If we require that this deviation be less than $\lambda/10$, then

$$\frac{(n')^2 L^3}{n_0 \lambda} < 0.3. \tag{6.47}$$

If this inequality is not satisfied, refraction errors may not be negligible. It is convenient to express equation 6.47 in terms of the fringe spacing $d_f = \lambda / n' L$:

$$\frac{\lambda L}{n_0 d_f^2} < 0.3. \tag{6.48}$$

To consider the analysis of interferograms when refraction is not negligible, the role of the imaging lens used to view or photograph the fringe pattern must be accounted for. Such a lens is shown in Figure 6.9. The interferogram is formed by imaging a plane $z = z_f$ onto a screen or photographic film. The probing ray will appear to have originated at point P, determined by projecting the exiting ray back to the focus plane $z = z_f$. The interferogram is formed when this ray interferes at point P' with the straight ray which passed through point P in the undisturbed test section during the reference holographic exposure. Using equations 6.43 and 6.44 and the geometry of Figure 6.9, we can determine the location of this ray:

$$y_i = y_0 + \frac{n' L}{n_0} \left(z_P - \tfrac{1}{2} L \right). \tag{6.49}$$

The optical pathlength difference determined at point P' by the fringe pattern is

$$\Delta \Phi = \Phi - n_0 \left(z_f + \overline{PQ} \right), \tag{6.50}$$

where $\overline{PQ} = \overline{PE}$ is the radius from point P to the point at which the probing ray exits from the test section. The distance $\overline{PQ}$ enters the analysis because, for an ideal lens or imaging system, the ray from E to P' will have an optical pathlength equal to that from Q to P'. Since

$$\overline{PQ} = \overline{PE} = [1+(y_0')^2]^{1/2}(L-z_f),$$

equations 6.44 and 6.50 can be combined to give

$$\Delta\Phi = n_0 L^2 \left(\frac{n'}{n_0}\right)^2 \left(\tfrac{1}{2}z_f - \tfrac{1}{6}L\right), \tag{6.51}$$

where terms smaller than $(n'/n_0)^2$ have been neglected.

If an observer follows the usual approach of counting fringes, the optical pathlength difference given by equation 6.51 will be incorrectly assigned to a straight horizontal ray $y = y_i$. Errors of this kind can be alleviated by optimum choice of the focus plane $z = z_f$, or corrected by computation. To do so, we first compute the optical pathlength difference along a ficticious straight ray $y = y_i$ through the test section. If equation 6.41 adequately represents the refractive index distribution near $y = y_i$, this optical pathlength difference is

$$\Delta\Phi_i = n_0 L \left(\frac{n'}{n_0}\right)(y_i - y_0).$$

Combining this with equation 6.49 gives

$$\Delta\Phi_i = n_0 L^2 \left(\frac{n'}{n_0}\right)^2 \left(z_f - \tfrac{1}{2}L\right). \tag{6.52}$$

Here $\Delta\Phi_i$ is the optical pathlength difference which would yield the correct value of refractive index at $y = y_i$ when the usual fringe interpretation procedure is followed; $\Delta\Phi$ is the optical pathlength difference measured experimentally at point P'. The deviation of $\Delta\Phi$ from $\Delta\Phi_i$ is

$$\Delta\Phi - \Delta\Phi_i = n_0 L^2 \left(\frac{n'}{n_0}\right)^2 \left(\tfrac{1}{3}L - \tfrac{1}{2}z_f\right). \tag{6.53}$$

This difference vanishes if $z_f = \tfrac{2}{3}L$. Hence, *to this order of approximation, refraction errors are eliminated by focusing on a plane $L/3$ from the exit plane of the test section.* This result has been derived by a more formal expansion procedure by Svensson,[34] Wachtell,[35] and Howes and Buchele.[36]

6.2 FRINGE INTERPRETATION

The result is also unaffected, to this order of approximation, by the presence of windows at the entrance and exit planes of the test section.[36] If for some reason another focus plane $z = z_f$ is used, equation 6.53 can serve to computationally correct for refraction errors:

$$[n(y_i) - n_0] \cdot L = \Delta\Phi_i$$

$$= \Delta\Phi - n_0 L^2 \left(\frac{n'}{n_0}\right)^2 \left(\tfrac{1}{3}L - \tfrac{1}{2}z_f\right).$$

Since $\Delta\Phi = N(y_i) \cdot \lambda$, where $N(y_i)$ is the fringe order number,

$$n(y_i) - n_0 = N(y_i) \cdot \frac{\lambda}{L} - n_0 L \left(\frac{n'}{n_0}\right)^2 \left(\tfrac{1}{3}L - \tfrac{1}{2}z_f\right). \tag{6.54}$$

This equation replaces equation 6.2 when refraction is not negligible. More precise approaches to correction for refraction error are discussed by Anderson et al.[37] and by Howes and Buchele.[36] Correction schemes for specific problems of boundary layer measurement have been developed.[38-40] When very strong refraction occurs, iterative schemes have been proposed.[41] An initial estimate of the field is computed by neglecting refraction. Ray tracing calculations based on this estimated field are used to predict a fringe pattern, which then is compared with the experimental interference fringes. The estimated refractive index distribution is modified to improve agreement, and the procedure is repeated until good agreement is attained. It has been noted that this procedure is unstable in some cases[38] and is therefore difficult to carry out by computer without operator intervention at each iterative step.

Refractive errors in radially symmetric refractive index distributions have been studied[42, 43] because they are of particular importance in plasma diagnostics. It has been found by computer simulation that refraction errors are generally quite small if equation 6.5 is used to invert data obtained by focusing on the center plane containing the axis of symmetry.

A serious problem that can arise due to refraction is *apparent ray crossing*. This can occur in a strongly refracting medium or a long test section when the refractive index is a nonlinear function of y. Two probing rays traversing a strongly refracting medium are shown in Figure 6.10. The upper ray is refracted more strongly than the lower ray. The imaging lens used to view or photograph the interference pattern is focused on the plane $z = z_f$. When the exiting rays are projected back to the focus plane, the upper ray, which entered at y_{01}, appears to originate at point P_1, which is below P_2, the apparent origin of the lower ray, which entered at y_{02}. The resulting fringe pattern is thus ambiguous and cannot be interpreted.

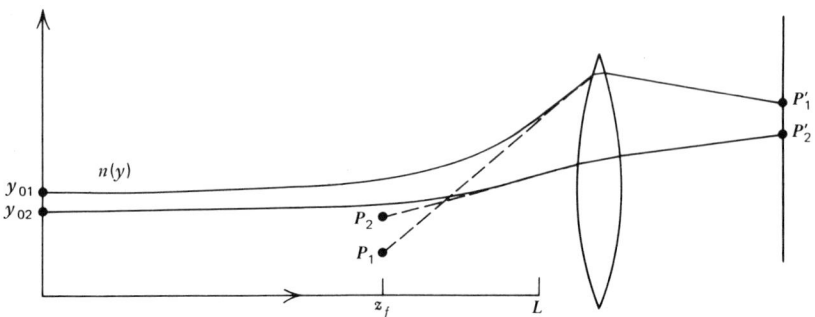

Figure 6.10 Apparent ray crossing due to refraction of the probing rays.

Howes and Buchele[36] show that focusing on the plane $z_P = L/2$ tends to minimize apparent ray crossing. If this phenomenon is suspected in an experiment, it is useful to analyze an approximate model of the object under study to aid in choosing a focus plane and analyzing the fringes—an approach illustrated by Beach et al.[39] This comment also applies to the problem of spurious fringes due to reflection and diffraction which occur very close to solid boundaries within the test section.[44]

6.3 SPECIAL TECHNIQUES: MULTIPLE-BEAM, MULTIPLE-WAVELENGTH, AND MULTIPLE-PASS INTERFEROMETRY

The use of multiple reference or reconstruction beams adds flexibility to the holographic interferometry of transparent objects. Phase difference amplification, shearing interferometry, and moiré interferometry are based on the use of multiple beams. These techniques, as well as multiple-wavelength and multiple-pass holographic interferometry, are discussed in this section as further background for the consideration of applications in Section 6.4.

Consider a single-exposure, *three-beam hologram* recorded with the system shown in Figure 6.11a. The object wave and reference waves, assumed for simplicity to have unit irradiance, have the following complex amplitudes in the hologram plane:

$$\mathbf{U}_o(x,y) = \exp[i\phi(x,y)],$$

$$\mathbf{U}_{R_1}(x,y) = \exp(i2\pi f y), \qquad \mathbf{U}_{R_2} = \mathbf{U}_{R_1}^* = \exp(-i2\pi f y),$$

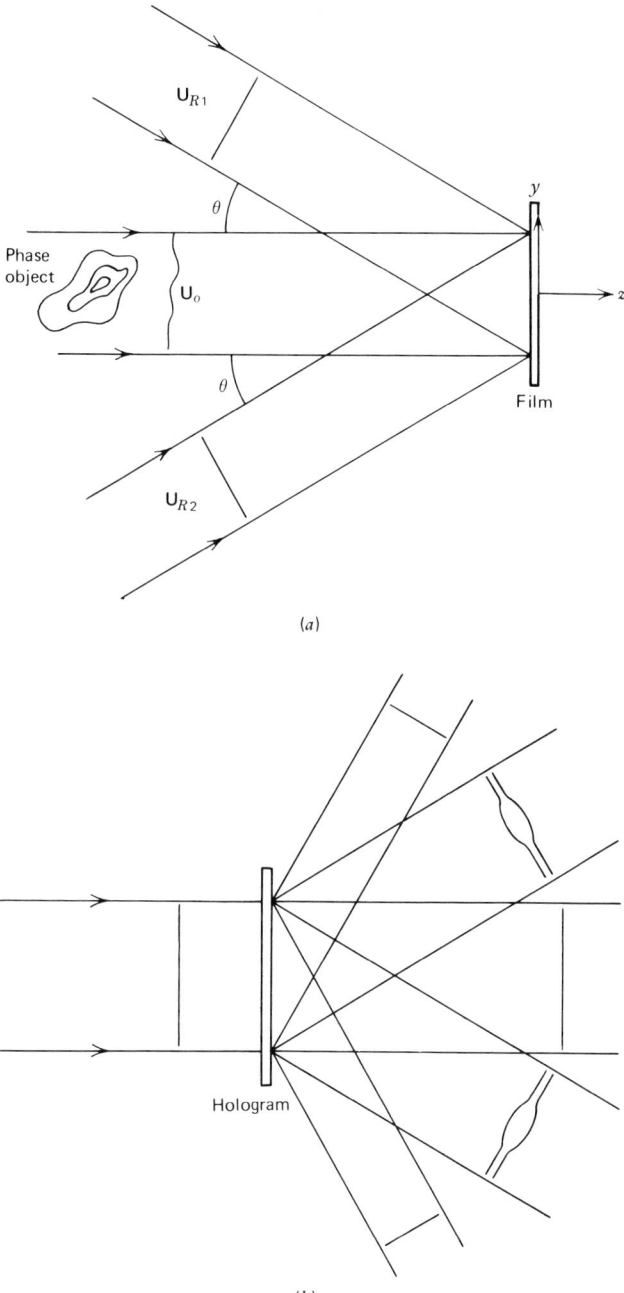

Figure 6.11 Dual-reference-wave holographic interferometer for three-beam, single-exposure interferometry for fringe doubling. (*a*) Recording the hologram. (*b*) Reconstruction by a plane wave travelling in the direction of the original object wave.

335

where $f = (\sin\theta)/\lambda$. A hologram formed by simultaneously exposing a film plate to all three beams will have a transmittance of the form

$$\begin{aligned}\mathbf{t}(x,y) = {} & k_0 + k_1 \big(\exp(-i2\pi fy) \cdot \{\exp[i\phi(x,y)] + \exp[-i\phi(x,y)]\} \\ & + \exp(i2\pi fy) \cdot \{\exp[i\phi(x,y)] + \exp[-i\phi(x,y)]\} \\ & + \exp(-i4\pi fy) + \exp(i4\pi fy)\big).\end{aligned} \quad (6.55)$$

If this hologram is illuminated by a plane wave traveling in the direction of the original object wave, as shown in Figure 6.11b, each diffracted wave of +1 or −1 order will have an irradiance proportional to

$$\begin{aligned}I &= |\exp[i\phi(x,y)] + \exp[-i\phi(x,y)]|^2 \\ &= 2\{1 + \cos[2\phi(x,y)]\}.\end{aligned} \quad (6.56)$$

Equation 6.56 represents an interferogram with phase difference amplification by a factor of 2. This technique, reported by De and Sevigny,[45] is similar to the method discussed in Section 5.5 in which the hologram is recorded with a single reference wave, but two conjugate reconstruction waves are used.

If the dual reference waves are oriented as shown in Figure 6.12 and a hologram is formed by a single exposure of the plate to all three waves, a similar analysis indicates that each diffracted wave of +1 or −1 order will have irradiance proportional to

$$I = 1 + \cos[\phi(x,y)]. \quad (6.57)$$

Equation 6.57 represents an ordinary linear interferogram; however, the interferogram is formed with a single exposure, which is advantageous in some applications. De and Sevigny[45] also described a self-aligning system using a grating to generate object and reference waves with the proper orientations. Politch et al.[46–48] have analyzed more general configurations in which two reference beams and one reconstruction beam are used. They give information regarding noise problems which can occur in superimposed holograms of the multibeam type.

Multiple-reference-wave, multiple-exposure holography also can be used to double the sensitivity of holographic interferometry.[49] In this case the desired fringes are displayed in a moiré pattern. Although this technique is more complicated than the single-exposure method of De and Sevigny, it has the advantage of being insensitive to small aberrations introduced by low-quality optical elements.

In *shearing interferometry* a fringe pattern is formed by the interference of two identical wavefronts which are displaced with respect to each other.

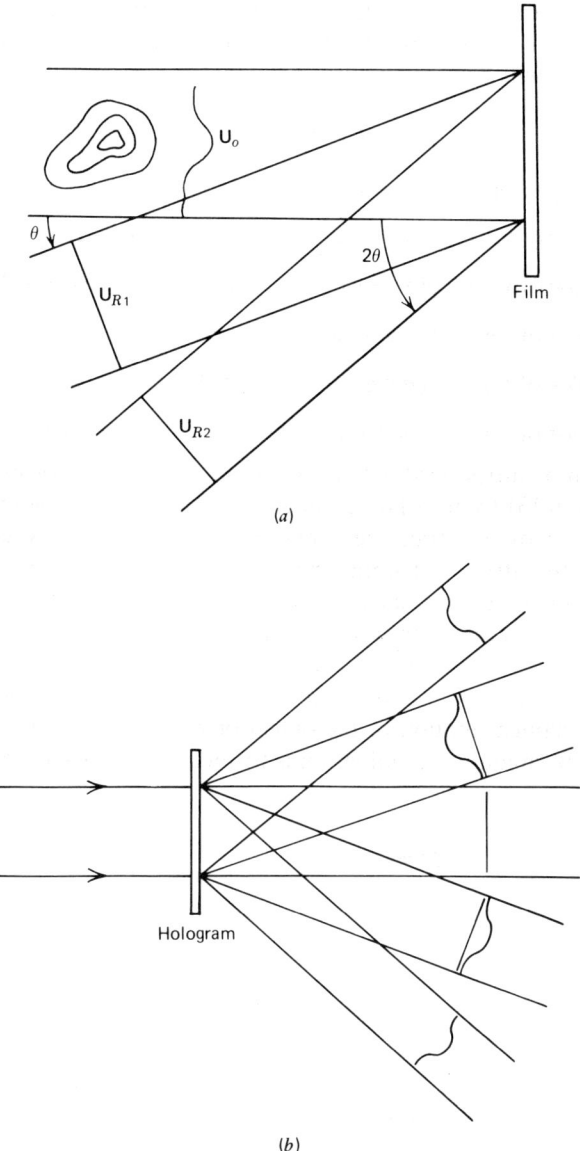

Figure 6.12 Dual-reference-wave holographic interferometer for three-beam, single-exposure interferometry with overlapping diffraction orders. (*a*) Recording the hologram. (*b*) Reconstruction by a plane wave traveling in the direction of the original object wave.

This can be done easily by holography. If the transparent object under study is a pure phase object, $\exp(i\phi)$, the fringe pattern of a shearing interferogram will be

$$I = 1 + \cos \Delta \phi. \tag{6.58}$$

Five basic types of shear can be defined:[50]

$\Delta \phi = \phi(x,y) - \phi(x, y + \Delta y)$	lateral shear Δy,
$\Delta \phi = \phi(x,y) - \phi(x, -y)$	reversion or folded shear,
$\Delta \phi = \phi(r,\varphi) - \phi(r, \varphi + \Delta \varphi)$	rotational shear $\Delta \varphi$,
$\Delta \phi = \phi(r,\varphi) - \phi(mr, \varphi)$	radial shear, ratio m,
$\Delta \phi = \phi(x,y,z) - \phi(x,y, z + \Delta z)$	longitudinal shear Δz.

Shear can be introduced with a simple off-axis, two-exposure interferometer (Figure 6.13) by translating either the object or the hologram between exposures, or by rotating them about the z axis. If desired, reference fringes can be introduced simultaneously by tilting either the mirror M or the hologram. Real-time shearing interferometry can be accomplished in an analogous manner. In practice this system is suitable for forming interferograms with lateral or rotational shear; other types of shear are introduced more conveniently with different systems which do not require precise mechanical motion of the object or hologram. In one such system, devised by Bryngdahl,[50] a hologram is formed by two waves, both of which

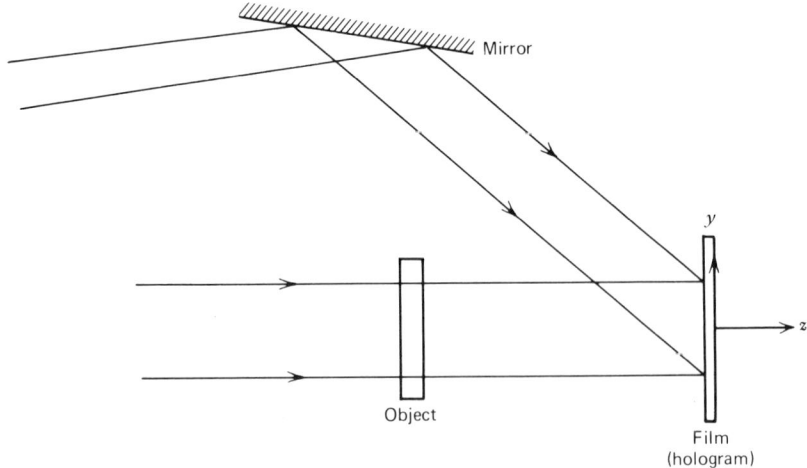

Figure 6.13 Two-exposure holographic shearing interferometer.

have passed through the phase object. Shear is introduced optically between the two. The object is then removed, and a second exposure of the hologram to the two waves is made. When the resulting hologram is illuminated by one of these waves, an interference pattern is formed between the other, undisturbed wave and a wavefront having the shape of the difference between the two object waves. This is illustrated by the system shown in Figure 6.14. The object is imaged onto the hologram plane by a telecentric system so that the wavefronts which form the hologram will be plane if the object is absent. A tilted flat glass plate is used to introduce a lateral shear between the two images of the object.

Let an unsheared and a sheared wave have complex amplitudes $\exp(i\phi)$ and $\exp[i(2\pi fx + \phi_s)]$, respectively, at the hologram plane. The hologram is first exposed to these two waves; then the object is moved to introduce shear, and a second exposure is made. The transmittance of this hologram is proportional to the irradiance

$$I = |\exp(-i\phi) + \exp[-i(2\pi fx + \phi_s)]|^2 \\ + |1 + \exp(-i2\pi fx)|^2. \tag{6.59}$$

When this hologram is illuminated by the wavefront with amplitude $\exp(i2\pi fx)$, the virtual image term has the irradiance

$$I = |1 + \exp[-i(\phi - \phi_s)]|^2 \\ = 2[1 + \cos(\phi - \phi_s)], \tag{6.60}$$

which is the desired interferogram. Similar methods are discussed by Vest

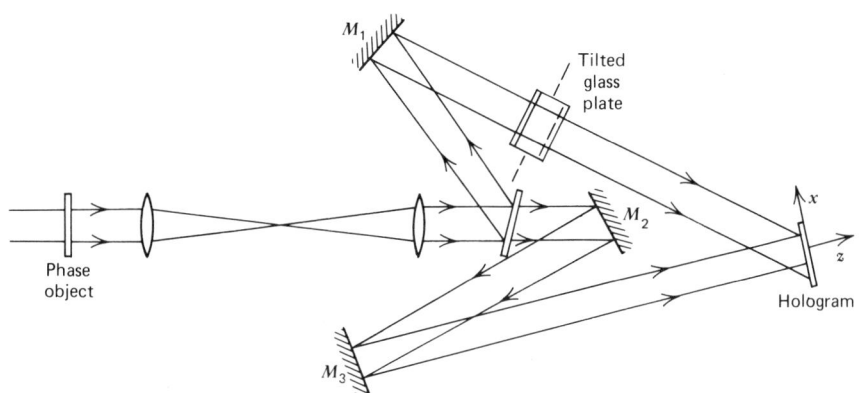

Figure 6.14 Holographic shearing interferometer due to Bryngdahl (ref. 50).

and Sweeney[51] and by Shamir.[52] Reference fringes can be introduced by tilting mirror M_3 between the two exposures. Radial shear can be introduced by a variant of this system in which the images of the object are formed at the hologram plane with two slightly different magnifications. Shearing interferometry can also be accomplished by using two reconstruction waves separated slightly in angular orientation to reconstruct a single-exposure hologram.[53]

In Section 5.5 and in the first part of the present section, multiple-reference or multiple-reconstruction beam techniques for doubling phase differences were discussed. In either procedure, if the object is imaged onto the hologram plane, the resulting interferogram can be considered to be due to a longitudinal shear about the hologram plane. This has been discussed and demonstrated experimentally by Bryngdahl,[54] using a single-exposure, off-axis hologram which is reconstructed by illuminating it with the original reference wave and its conjugate. Two first-order waves diffracted by a grating are used as the conjugate reconstruction waves. If the grating is replaced by a Gabor zone plate, a simple radial shear interferometer can be developed by applying similar principles.[55, 56]

Doi et al.[57] have developed a single-exposure holographic interferometer in which variable lateral shear can be introduced during reconstruction. Two reference waves, obtained by using a biprism (see Figure 6.15), are linearly polarized along orthogonal directions (e.g., 0° and 90°). The light passing through the object is polarized at 45°. A single exposure is made, and the resulting hologram is then replaced in its original position, illuminated by the two reference waves, and viewed through a polarizing screen oriented at 45°. Shear, which is lateral to first order, can be

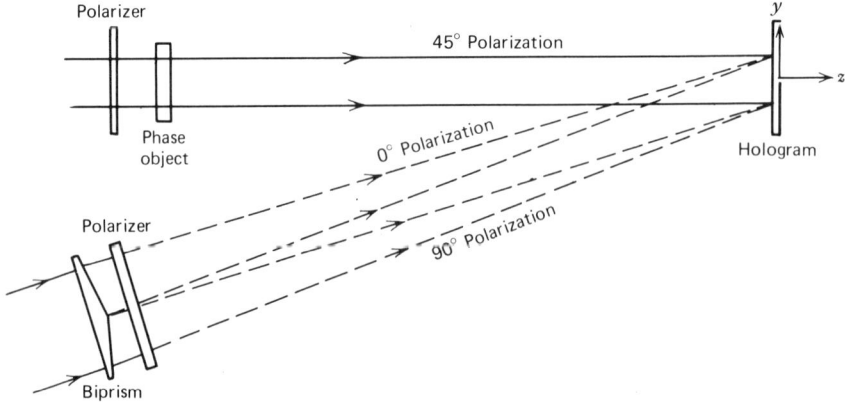

Figure 6.15 Single-exposure holographic shearing interferometer.

6.3 SPECIAL TECHNIQUES

introduced by rotating the hologram about the z axis. In another method based on polarization effects a plane parallel plate of birefringent material such as quartz is placed behind the transparent object in an ordinary off-axis holographic system.[58] The birefringent plate divides the object wave into two laterally sheared waves which are linearly polarized in the 0° and 90° directions, respectively. A single exposure is recorded, using a reference wave polarized at 45°. The two waves reconstructed with this hologram form a laterally sheared interferogram. If the birefringent plate is replaced by a birefringent lens, radial shear interferograms can also be formed.

We conclude this section by discussing two additional techniques for varying the sensitivity of holographic interferometry. The first is to simultaneously record holograms in different wavelengths;[59-62] the second is to pass the object wave through the object several times.[63]

Single-exposure, multiple-wavelength holography can be used to produce interferograms of either reduced or increased sensitivity. The technique, which is very similar to multiple-wavelength contouring (see Section 7.2), has been discussed and experimentally demonstrated by Weigl[59, 60] and by Afanaseva et al.[62] Consider a single-exposure hologram recorded with an off-axis system, as shown in Figure 6.16a. The object and reference waves both contain light of two wavelengths, λ_1 and λ_2, obtained either from two lines of a single laser or from two different lasers. The transmittance of this hologram is of the form

$$t(x,y) = k_0 + k_1\{|\exp[i\phi_1(x,y)] + \exp(i2\pi f_1 x)|^2 \\ + |\exp[i\phi_2(x,y)] + \exp(i2\pi f_2 x)|^2\}, \quad (6.61)$$

where $f_{1,2} = \sin\theta/\lambda_{1,2}$, and $\phi_{1,2}(x,y)$ is the phase distribution, in each wavelength, due to the phase object:

$$\phi_{1,2}(x,y) = \frac{2\pi}{\lambda_{1,2}} \int_L^{L+\Delta L} n(x,y,z)\,dz. \quad (6.62)$$

We assume that the transparent medium is *nondispersive*, that is, the refractive index $n(x,y,z)$ is independent of wavelength. For viewing, the hologram is illuminated simultaneously by two reconstruction waves of wavelength λ_3, as shown in Figure 6.16b. These waves have complex amplitudes $\exp(i2\pi f_a x)$ and $\exp(i2\pi f_b x)$, where $f_{a,b} = (\sin\theta_{a,b})/\lambda_3$. The reconstructed optical wave will have a complex amplitude at the hologram plane which includes the following important terms:

$$[\exp(-i2\pi f_a x) + \exp(-i2\pi f_b x)] \\ \cdot \{k_0 + k_1[4 + \exp(i\phi_1 - 2\pi f_1 x) + \exp(-i\phi_1 + 2\pi f_1 x) \\ + \exp(i\phi_2 - 2\pi f_2 x) + \exp(-i\phi_2 + 2\pi f_2 x)]\}. \quad (6.63)$$

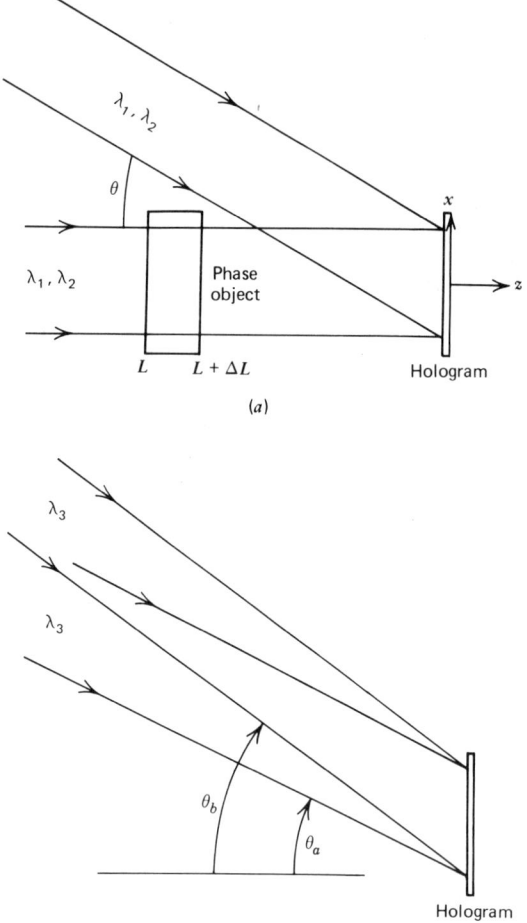

Figure 6.16 Single-exposure, multiple-wavelength holographic interferometer used for variable sensitivity measurements. (a) Recording the hologram with light of wavelengths λ_1 and λ_2. (b) Reconstruction with light of wavelength λ_3.

Among the terms in equation 6.63 are

$$\exp(i\phi_1) \cdot \exp[i2\pi(f_a - f_1)x] \\ + \exp(i\phi_2) \cdot \exp[i2\pi(f_b - f_2)x]. \qquad (6.64)$$

For a given reconstruction wavelength λ_3, angles θ_a and θ_b can be chosen so that $f_a = f_1$ and $f_b = f_2$; if these conditions are satisfied, both terms in

6.3 SPECIAL TECHNIQUES

equation 6.64 represent waves propagating in the z direction. The resulting interferogram has an irradiance distribution proportional to

$$I = 1 + \cos(\phi_2 - \phi_1).$$

Since $1/\lambda_2 - 1/\lambda_1 = (\lambda_1 - \lambda_2)/\lambda_1\lambda_2$, this interferogram will have a sensitivity approximately $(\lambda_1 - \lambda_2)/\lambda$ times that of an ordinary interferogram if λ_1 and λ_2 differ only slightly. Large *reductions* in sensitivity can be attained by this technique. A similar analysis shows that, if θ_a and θ_b are chosen so that $f_a = f_1$ and $f_b = -f_2$, the interferogram formed by waves traveling in the z direction has an irradiance distribution proportional to

$$I = 1 + \cos(\phi_2 + \phi_1).$$

This interferogram has a sensitivity approximately $(\lambda_1 + \lambda_2)/\lambda$ times that of an ordinary interferogram, so modest *increases* in sensitivity can also be attained.

Even greater variations in sensitivity are made possible by *nonlinear* multiple-wavelength holography, as demonstrated by Mustafin and Seleznev.[61] The study of *dispersive* media by multiple-wavelength holographic interferometry is very important in plasma diagnostics and will be discussed in Section 6.5.

Multiple-pass interferometry using holography also can serve to increase sensitivity. A simple system described by Weigl et al.[63] is shown in Figure 6.17. The phase object is placed in a resonant cavity formed by two parallel beamsplitters separated by a distance L. The light emerging from this cavity to illuminate the hologram consists of a series of waves which have traversed the object $1, 3, 5, \ldots$ times. If $2L > l_c$, where l_c is the coherence length of the laser light, these waves will not be mutually coherent. Furthermore, only one wave will be coherent with a given

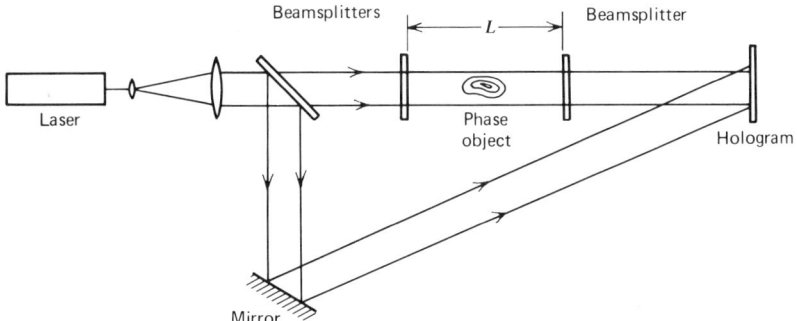

Figure 6.17 Multiple-pass holographic interferometer for increasing sensitivity.

reference wave. Hence, for example, if the length of the reference beam is chosen to match that of the beam which traverses the object three times, only that object wave will contribute to the hologram. If a two-exposure hologram is recorded with this system in the usual manner, a phase difference amplification by a factor of 3 will result. In a similar manner phase multiplication factors of 1,3,5,... can be achieved. The fraction of light in the object wave that carries the desired information decreases as the multiplication factor increases. In practice this limits the attainable multiplication factor. It should also be noted that high-precision optical elements are required even though this method is a two-exposure one.

6.4 APPLICATIONS

In this section we consider the application of holographic interferometry to aerodynamics and flow visualization, plasma diagnostics, heat and mass transfer measurements, and stress analysis of transparent objects.

For each of these topics the relation of the property of primary interest, such as density or temperature, to the index of refraction will be discussed, useful characteristics of holographic methods and relevant special techniques will be described, and examples of applications and references to the technical literature will be given.

6.4.1 Aerodynamics and Flow Visualization

Optical interferometry has been used for many years to study the flow of compressible gases in wind tunnels and shock tubes. The most common instrument for this purpose is the Mach-Zehnder interferometer, originally described by L. Zehnder and L. Mach in 1891–1892.[64, 65] A detailed discussion of this instrument and its application to aerodynamics is given by Ladenburg and Bershader.[66] More recently, holography has also become an accepted method of flow diagnostics.[67, 68] Because of its simplicity and state of development, the Mach-Zehnder interferometer is likely to remain the most common interferometer for analysis of compressible flows; however, the unique features of holographic interferometry make possible new types of measurements and extend the range of optical and environmental conditions in which interferograms can be recorded.

Interferometry is used to determine the distribution of *density*, the mass per unit volume, in a gas. Density, denoted by ρ, is related to the refractive index of the gas by the *Gladstone-Dale equation*:

$$n - 1 = K\rho, \qquad (6.65)$$

where K, the *Gladstone-Dale constant*, is a property of the gas. It is a weak function of the wavelength of light and is nearly independent of tempera-

TABLE 6.1 GLADSTONE-DALE CONSTANTS FOR AIR AT A TEMPERATURE OF 288°K

λ(nm)	K (m^3/kg)
356.2	0.2330×10^{-3}
380.3	0.2316
407.9	0.2304
447.2	0.2290
480.1	0.2281
509.7	0.2274
567.7	0.2264
607.4	0.2259
644.0	0.2255
703.4	0.2250
912.5	0.2239

Source: Reference 69.

ture and pressure under moderate physical conditions. Values of K for air as a function of wavelength are given in Table 6.1. Table 6.2 contains values of K for several common gases at wavelengths of 514.5 and 632.8 nm, which correspond to the dominant lines of argon ion and He-Ne lasers, respectively. The Gladstone-Dale constant of a mixture of gases can be calculated as a mass-weighted average of the values of K for the component gases:[69]

$$K = \sum_i a_i K_i, \qquad (6.66)$$

where a_i is the mass fraction, and K_i the Gladstone-Dale constant, of the ith component.

Figure 6.18 is a typical interferogram of a high-speed aerodynamic flow. The techniques described in Section 6.2 could be applied to determine the

TABLE 6.2 GLADSTONE-DALE CONSTANTS FOR GASES AT 300°K, 0.1013 MN/m^2

Gas	K (m^3/kg)	
	$\lambda = 514.5$ nm	$\lambda = 632.8$ nm
Ar	0.175×10^{-3}	0.158×10^{-3}
He	0.196	0.195
CO_2	0.229	0.227
N_2	0.240	0.238
O_2	0.191	0.189

Source: Values calculated from data in ref. 70.

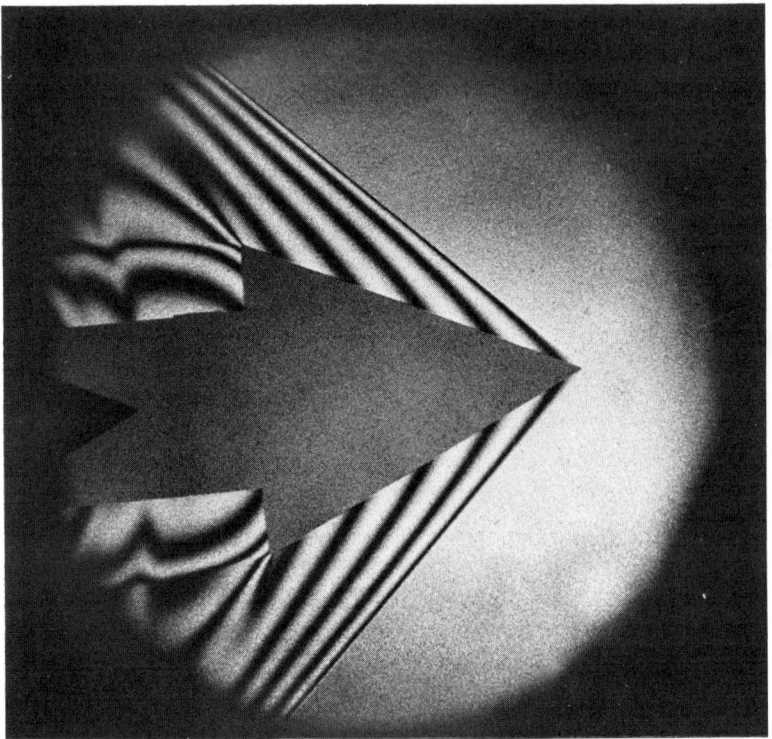

Figure 6.18 Two-exposure holographic interferogram of high-speed air flow past a cone. (ref. 74).

distribution of change of refractive index throughout the flow field. In this case the flow field is radially symmetric, so equation 6.5 would be used to determine $f(r) = n(r) - n_0$, where n_0 is the ambient refractive index. Equation 6.65 would then be applied to determine the density distribution in the flow field:

$$\rho(r) = \rho_0 + \frac{f(r)}{K}, \qquad (6.67)$$

where ρ_0 is the ambient density far from the test object.

The potential usefulness of holographic interferometry in aerodynamic diagnostics was illustrated in the early paper by Heflinger et al.,[71] which included a diffuse-illumination holographic interferogram of a bullet traveling through argon at a speed of 550 m/s. The hologram was recorded using a Q-switched ruby laser. After this work and that of Chau and

6.4 APPLICATIONS

Mullaney,[72] Smigielski and Royer,[73] Philbert and Surget,[74] and others demonstrated several advantages of holographic interferometry for aerodynamic investigations. These include its simplicity, the use of relatively low-quality optics (see Section 5.3), and the ability to obtain several interferograms with different viewing directions from a single hologram (see Section 5.3). Single-exposure holography is also useful because it provides a permanent record of the optical wavefront which has probed the test section.[75-77] This probing wavefront can be reconstructed later, and a shadowgraph, schlieren, or other optically filtered image can be formed.[67, 75, 78-81] Holography is useful for these types of flow visualization because the critical alignment of schlieren knife-edges and other optical components can be done at leisure on a stable optical bench rather than in the experimental apparatus. This is especially important when pulsed lasers are used to study transient phenomena. Such postprocessing is usually done using a continuous wave He-Ne laser. When a He-Ne laser ($\lambda = 632.8$ nm) is used to reconstruct a hologram recorded with a pulsed ruby laser ($\lambda = 693.4$ nm), care must be taken to minimize wavefront distortions caused by the wavelength difference if quantitative evaluation of filtered images is required.[82] If multiple-reference-beam or multiple-hologram techniques are used, both holographic interferometry and filtered-image flow visualization can be employed in the same experiment.

The development of reliable pulsed lasers with reasonably good coherence properties has been crucial to the application of holographic interferometry to gas dynamics. Much of the development of this technology as applied to holographic interferometry was carried out by a group at TRW.[71, 83, 84] Tanner,[85] Surget,[86] Fagot,[87] Veret,[88] Radley and Havener,[89] Hirth et al.,[90] Balozerov and Chernykh,[91] Reinheimer et al.,[92] Wortberg,[93] Law,[94, 95] Shatilov et al.,[96, 97] Berezkin et al.,[98] and others have continued the development of holographic interferometry as a tool for aerodynamic and shock tube studies.

The cancellation of phase errors due to low-quality optical elements, including test section windows, is an important attribute of holographic interferometry when large-field interferograms are required. This also makes it possible to use transparent test models to obtain interferograms in flow regions which are inaccessible by classical interferometry. Beamish et al.[100] demonstrated this by recording interferograms showing the interaction of a shock with the boundary layer over a model of a delta wing. The model, which is shown in Figure 6.19a, was manufactured from Plexiglas and mounted in 0.15×0.15 m test section of a supersonic free-jet wind tunnel. Diffuse-illumination interferograms were recorded using a pulsed ruby laser. The diffuse light passed through the transparent object while it was inserted in a Mach 4.0 flow. Figures 6.19b, 6.19c, and 6.19d are

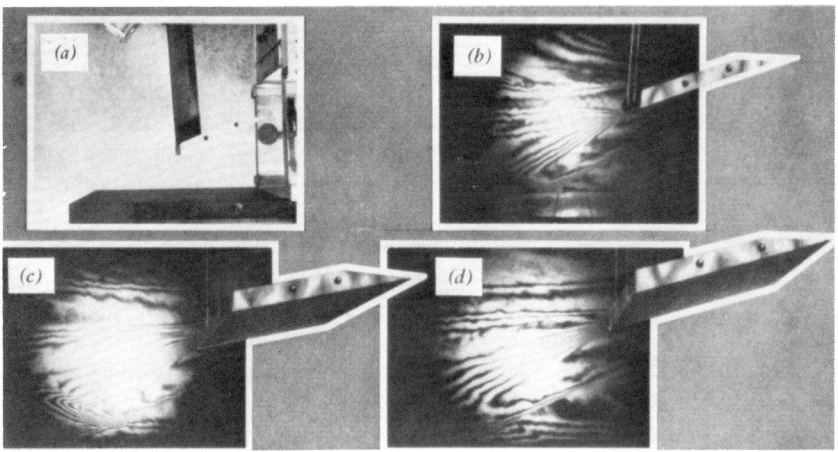

Figure 6.19 Holographic interferograms recorded with light passing through a transparent model of a wing (ref. 100). (a) The model mounted on a turn table in a wind tunnel. (b)–(d) Interferograms with wing at angles of attack 0°, 5°, and 10°, respectively. (Reprinted with permission from the American Institute of Aeronautics and Astronautics, *AIAA Journal*, Vol. 7, No. 10, p. 2042, Figure 2.)

interferograms obtained in this manner at various angles of attack. Kosakoski and Collins[101] have used a Lucite model to measure the three-dimensional density distribution in a transonic corner flow. It may prove feasible to apply fiber optical illumination systems to record holographic interferograms of flows in even less accessible regions.

In 1971 Matulka and Collins[30] reported the quantitative determination of a three-dimensional, asymmetric density field by holographic interferometry. This important development greatly extended the class of flows which can be studied quantitatively by interferometry. The object they studied was a tilted free air jet inside a test section, which is shown schematically in Figure 6.20. Three two-exposure, diffuse-illumination holograms were recorded simultaneously. These could be viewed from various directions. They were photographed to provide two-dimensional interferograms recorded at six viewing angles separated by 15° intervals. These interferograms were then analyzed by a method like that discussed in Section 6.2. Measurements of the type made by Matulka and Collins could be made using Mach-Zehnder interferometry to record several interferograms as the object is rotated. Holographic interferometry, however, can instantaneously capture the optical information for many directions, so its use is convenient for steady flows and essential for transient flows.

6.4 APPLICATIONS

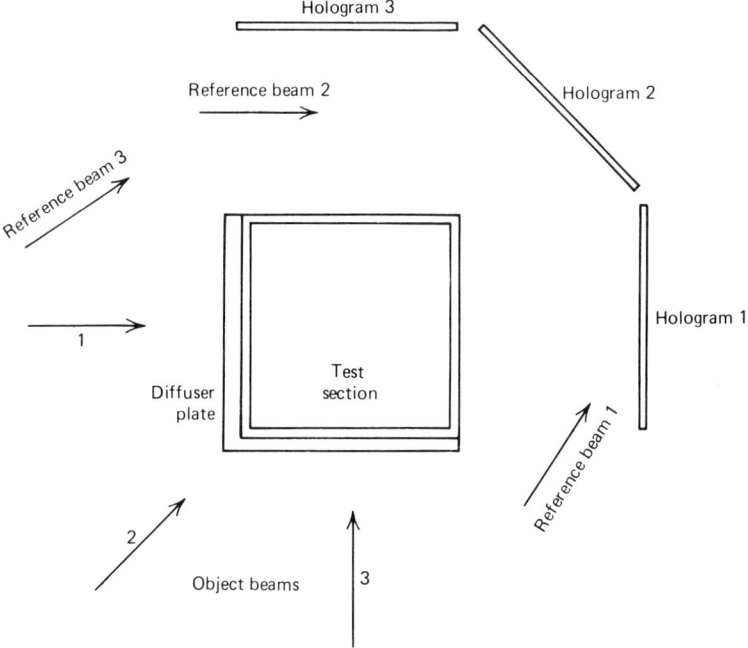

Figure 6.20 Optical arrangement for recording a holographic interferogram with a wide range of viewing angles (ref. 30).

Following the work of Matulka and Collins, other workers have reported measurements of asymmetric aerodynamic density fields. Pulsed laser holographic interferometry has been used to measure the asymmetric density distribution near right circular cones at various angles of attack.[29,102] Figure 6.21 shows results of measurements by Zien et al.[29] of the density change in a Mach 2.0 flow past a large cone of 15° half-angle at a 16.5° angle of attack. These tests were conducted in a large-scale wind tunnel. The results are believed to have maximum errors of less than 6 percent of the ambient density.

Kosakoski and Collins[101] studied the complicated density field in a transonic (Mach 0.937) corner flow near a model of an aircraft. Their experimental apparatus is shown schematically in Figure 6.22. The model was rotated about its axis, and separate holographic interferograms were recorded at approximately 11° intervals through a 180° range of viewing angles. The model was fabricated from transparent Lucite so that interferograms could be recorded in all viewing directions, as shown in Figure 6.23. The density distribution calculated from these data for one transverse

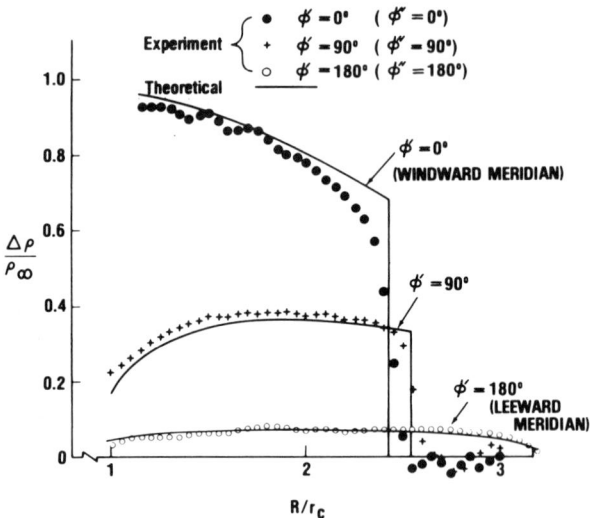

Figure 6.21 Measurements of an asymmetric density field made in a holographically instrumented, large-scale supersonic wind tunnel (ref. 29). $\Delta\rho$ is the density change in the flow field around a cone of radius r_c at 15° angle of attack, and R, ϕ' are polar coordinates in a plane normal to the axis of the cone. (Reprinted with permission from the American Institute of Aeronautics and Astronautics, *AIAA Journal*, Vol. 13, 1975, p. 842, Figure 5.)

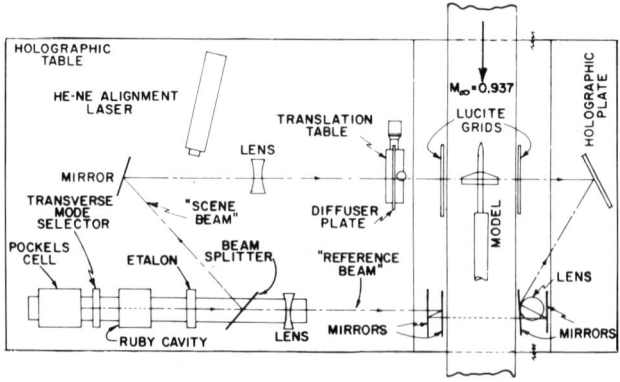

Figure 6.22 Holographic system used to measure the asymmetric density field in a transonic corner flow. (ref. 101). (Reprinted with permission from the American Institute of Aeronautics and Astronautics, *AIAA Journal*, Vol. 12, No. 6, p. 768, Figure 1.)

6.4 APPLICATIONS

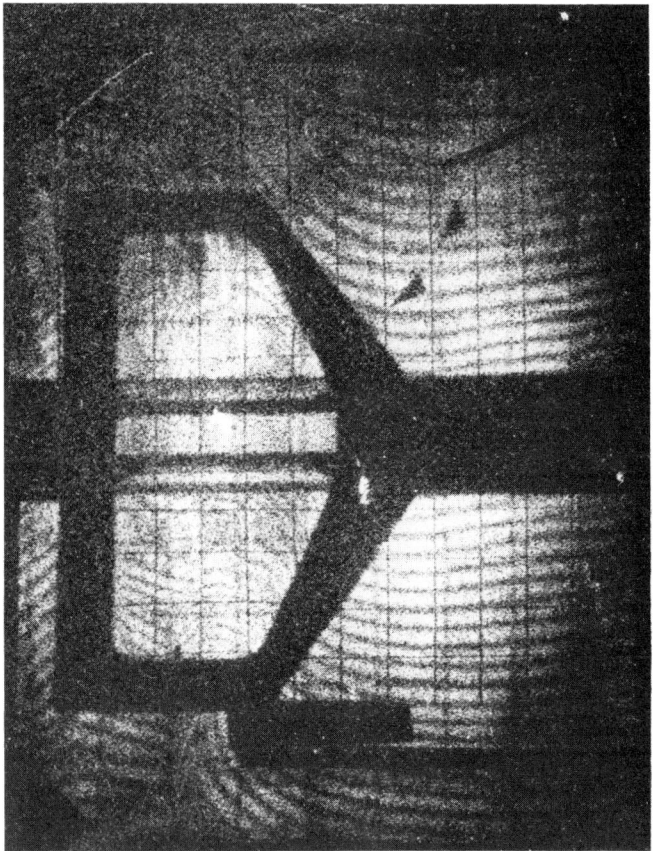

Figure 6.23 Holographic interferogram of density field around an aerodynamic model. (ref. 101). Note that part of the fringe pattern is formed with light passing through the transparent lucite model. (Reprinted with permission from the American Institute of Aeronautics and Astronautics, *AIAA Journal*, Vol. 12, No. 6, p. 769, Figure 3.)

plane is shown in Figure 6.24a. Figure 6.24b is a schematic drawing of the three-dimensional shock structure determined with the aid of the interferograms. The density changes are believed to be accurate within about 6 percent of the value of ambient density.

Holographic interferometry also has application to some flow visualization problems in low-speed fluid mechanics, such as the study of stratified flows.[103, 104] Figure 6.25 is a two-exposure interferogram of a cylinder of rectangular cross section moving slowly through density-stratified salt water. During the first exposure the cylinder was towed through a tank of

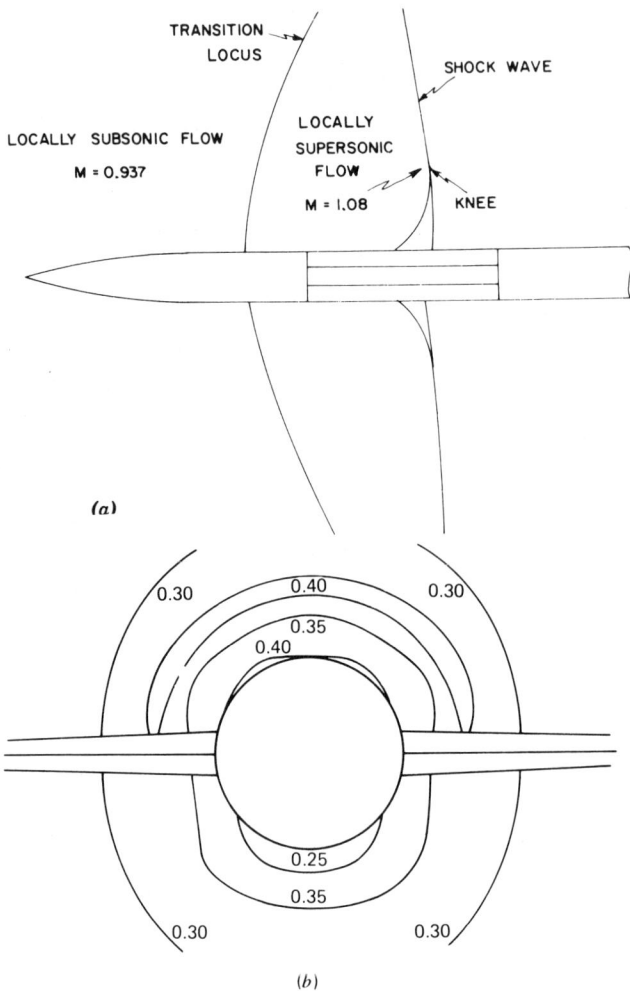

Figure 6.24 Density field in a transonic corner flow (ref. 101). (*a*) Schematic drawing of shock wave structure near the aerodynamic model. (*b*) Contour plot of density field in a transverse plane determined by holographic interferometry. (Reprinted with permission from the American Institute of Aeronautics and Astronautics, *AIAA Journal*, Vol. 12, p. 769, Figures 4 and 6.)

salt water with a linear density variation in the vertical direction. The tank was then emptied and refilled with pure water before the second exposure was recorded. In the absence of flow the interferogram would display a set of equally spaced, parallel, straight fringes corresponding to the surfaces of constant density. These surfaces are displaced because of the motion of the cylinder. The fringes in Figure 6.25 can be shown to be streamlines whose

6.4 APPLICATIONS

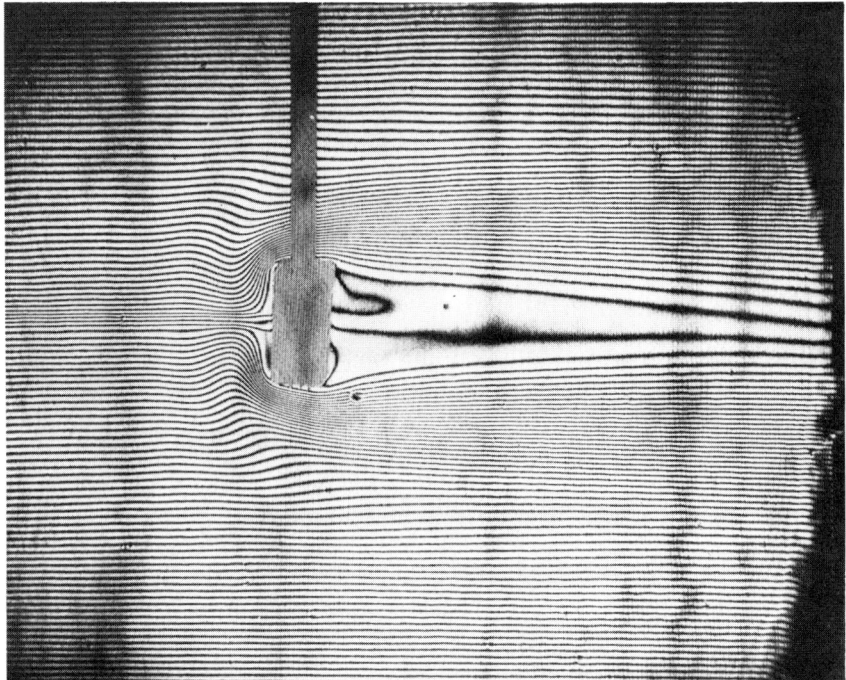

Figure 6.25 Holographic interferogram displaying streamlines of a density-stratified fluid as a rectangular bar is towed slowly from left to right. (ref. 104).

spacing is proportional to the velocity of the salt water. Experiments of this type have application to laboratory modeling of atmospheric phenomena. Witte[103] used this technique to visualize the wake of a model submarine. These applications illustrate the usefulness of phase error cancellation; the interferogram in Figure 6.25 was obtained with a large (1.22×0.15×0.31 m) test section made of ordinary plate glass.

6.4.2 Plasma Diagnostics

When matter is heated to a sufficiently high temperature, some electrons are excited to the point that they separate from the nucleus, thereby creating a *plasma*, that is, a collection of atoms, ions, and electrons. One important task in the experimental study of plasmas is to determine the spatial distribution of densities of these particles, particularly that of the electrons. The refractive index of a plasma is the sum of the refractive indices of the atoms, ions, and electrons weighted by their number densities. The refractive indices of the heavy particles (atoms and ions) are described by the Gladstone-Dale relation, equation 6.65, and are of the same order of magnitude. For example, the Gladstone-Dale constants associated with argon are $K_{ion} = \frac{2}{3} K_{atom}$. The values of K_{atom} and K_{ion} are only weakly dependent on the wavelength λ of the probing light.[105] The dependence of the refractive index on the number density of electrons, N_e (the number of electrons per unit volume), is quite different:[106]

$$n_e = \left(1 - \frac{N_e e^2 \lambda^2}{2\pi m_e c^2}\right)^{1/2},$$

where n_e is the refractive index of the electron gas, e is the charge and m_e the mass of an electron, and c is the speed of light. Evaluation of the constants in this equation gives

$$n_e = (1 - 8.92 \times 10^{-14} \lambda^2 N_e)^{1/2}, \tag{6.68}$$

where λ is given in centimeters and N_e in centimeters^{-3}. This equation can be approximated by

$$n_e - 1 = -4.46 \times 10^{-14} \lambda^2 N_e \tag{6.69}$$

if the deviation of n_e from unity is small.

Note that, unlike a neutral gas, the electron gas is very *dispersive*, that is, its refractive index is strongly dependent on λ. Further contrast between the contributions of heavy particles and electrons can be noted by evaluating equation 6.69 at a specific wavelength, say $\lambda = 546.3$ nm, and comparing it, as an example, with the Gladstone-Dale relation for argon expressed in terms of the number density of atoms, N_a:[107]

$$n_e - 1 = -13.33 \times 10^{-23} N_e, \tag{6.70}$$

$$n_a - 1 = +1.06 \times 10^{-23} N_a. \tag{6.71}$$

In addition to being much more dispersive, the contribution per electron is an order of magnitude greater than that per atom and is of the opposite sign. The electron gas therefore dominates the refractive index in moderately and highly ionized plasmas.

Because of the strong dispersion indicated by equation 6.69, measurements of refractive index at two different wavelengths enable one to determine the electron density directly:

$$n(\lambda_1) - n(\lambda_2) \cong n_e(\lambda_1) - n_e(\lambda_2)$$
$$= -4.46 \times 10^{-14} (\lambda_1^2 - \lambda_2^2) N_e. \tag{6.72}$$

This two-wavelength technique is discussed further below.

6.4 APPLICATIONS

Zaidel' and co-workers[108] and Jahoda and co-workers[109] described work on the application of holographic interferometry to plasma diagnostics that was carried out in 1966 and 1967. Since these initial efforts were reported, holographic interferometry has proved to be a valuable technique. Unlike many common diagnostic techniques such as spectroscopy and streak interferometry, it provides a full-field fringe pattern which is useful in forming a qualitative description of the plasma structure, as well as for measuring density distributions quantitatively. Phase error cancellation by two-exposure holographic interferometry can be very important because deposits often build up on test chamber windows. Phase error cancellation is especially important for studying plasmas confined in cylindrical tubes, since this is virtually impossible by classical interferomety. Pulsed-laser holography is compatible with the highly transient nature of many plasma events. Holographic interferograms can be recorded in a fairly wide variety of wavelengths, and good spatial resolution is possible. Nonlinear holographic interferometry provides a simple method of recording two-wavelength data for measuring electron number density.

Pulsed ruby lasers ($\lambda = 693.4$ nm) were used for most of the early holographic studies of plasmas. More recently, several different types of lasers have been used in order to extend the range of wavelengths of the probing radiation into the ultraviolet and infrared regions of the spectrum. One reason for this can be seen by examining equation 6.69. The change in refractive index due to a change in electron number density is proportional to λ^2. The fringe number, or shift, corresponding to a change in refractive index is proportional to λ^{-1}. Therefore the sensitivity of an interferometer to electron number density is proportional to λ, and longer wavelengths provide higher sensitivity. This is the opposite of the situation with nondispersive media, for which sensitivity is proportional to λ^{-1}. If single-wavelength, two-exposure holographic interferometry is used, the ratio of fringe shift due to electrons to that due to heavy particles is proportional to λ^2, so errors due to neglecting the heavy particles decrease rapidly with increasing wavelength. There are, however, restrictions on wavelength when electron number densities are high. Equation 6.68 indicates that there is a *critical electron number density* N_{cr}, at which the refractive index goes to zero:

$$N_{cr} = 1.12 \times 10^{13} \lambda^{-2}. \qquad (6.73)$$

Refraction will be very strong in regions of a plasma where N_e is close to N_{cr}. In regions where N_e exceeds N_{cr}, the probing light will not be transmitted by the plasma. Therefore in small, very dense plasmas with high gradients it may be necessary to use light with rather small wavelengths.

Wavelength-dependent transmittance of test chambers may also affect the choice of probing wavelength. An example of this is the study, by infrared holography, of a theta-pinch plasma in a quartz discharge tube reported by Kristal.[110] Quartz blocks the 10.6 μm light from CO_2 lasers, which are the most common source of coherent infrared radiation. This problem was circumvented by using a hydrogen fluoride (HF) chemical laser system with $\lambda = 3$ μm.

Some types of lasers used in plasma diagnostics are listed in Table 6.3, along with their wavelengths and the corresponding critical electron number densities. Light from ruby lasers is often frequency doubled, and holographic interferometry using frequency-tripled light from a neodynium (Nd) laser has been used for plasma diagnostics.[111] This method of decreasing the wavelength of high-power laser radiation is accomplished by passing the light through an optically nonlinear material, most commonly a potassium dihydrogen phosphate (KDP) crystal. The electric polarization of such materials responds nonlinearly to the electric field of the incident light, thereby generating harmonics.[112] Dreiden et al.[113] used 765.8 nm light for holographic interferometry of a plasma. This light was generated by stimulated Raman scattering of ruby laser light in nitrobenzene.

In 1966–1967 several aspects of holographic analysis of plasmas were reported in the Soviet literature and were later reviewed by Zaidel' et al.[108] In particular, both Gabor-type[119, 120] holography and off-axis[121] holography were used to study laser-induced spark discharges. In the United States, Jahoda et al.[109] applied diffuse-illumination interferometry to examine a theta-pinch plasma, and noted that wedge reference fringes could be introduced in order to make subfringe measurements. Two-exposure holographic interferometry is now used in a rather routine manner as a

TABLE 6.3 LASERS USED FOR HOLOGRAPHIC PLASMA DIAGNOSTICS

Type of laser	Wavelength	N_{cr}	Recording medium	Reference
Ruby	694 nm	2×10^{21}	Photographic film	
Frequency doubled	347 nm	9×10^{21}	Photographic film	114
Nd	1.06 μm	1×10^{21}	Bismuth film	115
Frequency doubled	530 nm	4×10^{21}		
Frequency tripled	354 nm	6×10^{21}	Photographic film	111
Yag: Nd^{3+}	1.06 μm	1×10^{21}	M_nB_i film	118
HF chemical	3 μm	1×10^{20}	Bismuth film	110
CO_2	10.6 μm	1×10^{19}	Bismuth film	116
CO_2	10.6 μm	1×10^{19}	Plexiglas	117

6.4 APPLICATIONS

plasma diagnostic technique.[122-130] Once the potential of holography for plasma diagnostics was recognized, various schemes were developed which utilize specific unique properties of holographic interferometry. A good example is its application to the two-wavelength method for determining electron number density.

As indicated by equation 6.72, if the spatial distribution of the refractive index of a plasma can be measured at two separate wavelengths, λ_1 and λ_2, N_e can be determined without error that is due to the refractive index of heavy particles. A system for doing this is shown schematically in Figure 6.26. Light from a pulsed ruby laser ($\lambda_1 = 694.3$ nm) passes through a frequency doubler to produce two colinear beams of wavelengths $\lambda_1 = 694.3$ nm and $\lambda_2 = 347.1$ nm, which enter a holographic interferometer. Since λ_1 and λ_2 differ by a factor of 2, they are not coherent with each other; therefore separate holograms are recorded in each wavelength during a single exposure with this system. Two sequential exposures are recorded on the film plate—one with the plasma ignited, and one with no plasma. The resulting hologram can be illuminated with a continuous wave He-Ne laser whose wavelength $\lambda_3 = 632.8$ nm is nearly equal to λ_1. Since the mean spatial frequency of the hologram recorded with λ_2 is twice that of the hologram recorded with λ_1, the wavefronts reconstructed from the λ_2 hologram will propagate at an angle of approximately θ with respect to the waves reconstructed from the λ_1 hologram (Figure 6.27a). The two interferograms are therefore spatially separated and can be photographed and analyzed individually.[114, 131] Figure 6.28 shows two interferograms of this

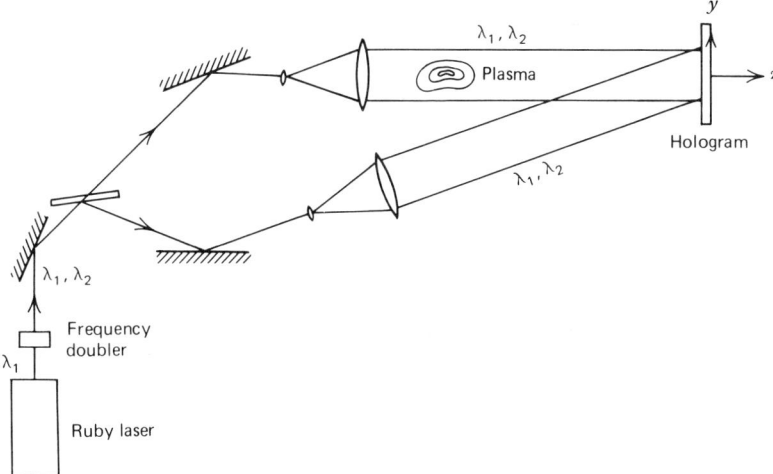

Figure 6.26 Holographic system for recording two-wavelength, two-exposure interferograms of dispersive plasmas. This system is used for measuring electron density distributions.

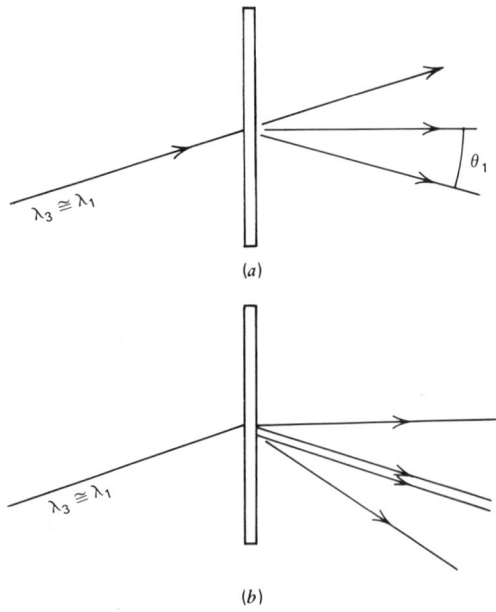

Figure 6.27 Reconstruction of waves from a two-wavelength holographic interferogram. (*a*) The hologram is linear. (*b*) The hologram is nonlinear.

type obtained from the same hologram. Note that the fringe shift is greater in the hologram recorded at the higher wavelength. The plasma was created by breakdown of a helium atmosphere due to a 0.2 J, 30 ns pulse of focused radiation from a ruby laser.[132] A similar approach was taken by Jeffries;[133] however, he used separate reference beams for each wavelength in order to accurately adjust the exposure levels. Two-exposure, two-wavelength holographic interferometry has been applied to the study of laser-induced air breakdown by Komissarova et al.[131, 134] Seftor[135] used the method to determine the distribution of N_e around exploding aluminum wires in a vacuum.

Nonlinear holography can be used to produce an interferogram with fringes which are contours of constant difference in refractive index $n(\lambda_1) - n(\lambda_2)$. By this method one avoids working with two separate interferograms; this is desirable because N_e is the difference between two large numbers and therefore is sensitive to small errors in fringe position. A single-exposure, two-wavelength hologram is recorded using a system like that shown in Figure 6.26. The exposure and development procedure are chosen so that a nonlinear hologram is produced (see Section 5.5). The

6.4 APPLICATIONS

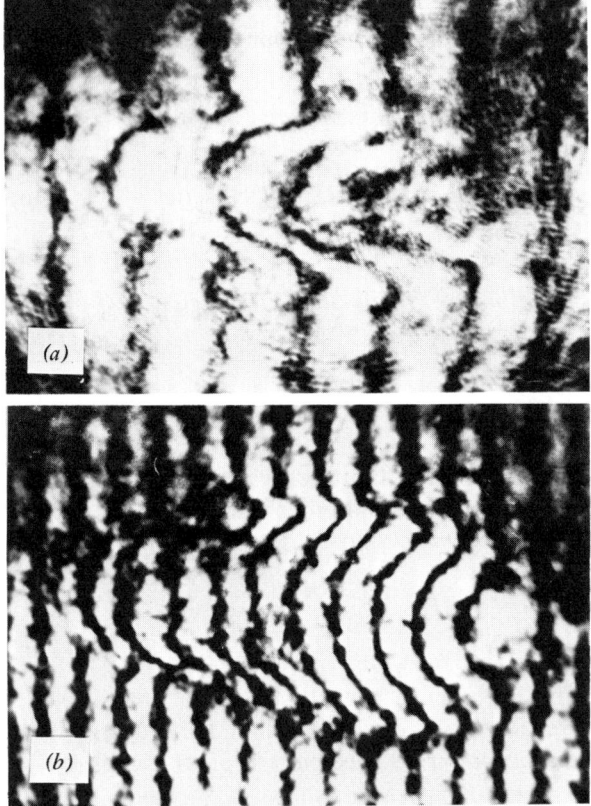

Figure 6.28 Holographic interferograms of a plasma recorded with light of different wavelengths (ref. 132). (a) $\lambda = 694.3$ nm. (b) $\lambda = 347.1$ nm.

transmittance of this hologram can be determined using equation 5.73:

$$t(x,y) = \sum_{n=-\infty}^{\infty} C_n \exp\left\{in\left[\frac{2\pi}{\lambda_1}\sin(\theta)y - \phi_{\lambda_1}(x,y)\right]\right\} + \sum_{n=-\infty}^{\infty} C_n$$

$$\times \exp\left\{in\left[\frac{2\pi}{\lambda_2}\sin(\theta)y - \phi_{\lambda_2}(x,y)\right]\right\}, \quad (6.74)$$

where ϕ_{λ_1} and ϕ_{λ_2} are the phase distributions due to the plasma at wavelengths λ_1 and λ_2, respectively. The desired interferogram is obtained by illuminating this hologram with a single plane wave of He-Ne laser light at

$\lambda_3 \cong \lambda_1$ (see Figure 6.27b). The reconstruction wave has the following complex amplitude at the hologram plane:

$$U_R(x,y) = \exp\left[-i\frac{2\pi}{\lambda_3}\sin(\theta)y\right].$$

Consider the following two terms from the series which result when $t(x,y)$ is multiplied by U_R:

$$C_1 \exp\left\{-i\left[2\pi\left(\frac{1}{\lambda_3}-\frac{1}{\lambda_2}\right)\sin(\theta)y + \phi_{\lambda_2}(x,y)\right]\right\}$$
$$+ C_2 \exp\left\{-i\left[2\pi\left(\frac{1}{\lambda_3}-\frac{2}{\lambda_1}\right)\sin(\theta)y + 2\phi_{\lambda_1}(x,y)\right]\right\}.$$

These terms represent the first-order wavefront recorded with light of wavelength λ_2 and the second-order wavefront recorded with light of wavelength λ_1, respectively. Since $\lambda_3 \cong \lambda_1 = 2\lambda_2$, this reduces to

$$C_1 \exp\left[\frac{i2\pi\sin(\theta)y}{\lambda_1}\right] \cdot \left\{\exp\left[-i\phi_{\lambda_2}(x,y)\right] + \exp\left[-i2\phi_{\lambda_1}(x,y)\right]\right\}$$

if $C_1 = C_2$. This expression shows that both of these wavefronts travel in the same direction (Figure 6.27b). The interferogram formed by these wavefronts displays an irradiance pattern proportional to

$$I = 1 + \cos\left[\phi_{\lambda_2}(x,y) - 2\phi_{\lambda_1}(x,y)\right]. \tag{6.75}$$

Since

$$\phi_{\lambda_1} = \frac{2\pi}{\lambda_1}\int n(\lambda_1)\,ds$$

and

$$\phi_{\lambda_2} = \frac{2\pi}{\lambda_2}\int n(\lambda_2)\,ds, \tag{6.76}$$

we see that for $\lambda_1 = 2\lambda_2$ the phase difference in equation 6.75 is

$$\phi_{\lambda_2} - 2\phi_{\lambda_1} = \frac{2\pi}{\lambda_2}\int\left[n(\lambda_2) - n(\lambda_1)\right]ds. \tag{6.77}$$

6.4 APPLICATIONS

Equation 6.77 indicates that the interference fringes are contours of constant difference in refractive index at wavelengths λ_1 and λ_2, which are precisely the data required for evaluation of equation 6.72. This technique was suggested by Ostrovskaya and Ostrovskii[136] and demonstrated by Ignatov et al.[137] for a laser-induced air breakdown. Radley[138] used this technique to determine the electron density distribution in a stationary plasma produced by focusing a 6.2 kW continuous wave CO_2 laser in air.

Since the time scales of some plasma events, as well as of probing pulses, are very short—on the order of nanoseconds or picoseconds—simple optical delay lines can be used to accomplish timing or sequencing for holographic interferometry. One of the early systems used by Zaidel' et al.[120] is shown schematically in Figure 6.29. This system was used to record time-resolved Gabor holograms of a plasma created by laser-induced air

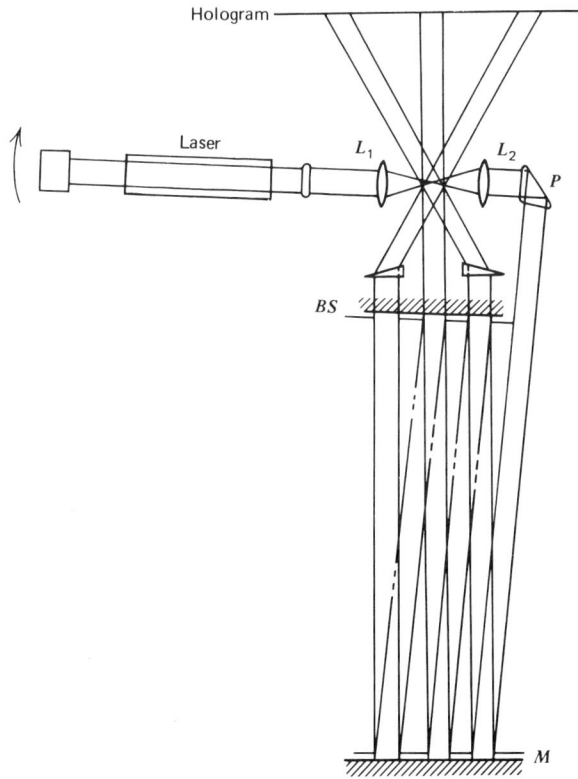

Figure 6.29 System for recording a sequence of time-resolved Gabor holograms of a plasma (Ref. 120).

breakdown. The plasma was produced at the point where light from the Q-switched ruby laser is focused between lenses L_1 and L_2. After this beam is re-expanded, it is directed by prism P into the holographic system to serve as diagnostic illumination. The optical delay $\tau = 2l/c$ results each time the beam travels from the beamsplitter BS to the mirror M and back to an aperture, where it enters the test region to form a hologram. The system shown in Figure 6.29 produces a sequence of three holograms separated by time intervals of $2l/c$; in the experiments reported in ref. 120 this delay was 40 ns. This approach has been adapted to the holographic interferometry of plasma.[121, 139]

A different approach to the sequencing of holographic interferograms was reported by Landry and McCarthy.[140] They used a four-cavity laser to produce four consecutive pulses. A beam directing prism selected different volumes of the lasing medium (ruby rod), each of which had its own Q-switch shutter. The interferograms were formed with diffuse illumination on a single photographic plate. Each hologram was recorded with a different reference beam direction, so that it could be reconstructed individually. The minimum time interval between exposures was controllable to within 25 ns, and the maximum interval was 1.25 ms. The system was used to study exploding bridge wires. The sequencing made it possible to observe a shock front moving at approximately 0.1 cm/μs.

Separate lasers can be used to obtain a sequence of holographic interferograms. This was done by Thomas et al.[141] for the analysis of a theta-pinch plasma. Armstrong and Forman[142] reported a system using two Pockles cells to produce two colinear but orthogonally polarized beams to obtain holograms with temporal separations of 0.2 to 100 μs. Real-time holographic interferometry has also been applied for time-resolved analysis of plasmas.[143, 144]

Kelly and Mix[145] have suggested that fringe visibility can be used to estimate plasma velocities. If the plasma moves a distance equal to a fraction of a fringe spacing during the exposure of a holographic interferogram, the visibility of the fringe will be reduced. If the pulse shape of the laser light is known accurately, the reduction in fringe visibility can be calculated as a function of plasma velocity. Similarly, a double-spiked pulse could be used.

With current technology temporal resolutions in the subnanosecond[146] and picosecond[111, 147] range have been reported. Spatial resolutions sufficient for studying plasmas with length scales of 100 μm, which give rise to fringe spacings of about 2 μm, are attainable with holographic interferometry.[148] Current research efforts are aimed at improving both spatial and temporal resolution[149] in order to extend the range of plasma phenomena accessible to holographic diagnostics.

6.4.3 Heat and Mass Transfer

Interferometry is used in heat and mass transfer experiments to determine the spatial distribution of temperature or concentration of chemical species. Let us first consider the relation between temperature and refractive index. The refractive index of a gas is given by the Gladstone-Dale relation, equation 6.65. In most circumstances the density of a gas can be calculated using the ideal gas equation of state:

$$\rho = \frac{MP}{RT}, \tag{6.78}$$

where P(Pa) is the pressure, M is the molecular weight of the gas, $R = 8.3143$ J/mol K is the universal gas constant, and T (°K) is the absolute temperature. Combining equations 6.65 and 6.78 yields

$$n - 1 = \frac{KMP}{RT}. \tag{6.79}$$

The slope of a curve of refractive index versus temperature is therefore

$$\frac{dn}{dT} = \frac{-KMP}{RT^2}. \tag{6.80}$$

If temperature changes in a given experiment are small, the right-hand side of equation 6.80 will be approximately constant, so a linear relation between change of refractive index and change of temperature results. For example, for air at 288°K and 0.1013 MPa, the Gladstone-Dale constant at $\lambda = 632.8$ nm is 0.226×10^{-3} m³/kg, and its molecular weight is 28.97. Therefore

$$\frac{dn}{dT} = -0.961 \times 10^{-6} K^{-1}.$$

Greater precision can be attained by using the following relation for the refractive index of air at 632.8 nm:[150]

$$n - 1 = \frac{0.292015 \times 10^{-3}}{1 + 0.368184 \times 10^{-2} T} \tag{6.81}$$

or, at 514.5 nm,

$$n - 1 = \frac{0.294036 \times 10^{-3}}{1 + 0.369203 \times 10^{-2} T}, \tag{6.82}$$

where T is in degrees Celsius. These equations are based on the Gladstone-Dale relation with wavelength dependence calculated according to Meggers and Peters,[151] and include small corrections introduced by Tilton.[152]

Refractive indices of liquids are related to density by the Lorentz-Lorenz equation:

$$\frac{n^2-1}{\rho(n^2+2)} = \bar{r}(\lambda), \qquad (6.83)$$

where $\bar{r}(\lambda)$ is the *specific refractivity* and is a function of the substance and wavelength of light. The Gladstone-Dale relation, equation 6.65, is a first-order approximation to equation 6.83 valid only for gases. The density of liquids is not a simple function of temperature, as in the case of gases, so empirical relations must be found between n and T. Schödel[153] has determined the slope dn/dT for several liquids at 298°K. These values, as reported by Hauf and Grigull,[154] are given in Table 6.4, along with measurements by Coumou et al.[155] Empirical relations for the refractive index of water are given by Tilton and Taylor;[156] however, these relations are quite complicated and include 13 empirical constants. A simpler, yet

TABLE 6.4 CHANGE IN REFRACTIVE INDEX OF LIQUIDS WITH TEMPERATURE AT 298°K

Liquid	$-dn/dT$ (°K^{-1})		Reference
	$\lambda = 546.1$ nm	$\lambda = 632.8$ nm	
Water	1.00×10^{-4}	0.985×10^{-4}	154
Methyl alcohol	4.05×10^{-4}	4.0×10^{-4}	154
Ethyl alcohol	4.05×10^{-4}	4.0×10^{-4}	154
Isopropyl alcohol	4.15×10^{-4}	4.15×10^{-4}	154
Benzene	6.42×10^{-4}	6.40×10^{-4}	154
Toluene	5.55×10^{-4}	5.55×10^{-4}	154
Nitrobenzene	4.68×10^{-4}	4.68×10^{-4}	154
c-Hexane	5.46×10^{-4}	5.43×10^{-4}	154
n-Hexane	5.43×10^{-4}	5.4×10^{-4}	154
n-Octane	4.76×10^{-4}		155
n-Decane	4.48×10^{-4}		155
n-Hexadecane	4.06×10^{-4}		155
Isooctane	4.87×10^{-4}		155
Acetone	5.31×10^{-4}	5.31×10^{-4}	154
Chloroform	5.98×10^{-4}	5.98×10^{-4}	154
Carbon tetrachloride	5.99×10^{-4}	5.98×10^{-4}	154
Carbon disulfide	7.96×10^{-4}	7.96×10^{-4}	154

quite accurate, equation for the refractive index of water at 632.8 nm[150] is

$$n - 1.3331733 = -(1.936T + 0.1699T^2) \times 10^{-5} \tag{6.84}$$

or, for 514.5 nm,

$$n - 1.337253 = -(2.8767T + 0.14825T^2) \times 10^{-5}, \tag{6.85}$$

where T is in degrees Celsius. These equations are a modification of the Dobbins and Peck[157] quadratic fit to the Tilton-Taylor data.

Murphy and Alpert[158] have demonstrated that the following relation gives values of dn/dT accurate to within about 2.0 percent:

$$\frac{dn}{dT} = -\frac{3}{2}\left[\frac{n(n^2-1)}{2n^2+1}\right]\beta, \tag{6.86}$$

where β is the coefficient of thermal expansivity of the liquid. Equation 6.86 is useful for estimating dn/dT when limited data are available.

The advantages and characteristics of holographic interferometry as applied to heat transfer measurements are the same as those discussed for aerodynamic measurements. The most useful characteristics are phase error cancellation, permitting the use of low-quality test section windows, and the application of multidirectional interferometry for measurement of asymmetric temperature fields. Zinnes[159] demonstrated the simplicity and usefulness of holographic interferometry for temperature measurement in a study of the coupling of conduction and natural convection from a vertical surface. Two-exposure interferograms were used to measure the temperature distribution in air near nonuniformly heated glass and ceramic plates in order to verify theoretical predictions. Aung and O'Regan[160] made detailed measurements of the thermal boundary layer near an isothermal vertical plate. Their results were compared with accepted analytical expressions and indicate that holographic interferometry can be used to make precision heat transfer measurements. The accuracy was found to be equal to that of Mach-Zehnder interferometry, although the fringes formed holographically have somewhat rougher profiles due to laser speckle and the effects of film grain and emulsion shrinkage. Holographic interferometers generally are less expensive and easier to adjust than Mach-Zehnder interferometers for applications of this type. Schmitt and Beer[161] have used holographic interferometry in a study of natural convection heat transfer from a vertical plate to carbon dioxide near its critical state. Real-time holographic interferometry has also been employed to study developing natural convection between parallel plates.[162]

The use of holographic interferometry for temperature measurement in a rather complex configuration was demonstrated by Panknin et al.[163] They studied the transition from laminar to turbulent flow in closely spaced tube bundles, which are used in heat exchangers. Three of the heated tubes are shown schematically in Figure 6.30a. The axis of the plane object wave was parallel to the 82-cm-long tube bundle. The end windows of the test section were manufactured of ordinary glass. The holographic exposure was recorded, using an argon ion laser, and developed *in situ* so that real-time interferometry could be applied. Figure 6.30b shows two interferograms corresponding to different tube spacings and flow Reynolds numbers. The tubes are 78 mm in diameter, and the interferometric sensitivity is about 1.5°K per fringe.

Another application to the study of forced convection was reported by Antonini et al.[164] They used real-time holographic interferometry to study convection from an oscillating heated wire.

The phase-error cancellation property permits one to record interferograms of gases contained in chambers of complicated shape. Several investigators have taken advantage of this property to record interferograms of the heated gas in various types of light bulbs.[165-167] Figure 6.31 is a two-exposure, diffuse-illumination interferogram of a 100 W domestic light bulb. Interferograms like this were used by Fitzgerald and Hörster[168] to estimate temperature distributions as a function of pressure in argon-filled lamps.

Temperature rises in irradiated media are indicative of the amount of radiant energy that has been absorbed. Hussmann[169] showed that holographic interferometry can be used to measure temperature profiles, and therefore energy deposition profiles, when a portion of an electron beam is absorbed by water. Absorption of laser radiation, gamma rays, and so on can also be measured by this calorimetric technique.

Sweeney and Vest[170] and Radulovic and Vest[171] have used multidirectional holographic interferometry to measure three-dimensional temperature distributions in asymmetric natural convection plumes. These

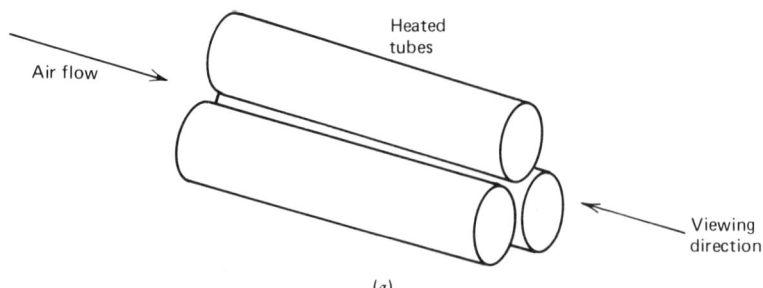

Figure 6.30 Holographic interferometry of the temperature field in air near closely spaced tube bundles. (a) Configuration of tubes. (b) Interferograms at two different Reynolds numbers and tube spacings (ref. 163).

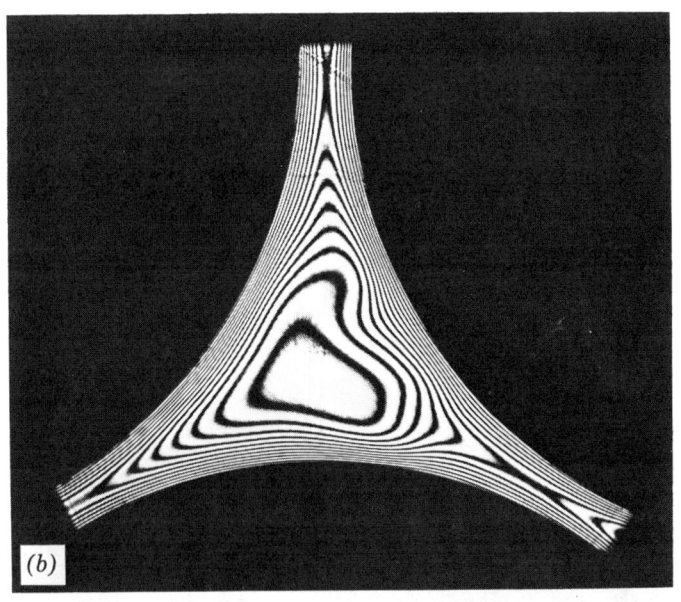

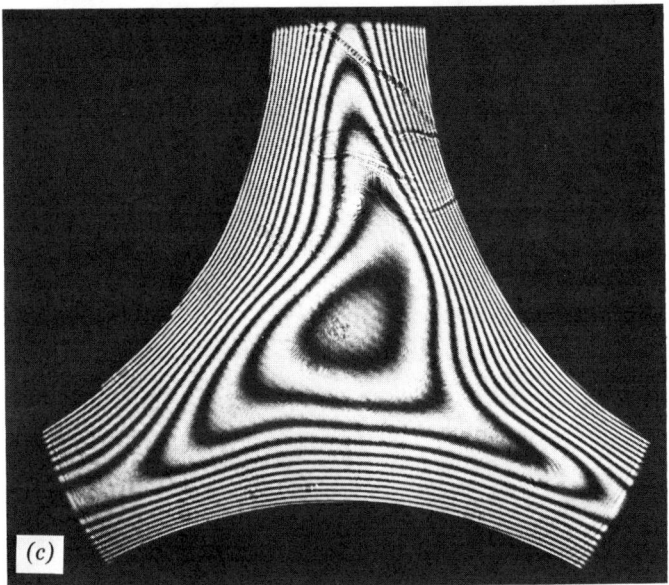

Figure 6.30 Cont.

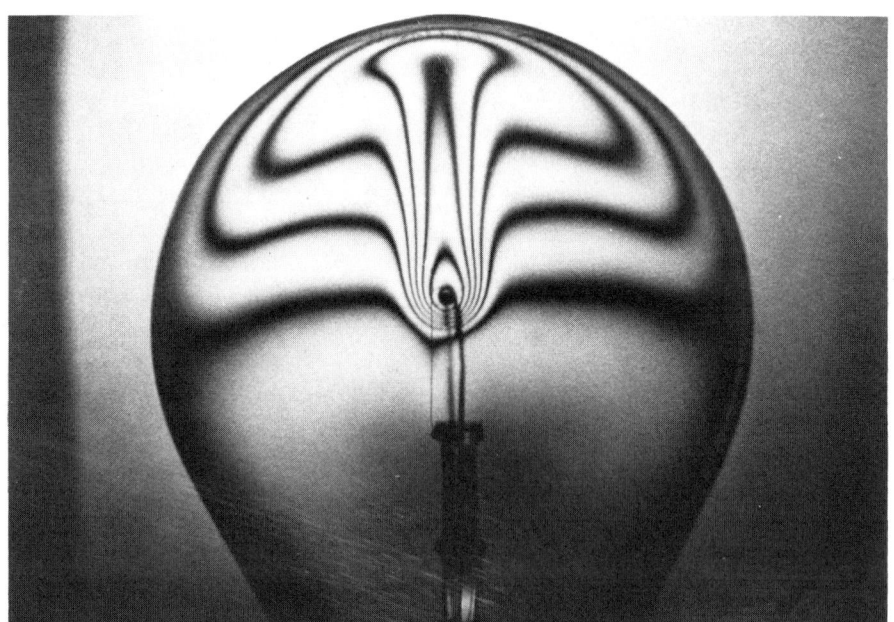
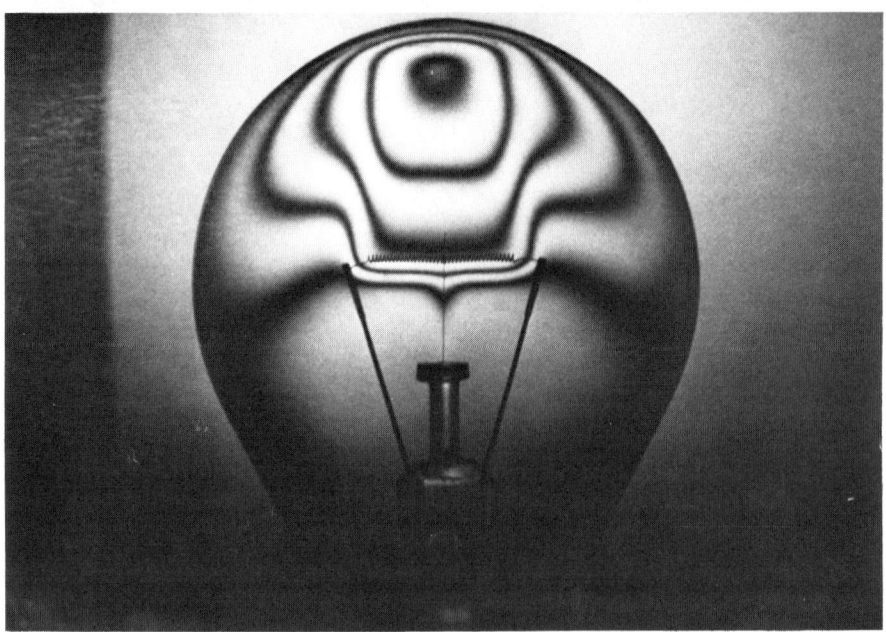

Figure 6.31 Interferograms of the temperature field in a 100 W domestic light bulb (ref. 168).

6.4 APPLICATIONS   369

measurements were made using an interferometer of the type depicted in Figure 5.22. This interferometer uses a phase grating to analyze the object wave into several plane waves which pass through the test section in different directions. The test section was a rectangular tank, manufactured from ordinary glass, and filled with distilled water. The thermal plumes rose above two electrically heated horizontal metal disks. Two views of a holographic interferogram recorded with this system are shown in Figures 6.32a and 6.32b. These interferograms were recorded with viewing directions separated by 23°. Data obtained from a sequence of interferograms

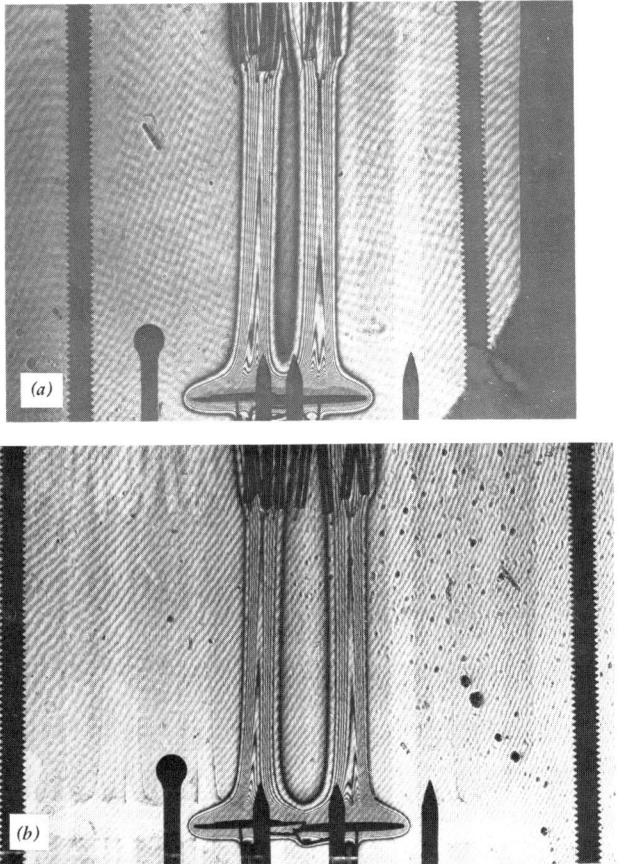

Figure 6.32 Multidirectional holographic interferometry of an asymmetric thermal plume rising in water above two heated disks (ref. 171). (a) Interferogram viewed from one direction. (b) Interferogram recorded when the same hologram is viewed from a different direction. The two directions differ by 23°. (c) The reconstructed temperature field in a horizontal plane 3.0 cm above the disks. The dots labelled t.j. denote location of thermocouple junctions.

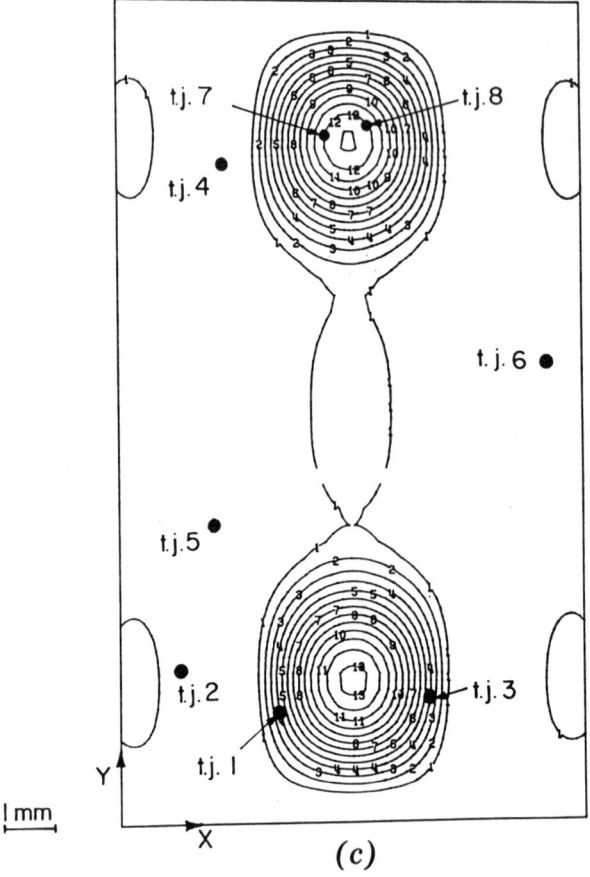

Figure 6.32 Cont.

of this type were analyzed by a method of the kind discussed in Section 6.2. Typical results of these computations are displayed in Figure 6.32c as isothermal contours in a horizontal plane 3 cm above the heated disks. Comparison with several thermocouple measurements indicated that the temperature fields were determined accurately to within about ±1°C by this method.

When interferometry is applied to *mass transfer* experiments, the objective is to determine the spatial distribution of concentration of each component present in a gaseous or liquid mixture. In the case of gases, equation 6.66 relates the Gladstone-Dale constant of a mixture to the mass fraction of each component. The mass fraction a_i of the ith component is equal to ρ_i/ρ, where ρ_i is the *mass concentration* of the ith component and

6.4 APPLICATIONS

ρ is the density of the mixture. For an *isothermal* binary mixture of components a and b, equations 6.65 and 6.66 can be combined to give

$$n - 1 = \rho_a K_a + \rho_b K_b.$$

If the diffusion of substance b into substance a is to be studied, the first exposure of a two-exposure hologram will be recorded while the test chamber is filled with pure substance a (e.g., air) at temperature T and pressure P. During this first exposure the refractive index n_0 will be uniform throughout the test chamber. The second exposure will be recorded during the mass transfer process. The resulting interferometric fringes can be interpreted by the usual methods (Section 6.2) to determine the change of refractive index, given by

$$\Delta n = n - n_0 = \rho_a K_a + \rho_b K_b - \rho_0 K_a \tag{6.87}$$

where ρ_0 is the original density of substance a. The relations governing mixtures of ideal gases can be used to re-express equation 6.87 in the form

$$\Delta n = \left(K_b - \frac{M_a}{M_b} K_a \right) \rho_b, \tag{6.88}$$

where M_a and M_b are the molecular weights of substances a and b, respectively. Equation 6.88 directly relates the concentration of substance b to the change of refractive index.

If both the temperature and the concentration change between holographic exposures, analysis of interferograms is more complicated; however, nonisothermal processes in multicomponent gases are technologically important. If the temperature of a binary ideal gas mixture changes from its initial value T_0 to T, equation 6.88 must be replaced by the more general expression

$$\Delta n = \frac{\rho_b}{M_b}(M_b K_b - M_a K_a) + \frac{M_a K_a P}{R}\left(\frac{1}{T} - \frac{1}{T_0} \right). \tag{6.89}$$

Equation 6.89 has two unknowns, T and ρ_b. It is possible to apply two-wavelength interferometry to obtain a second equation since the Gladstone-Dale constants depend on wavelength. This technique was developed for classical interferometers by El-Wakil[172] and adapted to holographic interferometry by Mayinger and Panknin.[173] Interferograms are recorded at two different wavelengths, λ_1 and λ_2. These frequencies should be separated as widely as possible because gases are only slightly dispersive. If we denote the changes of refractive index in each wavelength by Δn_{λ_1} and Δn_{λ_2}, then

$$\Delta n_{\lambda_1} = \frac{\rho_b}{M_b}\left(M_b K_{b_{\lambda_1}} - M_a K_{a_{\lambda_1}}\right) + \frac{M_a K_{a_{\lambda_1}} P}{R}\left(\frac{1}{T} - \frac{1}{T_0}\right), \quad (6.90)$$

$$\Delta n_{\lambda_2} = \frac{\rho_b}{M_b}\left(M_b K_{b_{\lambda_2}} - M_a K_{a_{\lambda_2}}\right) + \frac{M_a K_{a_{\lambda_2}} P}{R}\left(\frac{1}{T} - \frac{1}{T_0}\right). \quad (6.91)$$

We can eliminate ρ_b/M_b from equations 6.90 and 6.91 to obtain

$$\frac{\Delta n_{\lambda_1}}{M_b K_{b_{\lambda_1}} - M_a K_{a_{\lambda_1}}} - \frac{\Delta n_{\lambda_2}}{M_b K_{b_{\lambda_2}} - M_a K_{a_{\lambda_2}}}$$
$$= \left(\frac{K_{a_{\lambda_1}}}{M_b K_{b_{\lambda_1}} - M_a K_{a_{\lambda_1}}} - \frac{K_{a_{\lambda_2}}}{M_b K_{b_{\lambda_2}} - M_a K_{a_{\lambda_2}}}\right)\frac{M_a P}{R}\left(\frac{1}{T} - \frac{1}{T_0}\right). \quad (6.92)$$

Once Δn_{λ_1} and Δn_{λ_2} are determined from the two interferograms, the temperature distribution can be calculated using equation 6.92. After the distribution of T is known, the concentration ρ_b can be calculated using either equation 6.90 or 6.91. The two-wavelength method is susceptible to small errors because the Gladstone-Dale constants are only weak functions of wavelength. It is necessary to maintain careful registration between the two interferograms. This is accomplished easily if both interferograms are recorded simultaneously on the same holographic plate. Mayinger and Panknin[173] used an He-Ne laser ($\lambda_1 = 632.8$ nm) and an argon laser ($\lambda_2 = 457.9$ nm) to record separate holographic interferograms on a single plate, and applied the data to calculate both temperature and concentration profiles in a laminar boundary layer. The boundary layer was formed by natural convection of air past a heated vertical plate from which naphthalene evaporated.

The techniques of holographic interferometry developed for heat and mass transfer measurements should be applicable to the study of combustion processes.[174, 175] To date, however, very little appears to have been done in this area. The application of holographic interferometry to quantitative combustion diagnostics should be a fruitful topic for future development, as demonstrated by Varde's study of flame propagation in a stratified mixture.[176]

Holographic interferometry has also been used to determine diffusion coefficients for isothermal mass transfer in liquids.[177-180] Real-time interferometry was applied to determine concentration profiles as a function of time as various substances diffused into pure water. An analytical solution was known for the process, but involved a diffusion coefficient whose value was unknown. This coefficient was determined by fitting the interferomatic data, using a least-mean-square technique, to the known form of solution.

Pipman et al.[181] have used holographic interferometry to study mass diffusion at very low temperatures. They recorded interferograms which show concentration gradients due to interdiffusion of the helium isotopes ^{3}He and ^{4}He at 0.4°K.

A detailed description of a system for precise measurement of diffusion coefficients has been given by Durou.[182] It employs real-time holographic interferometry to observe changing fringe patterns over periods up to 24 h. To maintain the necessary thermal and mechanical stability, the system is enclosed and vibration isolated. Certain optical adjustments can be made by remote control, and the fringes are monitored by a television system which images the plane within the diffusion cell, where refraction errors are minimized according to equation 6.52. The isothermal diffusion coefficient for potassium chloride in water was measured with errors of less than 1 percent. An alternative, and perhaps simpler, technique for determining diffusion coefficients has been described by Bochner and Pipman.[183] They suggest the use of time-differential holographic interferometry, which permits a very direct determination of the diffusion coefficient if it can be assumed to be independent of concentration.

6.4.4 Stress Analysis

Holographic interferometry can be applied to stress analysis of transparent solids. The optical pathlength of light passing through a solid changes when the object is stressed. There are two reasons for this: the thickness of the model changes, and its refractive index is a function of its state of stress (*stress-optical effect*). We consider here the measurement of the transverse strain ϵ_z of a planar object. A thin test specimen is depicted in Figure 6.33; it has thickness t and is subjected to a tensile force F. The specimen is in a state of plane stress since all stresses lie in the x-y plane. Because of the Poisson effect, however, the material will have a strain ϵ_z in the z direction, where

$$\epsilon_z = \frac{\delta t}{t} \qquad (6.93)$$

is the change in thickness divided by the initial thickness. This transverse strain in an elastic material is related to the stress field:

$$\epsilon_z = \frac{-\nu}{E}(\sigma_1 + \sigma_2). \qquad (6.94)$$

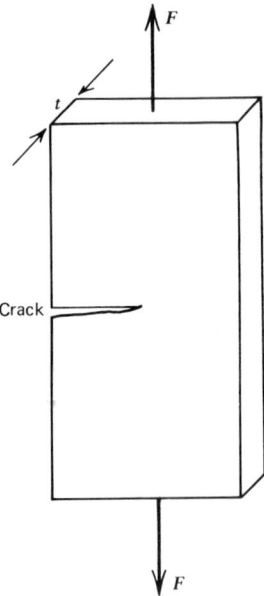

Figure 6.33 Thin specimen of transparent material subjected to a tensile load. This specimen is used to study the strain field near the tip of the crack which extends into the specimen from the left edge.

Here σ_1 and σ_2 are the principal stresses, which are mutually orthogonal and lie in the x-y plane. The objective is to determine the magnitude of the transverse strain at each point in the specimen.

Suppose that the specimen depicted in Figure 6.33 is placed in a holographic interferometer like that shown in Figure 5.9. The plane object wave traverses the object in the z direction. If two holographic exposures are recorded, one before stressing the specimen and one after the stress is applied, the difference in optical pathlength at any point between exposures is

$$\Delta\Phi = n(t + \delta t) - n_0 t - 1 \cdot \delta t, \tag{6.95}$$

where n is the refractive index of the stressed material, n_0 is its refractive index in its initial unstressed state, and the refractive index of the ambient air is 1.

The refractive index n is related to the state of stress by the two-dimensional *Maxwell-Neumann stress-optic law*:[184]

$$n_1 - n_0 = A\sigma_1 + B\sigma_2,$$
$$n_2 - n_0 = B\sigma_1 + A\sigma_2, \tag{6.96}$$

6.4 APPLICATIONS

where n_1 is the refractive index for light polarized in the direction of σ_1, and n_2 is the refractive index for light polarized in the direction of σ_2. Also, A and B are the stress-optic coefficients of the material. We are concerned here with materials of low stress-optic sensitivity so that $A \cong B$ and $n_1 \cong n_2$. For such optically isotropic materials equation 6.96 is replaced by

$$n - n_0 = A(\sigma_1 + \sigma_2). \tag{6.97}$$

Equations 6.93 and 6.97 can be combined with equation 6.95 to give the following expression for the optical pathlength change:

$$\Delta\Phi = \{\epsilon_z[n_0 - 1 + A(\sigma_1 + \sigma_2)] + A(\sigma_1 + \sigma_2)\}t.$$

The sum $(\sigma_1 + \sigma_2)$ is related to ϵ_z by equation 6.94, so

$$\Delta\Phi = \left[\epsilon_z\left(n_0 - 1 - A\frac{E}{\nu}\epsilon_z\right) - A\frac{E}{\nu}\epsilon_z\right]t.$$

The term $[A(E/\nu)\epsilon_z^2]$ is small compared with the other terms in this equation and can be neglected. Therefore

$$\Delta\Phi = \left[\left(n_0 - \frac{E}{\nu}A\right) - 1\right]t\epsilon_z. \tag{6.98}$$

The quantity $[n_0 - (E/\nu)A]$ can be considered to be the effective refractive index of the material. The optical pathlength change is related to fringe order number N by the relation $\Delta\Phi = N\lambda$, so

$$\epsilon_z = \frac{N\lambda}{\{[n_0 - (E/\nu)A] - 1\}t}. \tag{6.99}$$

Equation 6.99 permits a direct quantitative determination of transverse strain from an interferogram.

This method has been applied by Dudderar and O'Regan[185, 186] to experimental fracture mechanics. In particular, they measured the strain field near the tip of a crack in a tensile specimen like that shown in Figure 6.33. The specimens they studied were made from polymethylmethacrylate (PMM) and had thickness ranging from 0.7 to 13 mm. Figure 6.34 shows a typical interferogram from this study. The fringes are compared with theoretical contours of constant ϵ_z. This work was extended to a detailed study of stress intensity factors[187, 188] and effective crack lengths.[189] Soga et al.[190] have used two-exposure holographic interferometry to determine strains in optical windows installed in a high-pressure test chamber. Although the determination of transverse strain has been discussed here in

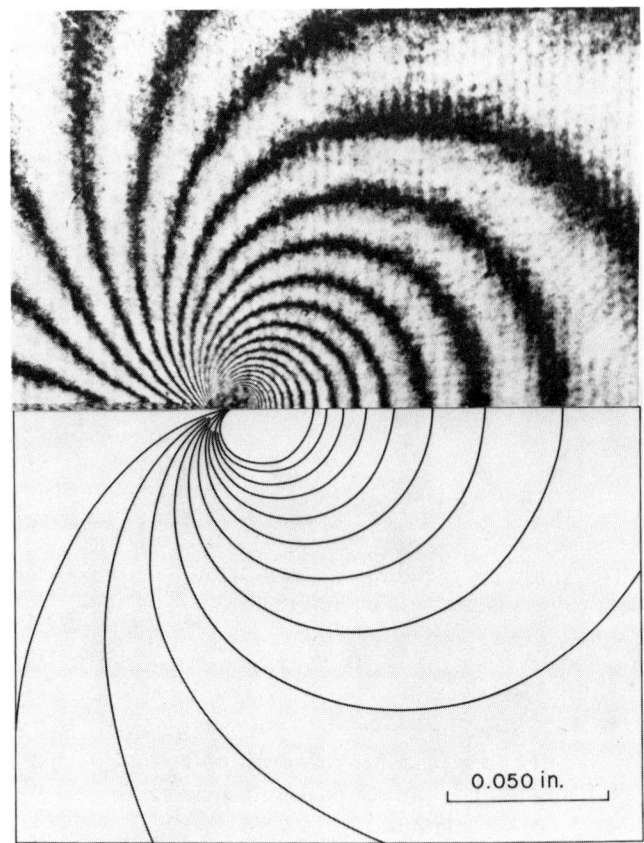

Figure 6.34 Comparison of interferogram and theoretical contours of constant transverse strain (ref. 186). The test specimen is a 3.2-mm-thick plate of PMM subjected to a tensile load equal to 96 percent of the load which causes fracture.

terms of two-exposure holographic interferometry, real-time interferometry can also be used. In fact, it may be advantageous in that the interferometer can be adjusted to minimize the number of unwanted fringes due to large displacement of the test specimen.[191] The disadvantage of the real-time approach is that the model must have very flat, parallel faces; however, with care problems caused by nonparallel faces can be minimized.[191]

No experimental investigations of radially symmetric or asymmetric stress fields by holographic interferometry have been reported to date; however, the general techniques for the analysis of interferograms discussed in Section 6.2 could be used to determine the sum of principal stresses at each point in transparent models of quite arbitrary shape and stress state. The application of holographic interferometry to stress analysis of birefringent models, for which A and B are not equal in equation 6.96, is discussed in Section 7.6.

REFERENCES

1. F. D. Bennett, W. C. Carter, and V. E. Bergdolt, Interferometric analysis of air flow about projectiles in free flight, *J. Appl. Phys.*, **23**, 453–469 (1952).
2. R. N. Bracewell, *The Fourier Transform and Its Applications*, McGraw-Hill, New York, 1965, pp. 262–266.
3. W. Hauf and U. Grigull, Optical methods in heat transfer, in *Advances in Heat Transfer*, Vol. 6, J. P. Hartnett and T. F. Irvine, Jr. (Eds.), Academic Press, New York, 1970, pp. 267–274.
4. R. Ladenburg, J. Winkler, and C. C. Van Voorhis, Interferometric study of faster than sound phenomena, Part I, *Phys. Rev.*, **73**, 1359–1377 (1948).
5. R. Barakat, Solution of an Abel integral equation for band-limited functions by means of sampling theorems, *J. Math. Phys.*, **43**, 325–331 (1964).
6. D. W. Sweeney, A comparison of Abel integral inversion schemes for interferometric applications, *J. Opt. Soc. Am.*, **64**, 559 (1974).
7. P. B. Gooderum and G. P. Wood, NACA TN 2173, 1950.
8. E. T. Whittaker and G. N. Watson, *A Course of Modern Analysis*, 4th ed., Cambridge University Press, Cambridge, 1962, p. 229.
9. O. H. Nestor and H. N. Olsen, Numerical methods for reducing line and surface probe data, *SIAM Rev.*, **2**, 200–207 (1960).
10. W. L. Barr, Method for computing the radial distribution of emitters in a cylindrical source, *J. Opt. Soc. Am.*, **52**, 885–888 (1962).
11. K. Bockasten, Transformation of observed radiances into radial distribution of the emission of a plasma, *J. Opt. Soc. Am.*, **51**, 943–947 (1961).
12. J. W. Bradley, Density determination from axisymmetric interferograms, *AIAA J.*, **6**, 1190–1192 (1968).
13. R. South, An extension to existing methods of determining refractive indices from axisymmetric interferograms, *AIAA J.*, **8**, 2057–2059 (1970).
14. S. I. Herlitz, A method for computing the emission distribution in cylindrical light sources, *Ark. Fys.*, **23**, 571–574 (1963).
15. R. N. Bracewell, Strip integration in radio astronomy, *Aust. J. Phys.*, **9**, 198–217 (1956).
16. I. D. Kulagin, L. M. Sorokin, and E. A. Dubrovskaya, Evaluation of some numerical methods for solving Abel's integral equation, *Opt. Spectrosc.*, **32**, 459–462 (1972).
17. I. M. Gel'Fand, M. I. Graev, and N. Ya. Vilenkin, *Generalized Functions*, Vol. 5, Academic Press, New York, 1966.
18. M. V. Berry and D. F. Gibbs, The Interpretation of optical projections, *Proc. Roy. Soc. (London)*, **A314**, 143–152 (1970).
19. D. Ludwig, The Radon transform on Euclidean space, *Commun. Pure Appl. Math.*, **19**, 49–81 (1966).

20. A. M. Cormack, Reconstruction of densities from their projections, with applications in radiological physics, *Phys. Med. Biol.*, **18**, 195–207 (1973).
21. C. M. Vest, Formation of images from projections: Radon and Abel transforms, *J. Opt. Soc. Am.*, **64**, 1215–1218 (1974).
22. P. P. B. Eggermont, *Three-Dimensional Image Reconstruction by Means of Two-Dimensional Radon Inversion*, Report 75-WSK-04, Technological University, Eindhoven, The Netherlands, 1975.
23. D. W. Sweeney and C. M. Vest, Reconstruction of three-dimensional refractive index fields from multidirectional interferometric data, *Appl. Opt.*, **12**, 2649–2664 (1973).
24. W. Alwang, L. Cavanaugh, R. Burr, and A. Hauer, *Optical Techniques for Flow Visualization and Fluid Flow Measurement in Aircraft Turbomachinery*, Item 1, Final Report PWA-3942, Pratt and Whitney Aircraft Co., Hartford, CT, 1970.
25. G. Golub, Numerical methods for solving least squares problems, *Numer. Math.*, **7**, 206–216 (1965).
26. G. N. Ramachandran and A. V. Lakshminarayanan, Three-dimensional reconstruction from radiographs and electron micrographs. Part III: Description of the convolution method, *Indian J. Pure Appl. Phys.*, **9**, 997–1003 (1971).
27. R. N. Bracewell and A. C. Riddle, Inversion of fan beam scans in radio astronomy, *Astrophys. J.*, **150**, 427–434 (1967).
28. J. W. Cooley and J. W. Tukey, An algorithm for the machine calculation of complex Fourier series, *Math. Comp.*, **19**, 297–301 (1965).
29. T.-F. Zien, W. C. Ragsdale, and W. C. Spring III, Quantitative determination of three-dimensional density field by holographic interferometry, *AIAA J.*, **13**, 841–842 (1975).
30. R. D. Matulka and D. J. Collins, Determination of three-dimensional density fields from holographic interferometry, *J. Appl. Phys.*, **42**, 1109–1119 (1971).
31. H.-G. Junginger and W. Van Haeringer, Calculation of three-dimensional refractive-index field using phase integrals, *Opt. Commun.*, **5**, 1–4 (1972).
32. Yu. P. Presnyakov, Calculation of a two-dimensional refractive index function, *Opt. Spectrosc.*, **40**, 69–70 (1976).
33. R. Gordon, *A Treatise on Reconstruction from Projections and Computerized Tomography* (to be published).
34. H. Svensson, The second-order aberrations in the interferometric measurement of concentration gradients, *Opt. Acta*, **1**, 25–32 (1954).
35. G. P. Wachtell, *Refraction Error in Interferometry of Boundary Layer in Supersonic Flow Along a Flat Plate*, Ph.D. Dissertation, Princeton University, 1951 (available from University Microfilms, Inc., Ann Arbor, MI).
36. W. L. Howes and D. R. Buchele, Optical interferometry of inhomogeneous gases, *J. Opt. Soc. Am.*, **56**, 1517–1528 (1966).
37. E. E. Anderson, W. H. Stevenson, and R. Viskanta, Estimating the refractive error in optical measurements of transport phenomena, *Appl. Opt.*, **14**, 185–188 (1975).
38. F. W. Schmidt and M. E. Newell, Evaluation of refraction errors in interferometric heat transfer studies, *Rev. Sci. Instrum.*, **39**, 592–595 (1968).
39. K. W. Beach, R. H. Muller, and C. W. Tobias, Light-deflection effects in the interferometry of one-dimensional refractive-index fields, *J. Opt. Soc. Am.*, **63**, 559–566 (1973).
40. Reference 3, pp. 230–267.

41. F. J. Weinberg, *Optics of Flames*, Butterworth, London, 1963, p. 232.
42. C. M. Vest, Interferometry of strongly refracting axisymmetric phase objects, *Appl. Opt.*, **14**, 1601–1606 (1975).
43. D. W. Sweeney, D. T. Attwood, and L. W. Coleman, Interferometric probing of laser produced plasmas, *Appl. Opt.*, **15**, 1126–1128 (1976).
44. F. R. McLarnon, R. H. Muller, and C. W. Tobias, Reflection effects in interferometry, *Appl. Opt.*, **14**, 2468–2472 (1975).
45. M. De and L. Sevigny, Three-beam holographic interferometry, *Appl. Opt.*, **6**, 1665–1671 (1967).
46. J. Politch and J. Ben-Uri, Considerations in multi-beam holography. Part I: Analysis in plane holography, *Optik*, **38**, 368–386 (1973).
47. J. Politch, J. Shamir, and J. Ben-Uri, Some characteristics of multi-beam holography, *Opt. Laser Technol.*, **3**, 226–228 (1971).
48. J. Politch, J. Shamir, and J. Ben-Uri, Improved polarization holography, *Appl. Phys. Lett.*, **16**, 496–498 (1970).
49. K. S. Mustafin and V. A. Seleznev, Three-beam holographic interferometry, *Opt. Spectrosc.*, **30**, 80–82 (1971).
50. O. Bryngdahl, Shearing interferometry by wavefront reconstruction, *J. Opt. Soc. Am.*, **58**, 865–871 (1968).
51. C. M. Vest and D. W. Sweeney, Holographic interferometry with both beams traversing the object, *Appl. Opt.*, **9**, 2810–2812 (1970).
52. J. Shamir, Reduced sensitivity phase object analysis by moiré techniques, *Appl. Opt.*, **12**, 271–274 (1973).
53. M. Marquet, M. H. Bourgeon, and J. C. Saget, Interférométrie par holographie, *Rev. Opt., Theor. Instrum.*, **45**, 501–506 (1966).
54. O. Bryngdahl, Longitudinally reversed shearing interferometry, *J. Opt. Soc. Am.*, **59**, 142–146 (1969).
55. J. C. Fouéré and C. Roychoudhuri, Holographic, radial and lateral shear interferometer, *Opt. Commun.*, **12**, 29–31 (1974).
56. J. C. Fouéré and D. Malacara, Holographic radial shear interferometry, *Appl. Opt.*, **13**, 2035–2039 (1974).
57. Y. Doi, T. Komatsu, and T. Fujimoto, Shearing interferometry by holography, *Jap. J. Appl. Phys.*, **12**, 1036–1042 (1973).
58. S. Mallick and M. L. Roblin, Shearing interferometry by wavefront reconstruction using a single exposure, *Appl. Phys. Lett.*, **14**, 61–62 (1969).
59. F. Weigl, A generalized technique of two-wavelength nondiffuse holographic interferometry, *Appl. Opt.*, **10**, 187–192 (1971).
60. F. Weigl, Two-wavelength holographic interferometry for transparent media using a diffraction grating, *Appl. Opt.*, **10**, 1083–1086 (1971).
61. K. S. Mustafin and V. A. Seleznev, Holographic interferometry with variable sensitivity, *Opt. Spectrosc.*, **32**, 532–535 (1972).
62. A. L. Afanaseva, K. S. Mustafin, and V. A. Seleznev, Polychromatic holographic interferometry with recording in a three-dimensional medium, *Opt. Spectrosc.*, **32**, 312–313 (1972).
63. F. Weigl, O. M. Friedrich, Jr., and A. A. Dougal, Multiple-pass nondiffuse holographic interferometry, *IEEE J. Quant. Electron.*, **QE-2**, 41–49 (1970).

64. L. Zehnder, Ein neuer Interferenzrefraktur, *Z. Instrumentenk.*, **11**, 275–285 (1891).
65. L. Mach, Uber einen Interferenzrefraktur, *Z. Instrumentenk.*, **12**, 89–93 (1892).
66. R. Ladenburg and D. Bershader, Interferometry, in *Physical Measurements in Gas Dynamics and Combustion*, R. W. Ladenburg (Ed.), Princeton University Press, Princeton, NJ, 1954, pp. 47–78.
67. J. D. Trolinger, *Laser Instrumentation for Flow Field Diagnostics*, AGARDograph 186, North Atlantic Treaty Organization, 7 Rue Ancelle, 92200 Neuilly Sur Seine, France, 1974.
68. J. D. Trolinger, Flow visualization holography, *Opt. Eng.*, **14**, 470–481 (1975).
69. W. Merzkirch, *Flow Visualization*, Academic Press, New York, 1974.
70. *International Critical Tables*, McGraw-Hill, New York, 1933.
71. L. O. Heflinger, R. F. Wuerker, and R. E. Brooks, Holographic interferometry, *J. Appl. Phys.*, **37**, 642–649 (1966).
72. H. H. M. Chau and G. J. Mullaney, An example of the application of pulsed light holography to aerodynamics, *Appl. Opt.*, **6**, 981 (1967).
73. P. Smigielski and H. Royer, Applications de l'holographie a l'aerodynamique, *L'Onde Electr.*, **48**, 223–225 (1968).
74. M. Philbert and J. Surget, Application de l'interférométrie en soufflerie, *Rech. Aérosp.*, No. 122, 55–60 (1968).
75. L. H. Tanner, Some applications of holography in fluid mechanics, *J. Sci. Instrum.*, **43**, 81–83 (1966).
76. L. D. Siebert and D. E. Geister, Pulsed holographic interferometry vs. schlieren photography, *AIAA J.*, **6**, 2194–2195 (1968).
77. W. G. Alwang, L. A. Cavanaugh, and D. Cain, The observation of three-dimensional shadowgraphlike images in holography of phase objects, *Appl. Opt.*, **8**, 1256–1257 (1969).
78. M. H. Horman, An application of wavefront reconstruction to interferometry, *Appl. Opt.*, **4**, 333–336 (1965).
79. A. F. Belozerov, A. N. Berezkin, A. I. Razumoskaya, and N. M. Spornik, Hologram study of gas flow in a ballistic wind tunnel, *Sov. Phys. Tech. Phys.*, **18**, 448–490 (1973).
80. P. Smigielski and A. H. Hirth, New holographic studies of high-speed phenomena, in *Proceedings of 9th International Congress on High-Speed Photography*, W. G. Hyzer and W. G. Chace (Eds.), Society of Motion Picture and Television Engineers, New York, 1970, pp. 321–326.
81. V. F. Ivanov, L. T. Mustafin, A. P. Shatilov, and E. S. Yushkov, Holographic formation of interferograms and shadow patterns of gas flow from a reflecting nozzle of a shock tube, *Sov. J. Quant. Electron.*, **4**, 1389–1390 (1975).
82. B. Ineichen, U. Kogelschatz, and R. Dändliker, Schlieren diagnostics and interferometry of an arc discharge using pulsed holography, *Appl. Opt.*, **12**, 2554–2556 (1973).
83. A. B. Witte and R. F. Wuerker, Laser holographic interferometry study of high-speed flowfields, *AIAA J.*, **8**, 581–583 (1970).
84. R. E. Brooks, L. O. Heflinger, and R. F. Wuerker, Pulsed laser holograms, *IEEE J. Quant. Electron.*, **QE-2**, 275–299 (1966).
85. L. H. Tanner, Holographic interferometer and fringe analyzer, and their use for the study of supersonic flow, *Opt. Laser Technol.*, **4**, 281–287 (1972).
86. J. Surget, Quantitative study of aerodynamic flows by holographic interferometry, *Rech. Aérosp.*, No. 154, 161–171 (1973).

87. H. Fagot, Visualization of gas flows by means of high-speed holography, *J. SMPTE*, **82**, 183 (1973).
88. C. Veret, Holographic interferometry applied to aerodynamics, *J. SMPTE*, **82**, 185 (1973).
89. R. J. Radley, Jr., and A. G. Havener, Application of dual hologram interferometry to wind-tunnel testing, *AIAA J.*, **11**, 1332–1333 (1973).
90. A. Hirth, P. Smigielski, and A. Stimpfling, Use of holography for visualization of the wake of projectiles in hypersonic flight at Mach 6, *Opt. Laser Technol.*, **3**, 195–199 (1971).
91. A. F. Belozerov and V. T. Chernykh, Production of optical image holograms with a Mach-Zehnder interferometer when a gas stream flows around the object, *Opt. Spectrosc.*, **28**, 552–553 (1970).
92. C. J. Reinheimer, C. E. Wiswall, R. A. Schmiege, R. J. Harris, and J. E. Dueker, Holographic subsonic flow visualization, *Appl. Opt.*, **9**, 2059–2065 (1970).
93. G. Wortberg, A holographic interferometer for gas dynamic measurements, in *Recent Developments in Shock Tube Research*, D. Bershader and W. Griffith (Eds.), Stanford University Press, Stanford, CA, 1973, pp. 267–276.
94. C. H. Law, Supersonic, turbulent boundary layer separation, *AIAA J.*, **12**, 794–797 (1974).
95. C. H. Law, Supersonic shock wave turbulent boundary-layer interaction, *AIAA J.*, **14**, 730–734 (1976).
96. A. P. Shatilov, L. T. Mustafina, B. F. Ivanov, A. F. Belozerov, and Ye. S. Yuskov, Recording of pulsed gasdynamic processes by holographic shearing interferometry, *Fluid Mech. Sov. Res.*, **5**, 154–162 (1976).
97. A. P. Shatilov, L. T. Mustafina, N. P. Mudrevskaya, V. F. Ivanov, and Ye. S. Yushkov, Holography of flow in a reflecting nozzle of a shock wave using a narrow reference beam, *Fluid Mech. Sov. Res.*, **5**, 163–170 (1976).
98. A. N. Berezkin, N. P. Mudrevskaya, L. T. Mustafina, and A. I. Razumovskaya, Shadow and interference patterns of low-density gas flow in hologram reconstruction, *Sov. Phys. Tech. Phys.*, **46**, 1724–1727 (1976).
99. A. F. Belozerov, A. V. En'Shin, V. G. Zaitsev, I. S. Zeilikovich, and N. M. Spornik, Use of hologram interferometer with reference beam formed from object beam in ballistic experiment. *Sov. Phys. Tech. Phys.* **21** 1160–1161 (1976).
100. J. K. Beamish, D. M. Gibson, R. H. Sumner, S. M. Zivi, and G. H. Humberstone, Wind-tunnel diagnostics by holographic interferometry, *AIAA J.*, **7**, 2041–2043 (1969).
101. R. A. Kosakoski and D. J. Collins, Application of holographic interferometry to density field determination in transonic corner flow, *AIAA J.*, **12**, 767–770 (1974).
102. R. C. Jagota and D. J. Collins, Finite fringe holographic interferometry applied to a right circular cone at angle of attack, *J. Appl. Mech.*, **39**, 897–903 (1972).
103. A. B. Witte, Holographic interferometry of a submarine wake in stratified flow, *J. Hydronaut.*, **6**, 114–115 (1972).
104. W. R. Debler and C. M. Vest, Visualization of a stratified flow by holographic interferometry, *Proc. Roy. Soc. (London)*, **A358**, 1–16 (1977).
105. R. A. Alpher and D. R. White, Optical refractivity of high temperature gases. II: Effects resulting from ionization of monatomic gases, *Phys. Fluids*, **2**, 162–169 (1959).
106. V. Ascoli-Bartoli, Plasma diagnostics based on refractivity, in *Physics of Hot Plasmas*, B. J. Rye and J. C. Taylor (Eds.), Plenum Press, New York, p. 405.

107. W. Merzkirch, *Flow Visualization*, Academic Press, New York, 1974, p. 70.
108. A. N. Zaidel', G. V. Ostrovskaya, and Yu. I. Ostrovskii, Plasma diagnostics by holography (a review), *Sov. Phys. Tech. Phys.*, **13**, 1153–1164 (1969).
109. F. C. Jahoda, R. A. Jeffries, and G. A. Sawyer, Fractional-fringe holographic plasma interferometry, *Appl. Opt.*, **6**, 1407–1410 (1967).
110. R. Kristal, Pulsed HF laser holographic interferometry, *Appl. Opt.*, **14**, 628–633 (1975).
111. D. T. Attwood, L. W. Coleman, and D. W. Sweeney, Holographic microinterferometry of laser produced plasma with frequency-tripled probe pulses, *Appl. Phys. Lett.*, **26**, 616–618 (1975).
112. F. Zernike and J. E. Midwinter, *Applied Nonlinear Optics*, Wiley-Interscience, New York, 1973.
113. G. V. Dreiden, A. N. Zaidel', Yu. I. Ostrovskii, and E. N. Shedova, Three-wavelength hologram diagnostics of an optical burst on a potassium target, *Sov. Phys. Tech. Phys.*, **18**, 972–974 (1974).
114. R. F. Wuerker, L. O. Heflinger, and R. A. Briones, Holographic interferometry with ultraviolet light, *Appl. Phys. Lett.*, **12**, 302–303 (1968).
115. J. N. Olsen, Picosecond infrared holography on bismuth film, *Appl. Phys. Lett.*, **24**, 220–222 (1974).
116. P. R. Forman, S. Humphries, Jr., and R. W. Peterson, Pulsed holographic interferometry at 10.6 μm, *Appl. Phys. Lett.*, **22**, 537–539 (1973).
117. E. M. Borkhudarov, V. R. Berezovskii, G. V. Geloshvili, M. I. Taklakishvili, T. Ya. Chelidze, and V. V. Chichinadze, Possible use of 10.6-μ holograms for plasma diagnostics, *Sov. Tech. Phys. Lett.*, **2**, 425–427 (1976).
118. B. M. Abakumov, N. D. Baikova, L. N. Gnatyuk, M. L. Gurari, and S. N. Marchenko, Recordings of holograms and holographic interferograms in MnBi films, *Sov. J. Quant. Electron.*, **5**, 873–874 (1975).
119. G. V. Ostrovskaya and Yu. I. Ostrovskii, Holographic investigation of a laser spark, *J.E.T.P. Lett.*, **4**, 83–85 (1966).
120. A. N. Zaidel', G. V. Ostrovskaya, Yu. I. Ostrovskii, and T. Ya. Chelidze, Time-resolved holography of laser-produced plasmas, *Sov. Phys. Tech. Phys.*, **11**, 1650–1652 (1967).
121. A. Kakos, G. V. Ostrovskaya, Yu. I. Ostrovskii, and A. N. Zaidel', Interferometry holographic investigation of a laser spark, *Phys. Lett.*, **23**, 81–83 (1966).
122. I. I. Ashmarin, Yu. A. Dykovskii, N. N. Degtyaranko, V. F. Elesin, A. I. Larkin, and I. P. Sapailo, Pulsed-holographic investigation of gas breakdown in front of a laser produced plasma, *Sov. Phys. Tech. Phys.*, **16**, 1881–1887 (1972).
123. J. L. Seftor, Holographic interferometry of exploding wires, *J. Appl. Phys.*, **44**, 4965–4969 (1973).
124. J. L. Seftor, Holographic interferograms of side-by-side exploding wires, *J. Appl. Phys.*, **45**, 5082–5083 (1974).
125. A. Bernard, A. Coudeville, A. Jolas, J. Laurspach, and J. de Mascureau, Experimental studies of the plasma focus and evidence for nonthermal processes, *Phys. Fluids*, **18**, 180–194 (1975).
126. R. Fedosejevs and M. C. Richardson, Subnanosecond microscopic holographic interferometry of plasmas produced by 1-nsec CO_2 laser pulses, *Appl. Phys. Lett.*, **27**, 115–117 (1975).
127. M. J. Forrest, P. D. Morgan, N. J. Peacock, K. Kuriki, M. V. Goldman, and T. Randolph, CO_2-laser-excited langmuir turbulence in a dense-plasma focus, *Phys. Rev. Lett.*, **37**, 1681–1684 (1976).

REFERENCES

128. M. J. Clauser, L. P. Mix, J. W. Poukey, J. P. Quintenz, and A. J. Toepfer, Enhanced deposition in electron beam targets due to beam stagnation, *Phys. Rev. Lett.*, **38**, 398–401 (1977).
129. D. B. Thomson, L. A. Jones, A. G. Bailey, and R. Engleman, Jr., Characteristics of high-density theta-pinches seeded with selected high-Z elements, *Plasma Phys. Suppl.*, pp. 209–213 (1976).
130. D. W. Swain, S. A. Goldstein, L. P. Mix, J. G. Kelly, and G. R. Hadley, The characteristics of a medium current relativistic electron-beam diode, *J. Appl. Phys.*, **43**, 1085–1093 (1977).
131. I. I. Komissarova, G. V. Ostrovskaya, L. L. Shapiro, and A. N. Zaidel', Two wavelength holography of a laser spark, *Phys. Lett.*, **29**, 262–263 (1969).
132. A. B. Ignatov, I. I. Komissarova, G. V. Ostrovskaya, and L. L. Shapiro, Holograph studies of a laser spark. III: Spark in hydrogen and helium, *Sov. Phys. Tech. Phys.*, **16**, 550–556 (1971).
133. R. A. Jeffries, Two-wavelength holographic interferometry of partially ionized plasmas, *Phys. Fluids*, **13**, 210–212 (1970).
134. I. I. Komissarova, G. V. Ostrovskaya, and L. L. Shapiro, Holographic studies of a laser spark. II: Double interferometry at long wavelength, *Sov. Phys. Tech. Phys.*, **15**, 827–833 (1970).
135. J. L. Seftor, Two-wavelength holographic interferometry of exploding wires, *J. Appl. Phys.*, **45**, 2903–2906 (1974).
136. G. V. Ostrovskaya and Yu. I. Ostrovskii, Two-wavelength hologram method for studying the dispersion properties of phase objects, *Sov. Phys. Tech. Phys.*, **15**, 1890–1892 (1971).
137. A. B. Ignatov, I. I. Komissarova, G. V. Ostrovskaya, and L. L. Shapiro, Two-frequency single exposure plasma holograph, *Sov. Phys. Tech. Phys.*, **16**, 315–319 (1971).
138. R. J. Radley, Jr., Two-wavelength holography for measuring plasma electron density, *Phys. Fluids*, **18**, 175–179 (1975).
139. I. I. Komissarova, G. V. Ostrovskaya, and L. L. Shapiro, Holographic investigation of laser sparks, *Sov. Phys. Tech. Phys.*, **13**, 1118–1121 (1969).
140. M. J. Landry and A. E. McCarthy, Use of the multiple cavity laser holographic system for EBW analysis, *Opt. Eng.*, **14**, 69–72 (1975).
141. K. S. Thomas, C. R. Harder, W. E. Quinn, and R. E. Siemon, Helical field experiments on a three-meter theta pinch, *Phys. Fluids*, **15**, 1658–1666 (1972).
142. W. T. Armstrong and P. R. Forman, Double-pulsed time differential holographic interferometry (for plasma diagnostics), *Appl. Opt.*, **16**, 229–232 (1977).
143. F. C. Jahoda, Submicrosecond holographic cine-interferometry of transmission objects, *Appl. Phys. Lett.*, **14**, 341–343 (1969).
144. W. Tiemann, Real-time holographic interferometry with microsecond resolution, *Siemens Forsch.- und Entwicklungsber.*, **5**, 163–170 (1976).
145. J. G. Kelly and L. P. Mix, Measurements of high-current relativistic electron diode plasma properties with holographic interferometry, *J. Appl. Phys.*, **46**, 1084–1090 (1975).
146. M. Richardson and A. Alcock, Subnanosecond interferometry of plasma filaments in a laser-produced spark, *Sov. J. Quant. Electron.*, **1**, 461–465 (1972).
147. P. Belland, C. DeMichelis, and M. Mattioli, Holographic interferometry of laser produced plasma using picosecond pulses, *Opt. Commun.*, **3**, 7–8 (1971).

148. D. T. Attwood and L. W. Coleman, Microscopic interferometry of laser-produced plasmas, *Appl. Phys. Lett.*, **24**, 408–410 (1974).

149. C. G. Murphy, Pulse clipping for holographic interferometry, *Appl. Opt.*, **15**, 23–24 (1976).

150. P. T. Radulovic, *Holographic Interferometry of Asymmetric Temperature or Density Fields*, doctoral dissertation, The University of Michigan, 1977.

151. W. F. Meggers and C. L. Peters, Measurements on the index of refraction of air for wavelengths from 2218 Å to 9000 Å, *Bull. Natl. Bur. Stands.*, **14**, 697–740 (1918–19).

152. L. W. Tilton, Standard conditions for precise prism refractometry, *J. Res. Natl. Bur. Stands.*, **14**, 393–418 (1935).

153. G. Schödel, *Kombinierte Wärmeleitung und Wärmestrahlung in Konvektionsfreien flussigheitsschichten*, Dissertation, Technische Hochschule München, Institut A für Thermodynamik, 1969.

154. Reference 3, p. 286.

155. D. J. Coumou, E. L. Mackor, and J. Hijmans, Isotropic light-scattering in pure liquids, *Trans. Faraday Soc.*, **60**, 1539–1547 (1964).

156. L. W. Tilton and J. K. Taylor, Refractive index of distilled water for visible radiation, at temperature 0 to 60°C, *J. Res. Natl. Bur. Stands.*, **20** (1938).

157. H. M. Dobbins and E. R. Peck, Change of refractive index of water as a function of temperature, *J. Opt. Soc. Am.*, **63**, 318–320 (1973).

158. C. G. Murphy and S. S. Alpert, Dependence of refractive index temperature coefficient on the thermal expansivity of liquids, *Am. J. Phys.*, **39**, 834–836 (1976).

159. A. E. Zinnes, The coupling of conduction with laminar natural convection from a vertical flat plate with arbitrary surface heating, *J. Heat Transfer*, **92**, Ser. C, 528–535 (1970).

160. W. Aung and R. O'Regan, Precise measurement of heat transfer using holographic interferometry, *Rev. Sci. Instrum.*, **42**, 1755–1759 (1971).

161. H. Schmitt and H. Beer, Holographic investigations regarding the heat transfer by natural convection at the critical point of carbon dioxide at small temperature differences, *Brennst.-Waerme-Kraft*, **29**, 15–23 (1977).

162. W. Aung, L. S. Fletcher, and V. Sernas, Developing laminar free convection between vertical flat plates with asymmetric heating, *Int. J. Heat Mass Transfer*, **15**, 2293–2308 (1972).

163. W. Panknin, M. Jahn, and H. H. Reinke, Forced convection heat transfer in the transition from laminar to turbulent flow in closely spaced circular tube bundles, Paper FC8.7, in *Proceedings of Fifth International Heat Transfer Conference* (September 3–7, 1974, Tokyo), Vol. 2, Japanese Society of Mechanical Engineers, pp. 325–329.

164. G. Antonini, G. Guiffant, and P. Perrot, The effect of transverse oscillations on heat transfer from a horizontal hot-wire to a liquid: Holographic visualization, *Int. J. Heat Mass Transfer*, **20**, 88–92 (1977).

165. N. J. Phillips and J. R. Coaton, Holography as a diagnostic technique in the study of lamps and lighting, *Light Res. Technol.*, **5**, 156–159 (1973).

166. G. M. Neumann, U. Mueller, and E. Schmidt, Holography applied to gas-filled incandescent lamps, 1, *Lichttechnik*, **27**, 214–218 (1975).

167. U. Mueller, G. M. Neumann, and E. Schmidt, Holography of gas-filled lamps, *Lichttechnik*, **27**, 282–283 (1975).

168. J. Fitzgerald and H. Hörster, The temperature distribution in gas-filled incandescent lamps, *Phillips Tech. Rev.*, **32**, 206–209 (1971).
169. E. K. Hussmann, A holographic interferometer for measuring radiation energy deposition profiles in transparent liquids, *Appl. Opt.*, **10**, 182–186 (1971).
170. D. W. Sweeney and C. M. Vest, Measurement of three-dimensional temperature fields above heated surfaces by holographic interferometry, *Int. J. Heat Mass Transfer*, **17**, 1443–1454 (1974).
171. P. T. Radulovic and C. M. Vest, Measurement of three-dimensional temperature fields by holographic interferometry, in *Applications of Holography and Optical Data Processing*, E. Marom and A. A. Friesem (Eds.), Pergamon Press, Oxford, 1977, pp. 241–249.
172. M. M. El-Wakil, An interferometric technique for measuring binary diffusion coefficients, *J. Heat Transfer*, **91**, Ser. C, 259–265 (1969).
173. F. Mayinger and W. Panknin, Holography in heat and mass transfer, Paper IL3, in *Proceedings of Fifth International Heat Transfer Conference* (September 3–7, 1974, Tokyo), Vol. 6, Japanese Society of Mechanical Engineers, pp. 28–43.
174. J. D. Trolinger, Holographic interferometry as a diagnostic tool in reactive flows, *Comb. Sci. Technol.*, **13**, 229–244 (1976).
175. H. J. Raterink and C. W. Lamberts, Holographic interferometry applied to high-speed flame research, in *Proceedings of 9th International Congress on High-Speed Photography*, W. G. Hyzer and W. G. Chace (Eds.), Society of Motion Picture and Television Engineers, New York, 1970, pp. 30–37.
176. K. S. Varde, An optical investigation of the combustion of a stratified mixture in a dual chamber confinement, *Can. J. Chem. Eng.*, **52**, 426–431 (1974).
177. G. E. Maddox and J. G. Becsey, Simple holographic interferometry of the Rayleigh type, *Rev. Sci. Instrum.*, **41**, 880–881 (1970).
178. J. G. Becsey, G. E. Maddox, N. R. Jackson, and J. A. Bierlein, Holography and holographic interferometry for thermal diffusion studies in solutions, *J. Phys. Chem.*, **74**, 1401–1403 (1970).
179. J. G. Becsey, N. R. Jackson, and J. A. Bierlein, Hologram interferometry for isothermal diffusion measurements, *J. Phys. Chem.*, **75**, 3374–3376 (1971).
180. V. Sanchez and J. Mahenc, Determination of Soret coefficient in binary liquid systems by holographic interferometry, *J. Chim. Phys.*, **73**, 485–490 (1976).
181. J. Pipman, S. Lipson, and J. Landau, Studies of phase changes in helium near absolute zero, *Laser Elektro-Optik*, **8**, 24–25 (1976).
182. C. Durou, Une installation d'interférométrie holographique destinée a l'étude des phénomènes de diffusion dans les liquids, *J. Phys. E (Sci. Instrum.)*, **6**, 1116–1120 (1973).
183. N. Bochner and J. Pipman, A simple method of determining diffusion constants by holographic interferometry, *J. Phys. D*, **9**, 1825–1830 (1976).
184. M. Nisida and H. Saito, A new interferometric method of two-dimensional stress analysis, *Exp. Mech.*, **4**, 366–376 (1964).
185. T. D. Dudderar, Applications of holography to fracture mechanics, *Exp. Mech.*, **9**, 281–285 (1969).
186. T. D. Dudderar and R. O'Regan, Measurement of the strain field near a crack tip in polymethylmethacrylate by holographic interferometry, *Exp. Mech.*, **11**, 49–56 (1971).
187. T. D. Dudderar and H. J. Gorman, The determination of mode I stress-intensity factors by holographic interferometry, *Exp. Mech.*, **13**, 145–149 (1973).

188. P. S. Theocaris, The determination of mode I stress-intensity factors by holographic interferometry: comments on previous discussion, *Exp. Mech.*, **15**, 150–152 (1975).
189. T. D. Dudderar and E. M. Doerries, A study of effective crack length using holographic interferometry, *Exp. Mech.*, **16**, 300–304 (1976).
190. N. Soga, D. Holcomb, and H. Spetzler, Determination of strains in optical windows of a high-pressure chamber by holographic interferometry and finite element analysis, *Rev. Sci. Instrum.*, **47**, 1453–1456 (1976).
191. R. Mark and R. O'Regan, Model interferometry with air holograms, *Exp. Mech.*, **12**, 332–334 (1972).

7
Related Measurement Techniques

7.1 INTRODUCTION

In this final chapter we discuss six coherent optical measurement techniques which either are extensions of holographic interferometry or are closely related to it in concept and practice. These are holographic contour generation, speckle photography and speckle interferometry, projected fringe techniques, techniques incorporating television systems, and holographic photoelasticity. Some of these techniques—in particular, speckle photography and speckle interferometry—can be used to measure the same quantities as holographic interferometry. An attempt is made to point out the relative merits of each technique and the manner in which they complement each other. The discussion of these methods is less detailed than that of holographic interferometry both by the necessity for brevity and by virtue of the fact that they are still in a stage of rapid development.

7.2 HOLOGRAPHIC CONTOUR GENERATION

Contour generation is the formation of an image of an object on which contours of constant elevation with respect to some plane are superimposed. Such an image provides an easily interpreted, full-field display of three-dimensional surface shapes. Applications include quality control, the recording of surface shape data for use in measurement of strain by holographic interferometry, the transfer of contour information from prototypes or models to numerically controlled machines, biomedical applications, and the measurement of mechanical wear. One method of generating surface contours optically is a form of holographic interferometry in which two holograms of an object are recorded, each with light of a different wavelength. The reconstructed image formed by such a hologram is a superposition of two images of the object, one of which is distorted and translated relative to the other. This gives rise to an image overlaid with interference fringes similar to those that would be formed if the object were physically deformed and translated between exposures. The experimental parameters such as wavelengths and system geometry can be chosen so that the fringes are the desired contours.

A simple system for *two-frequency holographic contouring*[1-3] is depicted in Figure 7.1. The object is illuminated by a point source located a distance S_o from a typical object point P. The reference wave originates at a point source a distance R_R from the center of the plate. Two sequential holographic recordings of the static object which is to be contoured are made on a single plate. During the first exposure the wavelength is λ_1, and during the second it is λ_2. After development the hologram is illuminated by a reconstruction wave, which is identical to the reference wave, of wavelength λ_2. The process can be analyzed using the nomenclature of Figure 7.1. This is identical to the nomenclature used in the holographic imaging equations 1.94 to 1.96, but we restrict the present analysis to the y-z plane, so $\beta_o = \beta_R = \beta_c = 0$.

Two holographic virtual images will be formed. The first, I_1, was recorded with λ_1 but reconstructed with λ_2. The second, I_2, was recorded and reconstructed with λ_2 and will therefore be identical to the original object. Suppose that we view these two images from the center of the hologram, $x = y = z = 0$. Inspection of equation 1.84 indicates that the light from image 1 of object point P will leave the hologram with phase

$$\phi_1 = k_1(S_o + R_o - R_R) + k_2(R_c),$$

and that the light from image 2 of object point P will leave the hologram

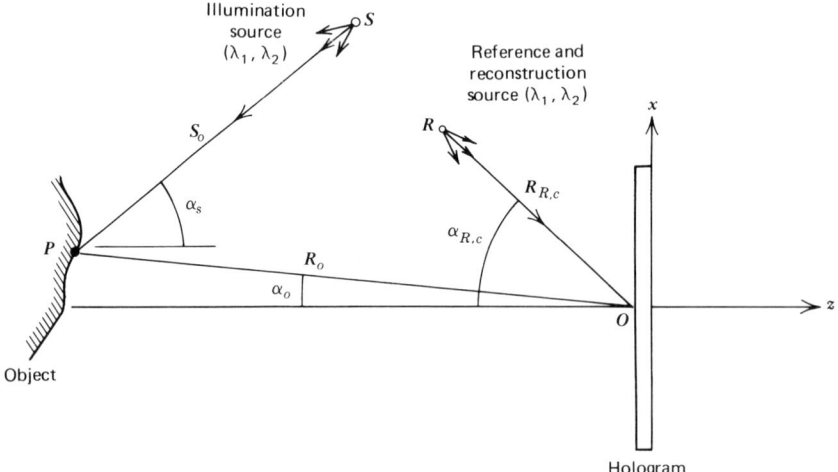

Figure 7.1 Optical system for two-wavelength holographic contouring.

7.2 HOLOGRAPHIC CONTOUR GENERATION

with phase

$$\phi_2 = k_2(S_o + R_o - R_R) + k_2(R_c).$$

Their difference $\Delta\phi = \phi_2 - \phi_1$ is

$$\Delta\phi = (S_o + R_o - R_R)(k_2 - k_1), \tag{7.1}$$

where $k_1 = 2\pi/\lambda_1$ and $k_2 = 2\pi/\lambda_2$. Bright fringes will thus be observed on the object surface at all points where

$$S_o + R_o = R_R + \frac{N}{1/\lambda_1 - 1/\lambda_2}, \quad N = 0, 1, 2, \ldots, \tag{7.2}$$

and dark fringes will be observed where $N = 0.5, 1.5, \ldots$. If we keep the observation point fixed at the origin, R_R is a constant, so equation 7.2 indicates that the fringes will represent the intersection of the object surface with a set of concentric ellipsoids with foci at S and O. Thus, although contours have been generated, they are not true depth contours, which would represent the intersection of the object surface with equally spaced parallel planes.

If we are to observe the interference fringes described by equation 7.2, the two images of point P must be separated in the direction transverse to the viewing direction by less than the resolution limit of the viewing system. To determine the actual separation of these images we must apply the holographic imaging equations, 1.94 to 1.96. In doing so, we will assume that both the illumination and reference waves are plane, that is, $R_s \to \infty$ and $R_R \to \infty$, because this will be shown later to be desirable. Image 2, which is recorded and reconstructed with λ_2, will be precisely coincident with the actual object position. Image 1, which is recorded with λ_1 and reconstructed with λ_2, will be distorted. The distance to this image of point P is given by equation 1.94:

$$R_{I_1} = \left(\frac{\lambda_1}{\lambda_2}\right) R_o. \tag{7.3}$$

Its x coordinate is calculated using equation 1.95:

$$x_{I_1} = R_{I_1} \sin \alpha_{I_1} = R_{I_1} \sin \alpha_c + \left(\frac{\lambda_2}{\lambda_1}\right) R_{I_1}$$
$$\times (\sin \alpha_o - \sin \alpha_R). \tag{7.4}$$

Combining equations 7.3 and 7.4 and noting that $x_o = R_o \sin \alpha_o = x_{I_2}$ yields

$$x_{I_1} - x_{I_2} = R_o \left(\frac{\lambda_1}{\lambda_2} \sin \alpha_c - \sin \alpha_R \right). \tag{7.5}$$

Equations 7.3 and 7.5 indicate that image 1 of point P is displaced both axially and transversally with respect to image 2 of point P.

The transverse separation of the two images requires that a very small viewing aperture be used, or that each object point be viewed from the direction along which its two images are aligned. This is quite inconvenient and makes interpretation difficult. A second difficulty with the method under consideration is that the fringe pattern obviously shifts if the observation point is moved to a different location on the hologram. Hence, if the virtual image is viewed or photographed through a system with a large aperture, the fringes will be blurred, that is, reduced in visibility. If distinct fringes are to be observed, the aperture must be small. Specifically, it must accept a cone of rays which intersects the hologram over an area small enough that the phase difference $\Delta \phi$ is less than a value such as $\pi/4$. This introduces very stringent requirements,[4] particularly for large values of $(\lambda_2 - \lambda_1)$, so that the contour pattern is likely to be dominated by laser speckle.

We have noted three difficulties with the two-frequency contouring scheme as described above:

1. The contouring surfaces are not plane.
2. The images are transversally displaced.
3. The value of $\Delta \phi$ varies rapidly with viewing direction.

The second and third difficulties can be alleviated by using the real-time method of Hildebrand and Haines,[1] in which the two images are manipulated by changing the reference source location until they closely overlap in some region of interest. Varner[5] has demonstrated a scheme in which a grating is introduced to automatically compensate for transverse displacements due to wavelength change. A third possibility is to introduce an imaging system between the object and the hologram; this approach is discussed below.

If the object, or its image, is very close to the film while the hologram is recorded, the transverse displacement indicated by equation 7.5 will tend to zero, thereby reducing one major difficulty. If a telecentric telescope like that in Figure 7.2 is used to image the object onto the hologram, the rays which form the image can be limited to those which deviate only slightly from the axial direction. With this system the light from each point on the object surface is mapped onto a very small region of the hologram, so difficulty 3 is greatly reduced. Specifically, fringes of adequate visibility

7.2 HOLOGRAPHIC CONTOUR GENERATION

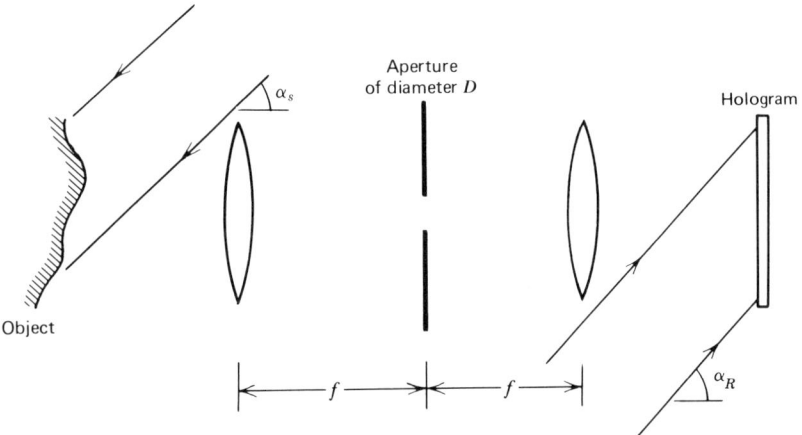

Figure 7.2 Apparatus for two-wavelength holographic contouring using a telecentric imaging system.

can be observed when the following criterion is satisfied:[6,4]

$$z_{max} \leq \frac{f/D}{|\sin \alpha_R|} \left[\frac{\lambda_1 \lambda_2}{2(\lambda_1 - \lambda_2)} \right] \tag{7.6}$$

for $\sin \alpha_R \geq D/2f$, where z_{max} is the maximum object depth measured from the hologram to the image of the object, f is the focal length of the lenses used in the telescope, and D is the diameter of the aperture in its center plane, as shown in Figure 7.2. If the system is arranged so that the object is illuminated along the z axis, $\alpha_R = 0$, the criterion becomes[4]

$$z_{max} \leq 4 \left(\frac{f}{D} \right)^2 \left(\frac{\lambda_1 \lambda_2}{\lambda_1 - \lambda_2} \right). \tag{7.7}$$

Plane contouring surfaces are formed if both the object and reference waves are collimated. This can be discerned by applying equation 7.2. Because all object points are observed along the z direction through the telecentric telescope, $R_o = -z$. Both the object and reference waves are plane, at any height x, so R_R and S_o are given, within an additive constant, by

$$R_R = -x \sin \alpha_R,$$
$$S_o = x \sin \alpha_s - z \cos \alpha_s.$$

The sign of the term $x \sin \alpha_s$ is positive because the image formed by the telescope is inverted. Equation 7.2 for the resulting contour surfaces is

$$x \sin \alpha_s - z \cos \alpha_s - z = -x \sin \alpha_R + N\left(\frac{\lambda_1 \lambda_2}{\lambda_1 - \lambda_2}\right),$$

which can be rearranged as

$$x(\sin \alpha_s + \sin \alpha_R) - z(1 + \cos \alpha_s) = N\left(\frac{\lambda_1 \lambda_2}{\lambda_1 - \lambda_2}\right),$$

$$N = 0, 1, 2, \ldots . \quad (7.8)$$

This is the equation of a family of planes with orientation

$$\frac{dx}{dz} = \frac{1 + \cos \alpha_s}{\sin \alpha_s + \sin \alpha_R}. \quad (7.9)$$

The difference in depth between two adjacent contours is

$$\Delta z = \frac{\lambda_1 \lambda_2}{(\lambda_1 - \lambda_2)(1 + \cos \alpha_s)}. \quad (7.10)$$

The contour surfaces can be made parallel to the hologram plane by choosing $\alpha_s = -\alpha_R$, as shown in Figure 7.2. The system shown in this figure was used to produce the contour maps in Figure 7.3.

As indicated by equation 7.10, contour intervals are determined primarily by the wavelength difference $(\lambda_1 - \lambda_2)$. Most experiments have been done using two lines from an argon ion laser, or from a He-Ne laser, which gives contour intervals on the order of $\Delta z = 10$ μm. Pulsed lasers have been used to produce holographic contour maps by exposing a hologram with a single laser pulse containing two frequencies. This technique was demonstrated by Heflinger and Wuerker,[7] who used a pulsed ruby laser with a resonant reflector of selected thickness as an output reflector. This yields two output lines such that the contour interval essentially equals the optical pathlength between the two surfaces of the reflector. These workers produced high-quality contour maps with 7.7 and 23 mm contour intervals. More recently, Henshaw and Ezekiel[8] have produced contour intervals ranging from 2.5 to 34 μm by using a pulsed xenon laser. Rapidly developing dye laser technology promises widely and simply variable contouring intervals in the future.

A second technique which can be used to generate contour maps is the *multiple-refractive-index method*.[9-11] This requires only a single-frequency

7.2 HOLOGRAPHIC CONTOUR GENERATION

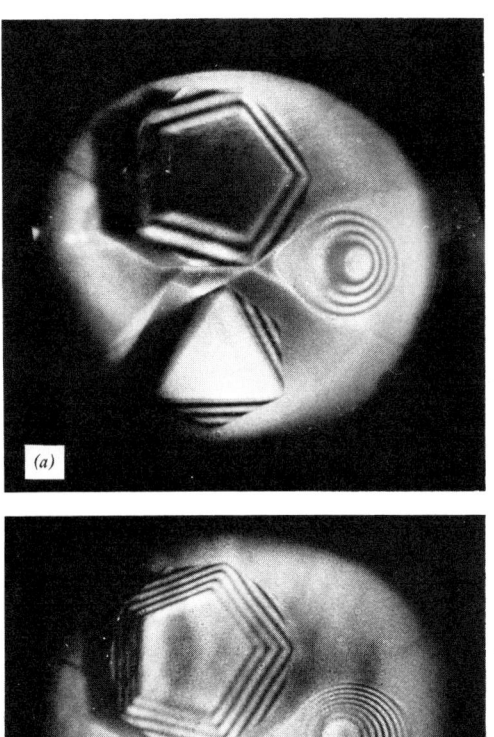

Figure 7.3 Contour maps produced with a system like that in Figure 7.2 (ref. 4): (a) $\lambda_1 - \lambda_2 = 0.03$ nm, contour interval = 6 mm; (b) $\lambda_1 - \lambda_2 = 0.06$ nm, contour interval = 3 mm.

laser. The object to be contoured is immersed in a glass tank containing a fluid having refractive index n_1. After a hologram of the object in the tank is recorded, the first fluid is replaced by a second one, which has refractive index n_2, and a second holographic recording is made on the same plate. The virtual image reconstructed when this hologram is illuminated contains contour fringes which are generated because of the optical pathlength differences in the two fluids. If the illumination wave is plane, the contour surfaces will be plane.

Figure 7.4 is a schematic diagram of a system used for multiple-index contouring. A beamsplitter is used so that the object can be illuminated and viewed in the normal direction. This is recommended because it minimizes aberrations caused by imperfections in the tank. For the same reason a telecentric telescope is used to select only viewing rays that are nearly normal to the tank surface. The hologram can be recorded anywhere to the right of the beamsplitter. Its location near the aperture was suggested by Zelenka and Varner[10] and results in a simple system with efficient utilization of light.

If z is the distance from the tank window to the object surface, the phase difference between light scattered to the image by a point in a fluid, and light scattered by the same point in a different fluid, is

$$\Delta\phi = 2\left(\frac{2\pi}{\lambda}\right)(n_1 - n_2)z,$$

so the contour interval will be

$$\Delta z = \frac{\lambda}{2|n_1 - n_2|}. \qquad (7.11)$$

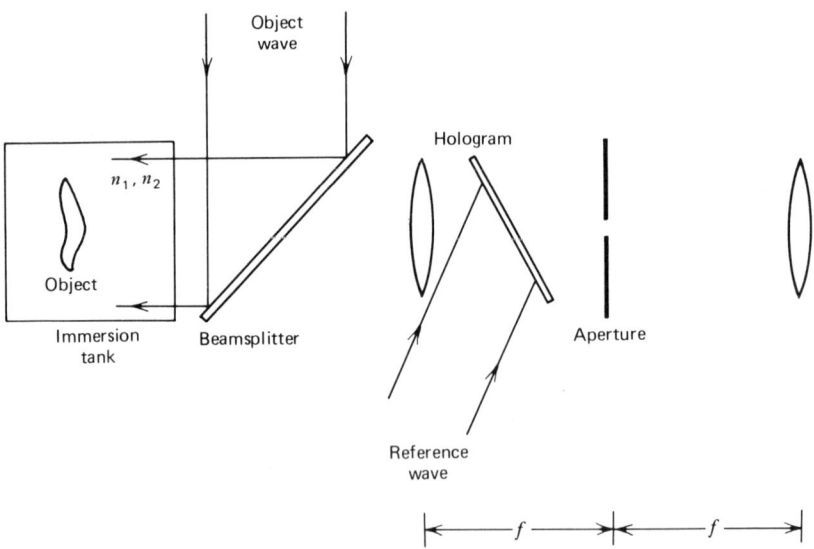

Figure 7.4 System for multiple-index holographic contouring.

7.2 HOLOGRAPHIC CONTOUR GENERATION

The value of Δz can be varied over a wide interval. If the tank is initially filled with air, $n_1 \cong 1$, and then is filled with water, $n_2 \cong 1.33$, for the second exposure, the contour interval will be $\Delta z \cong 1.5\lambda$. Combinations of various gases can be used to yield contour intervals of 300 μm and greater. Combinations of various liquids yield contour intervals in the range $1.5\lambda < \Delta z < 500\lambda$. Mixtures of gases or liquids having specific values of refractive index can be formed in order to obtain a desired contour interval. The refractive index of a mixture of gases is given by equation 6.65, where the Gladstone-Dale constant K of the mixture is given by equation 6.66. Liquids can also be mixed to obtain various values of n, but because of volume changes due to mixing, refractive indices cannot in general be added in a simple manner. Experimenters have found it convenient to use distilled water during the first exposure and a mixture of water with a small amount (approximately 1 percent) of ethyl alcohol during the second exposure. Figure 7.5, which was obtained using a

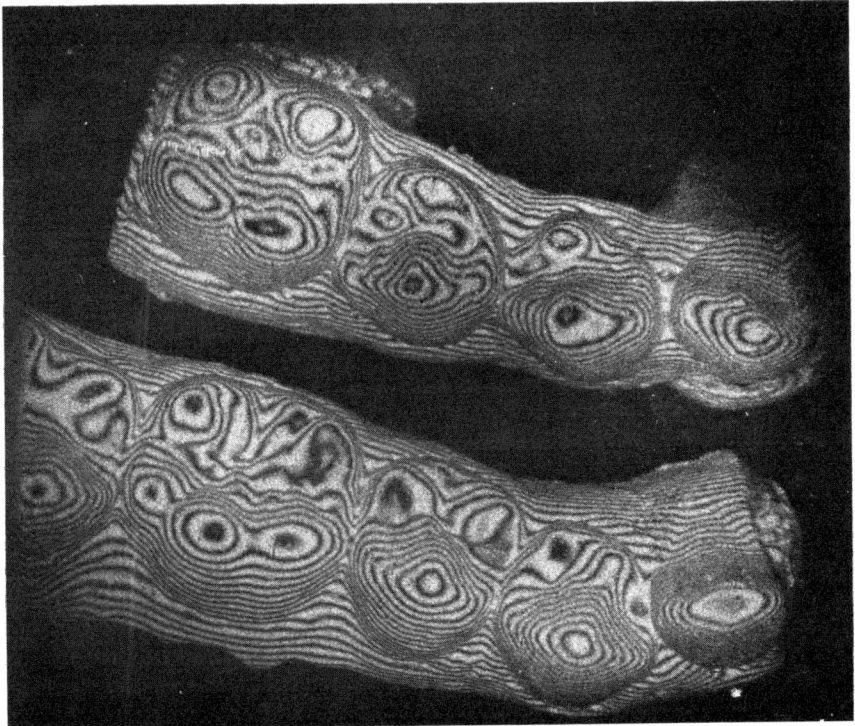

Figure 7.5 Contour map of a metal cast of a human lower dental arch recorded by the multiple-refractive-index method (ref. 12). Contour interval = 300 μm.

396 RELATED TECHNIQUES

water-ethyl alcohol mixture, is a contour map of metal casts of a human lower dental arch. The contour interval for this figure is approximately 300 μm.

7.3 SPECKLE PHOTOGRAPHY AND SPECKLE INTERFEROMETRY

In Section 1.4 we learned that whenever diffused laser light is imaged through a finite aperture a speckle pattern is formed. In most instances this is considered to be optical noise, which degrades images and limits the clarity of interference fringes. However, through the ingenious developments described in this section this phenomenon has been utilized, to the advantage of engineers and scientists, to form the basis of a class of measurement techniques.

7.3.1 Speckle Photography

Speckle photography is a technique for making moderate-sensitivity measurements of in-plane translation, strain, rotation, and vibration. It can also be used to measure out-of-plane rotation (tilt). Its principal attractions are the simplicity of the optical system and the relative ease of displaying and interpreting the results. Requirements for mechanical stability are much less severe than those for holographic interferometry. The sensitivity of the method can be varied during the readout process, and is generally less than that of holographic interferometry. The two methods are therefore complementary.

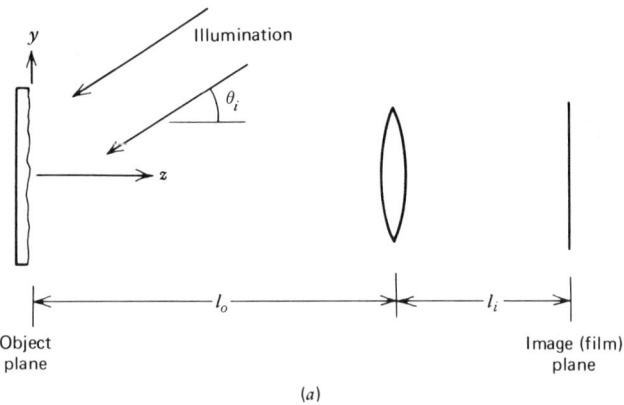

Figure 7.6 Two-exposure speckle photography for measurement of in-plane translation. (a) Recording system. (b) Processing system. (c) Fringe pattern formed in back focal plane of system (b). The object translated in the y direction between exposures.

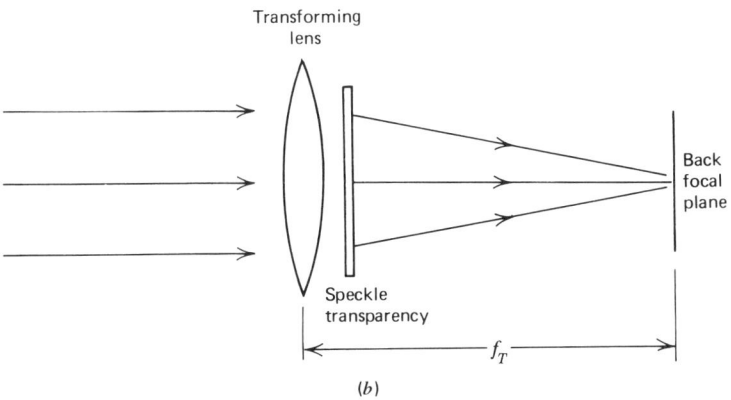

(b)

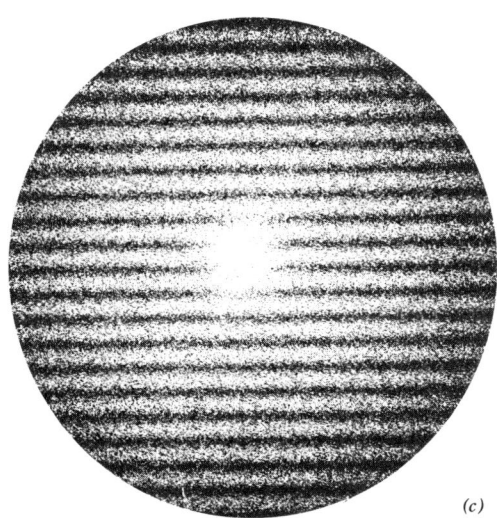

(c)

Figure 7.6 Cont.

To introduce the basic concepts of speckle photography we consider two experiments. The objective of the first is to measure the in-plane translation of a rigid, macroscopically flat object. In the second experiment the tilt of the same object is to be measured. The system for measuring in-plane translation is shown in Figure 7.6a. An image of the object surface is formed in the film plane by a lens of focal length f and diameter D. The object distance l_o and image distance l_i are related by equation 1.44. From the discussions in Section 1.4 we know that the image formed in the film plane will be modulated by a random pattern of speckles having a characteristic size b_s determined by the aperture of the lens:

$$b_s \cong 1.22\lambda\left(\frac{f}{D}\right).$$

If the object translates vertically by an amount L_y, the relative phase of the light in each of the various rays that contribute to the formation of each speckle will be unchanged. Hence the speckle pattern will simply translate as a whole in the film plane by an amount ML_y, where M is the lateral magnification of the imaging lens. Similarly, the speckle will translate an amount ML_x if the object translates L_x in the horizontal direction. The translation of the speckle pattern for these in-plane motions is independent of the angle of illumination θ_i.

To measure the *in-plane translation* of the object surface, the film is exposed twice—once before translation, and once after translation. On the assumption that the magnitude of the translation L is greater than the speckle size b_s, the developed film will contain a pair of identical speckle patterns separated by a distance ML. The separation ML of each speckle pair can be measured directly by microscopic examination of the film. Alternatively, by simple coherent optical processing of the developed film, the displacement can be displayed in the form of a fringe pattern. The film is placed in a converging spherical wave of laser light formed by a lens of focal length f_T, as shown in Figure 7.6b. Following the discussion of Section 1.3, we refer to this as the transforming lens. The irradiance distribution in the back focal plane of this lens consists of a bright central spot surrounded by a speckle pattern modulated by cosinusoidal fringes, as shown in Figure 7.6c. The bright central spot is formed by undiffracted light transmitted by the transparency; the modulated speckle pattern is generated by light diffracted by the speckle pattern recorded on the transparency. The cosinusoidal fringes are formed because each pair of corresponding speckles acts as a pair of identical sources of coherent light which form Young's fringes (see Section 1.2). Since all speckle pairs are separated by the same distance ML, all of the Young's fringes overlap to form the pattern shown in Figure 7.6c.

7.3 SPECKLE PHOTOGRAPHY AND SPECKLE INTERFEROMETRY

Interpretation of the fringes is straightforward. First, their orientation is normal to the in-plane translation **L**. Second, the magnitude of **L** can be determined by applying equation 1.16, which indicates that, if the speckles on the transparency are separated by a distance d_s, the fringes will have a spacing $d_f = \lambda f_T / d_s$. The in-plane translation of the object therefore is given by

$$L = \frac{\lambda f_T}{M d_f}, \qquad (7.12)$$

where λ is the wavelength of the laser light used to form the fringes, f_T is the focal length of the transforming lens, M is the magnification of the imaging system used to record the speckle photograph, and d_f is the fringe spacing.

Before discussing extensions of this method for measuring in-plane motions, let us consider how two-exposure speckle photography can be used to measure out-of-plane rotation, that is, *tilt*.[13,14] This requires only a small modification of the recording system. Instead of photographing the speckle pattern in the image plane, we move the film into the back focal plane of the lens, as shown in Figure 7.7. Again, a pattern of speckles of characteristic size $b_s = 1.22 \lambda (f/D)$ is formed in this plane. However, from the discussion in Section 1.3, it is clear that a given speckle in the film

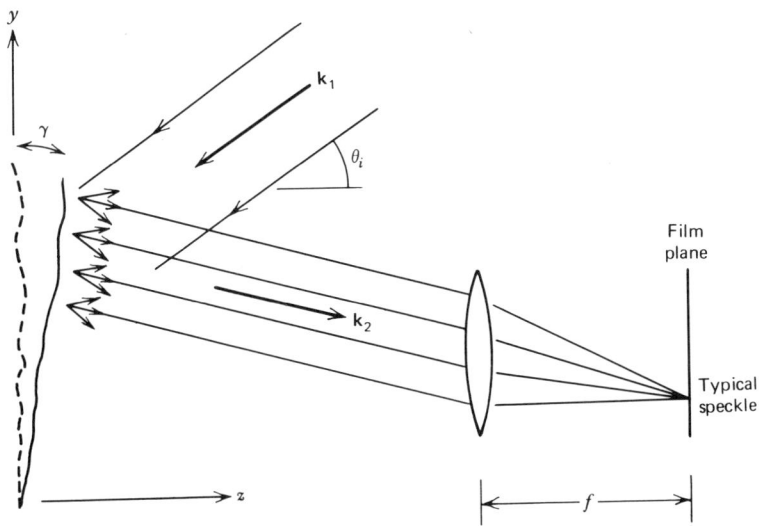

Figure 7.7 Two-exposure speckle photography for measurement of tilt. Light rays contributing to the formation of one speckle are shown.

plane is *not* associated with the light scattered by the neighborhood of a corresponding point on the object surface. Instead, each speckle is formed by all light scattered in a particular direction, as indicated in Figure 7.7. If the object translates in its own plane, $\mathbf{L} = \hat{\mathbf{i}} L_x + \hat{\mathbf{j}} L_y$, the speckle pattern remains stationary; however, if the object tilts by an angle γ, as in Figure 7.7, the speckle pattern in the back focal plane will translate. Each ray contributing to the formation of a single speckle leaves the object surface at a different location y, but all travel in nearly the same direction. When the object is tilted by an amount γ, there is a systematic change of relative phase among these rays. Using the relations developed in Chapter 2, we can write this phase shift as

$$\Delta\phi = \frac{2\pi}{\lambda}(\hat{\mathbf{k}}_2 - \hat{\mathbf{k}}_1) \cdot \mathbf{L}, \qquad (7.13)$$

where $\hat{\mathbf{k}}_1$ is the unit direction vector of the illumination, and $\hat{\mathbf{k}}_2$ is the unit direction vector of the rays contributing to the speckle under consideration. For the configuration in Figure 7.7, $\mathbf{L} = -\gamma y \hat{\mathbf{k}}$, and

$$k_{1x} = 0, \qquad k_{1y} = -\sin\theta, \qquad k_{1z} = -\cos\theta_i,$$

and

$$k_{2x} = 0, \qquad k_{2y} \cong 0, \qquad k_{2z} \cong 1;$$

therefore

$$\Delta\phi = \frac{2\pi}{\lambda}(1 + \cos\theta_i)\gamma y. \qquad (7.14)$$

Equation 7.14 describes a linear variation of phase with y, so the effect is as if all the rays contributing to a single speckle tilted by a small angle, $(1 + \cos\theta_i)\gamma$. The corresponding displacement in the focal plane can be calculated by using equation 1.45. If the angles involved are small, their tangents can be replaced by the angles themselves; then the translation d_s of the speckle in the back focal plane is

$$d_s = f(1 + \cos\theta_i)\gamma, \qquad (7.15)$$

where f is the focal length of the lens. If the angular aperture of the lens is not too large, the translation of the entire speckle pattern will be given by equation 7.15.

To measure tilt the film is placed in the back focal plane of the lens as in Figure 7.7 and is exposed twice—once before tilting the object, and once afterwards. The transparency formed by developing this film is then illuminated by a converging spherical wave formed by a lens of focal length f_T, as in Figure 7.6c. This will form a speckle pattern modulated by cosinusoidal fringes of spacing $d_f = \lambda f_T / d_s$, so the tilt of the object is given by

$$\gamma = \frac{\lambda f_T}{f(1+\cos\theta_i)d_f}. \quad (7.16)$$

Having described the basic concepts of speckle photography, we now consider several extensions and applications of the method.

In-plane vibrations can be measured by *time-averaged speckle photography*.[15-18] If the object in Figure 7.6a vibrates harmonically as a rigid body, $\mathbf{L}(t) = (\hat{\mathbf{i}}L_x + \hat{\mathbf{j}}L_y)\cos\omega t$, the speckles in the film plane translate by an amount $ML\cos\omega t$, where $L = (L_x^2 + L_y^2)^{\frac{1}{2}}$ is the amplitude. A time-average exposure therefore will record a pattern of *speckle streaks* of length $2ML$. The speckles spend most of each vibrational period near the two ends of the streaks; therefore, if the developed film is processed in the system of Figure 7.6b, the fringes observed in the back focal plane will be similar to those caused by an object translation of $2L$. Specifically, the form of the fringes that modulate the speckle pattern in the back focal plane is described by the square of a zero-order Bessel function:

$$J_0^2\left[\left(\frac{2\pi ML}{\lambda f_T}\right)x'\right],$$

where x' is a coordinate parallel to the direction of motion. The amplitude of the vibration is determined by measuring the distance d_f between the dark fringes to the left and those to the right of the bright central fringe. These fringes occur where the argument of J_0 equals ± 2.4, so

$$L = 0.76\left(\frac{\lambda f_T}{Md_f}\right), \quad (7.17)$$

where λ is the wavelength of the laser light used to form the fringes, f_T is the focal length of the transforming lens, and M is the magnification of the imaging system used to record the speckle photograph.

As may be expected from the analogous studies of vibration by time-average holographic interferometry, many extensions of this technique are possible, and several have been realized experimentally. For example, if the object translates in-plane with constant speed v during an exposure interval T, the fringes formed using the system of Figure 7.6b will be of the form[19]

$$\text{sinc}^2\left[\left(\frac{\pi M v T}{\lambda f_T}\right)x'\right].$$

Burch and Tokarski[20] showed that if N exposures are made when the object executes equal translations between each two exposures, the resulting pattern is similar to the diffraction from N equally spaced apertures. Archbold and Ennos[21] have conducted a series of experiments in which the object oscillates along two mutually orthogonal directions with different frequencies. The resulting fringe patterns were interpreted in terms of the Lissajou figures (trajectories) traced out by each speckle, while taking into account the portion of the period spent near each point in the trajectory. Patterns were presented and interpreted for four different rational frequency ratios. Irrational frequency ratios lead to a pattern similar to the diffraction pattern of a rectangular aperture, which is the product of two sinc functions. The interpretation of these time-averaged speckle patterns is similar to that based on the method of stationary phase in holographic interferometry (Section 4.3.3). Tiziani[14] demonstrated the application of vibration measurement by speckle photography to the analysis of a tuning fork designed for use in a quartz watch. He found the amplitude measurements to be accurate to within 3 percent.

If the object under study is strained or rotated during the experiment, displacement varies from point to point on the object surface. Hence a different method of analyzing speckle photographs must be devised.

There are two ways in which local in-plane displacements can be determined from a two-exposure speckle photograph recorded with a system like that in Figure 7.6a. First, the speckle photograph can be examined in one region at a time by illuminating it with a small beam, such as an unexpanded He-Ne laser beam, as shown in Figure 7.8. The speckle pattern on the transparency will diffract this beam into a diverging cone of angular extent[22]

$$\alpha \cong \left(\frac{f}{D}\right)^{-1}, \tag{7.18}$$

where f/D is the f number of the lens used to record the speckle photograph. If the small illuminated region contains pairs of identical speckles displaced by a distance d_s, the light in the resulting pairs of diffraction cones will interfere to form a diffraction pattern modulated by Young's fringes. If the fringes are viewed or photographed at a distance z_f from the transparency, their spacing will be $d_f = \lambda z_f / d_s$. Since $d_s = ML$, where M is the lateral magnification of the imaging system used to record

7.3 SPECKLE PHOTOGRAPHY AND SPECKLE INTERFEROMETRY

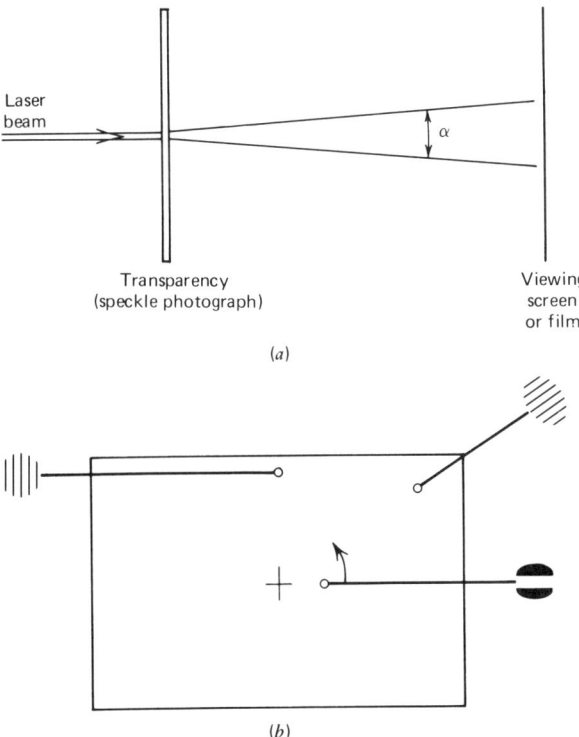

Figure 7.8 Formation of fringes by illuminating a two-exposure speckle photograph with a thin laser beam. (*a*) Optical system. (*b*) Appearance of fringes formed at various locations on a speckle photograph of a rigid plate which rotated about a normal axis through the center *O*.

the speckle photograph and L is the unknown displacement,

$$L = \frac{\lambda z_f}{M d_f}. \tag{7.19}$$

Equation 7.19 gives the displacement of points in a single small region of the object surface. The fringes are normal to the local in-plane displacement. Figure 7.8*b* is a schematic representation of the fringes which would be observed if the object had rotated about an axis normal to its surface.

An alternative method of analyzing speckle photographs results in an image of the object overlaid with fringes which are contours of constant vector displacement.[17,23-25] Such an image is produced by the simple spatial filtering system shown schematically in Figure 7.9*a*. The entire transparency is illuminated by a converging spherical wave of laser light. A small circular aperture is placed in the back focal plane of the first lens. Suppose that the aperture is placed at location y_a above the optical axis, as in Figure 7.9a. This aperture is located where a bright fringe would be if a thin beam had illuminated an object point which translated by an integral multiple of $L = \lambda f_T / M y_a$. The aperture would be located in a dark fringe associated with any object point which translated by $(N + \frac{1}{2})L$, where N is

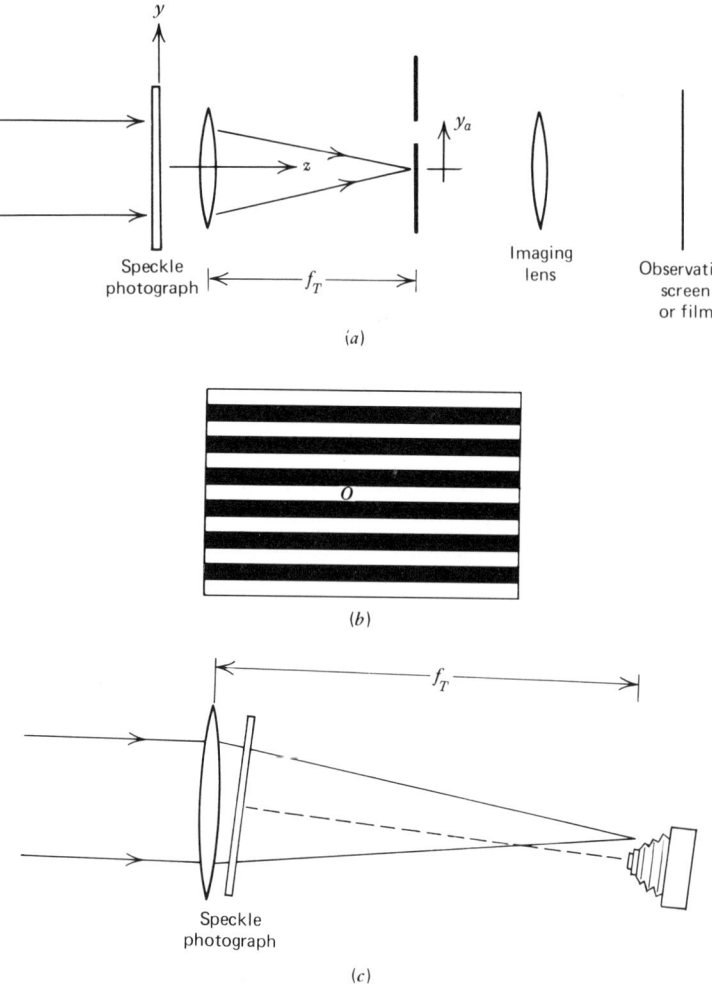

Figure 7.9 Whole-field analysis by speckle photography. (a) Optical system for forming fringes that are contours of constant y component of displacement. (b) Appearance of fringes in image of a rigid plate which rotated about a normal axis through the center O. (c) A simple form of the optical system using an ordinary camera.

an integer. Hence, when the entire film is illuminated in the system of Figure 7.9a, the image will be formed with light from only points whose y components of displacement are integer multiples of $\lambda f_T / M y_a$. The image is covered with bright fringes which are contours of constant y displacement, $L_y = N \lambda f_T / M y_a$, where $N = 0, \pm 1, \pm 2, \ldots$. For example, if a speckle photograph of an object which rotated about an axis normal to its surface were filtered in this manner, the image would be like that shown schematically in Figure 7.9b.

This filtering technique is quite flexible. If the aperture in the back focal plane is located at radial distance ρ_a from the optical axis and at an azimuthal angle ϕ_a, measured from the y axis, each bright fringe in the image is a contour connecting all points whose components of displacement in the direction of a line connecting the aperture and the optical axis are

$$L = \frac{N \lambda f_T}{M \rho_a}, \quad N = 0, \pm 1, \pm 2, \ldots . \quad (7.20)$$

A simple system for carrying out this type of spatial filtering is shown in Figure 7.9c. The imaging and filtering are done with an ordinary 35 mm camera.[26,27] The transparency and camera are mounted on a rigid arm which can be pivoted about the center of the transparency. The distance from the transforming lens to the camera is set so that undiffracted laser light comes to focus at the camera aperture. If this aperture is stopped down to a small diameter, it passes only light from a small region of the back focal plane of the transforming lens. This region is selected by pivoting the arm on which the camera and transparency are mounted. If the camera is focused on the transparency, an image with the desired contours of constant component of displacement can be photographed. Two images of this type are shown in Figure 7.10.

Duffy[24,28] introduced *double-aperture speckle photography*, which produces a final image with improved fringe quality and contrast. This improvement is achieved at the expense of increasing exposure time and fixing the sensitivity of the measurement a priori. In Duffy's system, which is shown in Figure 7.11a, two small apertures of diameter w_a are placed in front of the imaging lens used to record the speckle photograph. These apertures are located symmetrically about the optical axis and are separated by a distance d_a. The system can be an ordinary camera with the aperture plate mounted in front of the lens or, preferably, close to the iris diaphragm. The object is illuminated with laser light and is imaged onto the film plane. Large speckles of characteristic size $\lambda l_i / w_a$, determined by the diameter of the small apertures, are formed in the image (film) plane. This speckle is modulated by fine interference fringes whose spacing $\lambda l_i / d_a$

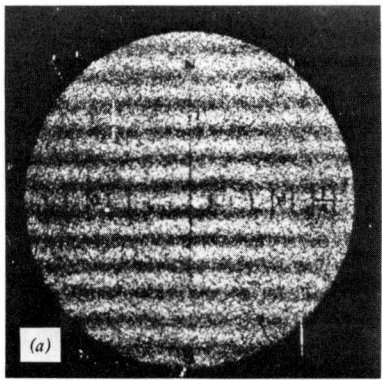 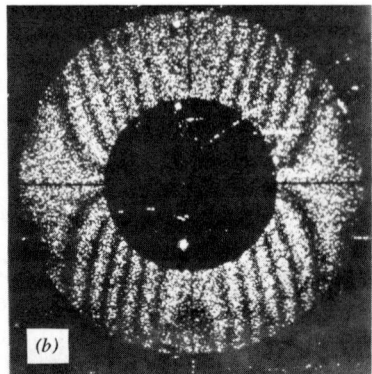

Figure 7.10 Displacement contours formed by a system like that in Figure 7.9c (ref. 26). (a) The object is a disk of 10 cm diameter which rotated about an axis normal to its surface. (b) The object is a ring (9 cm OD, 4 cm ID) which was loaded in diametral compression.

is determined by the separation of the apertures. A small portion of this speckle pattern is shown schematically in Figure 7.11c and is compared with the speckle pattern formed when the entire aperture, of diameter d_a, is used, as in Figure 7.6a.

When a two-exposure speckle photograph is recorded using a two-aperture system, the contrast of the fringes, which are visible in the regions of dark speckle, varies from point to point. The fringes formed during each exposure overlap in regions where the object either remained stationary or translated in the y direction by an integral multiple of $\lambda l_i / M d_a$. This results in high-contrast fringes on the developed transparency. In regions where the object translated in the y direction by $(N + \frac{1}{2})(\lambda l_i / M d_a)$, where N is an integer, bright fringes during the first exposure will overlap dark fringes during the second exposure, so fringe contrast on the developed transparency will be zero. When the transparency is placed into a processor like that in Figure 7.9a or 7.9c, regions with high-contrast fringes strongly diffract light in a direction such that it can be admitted by an aperture located at $y_a = \lambda f_T / w_a$. The image formed thus contains bright fringes which are contours connecting all points that translated in the y direction by an amount

$$L_y = N\left(\frac{\lambda l_i}{M w_a}\right), \qquad N = 0, \pm 1, \pm 2, \ldots . \tag{7.21}$$

The double-aperture method also can be used for real-time measurement of in-plane displacement. A speckle photograph is recorded in the system

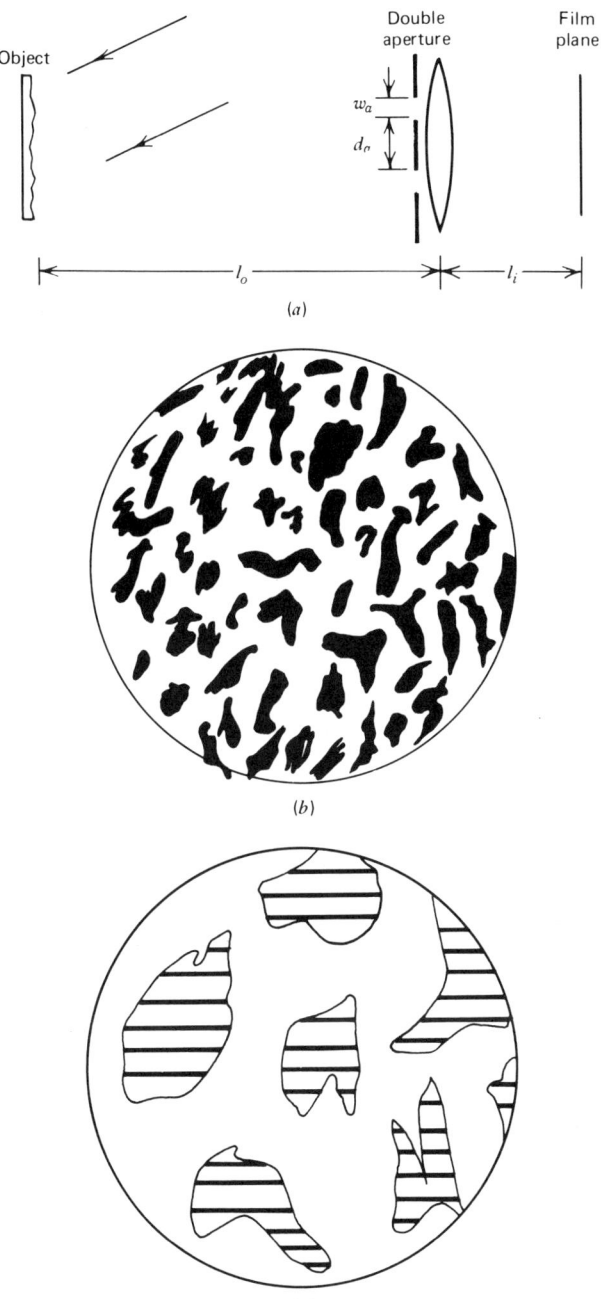

Figure 7.11 Double-aperture speckle photography. (*a*) System for recording speckle photographs. (*b*) Speckle formed if a large single aperture of diameter d_a is used. (*c*) Modulated speckle pattern formed using the system depicted in (*a*).

shown in Figure 7.11a. After the film is developed, it is carefully replaced in its initial position in the system, and the object is again illuminated and is viewed through the transparency. If the object is then translated or deformed, fringes which are contours of constant L_y can be observed in the image. The dark fringes are described by equation 7.21. Burch and Forno[29,30] have developed a *white-light* technique for measuring displacements normal to the viewing direction which is closely analogous to two-aperture speckle photography. The object is painted with retroreflective paint consisting of small glass spheres embedded in a matrix of opaque spheres. The object is then illuminated with incoherent light and photographed through a lens fitted with a parallel-slit aperture, which gives a ringing response to the light reflected by each glass sphere. This is analogous to the fine fringe pattern in two-aperture speckle photography. The photographs recorded by this method can be processed with coherent light, by the same methods used in speckle photography, to produce low-sensitivity contours of displacement. When possible, the surface can be painted or scribed with a grid tuned to the aperture separation of the camera in order to improve fringe quality.

So far we have discussed the use of speckle photography to measure lateral displacement and tilt. In the measurement of lateral displacement, the camera is focused on the object, so there is a mapping of information about object points onto corresponding image points. It is therefore possible to study deformations or rotations in which lateral displacement varies from point to point. Measurement of object tilt, on the other hand, requires that the speckle photograph be recorded in the back focal plane of the taking lens, so there is not a mapping of object points into image points. Deformations which result in variation of tilt from point to point cannot be measured by this method. This difficulty can be overcome, however, if it is known a priori that the object does not translate appreciably in the lateral direction.

To measure the distribution of tilt over a surface, the imaging system shown in Figure 7.6a is focused on a plane located a distance Δl_o in front of, or behind, the object surface. As explained above, if a region of the object surface tilts by a small angle γ, the rays contributing to the speckle effectively tilt by an angle $(1+\cos\theta_i)\gamma$. This results in a speckle displacement

$$d_s = M \Delta l_o (1 + \cos\theta_i)\gamma \qquad (7.22)$$

in the film plane. Consequently a two-exposure speckle photograph recorded in this manner can be processed by the system shown in Figure 7.8, or by the ones shown in Figure 7.9, to display the variation of tilt over the object surface. The sensitivity of this measurement is proportional to Δl_o,

7.3 SPECKLE PHOTOGRAPHY AND SPECKLE INTERFEROMETRY

the amount of defocusing; however, as Δl_o increases, the region of the object surface over which tilt is averaged also increases.

The measurement of local slopes is important in stress and strain analysis of plates and shells. Consider a flat plate, lying in the x-y plane. If it is bent slightly, it will have components of slope $\partial w/\partial x$ and $\partial w/\partial y$, where w is the deformation normal to the surface. The relation of these slopes to stress, strain, and bending moments is discussed in Chapter 4. By recording a two-exposure speckle photograph, using the full aperture, with the camera focused on a plane Δl_o in front of the surface, information regarding $\partial w/\partial x$ and $\partial w/\partial y$ can be recorded. Contours of these slopes can be displayed by processing the speckle photograph with the system shown in Figure 7.9a. If the filtering aperture is located on the x axis a distance ρ_{a_x} from the optical axis, the bright fringes will be contours of constant partial slope $\partial w/\partial x$:

$$\frac{\partial w}{\partial x} = \frac{N\lambda f_T}{2M\Delta l_o \rho_{a_x}}, \qquad N = 0, \pm 1, \pm 2, \dots. \tag{7.23}$$

Similarly, if the filtering aperture is located at ρ_{a_y} on the y axis, bright fringes will be contours of constant partial slope $\partial w/\partial y$:

$$\frac{\partial w}{\partial y} = \frac{N\lambda f_T}{2M\Delta l_o \rho_{a_y}}, \qquad N = 0, \pm 1, \pm 2, \dots. \tag{7.24}$$

In equations 7.23 and 7.24 it has been assumed that the object is illuminated by a source close to the camera, so that $\theta_i = 0$. This results in equal sensitivity to the two partial slopes. Point-by-point measurements of slope can also be made by illuminating the speckle photograph with a thin beam, as in Figure 7.8. Chiang and Juang[31] demonstrated this technique by forming partial slope contours on circular and triangular clamped plates subjected to concentrated central loads. Chiang and Lee[32] further used the method with pulsed-laser illumination to record contours of $\partial w/\partial x$ and $\partial w/\partial y$ on plates subjected to rapid transient stresses by impact loading.

Equations 7.22 and 7.15, which are used to determine tilt by speckle photography, are based on the use of collimated illumination. If the object is large, however, the use of collimated illumination is precluded. Gregory[33] has shown that tilt can be measured, even in the presence of translation, when the illumination wave is spherical. He found empirically that the trajectory of speckle motion is coincident with the curve traced by a ray which would be reflected to the observation plane if the surface consisted of a collection of tiny mirror facets. The plane in which speckle displacement is due to object tilt, and is independent of object translation, is the (virtual) plane containing the image of the source point in the mirror facets. Object tilt can be measured by focusing the camera on this plane.

Speckle motion recorded by a camera focused sharply on the object surface is independent of object tilt, regardless of the direction and shape of the illuminating wavefront. For practical purposes it is also independent of object translation in the observation direction; however, measurements of normal displacement with very low sensitivity, on the order of millimeters, can be made because of changes of magnification in the speckle image.[17] The requirement that the object surface be sharply in focus creates difficulties in practice when an object surface is not flat or must be viewed from an angle. Keeping such an object in focus, that is, increasing the depth of field, requires the use of a small aperture in the recording camera. This, however, increases the characteristic speckle size and decreases the sensitivity of the measurement. If the object is not in focus, speckle motion in the film plane will be due to a combination of lateral translation and tilt, and incorrect interpretation of the speckle photograph is likely to result.

Stetson[34] has analyzed the apparent motion of speckles due to arbitrary object displacement and deformation, and has concluded that by recording two two-exposure speckle photographs from the same observation direction, but with the camera focused on different planes, the object translation normal to that direction can be extracted. Furthermore, if such a measurement is made from three independent observation directions, the vector displacement, strains, shears, and rotations can, in theory, be determined as in holographic interferometry. The development of such a method would be quite useful because it could serve to measure deformations about two orders of magnitude greater than those which can be measured by holographic interferometry.

Holographic interferometry is most sensitive to deformations normal to the object surface, whereas speckle photography is most sensitive to deformations in the plane of the object surface. Thus it is natural to consider combining the two methods. Adams and Maddux[35] and Boone[36] have discussed such hybrid techniques for use in stress analysis. It is necessary to use image plane holography, in which the object is imaged onto the hologram,[35] or to physically attach the hologram to the object surface[36] and record a volume hologram.

Hudson and Setopoulos[37] have demonstrated a system for recording sequential pairs of speckle photographs in order to measure time-dependent deformations such as creep. A variety of applications have been reported. These include work in nondestructive testing,[38,39] metrology,[40,41] fracture mechanics,[42] and materials studies.[43]

There are two methods by which speckle photography can be applied to the study of *transparent objects*. The first method is potentially useful for

7.3 SPECKLE PHOTOGRAPHY AND SPECKLE INTERFEROMETRY

visualization and measurement in aerodynamics, heat transfer, and plasma diagnostics, and in the testing of weak optical elements. In this method the apparent displacement of speckles due to refraction of diffused light as it passes through the test medium is measured. The second method is applicable to both solid and fluid mechanics and is based on the measurement of speckle displacement caused by the motion of small scattering particles embedded in the test medium.

If a transparent test object is placed in front of a laser-illuminated diffuser, as shown in Figure 7.12, a speckle pattern can be formed in the film plane of a camera which images the diffuser. If the refractive index varies, either continuously or discontinuously, in the object, the rays contributing to formation of the speckle pattern in the film plane will be deflected by refraction (see Section 5.2). This causes a displacement of each speckle from the position it would occupy if the transparent object were not present. Let a two-exposure speckle photograph be recorded. During one exposure the test object is not present, and during the other exposure the object is present. This speckle photograph can be analyzed point by point to determine the speckle displacement. Alternatively, contours of constant displacement in a given direction can be produced by using the systems shown in Figure 7.9.[44,45]

To interpret the fringe systems generated by this method, the camera should have a small aperture so that all rays contributing to a particular speckle are deflected by essentially the same amount. Furthermore, it is preferable that all rays contributing to the image travel in nearly the same direction, so the camera should be placed a fairly large distance from the object. Alternatively, a telecentric system can be used to form the image. The problem of quantitative determination of the refractive index distribution in the test object from speckle deflection data is similar to the inversion problems associated with schlieren systems[46] and with plasma diagnostic techniques based on angular deflection of probing laser beams.[47] The inversion of speckle photography data should be feasible if

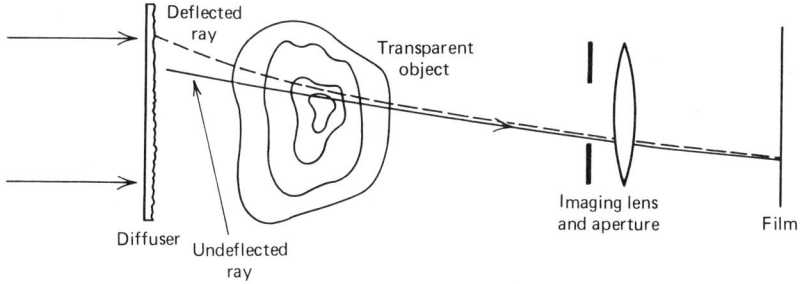

Figure 7.12 System for recording speckle photographs of refracting, transparent test objects.

careful account is taken of the role of the imaging system and apparent ray crossing (see Section 6.2.2). Other methods of using speckle photographs of transparent objects utilize white light illumination to produce color-schlieren-like images.[45,48]

Barker and Fourney[49,50] have shown that speckle photography can be used to measure internal material displacements of weakly scattering transparent objects. A planar region inside the test object is illuminated by a thin sheet of laser light. Small scattering centers, either naturally occurring or seeded into the material, scatter light out of the illuminated plane. The resulting laser speckle can be photographed by a camera focused on the illuminated plane. If the object is a solid, two-exposure speckle photographs can be recorded to determine the distribution of material displacement in the plane under study when the object is loaded mechanically. Such speckle photographs are processed and analyzed by the same methods used for measuring the surface deformation of opaque objects. Barker and Fourney[49] have demonstrated the method by measuring interior displacements of a Plexiglas cantilever beam and also the displacements in a Plexiglas cube subjected to a concentrated compressive load. If a double-pulsed laser is used, velocities can be measured in fluids[50] or in transiently stressed transparent solids.[49]

We conclude this discussion of speckle photography with a few comments regarding film and optical systems. The sensitivity of measurement by speckle photography is limited by the requirement that speckle motion in the film plane must exceed the characteristic speckle size, $b_s = 1.22\lambda(f/D)$. If $f/4$ is used as a typical relative aperture of the camera, $b_s = 3$ μm, so film resolution of about 300 lines/mm is required. Therefore a relatively high-resolution film is necessary. The film should have a high MTF (see Section 1.5) up to the maximum frequency of the speckle pattern, which is $f_{max} = [\lambda(f/D)]^{-1}$. High-resolution holographic emulsions, coated either on plates or on film, therefore are generally used for speckle photography.[22] Since film speed varies inversely with resolution, fairly long exposure times may be required, especially if the double-aperture method is used. Speckle photographs should be exposed and developed to achieve high contrast, so that the diffraction efficiency will be good. In fact, the best results are obtained by bleaching the speckle photograph, using one of the methods discussed in Section 1.5.[26,27]

The upper limit to measurable displacements or tilts is set by decorrelation of the two speckle patterns. This is evident because processing of speckle photographs requires formation of Young's fringes by diffraction of light by pairs of identical speckles. Translucent surfaces which undergo even a small tilt cannot be studied by speckle photography because the multiple reflections occurring in the material below the surface cause changes in the form of the speckle. Similarly, a loss of correlation will

7.3 SPECKLE PHOTOGRAPHY AND SPECKLE INTERFEROMETRY

result if strains are large enough to appreciably alter the microscopic structure of the opaque surface under study. In the absence of these problems, loss of correlation occurs when motions are sufficiently large that the aperture of the taking lens samples an appreciably different portion of the wavefront scattered by the object during the two exposures. When this occurs, only a fraction of the speckle pattern recorded on the photograph consists of pairs of identical speckles, so the fringes they generate are washed out by the light diffracted by the rest of the speckle. A rough estimate of the sensitivity range of speckle photography, based on experiments reported in the literature, is that motions resulting in speckle translations of the order of 15 to 200 μm *in the film plane* can be measured. This lower limit is roughly 5 times the size of speckles formed with an $f/4$ lens. In the upper limit the object translation equals about 10 percent of the diameter of the entrance pupil of a typical camera lens at $f/4$.

The practical sensitivity range in terms of actual object translation or tilt depends primarily on object size, magnification, and required depth of focus. In general the range extends to motions as large as 1.5 mm. Above this range direct measurement by ordinary double-exposure photography is practical. The lower limit of the range is about 5 μm, but depth-of-focus requirements in applications usually raise this limit to the order of 100 μm or greater. Hence in practice there is often a significant gap between the maximum displacement measurable by holographic interferometry and the smallest displacement measurable by speckle photography.

7.3.2 Speckle Interferometry

Speckle interferometry is a class of measurement techniques which involve the coherent addition (interference) of a speckle field and a plane reference wave or another speckle field. These techniques have a sensitivity comparable to that of holographic interferometry.

We begin our discussion by considering speckle interferometry in which a reference wave is used. Suppose that laser light is scattered by a rough surface toward an observation screen or photographic film. It was shown in Section 1.4 that the characteristic size of the speckle is $b_s = 1.22(\lambda z/D)$, where z is the distance from the object to the observation plane and D is the diameter of the object. If an imaging lens is used, $b_s = 1.22(\lambda f/D)$, where f/D is the f number of the lens. If we combine the diffused light with a coherent plane reference wave whose irradiance is equal to the mean irradiance of the speckle field, two changes in the speckle pattern occur. First, the size of a typical speckle is approximately doubled,[51] so $b_s = 2.44(\lambda z/D)$ or $b_s = 2.44(\lambda f/D)$. Second, there is a small change in the brightness distribution.[52]

The addition of the reference wave introduces a very important change in the behavior of the speckle pattern as the object translates toward, or away from, the observer. On the basis of the study of longitudinal statistics of speckle by Weigelt and Stoffregen,[53] speckle can be thought of as having a rather elongated structure to the right of the imaging lens, as shown in Figure 7.13a. Consequently, as the object is moved in an axial direction, very little change in the structure of the speckle pattern in the image plane occurs. In other words, small axial translation induces no change in the *relative* phase of the light scattered by each point on the surface. For this reason the speckle photography methods discussed in Section 7.3.1 are very insensitive to normal motions. Now suppose that a coherent reference wave traveling in the z direction is added to the speckle field at the image plane. The observed speckle pattern is now formed by interference of the speckle field with the reference wave. If the object moves a distance Δz,

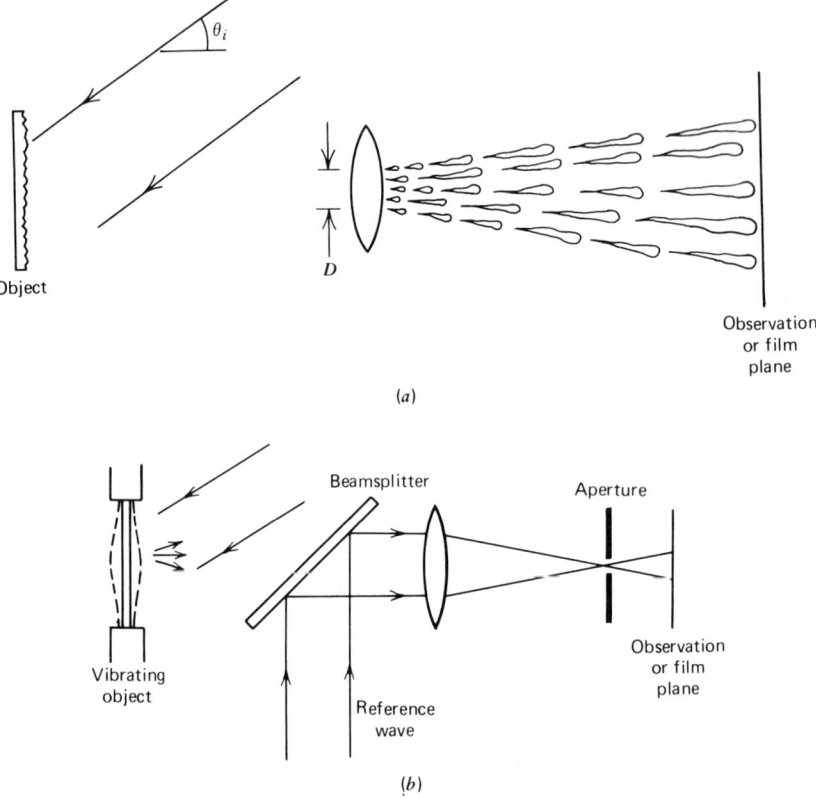

Figure 7.13 Vibration measurement by speckle interferometry. (*a*) Schematic representation of the longitudinal structure of speckle. (*b*) Speckle interferometry for observation of a vibrating surface.

7.3 SPECKLE PHOTOGRAPHY AND SPECKLE INTERFEROMETRY

the relative phase of these two fields will change by

$$\Delta\phi = \frac{2\pi}{\lambda}(1+\cos\theta_i)\Delta z. \tag{7.25}$$

The irradiance of the speckle pattern will therefore vary periodically as the object moves in the axial direction. If

$$\Delta z = \frac{N\lambda}{1+\cos\theta_i}, \quad N = 0, \pm 1, \pm 2, \ldots,$$

the speckle pattern will be identical to that of the original object with $\Delta z = 0$. If

$$\Delta z = \frac{(N+\frac{1}{2})\lambda}{1+\cos\theta_i}, \quad N = 0, \pm 1, \pm 2, \ldots,$$

there will be a contrast reversal, so that regions which were originally dark will now be bright, and vice versa. The speckle therefore appears to twinkle as the object is translated slowly in the axial direction. Speckle interferometers based on this effect can be used to observe the modal structures of vibrating surfaces.[54] Such a device is shown schematically in Figure 7.13b. The surface being studied is illuminated by laser light, and an image of it is formed using a variable aperture to control the speckle size. Light scattered from the nodal regions of the surface forms a distinct, stationary speckle pattern in the observation plane. The speckle irradiance at other image points, however, fluctuates periodically as the surface vibrates, and averages out to a relatively uniform value if it is observed visually or is recorded photographically with an exposure time greater than the vibrational period. This image is called a *speckle interferogram*. The observer can detect the nodal regions because they have a high speckle contrast. Regions which vibrate have very low-contrast speckle. A speckle interferogram of a vibrating rectangular plate is shown in Figure 7.14. The nodal region is quite obvious in this interferogram. Close inspection reveals some fluctuation of speckle contrast over the vibrating region of the surface. In fact, bands of constant speckle contrast are contours of constant vibrational amplitude. Ek and Molin[55] have shown that a normally illuminated object surface which vibrates sinusoidally,

$$\Delta z(x,y) = A(x,y)\sin\omega t, \tag{7.26}$$

gives rise to a speckle interferogram with contrast given by

$$\nu(x,y) = \frac{\{1+2\alpha J_0^2[(4\pi/\lambda)A(x,y)]\}^{1/2}}{1+\alpha}, \tag{7.27}$$

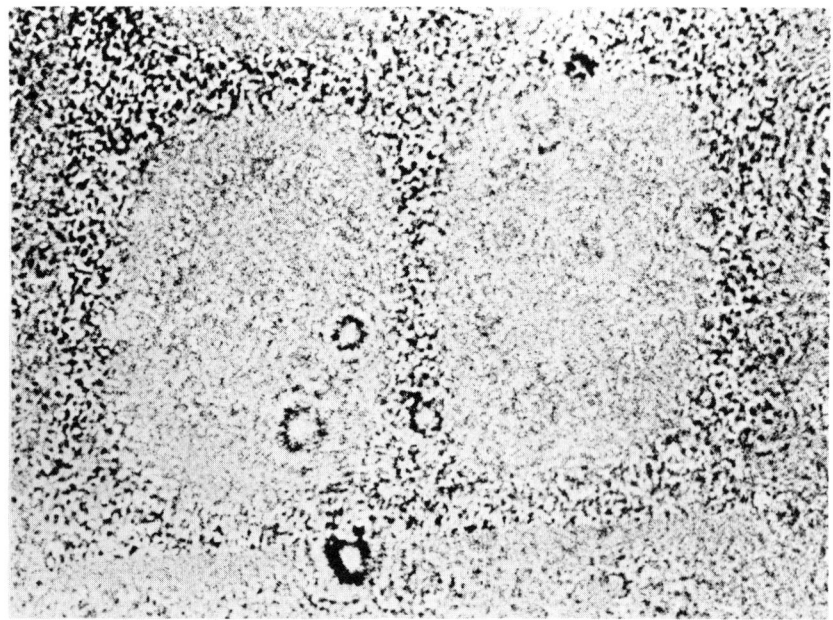

Figure 7.14 Speckle interferogram of a vibrating rectangular plate showing contours of constant vibrational amplitude (ref. 55).

where α is the ratio of reference wave irradiance to object wave irradiance. This contrast modulation is not large, and in practice it is difficult to detect more than two consecutive amplitude contours. The pattern is also rather difficult to photograph, although the detection of nodal areas by visual observation is generally quite easy.

Although the measurement of vibrational amplitudes by speckle interferometry is difficult, the technique can be used very effectively in conjunction with time-average holographic interferometry. The speckle interferometer can serve to identify resonance conditions and nodal patterns; then holographic interferometry can be used to record a high-quality fringe pattern for the quantitative evaluation of vibrational amplitude. The detailed design of a speckle interferometer for the observation of vibrating surfaces is given by Stetson.[56] This interferometer uses an external reference beam and can be inserted in the hologram plane of an off-axis holographic system. Once the desired resonance has been established under observation with the speckle interferometer, it can be removed and a time-average hologram can be recorded.

A second type of speckle interferometry involves the *coherent addition* of two speckle fields produced by illuminating the surface under study from

7.3 SPECKLE PHOTOGRAPHY AND SPECKLE INTERFEROMETRY

two different directions. Consider an object which is illuminated by two plane waves, as shown in Figure 7.15. An image of the object is formed using a lens with an aperture sufficiently small to yield a photographically resolvable speckle pattern. When two speckle fields having the same mean irradiance are coherently added, the resultant speckle field has the same statistical characteristics as the original fields.[57] The average speckle size, normalized brightness distribution, and contrast in the image plane will therefore be the same when both beams illuminate the object as when one beam illuminates it. At any point in the image plane, the irradiance of the speckle pattern formed by double illumination of the object can be represented by

$$I = a_1^2 + a_2^2 + 2a_1 a_2 \cos \Delta\phi, \tag{7.28}$$

where a_1 and a_2 are the real amplitudes of the light scattered from each illumination wave to the point, and $\Delta\phi$ is their phase difference. If the corresponding point in the object surface translates by a small amount **L**, the phase difference $\Delta\phi$ will change by an amount δ, which can be calculated by using equation 2.8. The phase change of light scattered from illumination wave 1 will be

$$\delta_1 = (\mathbf{k}_3 - \mathbf{k}_1)\cdot\mathbf{L}. \tag{7.29}$$

Similarly, for illumination wave 2,

$$\delta_2 = (\mathbf{k}_3 - \mathbf{k}_2)\cdot\mathbf{L}. \tag{7.30}$$

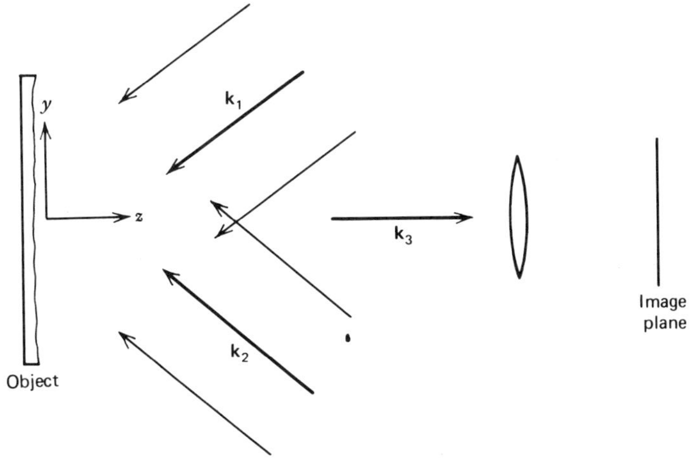

Figure 7.15 Dual-illumination-wave speckle interferometer.

The irradiance of the speckle pattern after displacement or deformation of the object will be

$$I = a_1^2 + a_2^2 + 2a_1 a_2 \cos(\Delta\phi + \delta), \tag{7.31}$$

where

$$\delta = (\mathbf{k}_2 - \mathbf{k}_1)\cdot\mathbf{L}, \tag{7.32}$$

and $\mathbf{k}_1$ and $\mathbf{k}_2$ are the propagation vectors of the two illumination waves. Note that δ is independent of the observation direction. According to equation 7.31, if $\delta = 2N\pi$, where N is an integer, the speckle pattern will be unchanged after the small object translation $\mathbf{L}$. However, if $\delta = (2N+1)\pi$, the speckle pattern after translation will be completely uncorrelated with the pattern before translation. Thus, if the object is slowly displaced or deformed, the speckle pattern in any region of the image will periodically lose and regain correlation. Three basic types of speckle interferometry have been devised to measure vector displacement based on this periodic change of speckle correlation with displacement: real-time, superposition, and two-exposure.[58] Each is discussed below.

Real-time speckle interferometry was described in a pioneering paper by Leendertz.[59] A photographic plate is placed in the image plane of the system shown schematically in Figure 7.15 and is exposed while the object is illuminated by both plane waves. The plate is developed and replaced precisely in its original position. This requires a high-precision plate holder, or the use of *in situ* processing, which is discussed in Section 4.3.4. The object is then reilluminated and viewed through the speckle mask (developed plate). The speckle mask contains a *negative* image, so the bright speckles in the scattered light illuminate black images of themselves, and only dark speckles illuminate the highly transparent portions of the plate. The amount of light transmitted by the mask is therefore low if the object does not change. However, if a region of the object is translated by an amount $\mathbf{L}$ so that $\delta = (2N+1)\pi$, the speckle pattern scattered to the plate will be uncorrelated with the image on the speckle mask, and light will be readily transmitted by the transparent portions of the mask. An observer viewing or photographing the object through the speckle mask will see light and dark fringes modulating a speckled image of the object. The dark fringes are contours of constant phase change $\delta = 2N\pi$, and the bright fringes are contours of constant phase change $\delta = (2N+1)\pi$. The fringes outline regions of the object which have the same component of displacement normal to the bisector of the two illumination directions and lying in the plane containing $\mathbf{k}_1$ and $\mathbf{k}_2$.

7.3 SPECKLE PHOTOGRAPHY AND SPECKLE INTERFEROMETRY

Usually the system is arranged so that the two illumination waves propagate in directions which are symmetrical about the surface normal, as in Figure 7.16a. A simple method[60] of realizing this illumination geometry is shown in Figure 7.16b. If θ is the angle between each illumination direction and the z axis, $(\mathbf{k}_2 - \mathbf{k}_1) \cdot \mathbf{L} = (2\pi/\lambda)(2\sin\theta)L_y$, so *bright fringes* are contours of constant displacement L_y:

$$L_y = \frac{(2N+1)\lambda}{4\sin\theta}. \tag{7.33}$$

This technique, for practical purposes, is sensitive only to a transverse

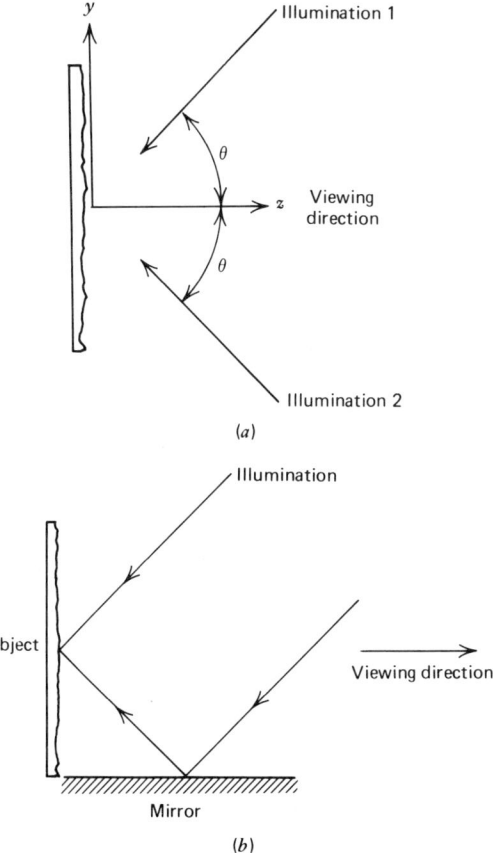

Figure 7.16 Speckle interferometry with symmetrically oriented dual illumination waves. (a) Illumination and viewing directions. (b) A scheme for realizing optical configuration (a).

component of displacement (L_y in this case). The sensitivity to this component can be greater than that readily attainable by holographic interferometry. The fringe visibility, however, is rather low; theoretically it is of the order of 14 percent.[59]

Leendertz[59] also showed that fringes can be formed by recording two separate photographs, one before and one after displacement, in the image plane of a system such as that shown in Figure 7.15. The aperture is adjusted to give relatively large speckle size. A positive contact print of one of the plates, say that exposed first, is made on a third plate. This contact print is placed in exact registration with the second plate, and the pair is illuminated from behind with incoherent light. The fringes of constant displacement component will then be seen as regions of high and low transmittance. This technique is referred to as *superposition speckle interferometry*.

Two-exposure speckle interferometry can be accomplished with the same system (Figure 7.15). The photographic plate or film is placed in the image plane and exposed twice, once before and once after displacement. The transmittance of the developed plate is proportional to the sum of two speckle patterns, I:

$$I = \left(a_1^2 + a_2^2 + 2a_1 a_2 \cos \Delta\phi\right)$$
$$+ \left[a_1^2 + a_2^2 + 2a_1 a_2 \cos(\Delta\phi + \delta)\right]. \quad (7.34)$$

In other words, the two speckle patterns are *incoherently added*. To see how the desired fringes can be produced, it is necessary to describe the statistics of the incoherent sum of two speckle patterns. If only a single exposure is recorded, the probability density distribution of irradiance is given by equation 1.49:

$$P_I(I) = \frac{1}{\langle I \rangle} \exp\left(\frac{-I}{\langle I \rangle}\right).$$

This function is plotted in Figure 1.22. There is a predominance of black speckles in such a field. Correspondingly, there is a predominance of transparent speckles in a negative photographic transparency of it. If two statistically identical but *uncorrelated* speckle patterns having equal mean irradiances are incoherently added, the probability density distribution of irradiance in the resultant speckle pattern is quite different:[52]

$$P_I(I) = \frac{4I}{\langle I \rangle^2} \exp\left(\frac{-2I}{\langle I \rangle}\right). \quad (7.35)$$

7.3 SPECKLE PHOTOGRAPHY AND SPECKLE INTERFEROMETRY

This function is plotted in Figure 7.17, which should be compared with Figure 1.22. The incoherent sum of two uncorrelated speckle patterns will have few dark speckles, that is, the speckle contrast will be low.

Regions of the doubly exposed plate in which the speckle patterns during the two exposures are uncorrelated will have relatively low-contrast speckles, whereas regions where the two patterns are well correlated will have high-contrast speckles, predominantly black. The difference in contrast can be enhanced by nonlinear photography.[22] The predominantly dark regions of good correlation will have high transmittance in the negative transparency, whereas the transmittance in regions of low correlation will be only moderate. Therefore, if the transparency is illuminated from behind with diffused white light, the fringes can be observed as relatively light and dark bands, which can be interpreted by using equation 7.32 or, in the case of symmetrical illumination, equation 7.33.

Regions of high correlation have greater speckle contrast and therefore diffract more light than regions of low correlation. This suggests a method of producing fringes of higher contrast. The photographic transparency of the two-exposure speckle pattern should be processed for enhanced contrast by overdevelopment[22] or bleaching (see Section 1.5). The transparency is then placed in a simple optical processor such as one of those in Figure 7.9. The aperture in these processors is replaced by a small opaque disk to block out the central spot of undiffracted light. The image is therefore formed with only diffracted light and will be brightest in regions of high correlation, and dark in regions of low correlation. Moderate-contrast fringes can be produced in this manner and can be interpreted by using equation 7.32 or 7.33.

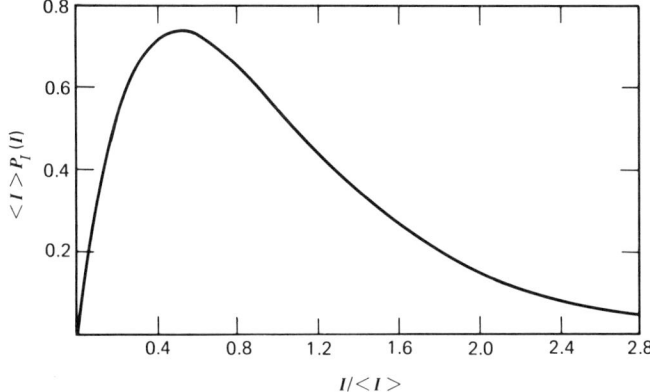

Figure 7.17 Probability density distribution of the irradiance of the incoherent sum of two speckle patterns having identical mean irradiances $\frac{1}{2}\langle I \rangle$.

Fringes of contrast approaching unity can be produced by two-exposure interferometry with two-beam illumination if the object is rigidly translated as well as deformed between exposures. The translation can be introduced intentionally, or can be an inadvertent result of mechanically loading the specimen. Suppose that, between exposures, the object in Figure 7.15 rigidly translated by an amount L_{T_y} in the y direction, and additionally has a local y displacement $v(x,y)$ due to deformation or rotation. If L_{T_y} is sufficiently large to cause a translation of the speckle image which is larger than the characteristic speckle size, interpretation of the speckle pattern will require consideration of both speckle photography and speckle interferometry.[17,61] By placing the developed transparency in an optical processor like that shown in Figure 7.9a, the fringes characteristic of speckle photography can be observed in the back focal plane. From them, the bulk, in-plane translation L_{T_y} can be determined by using equation 7.12. If the aperture shown in Figure 7.9a is placed at the position of the first *dark* speckle photography fringe, it will transmit only light which was not diffracted by the speckle pattern on the transparency. As we have seen, most of this light is scattered by the low-contrast speckle in regions of low correlation. Hence the image formed in this optical processor will contain the fringes of dual-illumination-wave speckle interferometry. If the aperture in Figure 7.9a is a thin slit parallel to the fringes in the focal plane, these speckle interferometry fringes can have very good contrast. The equation of *dark* fringes formed in this manner is equation 7.33, assuming that the object is illuminated symmetrically about the optical axis. If not, it can be discerned in general from equation 7.32. A more general analysis of two-exposure speckle interferometry when the speckle photography effect is also present is given by Stetson.[62]

As pointed out by Hung and Hovanesian,[58] illumination from multiple directions extends the technique of speckle interferometry so that all three components of surface deformation can be measured. The surface rotations and strains can be deduced from these components (see Section 4.2). Jones and Leendertz[63,64] have constructed a speckle interferometer of this type; however, the fringes are produced by electronic processing, a technique which is discussed in Section 7.5. Barker and Fourney[49] have used speckle interferometry to measure internal displacements of transparent solids with much greater sensitivity than can be achieved using the speckle photography technique discussed in Section 7.3.1.

We conclude this section with a brief discussion of *speckle shearing interferometry*. This technique, by which contours of derivatives of displacement can be displayed directly, is attractive for strain analysis because some of the error-prone numerical differentiations required to compute strains or bending moments from displacements are eliminated. In this method, which was developed independently by Leendertz and

7.3 SPECKLE PHOTOGRAPHY AND SPECKLE INTERFEROMETRY

Butters[65] and by Hung and Taylor,[66] a speckle pattern is formed by interference of two sheared images of the same object surface. A variety of devices can be used to introduce shear. Leendertz and Butters used a Michelson interferometer. Hung and Taylor[67] employed a lens with two small apertures in front of which glass plates could be tilted in order to vary the amount of shear. Hariharan used a shearing interferometer consisting of two diffraction gratings.[68] Here we discuss the technique in terms of a simple and versatile apparatus introduced by Hung et al.,[69] which is depicted in Figure 7.18. Note that this is essentially the same as the apparatus used for double-aperture speckle photography.

The object under study is assumed to be flat, and its surface lies in the x-y plane. It is illuminated by a plane, or nearly plane, wave of laser light with propagation vector $\mathbf{k}_1$. A double aperture, consisting of a plate with two holes of small diameter w_a separated by d_a and located symmetrically about the optical axis, is placed in front of the lens. The dotted line in Figure 7.18 denotes the plane in which a sharply focused image of the object would be formed if the full aperture of the lens were used. The film plane is placed a distance Δl_i behind the image plane. This produces two images of the object which are separated, that is, sheared, by a distance Δy_i, given by

$$\Delta y_i = \frac{d_a \Delta l_i}{l_i}. \tag{7.36}$$

These two images are in good focus because the small diameter w_a of the apertures results in a large depth of focus.

The light contributing to the speckle at a given point in the film plane comes from the neighborhoods of two object points separated by a

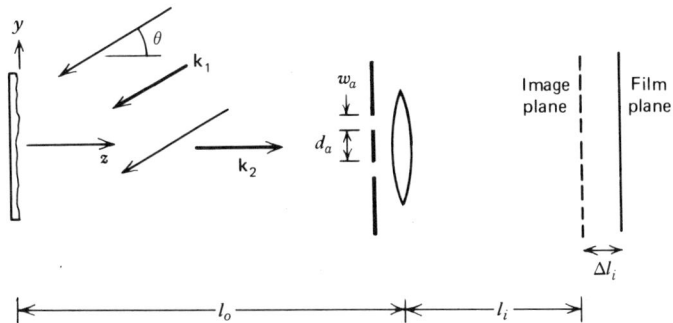

Figure 7.18 Double-aperture speckle shearing interferometer.

distance Δy_o, where

$$\Delta y_o = \frac{\Delta y_i}{M} = \frac{d_a \Delta l_i}{M l_i}, \qquad (7.37)$$

where M is the lateral magnification. The irradiance at a point in the film plane can be written as

$$I_1 = a^2(y) + a^2(y+\Delta y_o) + 2a(y)a(y+\Delta y_o)\cos\phi, \qquad (7.38)$$

where $a(y)$ represents the real amplitude in the film plane due to light scattered from the neighborhood of a point at y. This irradiance is recorded on the film during an initial exposure; then the object is deformed, and a second exposure is recorded. During the second exposure the irradiance in the film plane is

$$I_2 = a^2(y) + a^2(y+\Delta y_o) + 2a(y)a(y+\Delta y_o)\cos(\phi+\delta), \qquad (7.39)$$

where δ is the net phase change due to deformation of the object. The phase change of light scattered by the point initially at y is

$$\delta(y) = (\mathbf{k}_2 - \mathbf{k}_1)\cdot\mathbf{L}(y), \qquad (7.40)$$

where $\mathbf{L}(y)$ is the displacement of the object point initially at y. Similarly, if $\mathbf{L}(y+\Delta y_o)$ is the displacement of the point initially at $y+\Delta y_o$, then

$$\delta(y+\Delta y_o) = (\mathbf{k}_2 - \mathbf{k}_1)\cdot\mathbf{L}(y+\Delta y_o). \qquad (7.41)$$

If the object is viewed normally, along the z axis, and is illuminated as shown in Figure 7.18,

$$\delta(y) = \frac{2\pi}{\lambda}[v(y)\sin\theta + w(y)(1+\cos\theta)], \qquad (7.42)$$

where $v(y)$ and $w(y)$ are the y and z components of object deformation. Also, to first order,

$$\delta(y+\Delta y_o) = \frac{2\pi}{\lambda}\left\{\left[v(y) + \left(\frac{\partial v}{\partial y}\right)\Delta y_o\right]\sin\theta \right.$$

$$\left. + \left[w(y) + \left(\frac{\partial w}{\partial y}\right)\Delta y_o\right](1+\cos\theta)\right\}. \qquad (7.43)$$

7.3 SPECKLE PHOTOGRAPHY AND SPECKLE INTERFEROMETRY

Subtracting equation 7.42 from equation 7.43 gives the net phase change:

$$\delta = \frac{2\pi}{\lambda}\left[\left(\frac{\partial v}{\partial y}\right)\sin\theta + \frac{\partial w}{\partial y}(1+\cos\theta)\right]\Delta y_o. \tag{7.44}$$

The amplitude transmittance of the developed film will be proportional to the total irradiance $I = I_1 + I_2$:

$$I = \left[a^2(y) + a^2(y+\Delta y_o) + 2a(y)a(y+\Delta y_o)\cos\phi\right]$$
$$+ \left[a^2(y) + a^2(y+\Delta y_o) + 2a(y)a(y+\Delta y_o)\cos(\phi+\delta)\right], \tag{7.45}$$

where δ is given by equation 7.44. Equation 7.45 represents an incoherent addition of two speckle patterns. In regions where $\delta = 2N\pi$, for integer values of N, the speckle patterns will be well correlated and therefore will have high contrast and a preponderance of black speckles. In regions where $\delta = (2N+1)\pi$ the speckle pattern will have low correlation and low contrast.

Fringes of contrast correlation can be produced from two-exposure speckle shearing transparencies by any of the techniques described above. In particular, optical processing by spatial filtering in the systems shown in Figure 7.9 works well because the recorded speckle pattern has a strong spatial frequency component f_y corresponding to the spacing between the two apertures. High-contrast fringes can be formed by placing the aperture of the system shown in Figure 7.9a or 7.9c so that it accepts the bright spot of diffracted light which appears above or below the optical axis in the back focal plane of the transforming lens. The equation of bright fringes in the resulting image is

$$\left(\frac{\partial v}{\partial y}\right)\sin\theta + \frac{\partial w}{\partial y}(1+\cos\theta) = \frac{N\lambda M l_i}{d_a \Delta l_i}, \quad N = 0,1,2,\ldots. \tag{7.46}$$

In the system described by Hung et al.[68] the aperture plate actually has four small apertures, as shown in Figure 7.19a. A convenient shearing speckle interferometer can be constructed by attaching this aperture plate to an ordinary camera, preferably by separating the lens elements and placing it in the plane of the original camera aperture. If a double exposure is recorded with this system, the resulting speckle pattern will have strong spatial frequency components f_x and f_y. When this transparency is spatially filtered using one of the systems in Figure 7.9, the diffraction pattern in the transform plane will be as shown in Figure 7.19b. If the aperture is located so that it accepts light in the bright spot just above the optical axis, fringes described by equation 7.46 will be generated. If light from the bright spot

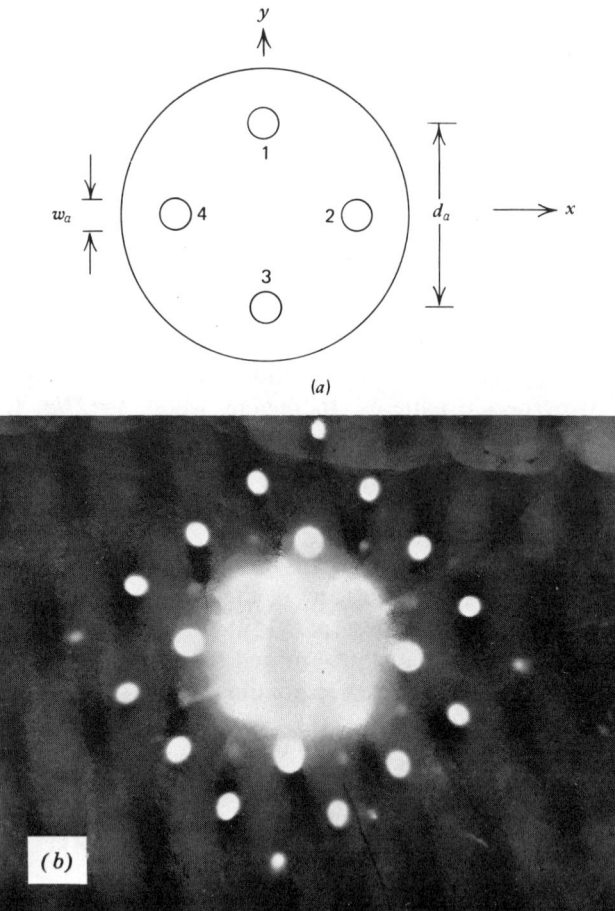

Figure 7.19 Four-aperture speckle shearing interferometry. (*a*) The aperture plate. (*b*) Diffraction pattern in the back focal plane of the transforming lens (ref. 69).

to the side of the axis is admitted, fringes described by the following equation will be formed:

$$\left(\frac{\partial u}{\partial x}\right)\sin\theta + \frac{\partial w}{\partial x}(1+\cos\theta) = \frac{N\lambda M l_i}{d_a \Delta l_i}, \qquad N = 0,1,2,\ldots. \quad (7.47)$$

Hence derivatives in both the x and y directions can be measured. Furthermore, if the aperture admits light from a bright spot lying along the axis at 45° to the x and y axes in the transform plane, derivatives of

7.3 SPECKLE PHOTOGRAPHY AND SPECKLE INTERFEROMETRY

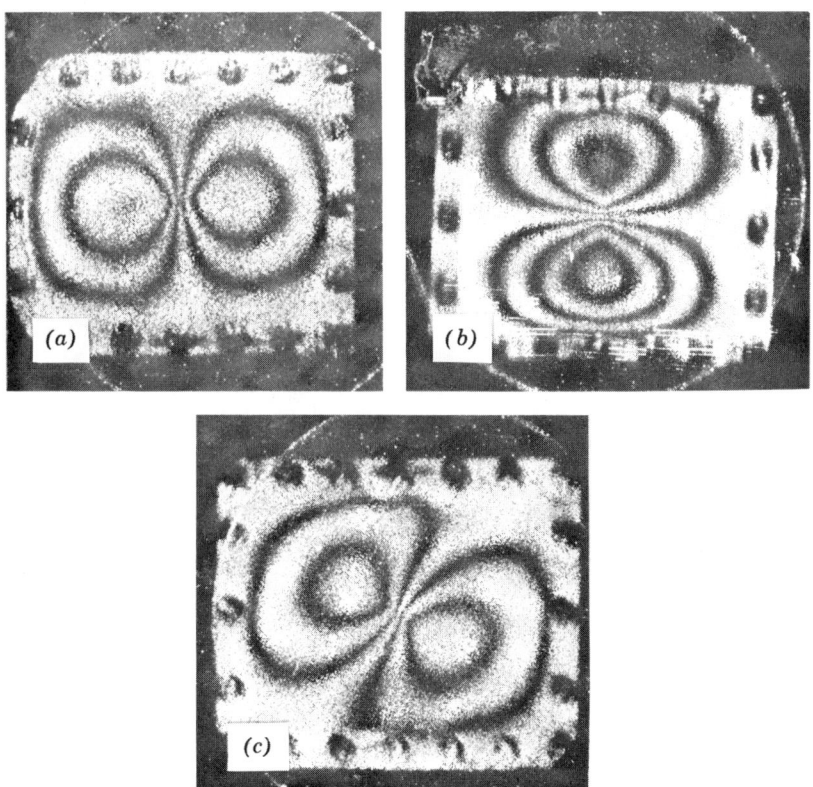

Figure 7.20 Four-aperture speckle shearing interferograms of a centrally loaded rectangular plate (ref. 69). The patterns display contours of (a) $\partial w/\partial x$, (b) $\partial w/\partial y$, (c) $\partial w/\partial x_{45°}$.

displacement along the corresponding direction on the object surface can be measured.

If we know a priori that either the in-plane rotation and shear or the normal deformation is negligible, the other quantity can be obtained directly from the fringe pattern. When interpreting speckle shearing interferograms, it must also be assumed that lateral translations, if any, are sufficiently small that speckle photography effects will be negligible. Figure 7.20 shows three speckle shearing interferograms obtained from a single transparency by the four-aperture method. The object is a clamped rectangular plate subjected to a concentrated normal load at its center. The interferograms display contours of $\partial w/\partial x, \partial w/\partial y$, and the derivative of w along a direction at 45° to the x axis. It is of interest to compare this figure with the holographic interferogram of a similar configuration in Figure 4.7a and the moiré patterns in Figures 4.7b and 4.7c. If $w=0$, the same technique can be used to determine in-plane strains ϵ_x, ϵ_y, and ϵ_{xy} (see Section 4.2). The technique is also applicable to measurement of vibrational amplitudes by time exposure.[68]

An alternative speckle shearing technique in which a full-aperture imaging system is used has been developed by Hung et al.[70] The coherently illuminated object is photographed in a misfocused plane as in Figure 7.18, but the full aperture of the taking lens is used. Two exposures are recorded, one before deformation of the object and one after deformation. The resulting transparency can be processed in a system similar to that in Figure 7.9a to provide contours of derivatives of w along an arbitrary direction in the x-y plane. This method is a speckle photography technique equivalent to that of Chiang and Juang.[31] However, Hung et al.[70] note that by using off-axis illumination of the object, derivatives of in-plane displacement can be measured as in speckle shearing interferometry.

The primary advantages of speckle interferometry are decreased mechanical stability and coherence requirements (for most systems), readily adjustable sensitivity, and ease of measuring in-plane displacements. In some cases, deformations are easily separated from rigid translations. The primary disadvantages are limited fringe resolution and clarity, and ambiguities of interpretation which can occur when the general nature of the deformation is not known a priori or when speckle photography and speckle interferometry effects coexist.

7.4 PROJECTED FRINGE TECHNIQUES

Projected fringe techniques constitute a class of moiré methods by which surface contour, deformation, and vibrational amplitude can be measured. The general topic of moiré metrology is outside the scope of this book; however, because projected fringe techniques are so closely analogous to holographic interferometry, and because they are complementary in that their sensitivity is low, they will be discussed briefly here.

Hildebrand and Haines,[1] Rowe and Welford,[71] and Brooks and Heflinger[72] suggested a contour generation scheme in which the object is simply illuminated by two nearly parallel, mutually coherent plane waves. These waves can be produced by two closely spaced point sources of laser light placed a large distance from the object. Alternatively, virtually any type of interferometer can be used to generate the nearly parallel plane waves. As an example, a Michelson interferometer is shown in Figure 7.21a. The two plane waves of equal irradiance propagate in directions specified by $\mathbf{k}_1$ and $\mathbf{k}_1'$, which differ by a small angle γ. The surfaces of

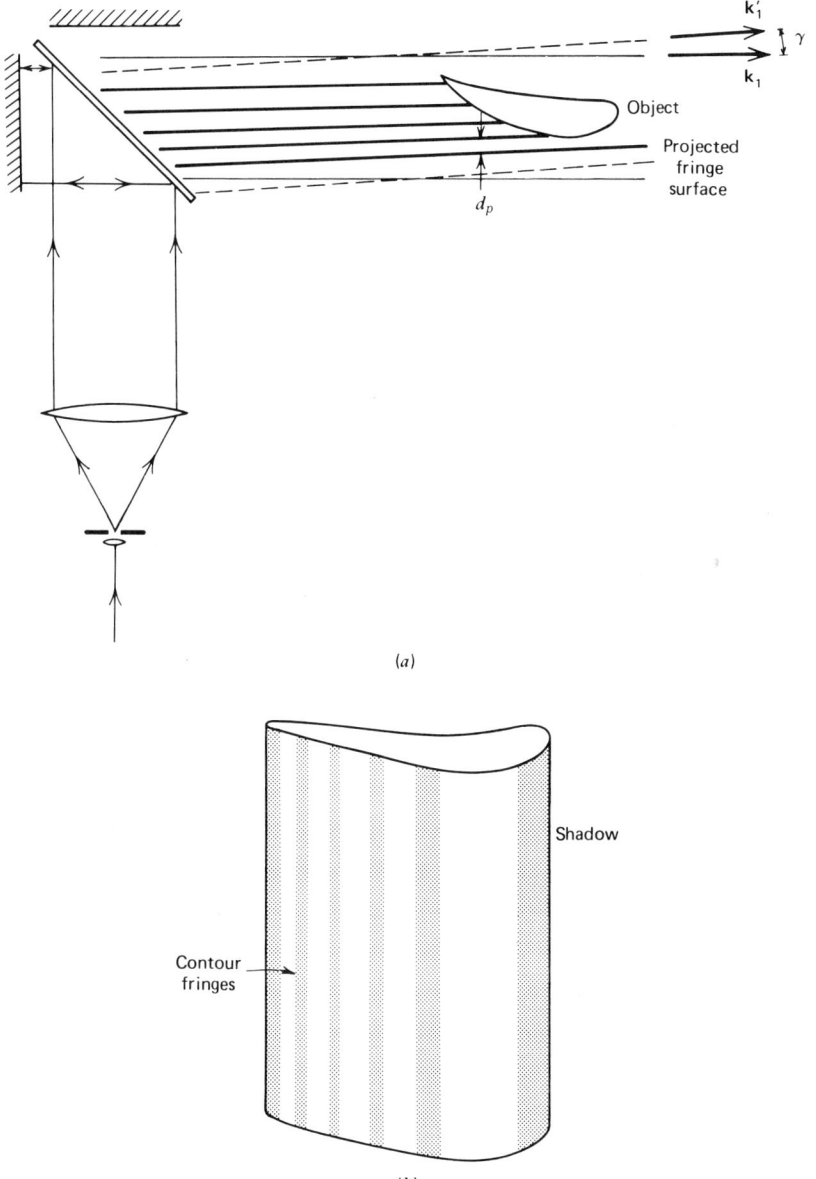

Figure 7.21 Contouring by the projected fringe method. (*a*) Projected fringe surfaces generated with a Michelson interferometer illuminate the object. (*b*) Sketch of the contour fringes on the object surface.

429

constant phase difference between these two waves are parallel planes, normal to the plane of the figure, which contain the bisectors of $\mathbf{k}_1$ and $\mathbf{k}'_1$. The surfaces along which the two waves differ in phase by π radians are termed *projected fringe surfaces*. Consecutive projected fringe surfaces are separated by a spacing d_p, given by

$$d_p = \frac{\lambda/2}{\sin(\gamma/2)}, \qquad (7.48)$$

where it is assumed that γ is small.

If an object is illuminated by the two plane waves, they will interfere and form dark fringes on its surface along its intersections with the projected fringe surfaces. If the object is photographed, a permanent record of the surface contour can be made, as shown schematically in Figure 7.21b. The contour interval normal to the projected fringes is equal to their spacing d_p, which is given by equation 7.48. This contour interval usually is large in comparison with holographic contour intervals. (See Section 7.1.) One difficulty with the present scheme is that in practice the illumination must strike the object at near-grazing incidence, so shadowing often occurs, as indicated in Figure 7.21b.

If a double-exposure photograph of an object illuminated by two coherent, nearly parallel plane waves is recorded, a moiré pattern is formed which is indicative of surface displacement or deformation between exposures. Similarly, a time-average photograph gives moiré fringes which are contours of constant vibrational amplitude. These applications to metrology were first demonstrated by Abramson[73] and later developed independently[74] and refined[75-77] by other workers. An equivalent method in which an incoherent projection of a fringe pattern is used was developed by Hovanesian and Hung[78] and by others.[77, 79] We will discuss these techniques assuming coherent illumination; however, the analysis is essentially the same for incoherent illumination (see refs. 77 and 78).

Figure 7.22 shows a set of projected fringe surfaces impinging on an object surface. The projected fringe surfaces are parallel to the unit vector $\mathbf{k}_1$. The unit vector $\mathbf{k}_2$ points from the object to the observer, and $\hat{\mathbf{n}}$ is the local unit normal to the intersection of the object surface with the plane defined by $\mathbf{k}_1$ and $\mathbf{k}_2$. The spacing of the projected fringes is d_p, which is specified by equation 7.48. The interference fringes on the object surface have the spacing

$$d_f = \frac{d_p}{-\hat{\mathbf{k}}_1 \cdot \hat{\mathbf{n}}} = \frac{d_p}{\cos\theta}. \qquad (7.49)$$

7.4 PROJECTED FRINGE TECHNIQUES

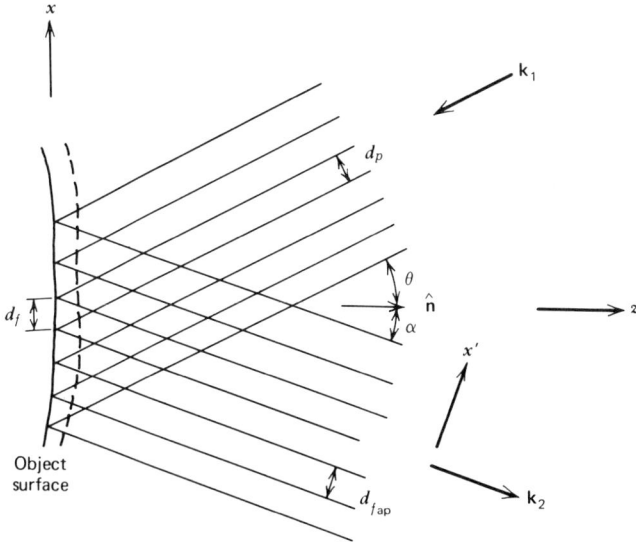

Figure 7.22 Nomenclature for analysis of projected fringe techniques.

Because of perspective, the observed fringes will have an apparent spacing $d_{f_{ap}}$:

$$d_{f_{ap}} = \frac{d_f}{\hat{\mathbf{k}}_2 \cdot \hat{\mathbf{n}}} = \frac{d_p \cos \alpha}{\cos \theta}. \tag{7.50}$$

If the surface translates or rotates in its own plane, no change in the fringe pattern will result. However, the normal component of displacement, $L_n = \mathbf{L} \cdot \hat{\mathbf{n}}$, causes each fringe to translate laterally by an amount $L_n(\tan\theta + \tan\alpha)$, as seen by the observer. If the surface is locally flat and lies in the x-y plane, the irradiance of the fringe pattern photographed is proportional to

$$I_1 = 1 + \cos\left(\frac{2\pi x}{d_{f_{ap}}}\right). \tag{7.51}$$

If the object is then displaced by an amount **L**, the irradiance of the fringe pattern photographed during a second exposure of the same film is

$$I_2 = 1 + \cos\left[\frac{2\pi x'}{d_{f_{ap}}} - \frac{2\pi L_n(\tan\theta + \tan\alpha)\cos\alpha}{d_{f_{ap}}}\right]. \tag{7.52}$$

These two patterns are *incoherently* superimposed, so the irradiance of the fringe pattern recorded on the double-exposure photograph is

$$I = I_1 + I_2.$$

After using several trigonometric identities, this summation can be written in the form

$$I = 2\left\{1 + \cos\left[\frac{\pi L_n(\tan\theta + \tan\alpha)(\cos\alpha)}{d_{f_{ap}}}\right]\right.$$
$$\left. \cdot \cos\left[\frac{2\pi x'}{d_{f_{ap}}} - \frac{\pi L_n(\tan\theta + \tan\alpha)(\cos\alpha)}{d_{f_{ap}}}\right]\right\}. \quad (7.53)$$

The first cosine term represents a slow modulation of the rapidly varying "carrier term" represented by the second cosine. Following the discussion in Section 4.2, we can say that the photographic image of the object will be covered by a moiré fringe pattern with *dark fringes* at locations where

$$\frac{\pi L_n(\tan\theta + \tan\alpha)(\cos\alpha)}{d_{f_{ap}}} = (2N+1)\pi, \quad (7.54)$$

with N being an integer. By combining equations 7.50 and 7.54, we obtain the formula for determining normal displacement from the moiré fringe pattern:

$$L_n = \frac{(2N+1)d_p}{(\tan\theta + \tan\alpha)\cos\alpha}, \quad (7.55)$$

where N is the order number of the dark moiré fringes. Note that the sign of the normal motion given by equation 7.55 is unknown.

Because this method does not depend on a matching of two microscopically identical wavefronts, as in holographic interferometry, it is possible to compare two physically different but macroscopically similar surfaces. For example, a manufactured part can be compared with a master part, or a damaged component with an undamaged one. This is illustrated in Figure 7.23, which is a two-exposure photograph of an automobile hubcap. During the first exposure an undamaged hubcap was photographed. For the second exposure it was replaced by a similar hubcap with a small dent. The dent is clearly outlined by the moiré fringe pattern, and its depth could be determined by applying equation 7.55. The photograph in this figure was made using an incoherent method. The fringes were projected by placing an amplitude grating in a 35 mm slide projector.[78]

7.4 PROJECTED FRINGE TECHNIQUES

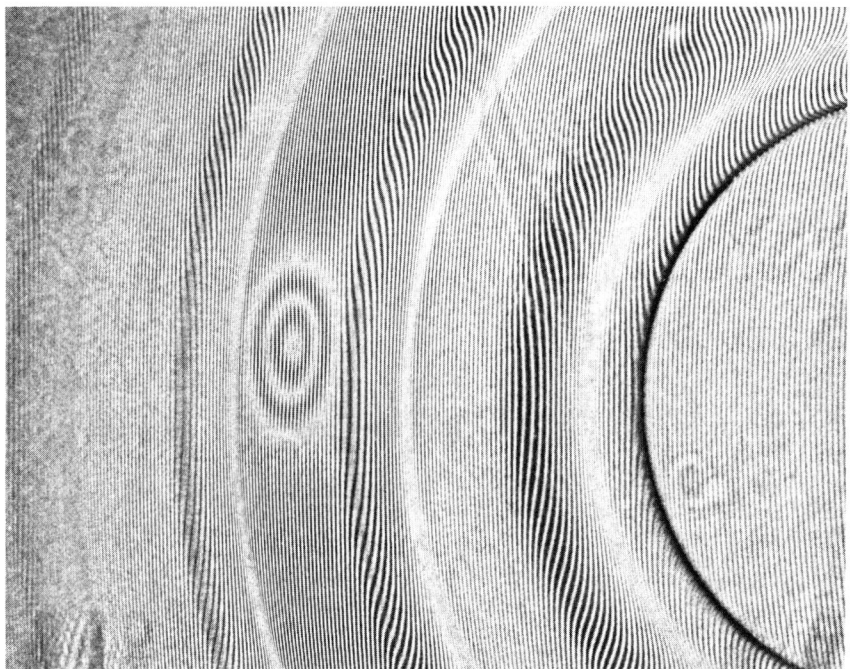

Figure 7.23 Two-exposure, projected fringe, moiré pattern showing a dent in an automobile hubcap (ref. 78).

Contours of constant *vibrational amplitude* can be formed by the projected fringe technique using time-exposure photography. This is illustrated by Figure 7.24, which is a photograph of a rectangular plate executing a harmonic, rigid-body, rocking motion about an axis oriented at 45° to its sides. If the normal motion of a flat object surface is

$$L_n(x,y)\cos\omega t, \tag{7.56}$$

the instantaneous irradiance in an image of the object illuminated by projected fringes is proportional to

$$I = 1 + \cos\left[\frac{2\pi x'}{d_{f_{\text{ap}}}} - \left(\frac{2\pi \tan\theta}{d_{f_{\text{ap}}}}\right)L_n(x,y)\cos\omega t\right]. \tag{7.57}$$

The exposure of a time-average photograph recorded over a period T, which is long compared with the vibrational period, is proportional to the

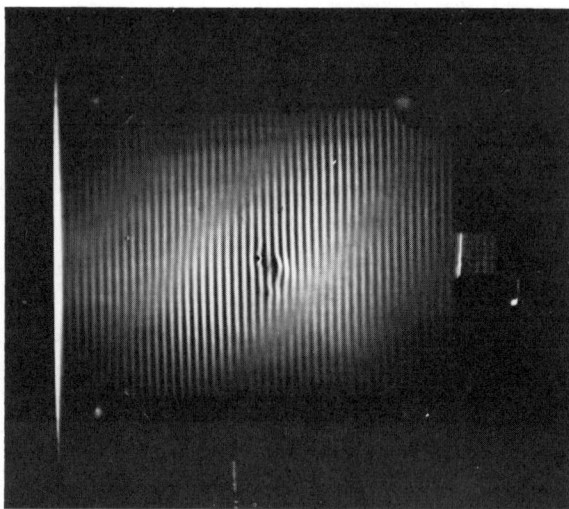

Figure 7.24 Time-exposure, projected fringe photograph of a plate executing a harmonic rigid rocking motion (ref. 74). The bands of reduced fringe visibility are contours of constant vibrational amplitude.

time-average irradiance $\langle I \rangle$:

$$\langle I \rangle = 1 + J_0\left[\left(\frac{2\pi \tan\theta}{d_{f_{\text{ap}}}}\right)L_n(x,y)\right]\cos\left(\frac{2\pi x'}{d_{f_{\text{ap}}}}\right). \quad (7.58)$$

This pattern consists of the basic high-frequency "carrier fringe" pattern, modulated by bands of decreased visibility. These bands are centered on regions where the amplitude is such that the Bessel function in equation 7.58 is a minimum. Using equation 7.58 and the asymptotic behavior of Bessel functions, it can be shown that the normal component of vibrational amplitude at the Nth band of low visibility is

$$L_n = \frac{d_p}{2\pi \sin\theta \cos\alpha}\left\{\left(2N - \tfrac{3}{4}\right)\pi - \tfrac{1}{8}\left[\left(2N - \tfrac{3}{4}\right)\pi\right]^{-1}\right\}. \quad (7.59)$$

This method of measuring vibrational amplitude also works in *real time*. If one visually observes a vibrating surface illuminated by projected fringes, the visual persistence of the eye will create an image like that shown in Figure 7.24.

Either two-exposure or time-average projected fringe photographs can be optically filtered to remove the carrier fringes, leaving distinct fringes which are contours of constant normal displacement. This can be done by placing a photographic transparency into either of the systems shown in Figure 7.9. Regions where the carrier fringe contrast is high diffract light, thereby yielding bright fringes in the final image. Dark fringes correspond to regions where the carrier fringe visibility is low because of the moiré effect. The visibility of the moiré fringes can also be enhanced by nonlinear photography.[76]

7.5 TECHNIQUES INCORPORATING TELEVISION SYSTEMS

Photographic recording materials have been used for all the coherent optical measurement techniques described in the preceding sections of this book. These materials have the disadvantage that the time and effort of development is not negligible, particularly in the case of real-time measurements. This is especially detrimental when repetitive inspection in an industrial environment is needed. For this reason it is natural to investigate the use of television systems to replace photographic recording materials. Such investigations were initiated by Butters and Leendertz[80] and by Macovski et al.[81] and have been followed by considerable development, especially by Butters and his colleagues. Unfortunately, the nearly instantaneous processing achievable with television systems is gained at the expense of drastically lowered spatial resolution. It is reduced from more than 1000 lines/mm for holographic emulsions to, typically, 500 resolution elements per line of a television image. Three-dimensional holographic interferometry is effectively ruled out by this low resolution. Television systems can be used, however, for speckle photography and interferometry, projected fringe techniques, and low-resolution image plane holographic interferometry. Indeed, the distinction between speckle and holographic methods is not always apparent when television systems are used.

First we consider *time-lapse methods*, the equivalent of two-exposure techniques in conventional holographic and speckle interferometry. A system for measuring *normal deformation* or tilt is shown in Figure 7.25.

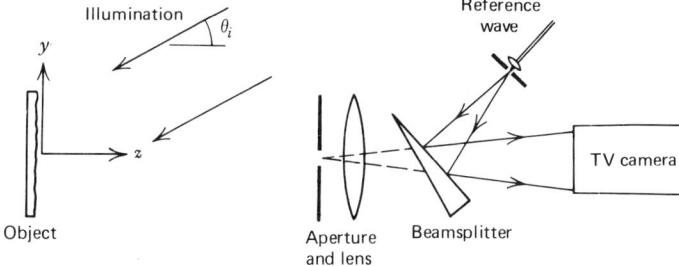

Figure 7.25 System for measuring normal deformation or tilt by speckle interferometry with electronic processing and television display.

Optically, this is a speckle interferometer. The aperture of the imaging lens must be stopped down until the speckle pattern is resolvable by the camera. To keep the spatial frequency of the speckle low, a spherical reference wave which diverges from an apparent source at the center of the aperture is used.[80, 82] From the discussion in Section 7.3.2 we know that, if a portion of the object surface translates in the normal direction by an amount

$$\Delta z = \frac{(N+\frac{1}{2})\lambda}{1+\cos\theta_i}, \qquad N = 0, \pm 1, \pm 2, \ldots, \qquad (7.60)$$

the speckle contrast at the corresponding region on the target of the camera is reversed from its initial value. The speckle contrast is unchanged near regions which translate by an amount

$$\Delta z = \frac{N\lambda}{1+\cos\theta_i}, \qquad N = 0, \pm 1, \pm 2, \ldots. \qquad (7.61)$$

A television image of the object in its initial configuration is recorded on a video disk. After the object is deformed, a second television image is recorded on a different track of the video disk. At playback the signal from the second track is *subtracted* from the signal from the first track. The resulting difference signal is then full-wave rectified and repetitively displayed on a television monitor. Regions where speckle contrast was unchanged between recordings give a zero difference signal and therefore appear as dark fringes in the image. Regions where speckle contrast was reversed appear as speckle-modulated bright fringes. The dark fringes therefore are contours of constant normal deformation specified by equation 7.61. It is assumed that the inaccuracy of this equation due to the use of a spherical reference wave is small. Continuous wave lasers have usually been employed for this technique, but Hughes[83] has reported preliminary experiments in which a pulsed laser was used, so the potential for examination of highly transient deformation exists.

In-plane deformation can be measured by adapting two-illumination-wave speckle interferometry to a television system of the type just discussed. Optically, the system is identical to that shown in Figure 7.15, with the imaging lens and film being replaced by the television camera.[80, 84] Again, the second of two consecutive speckle images is subtracted from the first, the resulting difference signal is full-wave rectified, and the final image is repetitively displayed. The equation of *dark fringes* produced by this system is

$$L_y = \frac{N\lambda}{2\sin\theta}, \qquad (7.62)$$

7.5 TECHNIQUES INCORPORATING TELEVISION SYSTEMS

where the angle θ is defined in Figure 7.16. A speckle interferogram produced in this manner is shown in Figure 7.26.

Another possibility is to apply television systems to projected fringe techniques, as reported by Macovski et al.[81] The advantage of this approach is that a large-aperture imaging system can be used because only the projected fringe pattern (and not the speckle) needs to be resolved by the television camera. This leads to efficient light utilization and good signal/noise ratios.

Macovski et al.[81] also devised a system which can be used with transparent and specularly reflecting objects. In this case neither speckle patterns nor projected fringes are used. The object is imaged onto the camera target, where a reference beam is added at a small offset angle. The irradiation on the target forms an off-axis, image plane hologram which is processed in essentially the same manner as the speckle interferograms described above. Another possibility is to use an in-line reference wave which is frequency shifted with respect to the object wave.[81] This system requires the use of a high-resolution device such as an image dissector, rather than an ordinary television camera.

Time-lapse systems can also be used to produce *contour maps* of objects simply by replacing the object by a flat surface during one recording if the projected fringe configuration is used.[81]

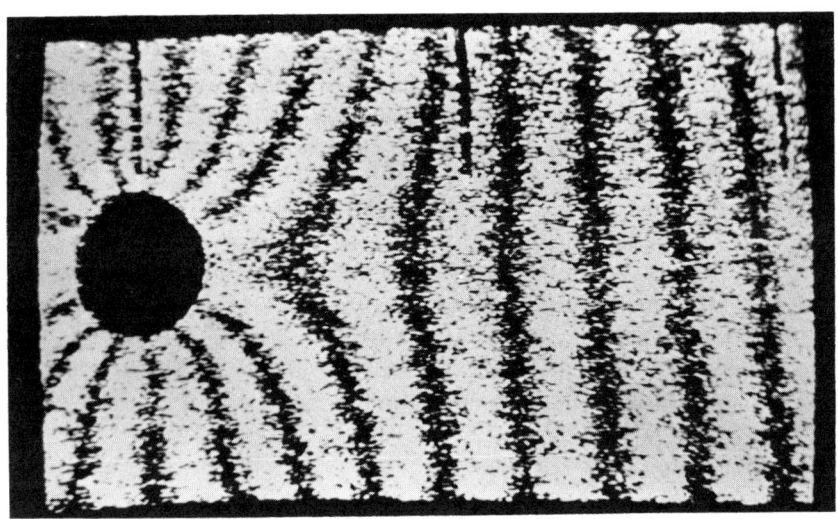

Figure 7.26 Speckle interferogram produced by electronic processing and display on a television monitor (ref. 84). The object is an aluminum bar, with a hole, loaded in uniaxial tension. (Reprinted by permission of the Council of the Institution of Mechanical Engineers from *Journal of Strain Analysis*, Vol. 9.)

Television systems are particularly well suited to measurement of *vibrational amplitudes* because no recording step is necessary. Most often it is the normal component of motion that is of interest in vibration analysis, so an in-line reference wave is used, as in Figure 7.25. The irradiance of the target of a television camera is integrated over the scan period, which is usually $\frac{1}{30}$ s. If the period of a sinusoidally vibrating object under study is shorter than this, the irradiance recorded by the camera will be a speckle pattern with contrast variation given by

$$\nu(x,y) = \left\{ 1 + 2\alpha J_0^2 \left[\frac{2\pi}{\lambda} (1 + \cos\theta_i) A(x,y) \right] \right\}^{1/2} \cdot (1+\alpha)^{-1}. \quad (7.63)$$

In this equation, which can be inferred from equations 7.25 and 7.27, $A(x,y)$ is the vibrational amplitude to be measured, and α is the ratio of reference-to-object-wave irradiance. The temporally varying signal produced by scanning the vidicon is high-pass filtered to eliminate variations of spatial frequency lower than that of the speckle pattern. This filtered signal is fed to a television monitor along with the unfiltered signal, which is used as an external synchronization source. Rectification of the signal is done by the monitor's cathode ray tube.[82] Regions of low visibility, which occur near the zeros of the Bessel function in equation 7.63, become dark fringes in the image displayed by the monitor.

Published results seem to indicate that approximately four fringes can be resolved by this method. Since many more fringes can be resolved by time-lapse methods, a larger number of fringes can be observed when stroboscopic or pulsed-laser techniques are adapted to television systems. Biedermann et al.[85] have reported a system in which the use of a double-slit aperture[86] in the imaging system made possible observations of up to 15 fringes.

Høgmoen and Løkberg[87] have adapted the concept of *frequency-translation holography* (see Section 4.3.5) to electronic speckle interferometry of vibrating objects. When the signal was read out by photoelectronic means, amplitudes as small as 0.01 nm were measurable. With visual observation, amplitudes as small as 2 nm could be measured. The same authors have used modulated reference electronic speckle interferometry to map the contours of constant relative phase of vibrating surfaces.[88]

7.6 HOLOGRAPHIC PHOTOELASTICITY

Photoelasticity is a branch of experimental stress analysis in which the state of stress in certain transparent materials is determined by its effect on the polarization of light transmitted through them. Test specimens are

7.6 HOLOGRAPHIC PHOTOELASTICITY

manufactured from materials which are stress birefringent, that is, noncrystalline materials whose refractive indices become double valued when they are stressed. In an unstressed state the refractive index of such a material is isotropic and has a value n_0. If the material is in a state of plane stress, the refractive index is n_1 along the direction of principal stress σ_1, and is n_2 along the direction of principal stress σ_2. By convention $\sigma_1 \geqslant \sigma_2$, and consequently $n_1 \leqslant n_2$. The refractive index is related to the principal stresses by the Maxwell-Neumann stress-optic law, equation 6.96. In classical photoelasticity[89] the difference between principal stresses $(\sigma_1 - \sigma_2)$ and their direction can be determined by rather simple methods. The sum of principal stresses $(\sigma_1 + \sigma_2)$ can also be determined by the use of an interferometer such as the Mach-Zehnder,[90] which places stringent requirements on the optical quality of the test specimen. The application of holography to photoelasticity, which was initiated by Fourney[91] and by Hovanesian et al.,[92] makes possible simultaneous recording of information about both $(\sigma_1 - \sigma_2)$ and $(\sigma_1 + \sigma_2)$ and alleviates much of the experimental difficulty associated with the use of classical interferometry. We use the term *holographic photoelasticity* to encompass the various techniques by which photoelasticity can be combined with holography or holographic interferometry. The early work in this area was described by Fourney[93] and by Holloway and Johnson.[94]

Before discussing holographic photoelasticity, it is useful to describe a few optical elements used to vary the state of polarization of light. In Section 1.2 it was noted that coherent light generally is elliptically polarized. Various optical elements can be used to convert it to linearly or circularly polarized light. In fact, most cw lasers have a Brewster window attached to the plasma tube so that the output is linearly polarized in the vertical direction. The polarization of light can be altered by passing it through an anisotropic material which is naturally birefringent. Such materials, by virtue of their crystaline structures, have different refractive indices in two orthogonal directions normal to the direction of propagation. These axes are referred to as the ordinary and extraordinary (or slow and fast) axes. The refractive indices along these directions are denoted by n_o and n_e, respectively. A *quarter-wave plate* is made from a crystal of this kind and has a thickness t such that $(n_o - n_e)t = \lambda/4$. If a quarter-wave plate is oriented with its axes at 45° to the direction of linear polarization, as in Figure 7.27, the two orthogonal components of the emerging light have equal amplitudes and differ in phase by $\pi/2$. Hence the wave is circularly polarized. If the plate is oriented at any angle other than 45°, the light will be elliptically polarized.

Circularly or elliptically polarized light can be linearly polarized by passing it through an *analyzer*, for example, a sheet of dichroic material

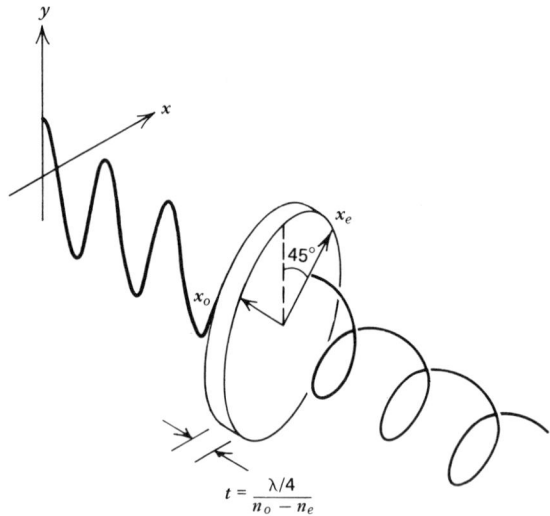

Figure 7.27 A quarter-wave plate oriented as shown here converts linearly polarized light to circularly polarized light.

like Polaroid, which transmits only one component of polarization and absorbs the other. Hence a combination of a quarter-wave plate and an analyzer can be used to rotate the axis of polarization of a laser beam; this can also be accomplished by passing the light through a *half-wave plate*, which is a birefringent crystal plate of thickness t such that $(n_o - n_e)t = \lambda/2$.

The analysis of holographic photoelasticity can be simplified if we note that a linearly polarized reference wave interferes only with the corresponding polarization component of the object wave. If a vertically polarized reference wave is used, a hologram of just the vertical component of the object wave is formed. Thus the holographic process acts as an analyzer. The image reconstructed using this hologram will have a real amplitude and phase distribution identical to that of the vertical component of the object wave. (This statement is true regardless of the polarization state of the reconstruction wave.) If the reference wave is circularly polarized, all polarization components of the object wave contribute to the hologram, because the vertical component of the reference wave interferes with the vertical component of the object wave, and the horizontal component of the reference wave interferes with the horizontal component of the object wave. Therefore each component contributes to the grating-like pattern which is the hologram. The object wave reconstructed from this hologram will be a coherent superposition of two waves having distribu-

7.6 HOLOGRAPHIC PHOTOELASTICITY

tions of real amplitude and phase identical to those of the two components of the object wave. (This statement is true regardless of the polarization state of the reconstruction wave.) We also note in passing that the true polarization of the object wave can be reconstructed by using two spatially separate reference waves which are linearly polarized in orthogonal directions. If the same two waves are used as reconstruction waves, the resulting image will have the same polarization state as the original object wave.[95-97] Alternatively, a single diffused and randomly polarized reference wave can be used.[98] Accurate reconstruction of a state of polarization places strict requirements on alignment, stability, and film characteristics.

Let us now consider *two-exposure holographic photoelasticity*, which is sometimes referred to as stress-holointerferometry. Figure 7.28 is a schematic diagram of a system used for recording the hologram. Quarterwave plates have been used to circularly polarize the reference and object waves. The test object is a specimen cut or cast in the form of a thin plate and is assumed to be loaded in a state of plane stress, that is, the principal stresses lie in the x-y plane, which is parallel to the faces of the specimen, and there is no stress in the z direction. The magnitude, sense, and orientation of the principal stresses σ_1 and σ_2 vary from point to point across the test specimen. Figure 7.29 shows the relative orientations of the polarization of the laser beam, the axes of the quarter-wave plate, and the principal stresses at a typical location on the specimen.

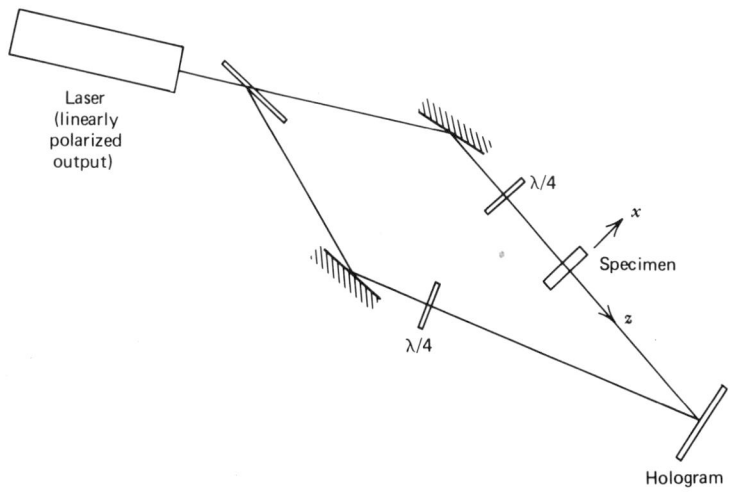

Figure 7.28 System for recording a two-exposure hologram of a stress-birefringent specimen in a state of plane stress.

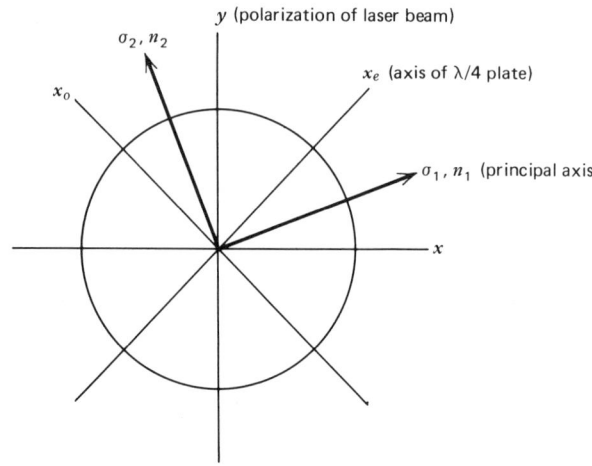

Figure 7.29 Relative orientation of the polarization axis of the laser beam, the quarter-wave plate, and the principal axes.

Two exposures are made: one with the specimen in an unstressed state, and one with the specimen loaded. If this hologram is developed and reilluminated with a plane wave of the same orientation as the reference wave, the virtual image will be formed by coherent superposition of four reconstructed waves. The first two have distributions of real amplitude and phase corresponding to the polarization components of the object wave in the σ_1 and σ_2 directions, respectively. Their complex amplitudes at the hologram plane are

$$\mathbf{U}_{o1} = \exp(-i\phi_1), \qquad (7.64a)$$

$$\mathbf{U}_{o2} = \exp(-i\phi_2). \qquad (7.64b)$$

For simplicity we have assumed a unit amplitude. Clearly, ϕ_1 and ϕ_2 are equal because the test specimen is not birefringent when it is unstressed; its refractive index n_0 is isotropic. The second two reconstructed waves have these complex amplitudes:

$$\mathbf{U}'_{o1} = \exp[-i(\phi_1+\Delta\phi_1)], \qquad (7.65a)$$

$$\mathbf{U}'_{o2} = \exp[-i(\phi_2+\Delta\phi_2)], \qquad (7.65b)$$

where $\Delta\phi_1$ and $\Delta\phi_2$ are the phase changes due to stressing the object. Since $\mathbf{U}_{o1} = \mathbf{U}_{o2}$, the irradiance of the reconstructed wave which forms the virtual

7.6 HOLOGRAPHIC PHOTOELASTICITY

image is proportional to

$$I = |2\mathbf{U}_{o1} + \mathbf{U}'_{o1} + \mathbf{U}'_{o2}|^2$$
$$= |2\exp(-i\phi_1) + \exp[-i(\phi_1 + \Delta\phi_1)] + \exp[-i(\phi_2 + \Delta\phi_2)]|^2. \tag{7.66}$$

When the quantity inside the bars of equation 7.66 is multiplied by its complex conjugate, the resulting exponential terms can be combined to show that the irradiance of the reconstructed wave is proportional to

$$I = 3 + 2\cos\Delta\phi_1 + 2\cos\Delta\phi_2 + \cos(\Delta\phi_1 - \Delta\phi_2)$$
$$= 2\left[1 + 2\cos\left(\frac{\Delta\phi_1 + \Delta\phi_2}{2}\right)\cos\left(\frac{\Delta\phi_1 - \Delta\phi_2}{2}\right) + \cos^2\left(\frac{\Delta\phi_1 - \Delta\phi_2}{2}\right)\right]. \tag{7.67}$$

The phase shift $\Delta\phi_1$ is caused by the effect of stress on refractive index n_1 and by the change in thickness of the object due to the Poisson effect (see Section 6.4.4):

$$\Delta\phi_1 = \frac{2\pi}{\lambda}\left[n_1 t - n_0 t_0 - n(t - t_0)\right]$$
$$= \frac{2\pi}{\lambda}\left[(n_1 - n_0)t + (n_0 - n)(t - t_0)\right]. \tag{7.68}$$

Here t_0 is the initial thickness of the specimen, n_0 is its initial refractive index, t is its thickness while it is stressed, and n is the refractive index of the medium which surrounds the test specimen. From equation 6.96,

$$n_1 - n_0 = A\sigma_1 + B\sigma_2.$$

The change in thickness of the specimen is

$$t - t_0 = \epsilon_z t_0, \tag{7.69}$$

where ϵ_z is the normal strain in the z direction. Combining equations 7.69 and 6.94 gives

$$t = t_0\left[1 - \frac{\nu}{E}(\sigma_1 + \sigma_2)\right], \tag{7.70}$$

where ν is the Poisson ratio, and E the modulus of elasticity of the test specimen. (Properties of various stress-birefringent materials are given in ref. 99.) Combining equations 7.70 and 7.68 results in

$$\Delta\phi_1 = \frac{2\pi}{\lambda}\left\{(A\sigma_1 + B\sigma_2)\left[1 - \frac{\nu}{E}(\sigma_1 + \sigma_2)\right]t_0 - (n_0 - n)\frac{\nu}{E}(\sigma_1 + \sigma_2)t_0\right\}.$$

For most stress-birefringent materials $\nu/E \sim 10^{-4}$ MPa^{-1}, and the maximum elastic stress is of the order of 10 MPa; therefore, inside the square brackets we neglect the term $(\nu/E)(\sigma_1 + \sigma_2)$, and obtain

$$\Delta\phi_1 = \frac{2\pi}{\lambda}\left[(A\sigma_1 + B\sigma_2) - (n_0 - n)\frac{\nu}{E}(\sigma_1 + \sigma_2)\right]t_0. \tag{7.71}$$

Similarly,

$$\Delta\phi_2 = \frac{2\pi}{\lambda}\left[(B\sigma_1 + A\sigma_2) - (n_0 - n)\frac{\nu}{E}(\sigma_1 + \sigma_2)\right]t_0. \tag{7.72}$$

Subtracting equation 7.71 from equation 7.72 yields

$$\Delta\phi_1 - \Delta\phi_2 = \frac{2\pi t_0}{\lambda}[(A - B)(\sigma_1 - \sigma_2)]. \tag{7.73}$$

Similarly,

$$\Delta\phi_1 + \Delta\phi_2 = \frac{2\pi t_0}{\lambda}\left[(A + B) - 2\frac{\nu}{E}(n_0 - n)\right](\sigma_1 + \sigma_2). \tag{7.74}$$

Substitution of equations 7.73 and 7.74 into equation 7.67 reveals that the irradiance of the reconstructed object wave is proportional to

$$I = 1 + 2\cos\left\{\frac{\pi t_0}{\lambda}\left[(A + B) - 2\frac{\nu}{E}(n_0 - n)\right](\sigma_1 + \sigma_2)\right\}$$
$$\cdot \cos\left\{\frac{\pi t_0}{\lambda}[(A - B)(\sigma_1 - \sigma_2)]\right\}$$
$$+ \cos^2\left\{\frac{\pi t_0}{\lambda}[(A - B)(\sigma_1 - \sigma_2)]\right\}. \tag{7.75}$$

It is convenient to introduce the following definitions:

$$A' = A - \frac{\nu}{E}(n_0 - n),$$
$$B' = B - \frac{\nu}{E}(n_0 - n), \tag{7.76}$$
$$C = A - B.$$

7.6 HOLOGRAPHIC PHOTOELASTICITY

Combining the preceding equations, we have

$$I = 1 + 2\cos\left[\frac{\pi t_0}{\lambda}(A' + B')(\sigma_1 + \sigma_2)\right]$$
$$\cdot \cos\left[\frac{\pi t_0}{\lambda} C(\sigma_1 - \sigma_2)\right]$$
$$+ \cos^2\left[\frac{\pi t_0}{\lambda} C(\sigma_1 - \sigma_2)\right]. \quad (7.77)$$

Equation 7.77 is the fundamental equation of two-exposure holographic photoelasticity. It indicates that the image reconstructed from a two-exposure hologram recorded with circularly polarized light will be modulated by a complicated fringe pattern.

It is desirable to produce two families of fringes: the *isochromatic fringes*, which are contours of constant values of $(\sigma_1 - \sigma_2)$, and the *isopachic fringes*, which are contours of constant values of $(\sigma_1 + \sigma_2)$. In classical photoelasticity a *bright isochromatic fringe* is the locus of points for which

$$\frac{\pi t}{\lambda} C(\sigma_1 - \sigma_2) = N\pi, \quad N = 0, \pm 1, \pm 2, \ldots, \quad (7.78)$$

and a *dark isochromatic fringe* is the locus of points for which

$$\frac{\pi t}{\lambda} C(\sigma_1 - \sigma_2) = (2N+1)\frac{\pi}{2}. \quad (7.79)$$

Thus the third term of equation 7.77 represents a classical isochromatic fringe pattern, but it is not directly separable from the combined isochromatic-isopachic pattern represented by the first two terms. If the spacings of the isochromatic and isopachic fringes are sufficiently different, and if they are nearly orthogonal, a relatively direct interpretation of the fringe pattern is possible. Inspection of equations 7.79 and 7.77 indicates that a dark isochromatic fringe becomes a half-tone $(I=1)$ fringe. Also, as an isopachic fringe crosses an isochromatic fringe, its irradiance changes from light to dark, or vice versa,[92, 93] so the isopachic fringes are easily recognized. As demonstrated by Sanford and Durelli,[100] these guidelines are not generally sufficient for quantitative determination of isochromatic and isopachic fringes, especially when the two families of fringes are nearly parallel. (Figure 7.30 shows a typical fringe pattern and confirms this difficulty.) For this reason, several schemes have been developed for producing separate isopachic and isochromatic fringe systems, or otherwise improving fringe interpretation. In the rest of this section, a few of these schemes will be described.

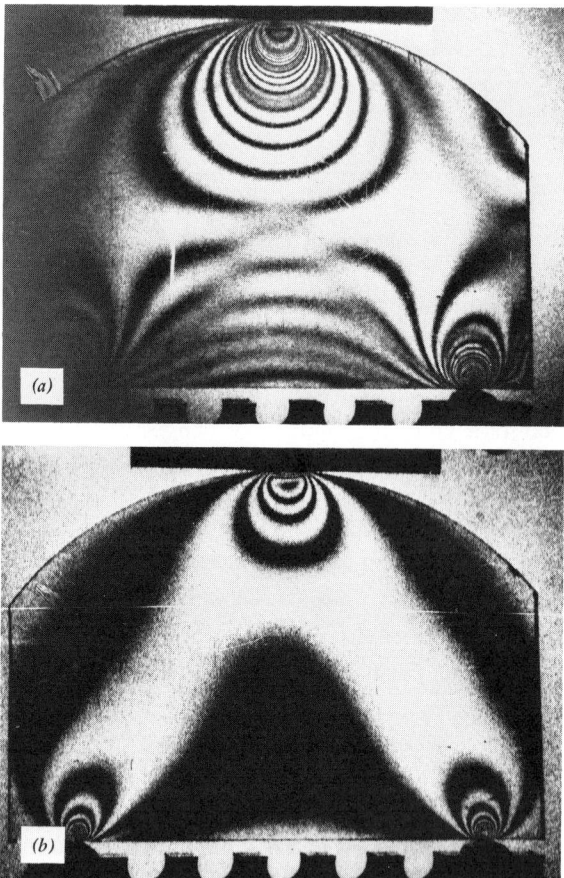

Figure 7.30 Photoelastic fringes (ref. 108). (a) Two-exposure holographic photoelasticity. (b) Pure isochromatic fringe pattern of the same test specimen.

The fringe system represented by equation 7.77 can be simplified by eliminating the third term. This can be accomplished by using an object wave which has been diffused by a volume diffuser, such as a plate of opal glass, which gives a random point-to-point variation of polarization.[101, 102] In the particular system shown in Figure 7.31 the diffuser is placed behind the test specimen, which is imaged onto photographic film or a viewing screen. The random irradiance distribution (speckle pattern) formed by light polarized along the σ_1 direction will be uncorrelated with that formed by light polarized along the σ_2 direction. Therefore the irradiance pattern of the image formed by a two-exposure hologram is the sum of the

7.6 HOLOGRAPHIC PHOTOELASTICITY

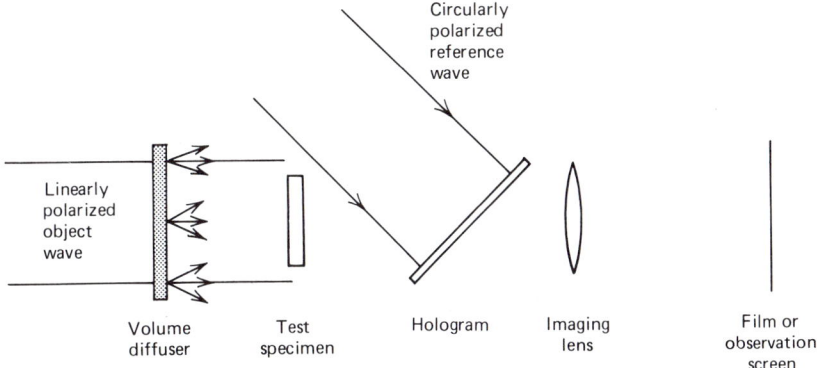

Figure 7.31 Recording system for holographic photoelasticity using a randomly polarized object wave.

irradiances of these two polarization components. Using the nomenclature of equations 7.64 and 7.65, we find that the irradiance is proportional to

$$I = |U_{o1} + U'_{o1}|^2 + |U_{o2} + U'_{o2}|^2. \tag{7.80}$$

Completing the analysis by using equations 7.73, 7.74, and 7.76, we can express the irradiance distribution as

$$I = 1 + \cos\left[\frac{\pi t_0}{\lambda}(A' + B')(\sigma_1 + \sigma_2)\right]$$
$$\cdot \cos\left[\frac{\pi t_0}{\lambda}C(\sigma_1 - \sigma_2)\right] \tag{7.81}$$

This fringe pattern is identical to that produced by Nisida and Saito,[90] using a Mach-Zehnder interferometer.

If the loading under study is static, it is possible to perform a second experiment to produce either an isochromatic or an isopachic fringe pattern. A single-exposure hologram of the test specimen in the stressed state gives the isochromatic fringe pattern. In this case the irradiance of the reconstructed object wave will be proportional to

$$I = |U'_{o1} + U'_{o2}|^2$$
$$= 2[1 + \cos(\Delta\phi_1 - \Delta\phi_2)]$$
$$= 4\cos^2\left(\frac{\Delta\phi_1 - \Delta\phi_2}{2}\right).$$

Using equations 7.73 and 7.76 and dropping the multiplicative constant 4, we find that the irradiance is proportional to

$$I = \cos^2\left[\frac{\pi t_0}{\lambda} C(\sigma_1 - \sigma_2)\right]. \tag{7.82}$$

The image reconstructed with this single-exposure hologram contains only bright and dark isochromatic fringes, described by equations 7.78 and 7.79, respectively. By using two spatially separate reference waves, the isochromatic pattern, equation 7.82, and the combined isochromatic-isopachic pattern, equation 7.77, can be recorded on a single hologram plate.[103] One reference wave, U_{R_1}, is used for two holographic exposures: the first with the specimen unloaded, and the second with the specimen loaded. The other reference wave, U_{R_2}, is used only during the second exposure with the specimen loaded. Reconstruction of this hologram with U_{R_1} gives the combined pattern, and reconstruction with U_{R_2} gives the isochromatic pattern alone.

A separate isopachic pattern also can be produced by modifying the optical system shown in Figure 7.28.[104, 105] The modified system is shown schematically in Figure 7.32. A circularly polarized wave passes through the test specimen and then through an optical rotator, either a passive or electro-optical type, which rotates the polarization axes by 90°. After

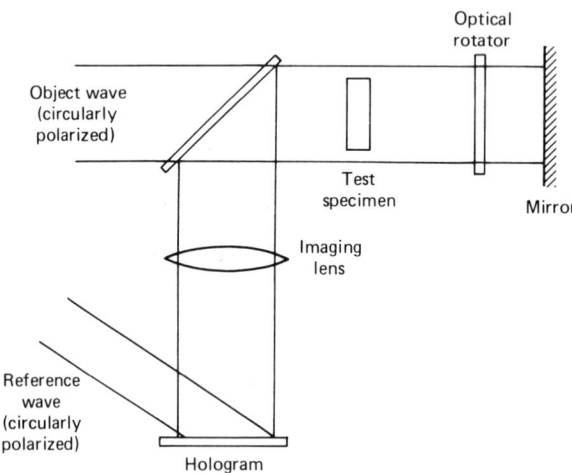

Figure 7.32 A scheme for producing isopachic fringes with a stress-birefringent test specimen.

7.6 HOLOGRAPHIC PHOTOELASTICITY

reflection by a normally oriented front surface mirror, the wave passes back through the rotator and traverses the test specimen a second time. Light which is polarized in the σ_1 direction after the first pass through the specimen is rotated to the σ_2 direction before the second pass through the specimen, and vice versa. Thus light of each polarization component undergoes the same total phase change. If a two-exposure hologram is recorded with this system, the irradiance of the reconstructed object wave will be proportional to

$$I = |\exp(-i\phi) + \exp[-i(\phi+\Delta\phi)]|^2$$
$$= 2(1+\cos\Delta\phi)$$
$$= 4\cos^2\left(\frac{\Delta\phi}{2}\right), \qquad (7.83)$$

where

$$\Delta\phi = \frac{2\pi t_0}{\lambda}\left[(A+B) - 2\frac{\nu}{E}(n_0-n)\right](\sigma_1+\sigma_2). \qquad (7.84)$$

Combining equations 7.83 and 7.84 and dropping the factor 4, we see that the reconstructed object wave irradiance is proportional to

$$I = \cos^2\left[\frac{\pi t_0}{\lambda}(A'+B')(\sigma_1+\sigma_2)\right]. \qquad (7.85)$$

This represents an isopachic fringe pattern with twice the usual sensitivity (because of the double pass through the specimen). A dual-reference-wave system can also be used to produce separately this isopachic pattern and the isochromatic pattern described by equation 7.82.

An alternative approach to separating the isopachic and isochromatic fringes is to apply *sandwich holography* to photoelastic analysis.[107] The sandwich hologram is formed as described in Section 5.3, using circularly polarized reference and object waves. The sandwich is formed from two holograms, one of the unstressed object and one of the stressed object. Because the holograms are formed with nondiffuse light, they can be viewed conveniently by illuminating them with white light incident at the same angle as the reference wave. By slightly translating one hologram with respect to the other, the isopachic fringes can be displayed as a finite-fringe interferogram with straight, horizontal reference fringes. (See Section 5.3.) Only the isopachic fringes are affected by this small translation. To see this, we note that the effect of the translation is to add a linear phase variation $2\pi f_y y$ to U'_{o1} and U'_{o2} in equation 7.66, so the image irradiance distribution corresponding to equation 7.67 is

$$I = 2\left[1 + 2\cos\left(\frac{\Delta\phi_1 + \Delta\phi_2 + 4\pi f_y y}{2}\right)\cos\left(\frac{\Delta\phi_1 - \Delta\phi_2}{2}\right)\right.$$
$$\left. + \cos^2\left(\frac{\Delta\phi_1 - \Delta\phi_2}{2}\right)\right]. \quad (7.86)$$

After $\Delta\phi_1$ and $\Delta\phi_2$ are evaluated in the usual manner, it is found that the irradiance distribution of the image is

$$I = 1 + 2\cos\left[\frac{\pi t_0}{\lambda}(A' + B')(\sigma_1 + \sigma_2) + 2\pi f_y y\right]$$
$$\cdot \cos\left[\frac{\pi t_0}{\lambda}C(\sigma_1 - \sigma_2)\right]$$
$$+ \cos^2\left[\frac{\pi t_0}{\lambda}C(\sigma_1 - \sigma_2)\right]. \quad (7.87)$$

Essentially, we have forced the fringes to fulfill the condition of having a relatively high spatial frequency and to be nearly normal to the isochromatic fringes in regions where they are vertical. Such a pattern is shown schematically in Figure 7.33. If there are regions where the isochromatics are predominantly horizontal, the holograms can be translated in the x direction relative to each other in order to introduce vertical reference fringes.

Hovanesian[108] has shown that a very simple real-time holographic interferometer can serve to clearly separate isochromatic and isopachic fringes. An off-axis holographic system with a linearly polarized reference wave is

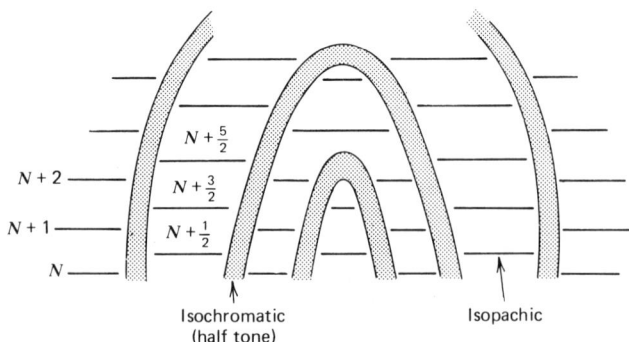

Figure 7.33 Sketch of fringe pattern formed by sandwich hologram photoelasticity.

7.6 HOLOGRAPHIC PHOTOELASTICITY

used. The object wave is circularly polarized by a quarter-wave plate before traversing the test specimen. After traversing the specimen, the object wave passes through a second quarter-wave plate. The fast axes of these two plates are oriented at 90° to each other. This configuration, which was used earlier by Fourney and Mate,[97] eliminates errors caused by large reference wave offset angles in systems which use circularly polarized reference and object waves. Two holographic exposures are made: one with the specimen unstressed, and one with it stressed. The hologram is developed and replaced in its initial position. The object is then re-illuminated and viewed through a linear polarizer simultaneously with the reconstructed object wave. The nature of the resulting real-time holographic interferogram can be varied by adjusting the ratio of reconstruction to object wave irradiance and the orientation of the polarizer. If the beam ratio is set so that the ratio of the reconstructed object wave irradiance to the real object wave irradiance is 4, the Nisida-Saito pattern described by equation 7.81 results. As this ratio is reduced, the isopachic fringes attain good visibility and are readily separated from the isochromatic fringes. If the reference wave is turned off, the remaining system is a classical circular polariscope, so isochromatic fringes can be formed. Ajovalasit[109, 110] has described techniques by which real-time holographic interferometry can be used to determine the *absolute retardations*, $\Delta\phi_1$ and $\Delta\phi_2$, if the directions of the principal stresses are known.

As mentioned earlier, it is possible to record the state of polarization of an object wave by using spatially separate object waves of different linear polarization. Other methods involve the use of a randomly polarized diffused reference wave,[98] or a reference wave generated by a mosaic of thin, parallel, strip polarizers of alternating orientation.[111] Thus a hologram of an appropriately illuminated stressed object can be recorded. The reconstructed object wave can then be examined by classical photoelastic techniques.[111, 112] This is potentially useful for recording transient stress patterns by pulsed-laser holography.

Holographic methods can be applied to the frozen-stress method of measuring three-dimensional stress fields. Here a test specimen manufactured from an appropriate stress-birefringent polymer is mechanically stressed while at an elevated temperature. The test specimen is then cooled to room temperature while remaining under mechanical load. Certain polymers have the property that the birefringence persists at room temperature after the load has been removed. The stresses are said to be "frozen" in the specimen, and could be studied by physically slicing the specimen into thin parallel sections, each of which can be subjected to photoelastic and interferometric analysis.[89] Dhir and Peterson[113] have shown that significant experimental simplification results if the individual sections are studied by holographic photoelasticity while immersed in a tank of liquid whose refractive index matches the initial refractive index of the polymer.

REFERENCES

1. B. P. Hildebrand and K. A. Haines, Multiple-wavelength and multiple-source holography applied to contour generation, *J. Opt. Soc. Am.*, **57**, 155–162 (1967).
2. K. Haines and B. P. Hildebrand, Contour generation by wavefront reconstruction, *Phys. Lett.*, **19**, 10–11 (1965).
3. B. P. Hildebrand and K. A. Haines, The generation of three-dimensional contour maps by wavefront reconstruction, *Phys. Lett.*, **21**, 422–423 (1966).
4. A. A. Friesem and U. Levy, Fringe formation in two-wavelength contour holography, *Appl. Opt.*, **15**, 3009–3020 (1976).
5. J. R. Varner, Simplified multiple-frequency holographic contouring, *Appl. Opt.*, **10**, 212–213 (1971).
6. J. S. Zelenka and J. R. Varner, A new method for generating depth contours holographically, *Appl. Opt.*, **7**, 2107–2110 (1968).
7. L. O. Heflinger and R. F. Wuerker, Holographic contouring via multifrequency lasers, *Appl. Phys. Lett.*, **15**, 28–30 (1969).
8. P. D. Henshaw and S. Ezekiel, High resolution holographic contour generation using a pulsed multicolor ion laser, *Appl. Opt.*, **12**, 2550–2552 (1973).
9. T. Tsuruta, N. Shiotake, J. Tsujiuchi, and K. Matsuda, Holographic generation of contour map of diffusely reflecting surface by using immersion method, *Jap. J. Appl. Phys.*, **6**, 661–662 (1967).
10. J. S. Zelenka and J. R. Varner, Multiple-index holographic contouring, *Appl. Opt.*, **8**, 1431–1434 (1969).
11. E. S. Marrone and W. B. Ribbens, Dual index holographic contour mapping over a large range of contour spacings, *Appl. Opt.*, **14**, 23–24 (1975).
12. J. R. Varner, Holographic and moiré surface contouring, Chapter 5 in *Holographic Nondestructive Testing*, R. K. Erf (Ed.), Academic Press, New York, 1974, pp. 106–147.
13. H. J. Tiziani, A study of the use of laser speckle to measure small tilts of optically rough surfaces accurately, *Opt. Commun.*, **5**, 271–276 (1972).
14. H. J. Tiziani, Analysis of mechanical oscillations by speckling, *Appl. Opt.*, **11**, 2911–2917 (1972).
15. H. J. Tiziani, Application of speckling for in-plane vibration analysis, *Opt. Acta*, **18**, 891–902 (1971).
16. U. Köpf, Ein koharent-optisches Verfahren zur messung mechanischer Schwingungen, *Optik*, **33**, 517–521 (1971).
17. E. Archbold and A. E. Ennos, Displacement measurement from double-exposure laser photographs, *Opt. Acta*, **19**, 253–271 (1972).
18. U. Köpf, Ein koharent-optisches Verfahren zur beruhrungslosen messung mechanischer Verformungen und Schwingungen, *Optik*, **35**, 144–151 (1972).
19. S. Mallick, Speckle-pattern interferometer to determine object deformations, *Nouv. Rev. Opt. Appl.*, **4**, 267–272 (1973).
20. J. M. Burch and J. M. J. Tokarski, Production of multiple beam fringes from photographic scatterers, *Opt. Acta*, **15**, 101–111 (1968).
21. E. Archbold and A. E. Ennos, Two-dimensional vibrations analyzed by speckle photography, *Opt. Laser Technol.*, **7**, 17–21 (1975).
22. E. Archbold, J. M. Burch, and A. E. Ennos, Recording of in-plane surface displacement by double-exposure speckle photography, *Opt. Acta*, **17**, 883–898 (1970).

23. W. Fink and P. A. Büger, Eine Methode zur kontaktlosen messung kleiner Verschiebungen rauher Oberflachen, *Z. Angew. Phys.*, 2/3, 176–178 (1970).
24. D. E. Duffy, Measurement of surface displacement normal to the line of sight, *Exp. Mech.*, 14, 378–384 (1974).
25. P. R. Khetan and F. P. Chiang, Strain analysis by one-beam laser speckle interferometry. 1: Single aperture method, *Appl. Opt.*, 15, 2205–2215 (1976).
26. G. Cloud, Practical speckle interferometry for measuring in-plane deformation, *Appl. Opt.*, 14, 878–884 (1975).
27. K. A. Stetson, A review of speckle photography and interferometry, *Opt. Eng.*, 14, 482–489 (1975).
28. D. E. Duffy, Moiré gauging of in-plane displacement using double aperture imaging, *Appl. Opt.*, 11, 1778–1781 (1972).
29. J. M. Burch and C. Forno, A high sensitivity moiré grid technique for studying deformation in large objects, *Opt. Eng.*, 14, 178–185 (1975).
30. C. Forno, White-light speckle photography for measuring deformation, strain, and shape, *Opt. Laser Technol.*, 7, 217–221 (1975).
31. F. P. Chiang and R. M. Juang, Laser speckle interferometry for plate bending problems, *Appl. Opt.*, 15, 2199–2204 (1976).
32. F. P. Chiang and C. H. Lee, Dynamic laser speckle interferometry applied to transient flexure problem, *Appl. Opt.*, 16, 3085–3087 (1977).
33. D. A. Gregory, Basic physical principles of defocused speckle photography: a tilt topology inspection technique, *Opt. Laser Technol.*, 8, 201–213 (1976).
34. K. A. Stetson, Problems of defocusing in speckle photography: its connection to hologram interferometry, and its solutions, *J. Opt. Soc. Am.*, 66, 1267–1271 (1976).
35. F. D. Adams and G. E. Maddux, Synthesis of holography and speckle photography to measure 3-D displacements, *Appl. Opt.*, 13, 219 (1974).
36. P. M. Boone, Use of reflection holograms in holographic interferometry and speckle correlation for measurement of surface displacement, *Opt. Acta*, 22, 579–590 (1975).
37. R. R. Hudson and D. D. Setopoulos, Speckle interferometric method for the determination of time-dependent displacements and strains, *Strain*, 11, 126–129 (1975).
38. L. C. de Backer, In-plane displacement measurement by speckle interferometry, *Nondestr. Test.*, 8, 177–180 (1975).
39. E. Archbold and A. E. Ennos, Laser photography to measure deformation of weld cracks under load, *Nondestr. Test.*, 8, 181–184 (1975).
40. A. R. Luxmoore, F. A. A. Amin, and W. T. Evans, In-plane strain measurement by speckle photography: a practical assessment of the use of Young's fringes, *J. Strain Anal.*, 9, 26–34 (1974).
41. G. Torbillon, Sur quelques methodes de speckle en métrologie, *Opt. Acta*, 24, 877–892 (1977).
42. W. T. Evans and A. R. Luxmoore, Measurement of in-plane displacements around crack tips by a laser speckle method, *Eng. Fracture Mech.*, 6, 735–743 (1974).
43. F. A. A. Amin and A. R. Luxmoore, The measurement of crystal length changes by a laser-speckle method, *J. Inst. Met.*, 101, 208–211 (1973).
44. U. Köpf, Application of speckling for measuring the deflection of laser light by phase objects, *Opt. Commun.*, 5, 347–350 (1972).
45. S. Mallick and M. L. Roblin, Speckle-pattern interferometry applied to the study of

phase objects, *Opt. Commun.*, **6**, 45–49 (1972).
46. W. Merzkirch, *Flow Visualization*, Academic Press, New York, 1974, pp. 147–148.
47. A. M. Hunter II and P. W. Schreiber, Determination of refractive index in inhomogeneous axisymmetric media from refraction angle measurements, *Appl. Opt.*, **14**, 25–26 (1975).
48. S. Debrus, M. Francon, C. P. Grover, M. May, and M. L. Roblin, Ground glass differential interferometer, *Appl. Opt.*, **11**, 853–857 (1972).
49. D. B. Barker and M. E. Fourney, Displacement measurements in the interior of 3-D bodies using scattered-light speckle patterns, *Exp. Mech.*, **16**, 209–214 (1976).
50. D. B. Barker and M. E. Fourney, Measuring fluid velocities with speckle patterns, *Opt. Lett.*, **1**, 135–137 (1977).
51. A. E. Ennos, Speckle interferometry, Chapter 6 in *Laser Speckle and Related Phenomena*, J. C. Dainty (Ed.), Springer-Verlag, Berlin, 1975, pp. 203–253.
52. J. M. Burch, Interferometry with scattered light, in *Optical Instruments and Techniques*, J. Home Dickson (Ed.), Oriel Press, Newcastle-upon-Tyne, 1970, pp. 213–219.
53. G. P. Weigelt and B. Stoffregen, The longitudinal correlation of a three-dimensional speckle intensity distribution, *Optik*, **48**, 399–408 (1977).
54. E. Archbold, J. M. Burch, A. E. Ennos, and P. A. Taylor, Visual observation of surface vibration nodal patterns, *Nature*, **222**, 263–265 (1969).
55. L. Ek and N.-E. Molin, Detection of the nodal lines and the amplitude of vibration by speckle interferometry, *Opt. Commun.*, **2**, 419–424 (1971).
56. K. A. Stetson, New design for laser image-speckle interferometer, *Opt. Laser Technol.*, **2**, 179–181 (1970).
57. J. W. Goodman, Statistical properties of laser speckle patterns, in *Laser Speckle and Related Phenomena*, J. C. Dainty (Ed.), Springer-Verlag, Berlin, 1975, pp. 9–75.
58. Y. Y. Hung and J. D. Hovanesian, Full-field surface-strain and displacement analysis of 3-D objects by speckle interferometry, *Exp. Mech.*, **12**, 454–460 (1972).
59. J. A. Leendertz, Interferometric displacement measurement on scattering surfaces utilizing speckle effect, *J. Phys. E.: Sci. Instrum.*, **3**, 214–218 (1969).
60. E. Archbold and A. E. Ennos, Applications of holography and speckle photography to the measurement of displacement and strain, *J. Strain Anal.*, **9**, 10–16 (1974).
61. J. N. Butters and J. A. Leendertz, A double-exposure technique for speckle pattern interferometry, *J. Phys. E.: Sci. Instrum.*, **4**, 277–279 (1971).
62. K. A. Stetson, Analysis of double-exposure speckle photography with two-beam illumination, *J. Opt. Soc. Am.*, **64**, 857–861 (1974).
63. R. Jones and J. A. Leendertz, Elastic constant and strain measurements using a three beam speckle pattern interferometer, *J. Phys. E.: Sci. Instrum.*, **7**, 653–657 (1974).
64. R. Jones, The design and application of a speckle pattern interferometer for total plane strain field measurement, *Opt. Laser Technol.*, **8**, 215–219 (1976).
65. J. A. Leendertz and J. N. Butters, An image-shearing speckle-pattern interferometer for measuring bending moments, *J. Phys. E.: Sci. Instrum.*, **6**, 1107–1110 (1973).
66. Y. Y. Hung and C. E. Taylor, Measurement of slopes of structural deflections by speckle-shearing interferometry, *Exp. Mech.*, **14**, 281–285 (1974).
67. Y. Y. Hung and C. E. Taylor, Speckle-shearing interferometric camera—a tool for measurement of derivatives of surface-displacement, *Proc. SPIE*, **41**, 169–175 (1973).
68. P. Hariharan, Speckle-shearing interferometry: a simple approach, *Appl. Opt.*, **14**, 2563 (1975).

69. Y. Y. Hung, R. E. Rowlands, and I. M. Daniel, Speckle-shearing interferometric technique: a full-field strain gauge, *Appl. Opt.*, **14**, 618–622 (1975).
70. Y. Y. Hung, I. M. Daniel, and R. E. Rowlands, Full-field optical strain measurement having post-recording sensitivity and direction sensitivity, *Exp. Mech.*, **18**, 56–60 (1978).
71. S. H. Rowe and W. T. Welford, Surface topography on non-optical surfaces by projected interference fringes, *Nature*, **216**, 786–787 (1967).
72. R. E. Brooks and L. O. Heflinger, Moiré gauging using optical interference patterns, *Appl. Opt.*, **8**, 935–939 (1969).
73. N. Abramson, Interferometric measurement of differences between two conditions, *Laser Focus*, **4**, 26–28 (1968).
74. C. M. Vest and D. W. Sweeney, Measurement of vibrational amplitude by modulation of projected fringes, *Appl. Opt.*, **11**, 449–454 (1972).
75. A. J. MacGovern, Projected fringes and holography, *Appl. Opt.*, **11**, 2972–2974 (1972).
76. J. Shamir, Moiré gauging by projected interference fringes, *Opt. Laser Technol.*, **5**, 78–86 (1973).
77. P. Benoit, E. Mathieu, J. Hormière, and A. Thomas, Characterization and control of three dimensional objects using fringe projection techniques, *Nouv. Rev. Opt*, **6**, 67–86 (1975).
78. J. D. Hovanesian and Y. Y. Hung, Moiré contour-sum, contour-difference, and vibration analysis of arbitrary objects, *Appl. Opt.*, **10**, 2734–2738 (1971).
79. B. Dessus and M. Leblanc, "Fringe method" and its application to the measurement of deformations, vibrations, contour lines, and differences of objects, *Opt. Elect.*, **5**, 369–391 (1973).
80. J. N. Butters and J. A. Leendertz, Holographic and video techniques applied to engineering measurements, *Meas. Control*, **4**, 349–354 (1971).
81. A. Macovski, S. D. Ramsey, and L. F. Schaefer, Time-lapse interferometry and contouring using television systems, *Appl. Opt.*, **10**, 2722–2727 (1971).
82. H. M. Pederson, O. J. Lökberg, and B. M. Förre, Holographic vibration measurement using a TV speckle interferometer with silicon target vidicon, *Opt. Commun.*, **12**, 421–426 (1974).
83. R. G. Hughes, The determination of vibration patterns using a pulsed laser with holographic and electronic speckle interferometry techniques, in *The Engineering Uses of Coherent Optics*, E. R. Robertson (Ed.), Cambridge University Press, Cambridge, 1976, pp. 199–218.
84. D. Denby and J. A. Leendertz, Plane-surface strain examination by speckle pattern interferometry using electronic processing, *J. Strain Anal.*, **9**, 17–25 (1974).
85. K. Biedermann, L. Ek, and L. Ostlund, A TV speckle interferometer, in *The Engineering Uses of Coherent Optics*, E. A. Robertson (Ed.), Cambridge University Press, Cambridge, 1976, pp. 219–221.
86. K. Biedermann and L. Ek, A recording and display system for hologram interferometry with low resolution imaging devices, *J. Phys. E.: Sci. Instrum.*, **8**, 571–691 (1975).
87. K. Høgmoen and O. J. Løkberg, Detection and measurement of small vibrations using electronic speckle pattern interferometry, *Appl. Opt.*, **16**, 1869–1875 (1977).
88. O. J. Løkberg and K. Høgmoen, Vibration phase measurement using electronic speckle pattern interferometry, *Appl. Opt.*, **15**, 2701–2704 (1976).
89. J. W. Dally and W. F. Riley, *Experimental Stress Analysis*, 2nd ed., McGraw-Hill, New York, 1978.

90. M. Nisida and H. Saito, A new interferometric method of two-dimensional stress analysis, *Exp. Mech.*, **4**, 366–376 (1964).
91. M. E. Fourney, Application of holography to photoelasticity, *Exp. Mech.*, **8**, 33–38 (1968).
92. J. D. Hovanesian, V. Brcic, and R. L. Powell, A new stress-optic method: stress-holo-interferometry, *Exp. Mech.*, **8**, 362–368 (1968).
93. M. E. Fourney, Advances in holographic photoelasticity, in *Applications of Holography in Mechanics*, W. G. Gottenberg (Ed.), American Society of Mechanical Engineers, New York, 1971, pp. 17–38.
94. D. C. Holloway and R. H. Johnson, Advancements in holographic photoelasticity, *Exp. Mech.*, **11**, 57–63 (1971).
95. M. E. Fourney, A. P. Waggoner, and K. V. Mate, Recording polarization effects via holography, *J. Opt. Soc. Am.*, **58**, 701–702 (1968).
96. O. Bryngdahl, Polarizing holography, *J. Opt. Soc. Am.*, **58**, 702 (1968).
97. M. E. Fourney and K. V. Mate, Further applications of holograph to photoelasticity, *Exp. Mech.*, **10**, 177–186 (1970).
98. C. N. Kurtz, Holographic polarization recording with an encoded reference beam, *Appl. Phys. Lett.*, **14**, 59–61 (1969).
99. Reference 89, p. 486.
100. R. J. Sanford and A. J. Durelli, Interpretation of fringes in stress-holo-interferometry, *Exp. Mech.*, **11**, 161–166 (1971).
101. H. Kubo and R. Nagata, Holographic photoelasticity with depolarized object wave, *Jap. J. Appl. Phys.*, **15**, 641–644 (1976).
102. J. Ebbeni, J. Coenen, and A. Hermanne, New analysis of holophotoelastic patterns and their application, *J. Strain Anal.*, **11**, 11–17 (1976).
103. A. Assa and A. A. Betser, The application of holographic multiplexing to record separate isopachic- and isochromatic-fringe patterns, *Exp. Mech.*, **14**, 502–504 (1974).
104. H. H. M. Chau, Holographic interferometer for isopachic stress analysis, *Rev. Sci. Instrum.*, **39**, 1789–1792 (1968).
105. R. O'Regan and T. D. Dudderar, A new holographic interferometer for stress analysis, *Exp. Mech.*, **11**, 241–247 (1971).
106. K. Gasvik, Separation of the isochromatic-isopachics pattern by use of retarders in holographic photoelasticity, *Exp. Mech.*, **16**, 146–150 (1976).
107. H. Uozato and R. Nagata, Holographic photoelasticity by using dual hologram method, *Jap. J. Appl. Phys.*, **16**, 95–100 (1977).
108. J. D. Hovanesian, Variable isochromatic/isopachic-fringe visibility in photoelasticity, *Exp. Mech.*, **14**, 233–236 (1974).
109. A. Ajovalasit, A single hologram technique for the determination of absolute retardations in holographic photoelasticity, *J. Strain Anal.*, **10**, 148–152 (1975).
110. A. Ajovalasit and A. Bardi, Holographic photoelasticity: determination of absolute retardations by a single hologram, *Exp. Mech.*, **16**, 273–275 (1976).
111. H. Kubo and R. Nagata, Further consideration of photoelasticity using polarization holography (a method in holographic photoelasticity), *Opt. Acta*, **23**, 519–528 (1976).
112. H. Kubo and R. Nagata, Application of polarization holography by the Kurtz's method to photoelasticity, *Jap. J. Appl. Phys.*, **15**, 1095–1100 (1976).
113. S. K. Dhir and H. A. Peterson, An application of holography to complete stress analysis of photoelastic models, *Exp. Mech.*, **11**, 560–564 (1971).

Author Index

Abramson, N., 90, 225, 227, 430
Adams, F. D., 410
Afanaseva, A. L., 341
Ågren, C.-H., 240
Ahlberg, J. H., 161
Ajovalasit, A., 451
Aleksandrov, E. B., 69, 169
Aleksoff, C. C., 210, 215
Allaire, R. A., 237, 238
Alpert, S. S., 365
Alwang, W., 325
Amadesi, S., 222
Anderson, E. E., 333
Antonini, G., 366
Aprahamian, R., 237, 242, 243
Archbold, E., 209, 231, 402
Armstrong, W. T., 362
Aung, W., 365

Balozerov, A. F., 347
Barakat, R., 318
Barker, D. B., 412, 422
Barnett, N. E., 187
Barr, W. L., 320
Beach, K. W., 334
Beamish, J. K., 347
Beer, H., 365
Bellani, V. F., 99
Bennett, F. D., 319
Berezkin, A. N., 347
Berry, M. V., 323
Bershader, D., 344
Biedermann, K., 50, 103, 201, 202, 438
Bjelkhagen, H., 225, 227, 231, 232
Blodgett, J. A., 231
Bochner, N., 373

Bockasten, K., 320
Bonch-Bruevich, A. M., 69, 150, 169
Boone, P. M., 103, 143, 164, 236, 410
Borissov, M., 239
Born, M., 262
Bracewell, R. N., 320, 328
Bradley, J. W., 320
Brandt, G. B., 160
Briers, J. D., 103
Brooks, R. E., 57, 275, 428
Bryngdahl, O., 298, 304, 338, 340
Buchele, D. R., 332, 333, 334
Burch, J. M., 57, 58, 280, 402, 408
Burchett, O. J., 221
Butters, J. N., 225, 423, 435

Champagne, E. B., 227, 230
Chang, T. Y., 89
Chau, H. H. M., 346
Chen, Y., 242
Chernykh, V. T., 347
Chiang, F. P., 409, 428
Chomat, M., 239
Collier, R. J., 57
Collins, D. J., 329, 348, 349
Coumou, D. J., 364

Dainty, J. C., 30
De, M., 336
Decker, A. J., 217
Dhir, S. K., 82, 451
Dobbins, H. M., 365
Doherty, E. T., 57
Doi, Y., 340
Dreiden, G. V., 356
Dubas, M., 174, 230

457

Dudderar, T. D., 375
Duffy, D. E., 405
Durelli, A. J., 445
Durou, C., 373

Ek, L., 103, 415
El-Wakil, M. M., 371
Ennos, A. E., 69, 74, 150, 209, 402
Evensen, D. A., 237
Ezekiel, S., 392

Fagot, H., 347
Fitzgerald, J., 366
Forman, P. R., 362
Forno, C., 408
Fourney, M. E., 412, 422, 439, 451
Froehly, C., 143
Fryer, P. A., 205
Fuchs, von P., 231

Gabor, D., 37, 58
Gates, J. W. C., 280
George, N., 279
Gibbs, D. F., 323
Gilbert, J. A., 83, 89, 234
Goldberg, J. L., 236, 237
Golub, G. H., 79
Gooderum, P. B., 319
Goodman, J. W., 24, 33, 283
Gordon, R., 329
Gottenberg, W. G., 234, 242
Graube, A., 52
Gregory, D. A., 409
Grigull, U., 318, 364
Grünewald, K., 220
Gusev, O. B., 278

Haines, K. A., 57, 120, 142, 169, 233, 390, 428
Hariharan, P., 226, 423
Harper, J. S., 301
Hauf, W., 318, 364
Havener, A. G., 347
Hecht, E., 15
Heflinger, L. O., 57, 275, 346, 392, 428
Hegedus, Z. S., 226
Henshaw, P. D., 392
Herlitz, S. I., 320
Hildebrand, B. P., 57, 120, 142, 169, 233, 390, 428

Hirth, A., 437
Hockley, B. S., 225
Høgmoen, K., 438
Hörster, H., 366
Holloway, D. C., 439
Horman, M. H., 57
Hovanesian, J. D., 422, 430, 439, 450
Howes, W. L., 332, 333, 334
Hsu, T. R., 237
Hudson, R. R., 410
Hughes, R. G., 436
Humberstone, G. N., 231
Hung, Y. Y., 422, 423, 425, 428, 430
Hunter, A. R., 236
Hussmann, E. K., 366

Ignatov, A. B., 361

Jahoda, F. C., 355, 356
Jeffries, R. A., 358
Johansson, S., 50
Johnson, R. H., 439
Jones, R., 422
Juang, R. M., 409, 428
Junginger, H.-G., 329

Kallard, T., 45
Katayanagi, T. E., 221
Kellie, T. F., 201, 202
Kelly, J. G., 362
Kersch, L. A., 230
Khanna, S. M., 232
King, P. W., 83, 89, 242
Kirstal, R., 356
Klappert, W.R., 230
Klyberg, G., 143
Köpf, U., 75
Kohler, H., 103
Komissarova, I. I., 358
Konstantinov, V. B., 278
Kosakoski, R. A., 349
Kulagen, I. D., 321
Kurtz, R. L., 103

Ladenburg, R., 318, 344
Lakshminarayanan, A. V., 328
Landry, M. J., 103, 362
Larinov, N. P., 231
Larminant, P. M. de, 230
Lashkari, M., 239

Law, C. H., 347
Lee, C. H., 409
Leendertz, J. A., 418, 420, 422, 423, 435
Leith, E. N., 40, 275
Levitt, J. A., 216
Lewak, R., 237
Listovets, V. S., 207
Liu, H. K., 103
Løkberg, O. J., 438
Lohmann, A. W., 298
Luxmoore, A. D., 218

McCarthy, A. E., 362
Mach, L., 344
Machado Gama, M. A., 143
Macovski, A., 435, 437
Maddux, G. E., 410
Mandel, L., 15
Mate, K. V., 451
Matsumoto, T., 79, 97, 98, 301, 305
Matulka, R. D., 329, 348, 349
Mayer, M. D., 221
Mayinger, F., 371, 372
Meggers, W. F., 364
Mendenhall, F. T., 241
Miler, M., 207, 239
Mix, L. P., 362
Molin, N.-E., 142, 188, 189, 201, 202, 415
Morton, T. M., 236
Mottier, F. M., 217
Moyer, R. G., 237
Mullaney, G. J., 347
Murphy, C. G., 365
Mustafin, K. S., 343

Nestor, O. H., 320, 321
Neumann, D. B., 215, 227
Nisida, M., 447, 451

Olsen, H. N., 320, 321
O'Regan, R., 365, 375
Orr, L. W., 72
Ostrovskaya, G. V., 361
Ostrovskii, Yu. I., 207, 361
Ovechkin, A. P., 103

Palmer, D. A., 58
Panknin, W., 365, 371, 372
Pearce, I. K., 217
Peck, E. R., 365

Penn, R. C., 227
Pennington, K. S., 57, 301
Peters, C. L., 364
Peterson, H. A., 451
Philbert, M., 347
Pipman, J., 373
Politch, J., 336
Powell, R. L., 57
Presnyakov, Yu. P., 329
Prikryl, I., 140
Pryputniewicz, R. J., 177

Radley, R. J., Jr., 347, 361
Radulovic, P. T., 366
Ramachandran, G. N., 328
Redman, J. D., 217
Reinheimer, C. J., 347
Reinsch, C., 79
Robertson, E. R., 242
Rowe, S. H., 428
Royer, H., 347

Saito, H., 164, 236, 447, 451
Sampson, R. C., 234
Sanford, R. J., 445
Schaefer, R. J., 231
Schmitt, H., 365
Schödel, G., 364
Schott, D., 231
Schumann, W., 169, 174, 230
Sciammarella, C. A., 83, 89, 234
Seftor, J. L., 358
Seleznev, V. A., 343
Setopoulos, D. D., 410
Sevigny, L., 336
Shames, I. H., 149
Shamir, J., 340
Shatilov, A. P., 347
Sikora, J. P., 82, 241
Smigielski, P., 347
Smith, H. M., 53, 202
Soga, N., 375
Sollid, J. E., 69, 161
Sona, A., 99
South, R., 320
Stavroudis, O. N., 264
Steel, W. H., 143
Stetson, K. A., 57, 109, 120, 140, 142, 165, 169, 174, 185, 188, 189, 195, 216, 217, 240, 241, 242, 410, 416, 422

Stevenson, W. H., 201, 202
Stoffregen, B., 414
Strope, D. H., 189, 244
Surget, J., 347
Svensen, H., 332
Sweeney, D. W., 318, 325, 340, 366

Takashima, M., 301
Tanner, L. H., 281, 347
Taylor, C. E., 423
Taylor, J. K., 364
Taylor, L. H., 160
Taylor, P. A., 240, 241
Thomas, K. S., 362
Tilton, L. W., 364
Tiziani, H. J., 402
Tokarski, J. M. J., 402
Tolansky, S., 302
Tonndorf, J., 232
Toyooka, S., 301
Tsuruta, T., 143
Tuschak, P. A., 237, 238

Upatnieks, J., 40, 275

Van Haeringer, W., 329
Varde, K. S., 372
Varner, J. R., 390, 394
Velzel, C. H. F., 301
Verbiest, R., 143, 164, 236
Veret, C., 347
Vest, C. M., 325, 339, 366
Vikram, C. S., 209

Vlasov, N. G., 222

Wachtell, G. P., 332
Wakashima, H., 240
Wall, M. R., 182
Wallach, J., 150
Walles, S., 62, 64, 143
Waters, J. P., 221
Watrasiewicz, B. M., 182
Wei, R. P., 230
Weigelt, G. P., 414
Weigl, F., 341, 343
Weingarten, V. I., 239
Welford, W. T., 143, 428
Wendendal, P. R., 232
Wictorin, L., 232
Williams, R., 237
Wilson, A. D., 150, 188, 189, 234, 235, 237, 244
Wise, C. M., 103
Witte, A. B., 353
Wolf, E., 15, 262
Wood, G. P., 319
Wortberg, G., 347
Wuerker, R. F., 57, 275, 392

Zaidel', A. N., 355, 356
Zajac, A., 15
Zehnder, L., 344
Zelenka, J. S., 394
Zien, T.-F., 349
Zinnes, A. E., 365
Zivi, S. M., 231

Subject Index

Abel transform, 267, 316
Absolute retardation, in photoelasticity, 451
Accuracy, of displacement measurement, 77
Airy distribution, 305
Amplitude object, 20
Amplitude transmittance, 19
Analyzer, polarization, 439
Anistropic materials, 236
Apparent ray crossing, 333
Aperture, 26
Aperture stop, 26

Beam, 152
  stresses and strains, 153
  vibration of, 237
Beamsplitter, 9, 42
  variable, 42
Beating, of fringe patterns, 161
  of optical waves, 88
Bleaching, of holograms, 52
Bleaching procedures, dry, 52
  wet, 52
Bouguer's formula, 263
Bragg condition, 200

Central section theorem, 328
Characteristic function, in statistics, 187
  of vibration, 180
Coherence, spatial, 13
  of lasers, 14
  temporal, 11
Coherence length, 11
Collimated wave, 5
Combustion, 372
Complete localization, curve of, 174
Complex amplitude, 16

Concrete, swelling of, 236
Condition, of matrix, 79
Conjugate image, holographic, 55
Conjugate wave, 24
Contour generation, 387
Contouring, multiple-refractive index, 392
  with television system, 437
  two-frequency, 388
Cubic spline functions, 157

Deformation gradient matrix, 172
Derotator, optical, 242
Development centers, 45
Differential interferometry, 244
Differentiation, by moiré effect, 161
  optical, 161
Diffraction efficiency, of holograms, 49
Diffraction theory, scaler, 16
Diffuser, depolarization by, 279
  ground glass, 278
  opal glass, 278
  surface, 278
  volume, 278
Displacement, electric, 2
Displacement measurement, *see* Fringe interpretation
Division of wavefront, spatial, 60
  temporal, 60

Electromagnetic wave, 2
Electron number density, critical, 355
Emulsion, photographic, characteristics of, 47
Emulsion distortion, effects of, 200
Ensemble average, 32
Entrance pupil, 26

Ergodic theorem, and holographic interferometry, 187
Error analysis, of holographic interferometry, 77
Exit pupil, 26

Finite difference, backward, 156
  central, 156
  differentiation, 156
  forward, 156
Flexural rigidity, of plates, 155
Fluid velocity, measurement by speckle photography, 412
f number, 27
Focal length, 26
Focal plane, 26
Focal point, 26
Fourier optics, 24
Frequency, circular, 2
  of light, 16
  spatial, 18
Frequency shifting, devices for, 88
Frequency-translated holography, 211, 212
  with television system, 438
Frieser plateau, 50
Fringe control, 227
  in nondestructive testing, 230
  real-time, 227
  two-exposure, 229
Fringe interpretation, coplaner displacement, 72
  by holodiagram, 90
  multiple-hologram analysis, 73
  normal displacement, 69
  single-hologram analysis, 76
  vector displacement, 69
Fringe laminae, 135
Fringe localization, 107, 284
  with collinated illumination, 111
  curve of, 113
  equations of, 113, 123, 138
  role of aperture, 108
  sharpness of, 113
  with slit aperture, 117
  with spherical illumination, 122
  strain measurement and, 169
Fringe locus function, 174
Fringe order, sign ambiguity in, 312
Fringes, of equal inclination, 125
Fringe parallax, 121

Fringe sharpening, 305
Fringe surface, 135
Fringe vector, 174
Frozen-stress photoelasticity, 451

Geometric optics, 256
Gladstone-Dale constant, 344
  of air, 345
  of gases, 345
  of gas mixtures, 345
Gladstone-Dale equation, 344
Grating, sinusoidal amplitude, 20
  sinusoidal phase, 24, 282

Haidinger fringes, 125
Half-wave plate, 440
Holodiagram, 90
  coherence requirements and, 44, 97
  moiré analogy, 93
Hologram, Gabor, 38
  development of, 52
  in-line, 38
  phase, 52
Holographic interferometry, 57
  in aerodynamics, 344
  comparison of components by, 230
  diffuse-illumination, 275
  discovery of, 57
  dual-hologram, 273
  dual-reference-beam, 273
  with electronic phase detection, 87
  error reduction, 82
  for flow visualization, 351
  for heat transfer measurements, 363
  heterodyne, 87
  at high temperature, 237
  in hostile environment, 237
  in industrial environment, 225
  key properties of, 60
  for mass transfer measurements, 370
  in mechanics, 233
  medical and dental applications, 231
  microwave, 57
  multidirectional, 321, 366
  with multiple illumination beams, 85
  multiple-object-wave, 236
  multiple-pass, 343
  multiple-wavelength, 341, 371
  nonlinear, 295
  pulsed laser, 244

real-time, 197
real-time time-average, 199
of rotating objects, 241
by scanning, 99, 103
stroboscopic, 203, 217
three-exposure, 75
time-average, 178
time sequencing of, 361
two-exposure, 58
Holographic subtraction, 86
Holography, Leith-Upatnieks, 40
multiple-wavelength, 341, 371
nonlinear, 358
off-axis, 40
single-wave, 227
three-beam, 334
ultrasonic, 57
Homogeneous deformation, 140, 175
Homologous rays, 143
Hunter-Driffield curve, 46

Ideal gas, equation of state, 363
Imaging equation, holographic, 53
thin-lens, 26
Impact loading, 244
Impact parameter, 263
Independent motions, 188
Induction, magnetic, 2
Influence coefficients, 241
In-plane translation, measurement by speckle photography, 398
In situ processing, 201, 202
Intensity, magnetic, 2
Interference, 6
of diffuse waves, 60
Interferogram, 57
finite-fringe, 269
hologram-moiré, 273
holographic, 57
infinite fringe, 269
Interferometer, 8
division of amplitude, 8
division of wavefront, 8
Michelson, 9
scatter plate, 280
single-path, 267
Interferometry, diffuse illumination, 275
dual-hologram, 273
dual-reference-beam, 273
measurement by speckle photography, 412

multidirectional, 321
multiple-beam, 302
multiple-pass, 343
shearing, 336
Inversion formula, for asymmetric objects, 323
for radially symmetric objects, 316
Irradiance, 6
Irrationally related modes, 189
Isochromatic fringes, 445
Isopachic fringes, 445

Laser light, characteristics of, 5
Laser speckle, *see* Speckle
Latent image, 46
Lens, transforming, 29
Light, 1
Liquid gate, 201
Localization, fringe, *see* Fringe localization
Lorentz-Lorenz equation, 364

Maxwell-Neumann equation, 374
Mirror, front-surface, 42
Modulation depth, 24
Modulation transfer function, of film, 50
Modulus of elasticity, 152
Monobath developing, 202
MTF, of film, 50
Multiple-hologram analysis, 73
Multiple vibration modes, 187
Musical instruments, vibration measurement of, 240

Newton's law of reflection, 257
Nodal fringe, 181
Nondestructive testing, 217
of circuit boards, 223
of composite cylinders, 221, 222
of concrete, 219
by direct mechanical stressing, 218
excitation methods, 218
of fiber-resin composites, 220
of honeycomb panels, 220
of paintings, 222
of pneumatic tires, 221
of reinforced plastics, 219
of rocket propellants, 221, 222
by thermal stressing, 222
by vibrational excitation, 224
Nonhomogeneous medium, 256

Nonsinusoidal vibration, 185
Norm, of matrix, 78
  of vector, 78
Normal mode analysis, 240
Normal mode function, 240

Object motion compensation, 215, 225
Object wave, holographic, 36
Observer-projection theorem, 133
Optical disturbance, 16
Optical pathlength, 264
Optical pathlength difference, 266, 312
Overdetermined system, error reduction by, 82

Pathlength, matching in holographic systems, 44
Pathlength error, cancelation of, 267
Phase hologram, 52
Phase modulation, of holograms, 214
Phase object, 20, 264, 312
  asymmetric, 321
  radially symmetric, 267, 315
  two-dimensional, 315
Phase shift amplification, 298
Phase shifter, gas cell, 86
  glass plate, 86
Phasor sum, 193
Photoelasticity, 438
  frozen-stress, 451
  holographic, 439
    two-exposure, 441
Photographic emulsions, characteristics of, 47
Plane wave, 3
Plasma, 353
Plate, 154
  deformation of, 234, 235, 236
  stresses and strains, 154
  vibration of, 237
Plate holder, 44
  for real-time interferometry, 200, 201
Poisson's ratio, 152
Polarization, circular, 5
  elliptical, 4
  linear, 3, 5
  plane, 5
Polarization effects, in interferometry, 202
Primary fringes, holographic, 90
Primary image, holographic, 54

Probability density function, of motion, 187
Projected fringe surface, 430
Projected fringe techniques, 428
  real-time, 434
Propagation vector, 3
Pseudoscopic image, 45
Pulsed lasers, for plasma diagnostics, 356

Quarter-wave plate, 5, 439

Radon transform, 323
Rationally related modes, 189
Ray crossing, apparent, 333
Ray equation, 256
Real image, 26
  holographic, 45
  use in holographic interferometry, 97
Reconstruction, holographic, 38
Reconstruction wave, holographic, 38
Recording media, non-silver-halide, 53
Redundant measurements, error reduction by, 82
Reference fringes, 269, 312
Reference-to-object-beam ratio, 50
Reference wave, holographic, 37
Reflection, Newton's law of, 257
Refraction, 258
  in radially symmetric medium, 262
  Snell's Law of, 258
  in stratified medium, 261
Refraction errors, 329
  elimination of, 332
Refractionless limit, 264, 312
Refractive index, 256
  of electron gas, 354
  of gas mixtures, 371
  temperature dependence of, 363, 364
Refractivity, specific, 364
Retroreflective paint, 227
Retroreflective viewing, 132
Rigid-body motion, fringe localization due to, 131
Rigid-body translation, fringe localization due to, 124
Rotation, 149
Rotation matrix, 172

Sandwich holography, 225, 273
  for photoelasticity, 449
Scatter plate, 280

SUBJECT INDEX

Secondary fringes, holographic, 93
Sensitivity, of photographic emulsions, 49
Separable motions, characteristic function of, 183, 188
Shear, optical, types of, 338
Shear modulus of elasticity, 152
Shells, vibration of, 239
Single-hologram analysis, 76
Sinusoidal motion, characteristic function of, 180
Slit aperture, 117
Slopes, measurement by speckle photography, 409
Snell's law, 258
Soft spring, vibration of, 185
Spatial filtering, 26
Spatial frequency, 18
Specific refractivity, 364
Speckle, contrast of, 33
  laser, 30
  probability density, 33
  statistics of, 33
  width, 35
Speckle fields, coherent addition of, 416
  incoherent addition of, 420
Speckle interferogram, 415
Speckle interferometry, 413
  real-time, 418
  shearing, 422
  superposition, 420
  two-exposure, 420
Speckle photography, 396
  double-aperture, 405
  sensitivity range of, 413
  with television system, 435
  time-averaged, 401
  of transparent objects, 410
  white-light, 408
Speckle reference beam, 216
Speed of light, 2, 256
Spherical wave, 3
Stability, of holographic systems, 44
Stationary phase, method of, 195
Stiff spring, vibration of, 185

Strain, in-plane, measurement of, 150
  normal, 147, 152
  shear, 147
Strain matrix, 172
Stress analysis, 373
Stress-optical effect, 373
Stress, plane, 150
Stress waves, 242

Tables, optical, 44
Temporal filtering, holographic, 212
Temporally dependent motion, 189
Temporally modulated holography, 210
  in nondestructive testing, 224
Thin-beam reconstruction, 99, 277
Thin lens, 26
Tilt, measurement by speckle photography, 399
Torsion bar, relaxation of, 236
Transforming lens, 29
Transmittance, amplitude, 47
  intensity, 47
Transpose, of matrix, 172
Turbine fan, vibration of, 242

Ultrasonic vibrations, 237

Vibration, measurement by projected fringes, 433
  measurement by speckle interferometry, 415
  measurement with television system, 438
  sinusoidal, 178
Virtual image, holographic, 26, 45
Visibility, fringe, 11, 115

Wavefront, 3
Wave number, 2
Wedge fringes, 269

Young's fringes, 8

Zero-motion fringe, identification of, 75